AF733285

BIRKHÄUSER

Mechanics, Structure and Evolution of Fault Zones

Edited by
Yehuda Ben-Zion
Charles Sammis

Birkhäuser
Basel · Boston · Berlin

Reprint from Pure and Applied Geophysics
(PAGEOPH), Volume 166 (2009) No. 10/11

Editors
Yehuda Ben-Zion
Charles Sammis
Department Earth Sciences
University of Southern California
College of Letters, Arts & Sciences
Los Angeles, CA 90089-0740
USA
Email: benzion@usc.edu
sammis@usc.edu

Library of Congress Control Number: 2009935412

Bibliographic information published by Die Deutsche Bibliothek:
Die Deutsche Bibliothek lists this publication in the Deutsche Nationalbibliografie; detailed bibliographic data is available in the Internet at <http://dnb.ddb.de>

ISBN 978-3-0346-0137-5 Birkhäuser Verlag AG, Basel · Boston · Berlin

This work is subject to copyright. All rights are reserved, whether the whole or part of the material is concerned, specifically the rights of translation, reprinting, re-use of illustrations, recitation, broadcasting, reproduction on microfilms or in other ways, and storage in data banks. For any kind of use, permission of the copyright owner must be obtained.

© 2010 Birkhäuser Verlag AG
Basel · Boston · Berlin
P.O. Box 133, CH-4010 Basel, Switzerland
Part of Springer Science+Business Media
Printed on acid-free paper produced from chlorine-free pulp. TCF ∞
Cover graphic: Based on a graphic from the article "Structural Properties and Deformation Patterns of Evolving Strike-slip Faults: Numerical Simulations Incorporating Damage Rheology" by Y. Finzi, E. H. Hearn, Y. Ben-Zion, and V. Lyakhovsky

Printed in Germany

ISBN 978-3-0346-0137-5 e-ISBN 978-3-0346-0138-2

9 8 7 6 5 4 3 2 1 www.birkhauser.ch

Contents

Pure appl. geophys. 166 (2009) 1533–1536
0033–4553/09/101533–4
DOI 10.1007/s00024-009-0509-y

© Birkhäuser Verlag, Basel, 2009

Pure and Applied Geophysics

Mechanics, Structure and Evolution of Fault Zones

YEHUDA BEN-ZION and CHARLES SAMMIS

1. Introduction

The brittle portion of the Earth's lithosphere contains a distribution of joints, faults and cataclastic zones that exist on a wide range of scale-lengths and usually have complex geometries including bends, jogs, and intersections. The material around these complexities is subjected to large stress concentrations, which lead during continuing deformation to the generation of new fracture and granular damage with an associated evolution of the elasticity, permeability and geometry of the actively deforming regions. The evolving material and geometrical properties within and around the fault zone can lead, in turn, to evolution of various aspects of earthquake and fault mechanics.

The geometrical complexity, material heterogeneities and wide range of effective space-time scales make the quantification of properties and processes associated with fault zones extremely challenging. Many fundamental questions concerning the mechanics, structure and evolution of fault zones remain unanswered despite considerable research spanning over 100 years. The fourteen papers in this volume provide recent theoretical and observational perspectives on fault zones. Topics include damage and breakage rheologies, development of instabilities, fracture and friction, dynamic rupture experiments, analysis of geodetic data, and geological and laboratory characterizations of fault traces, fault surfaces, fracture zones, and particles in rock samples from fault zone environments.

FINZI *et al.* present computer simulations of the evolution of geometrical and elastic properties of large strike-slip fault zones and associated deformation fields, using a model with a seismogenic upper crust governed by a continuum damage rheology. The fault zones form initially as complex segmented structures and evolve overall with continuing deformation toward contiguous simpler structures. The simulations generally produce flower-type structures, consisting of a broad damage zone in the top few kilometers and highly localized damage deeper. However, at persisting geometrical complexities significant damage extends throughout the seismogenic zone. DE JOUSSINEAU and AYDIN discuss observed geometrical properties of surface traces of strike slip faults in the Valley

Department of Earth Sciences, University of Southern California, Los Angeles, CA 90089-0740, USA.

of Fire State Park, Nevada, and elsewhere. Analyzing the evolution of trace complexity with cumulative slip, they find that the number of fault steps can be described as a negative power law of the total slip. They also show that the mean segment length and mean step dimensions can be described as positive power laws of the fault offset. In agreement with previous studies, the results indicate an overall coalescence and smoothing with progressive fault growth.

GRIFFITH *et al.* analyze the effects of fracture damage on the effective response of fault zone rocks using finite-element calculations corresponding to uniaxial compression tests. The examined model configurations incorporate digitized fracture maps from the Bear Creek fault zones in the Sierra Nevada batholith, and include fracture-induced material contrast between damaged tabular zones and stiffer host rocks. The damage zone is shown to be strongly anisotropic and associated with strain-weakening behavior. The contrast of elastic moduli between the damage zone and host rock can be larger than contrasts associated with juxtaposition of different rock lithologies. SAMMIS *et al.* review recent laboratory results on dynamic rupture between transparent brittle polymers with and without pre-existing damage in the form of array of off-fault cracks. Multiple modes and speeds of rupture are generated depending on the contrast of elastic stiffness and damage across the fault. Asymmetric dissipation on the network of pre-existing cracks in tension and compression can overcome other dynamic effects and arrest the rupture propagation in one direction. The scaling of laboratory results to natural earthquakes is discussed using theoretical models for dynamic rupture and damage generation.

LOCKER *et al.* measure the strength and matrix permeability of core samples from drillholes that cross the Nojima fault that ruptured during the 1995 M = 7.2 Kobe earthquake in Japan. They find a simple fault zone structure comprised of a relatively weak fine-grained core with low permeability that impedes flow normal to the fault, surrounded by a damage zone of fractured rock that provides a high-permeability conduit for vertical and horizontal flow in the fault plane. MOORE *et al.* examine in a companion paper the microstructures and alteration features that correspond to the changes of strength and permeability in these same core samples. The strength decreases regularly with increased fracturing and fragmentation. The permeability increases with increasing fracturing until the rock looses cohesion, at which point the permeability decreases by 1–2 orders of magnitude. Alteration in the fragmented rock further decreases the permeability. NGUYEN and EINAV use a breakage mechanics model to discuss the evolution of stress-strain curves, grain size distribution and related properties during confined comminution of rocks. The results shed light on the permeability evolution and energy dissipation associated with cataclasite zones. The theoretical results fit well observational results of laboratory fracturing experiments with several types of sandstones. The energy considerations highlight the role of changes to the stored strain energy with changes of the internal configurations of particles in the cataclastic material.

ROCKWELL *et al.* present laboratory characterizations of the chemistry and particle-size distributions of pulverized granite at different distances from the San Andreas and Garlock faults. The results show little weathering other than in the less than 1 micron fraction. The particle-size statistics generally follow a pseudo-power law with a continuously changing exponent. The San Andreas samples have finer grains than those collected from the Garlock, and the distributions from both faults tend to become finer with increasing proximity to the fault core. The rock pulverization appears to be associated with dynamic shattering under locally tensile stress, with a fracturing process that accounts for a very small fraction of the earthquake energy budget. DOR *et al.* analyze microstructure properties of damaged rocks in young sedimentary formations near the San Andreas fault in the Mojave. The rock samples indicate pervasive fracturing over a zone that is about 100–200 m from the fault. The fracture fabric is characterized by preferred orientations that vary with distance from the fault. The results imply a shallow damage generation process consistent overall with deformation under compression and perhaps also local dynamic tension very close to the fault.

Beeler discusses theoretically constitutive relations for seismic and aseismic slip on a fault governed by rate- and state-dependent friction. The study attempts to connect physical processes with the state of the sliding surface, and to account for strong weakening at high-slip rates as well as the strength near the brittle-ductile transition zone. DIETERICH and SMITH simulate stress evolution and seismicity on a rough fault governed by rate-state friction, which they compare with recent observations of the spatial distribution of near-fault seismicity. CANDELA *et al.* provide observational results on fault roughness with a vertical resolution from 0.5 mm to 20 mm, using LIDAR in the field and a laser profilimeter in the laboratary for less than 1 micron. They evaluate different signal processing techniques and use the most reliable to measure the Hurst scaling parameter of the fault surfaces parallel and perpendicular to the sliding direction.

HSU *et al.* use GPS data recorded around the time of the 1999 Chi-Chi earthquake in Taiwan to derive spatio-temporal variations of slip on the Chelungpu fault. In agreement with expectations based on rate-state friction, the results indicate coseismic slip on a shallow patch that was locked before the earthquake and temperature-induced transition to localized stable sliding at depth. Azimuthal differences between the pre- and post-seismic GPS velocities are assumed to reflect very low shear stress on the creeping décollement beneath central Taiwan. REGENAUER-LIEB *et al.* discuss a general thermo-dynamic framework in which brittle and ductile deformation processes may give rise to positive feedback resulting in shear instability. In this context, instabilities thought to produce landslides and icequakes might also be relevant to earthquake nucleation at the brittle-ductile transition in the crust.

As illustrated by the detailed and diverse results in this volume, fault zones have complex structure and mechanics that reflect and affect a variety of physiochemical processes. Many unanswered questions remain about the properties and processes associated with fault zones, especially concerning the interaction and feedback between different components of the fault zone structure and dynamics. Additional multi-disciplinary studies of the type exemplified by the papers in this volume are required to clarify further the mechanics, structure and evolution of fault zones.

Acknowledgements

We thank the authors of the papers for their contributions and the referees for critical reviews that improved the scientific quality of the volume. The latter include Jean-Paul Ampuero, Jean-Philippe Avouac, Nick Beeler, Harsha Bhat, Ronald Biegel, Emily Brodsky, Jim Dieterich, Giulio Di Toro, Ory Dor, Jim Evans, Nick Hayman, Boris Kaus, Yann Klinger, Nadia Lapusta, Dave Lockner, Tom Mitchell, Mike Oskin, Ruey-Juin Rau, Francois Renard, Tom Rockwell, Jim Savage, Ram Weinberger and Teng-fong Wong.

Pure appl. geophys. 166 (2009) 1537–1573
0033–4553/09/101537–37
DOI 10.1007/s00024-009-0522-1

© Birkhäuser Verlag, Basel, 2009

Pure and Applied Geophysics

Structural Properties and Deformation Patterns of Evolving Strike-slip Faults: Numerical Simulations Incorporating Damage Rheology

YARON FINZI,[1] ELIZABETH H. HEARN,[1] YEHUDA BEN-ZION,[2] and VLADIMIR LYAKHOVSKY[3]

Abstract—We present results on evolving geometrical and material properties of large strike-slip fault zones and associated deformation fields, using 3-D numerical simulations in a rheologically-layered model with a seismogenic upper crust governed by a continuum brittle damage framework over a viscoelastic substrate. The damage healing parameters we employ are constrained using results of test models and geophysical observations of healing along active faults. The model simulations exhibit several results that are likely to have general applicability. The fault zones form initially as complex segmented structures and evolve overall with continuing deformation toward contiguous, simpler structures. Along relatively-straight mature segments, the models produce flower structures with depth consisting of a broad damage zone in the top few kilometers of the crust and highly localized damage at depth. The flower structures form during an early evolutionary stage of the fault system (before a total offset of about 0.05 to 0.1 km has accumulated), and persist as continued deformation localizes further along narrow slip zones. The tectonic strain at seismogenic depths is concentrated along the highly damaged cores of the main fault zones, although at shallow depths a small portion of the strain is accommodated over a broader region. This broader domain corresponds to shallow damage (or compliant) zones which have been identified in several seismic and geodetic studies of active faults. The models produce releasing stepovers between fault zone segments that are locations of ongoing interseismic deformation. Material within the fault stepovers remains damaged during the entire earthquake cycle (with significantly reduced rigidity and shear-wave velocity) to depths of 10 to 15 km. These persistent damage zones should be detectable by geophysical imaging studies and could have important implications for earthquake dynamics and seismic hazard.

Key words: Damage rheology, fault zone structure, strike-slip fault evolution, compliant zones, fault stepovers.

1. *Introduction*

Understanding the geometrical and mechanical properties of fault zones is important for many geoscience fields, including earthquake mechanics, crustal hydrology and mineral exploration. Since crustal faults generally grow and evolve through repeated earthquake ruptures, there are fundamental feedback mechanisms between the earth-

[1] Department of Earth and Ocean Sciences, University of British Columbia, Vancouver B.C. V6T 1Z4, Canada. E-mail: yfinzi@eos.ubc.ca; ehearn@eos.ubc.ca
[2] Department of Earth Sciences, University of Southern California Los Angeles, CA 90089-0740, U.S.A. E-mail: benzion@usc.edu
[3] Geological Survey of Israel, Jerusalem 95501, Israel. E-mail: vladi@geos.gsi.gov.il

quakes sustained by a fault and its structural evolution. Observational and theoretical studies indicate that the temporal and frequency-size statistics of earthquakes change as faults evolve with cumulative slip, from disordered structures to more regular mature fault zones (e.g., WESNOUSKY, 1994; BEN-ZION, 1996; STIRLING *et al.*, 1996; LYAKHOVSKY *et al.*, 2001). The evolution of the permeability structures around large faults affects the fluid flow properties of the crust and deposition of minerals (e.g., MICKLETHWAITE and COX, 2004). The evolving fluid flow regime influences in turn the mechanics of earthquakes and faults (e.g., HICKMAN *et al.*, 1995 and references therein). Fault stepovers and other geometrical heterogeneities affect the initiation, propagation and termination of earthquakes (e.g., SIBSON, 1985; KING, 1986; HARRIS and DAY, 1999; OGLESBY *et al.*, 2003; WESNOUSKY, 2006). Contrasts of elastic and permeability properties across faults affect the mode and properties of dynamic ruptures, seismic radiation and aseismic slip (e.g., BEN-ZION and ANDREWS, 1998; RUDNICKI and RICE, 2006; YAMASHITA, 2007; AMPUERO and BEN-ZION, 2008; DUNHAM and RICE, 2008).

Detailed mapping of several exhumed fault zones (e.g., CHESTER *et al.*, 1993; EVANS *et al.*, 2000; SIBSON, 2003) and additional observations summarized by BEN-ZION and SAMMIS (2003) indicate that the internal structure of fault zones evolves from an early stage associated with distributed deformation and band-limited fractal structures at several hierarchies, through localization to principal slip zones, to a mature stage characterized by large-scale faults with tabular damage zones and narrow cores of ultra-cataclasites. However, the ranges of conditions over which such evolution takes place, and the coupling between the evolving structures and distributions of crustal stress and strain, are not well understood. Various studies attempted to model changes of fault properties with ongoing deformation. For example, OLSON and POLLARD (1989) modeled the evolution of joints based on linear elastic fracture mechanics. COWIE *et al.* (1993) simulated evolving geometrical properties of fault networks using a scalar elastic field on a lattice model with spring-like elements. ANDREWS (2005), BEN-ZION and SHI (2005) and TEMPLETON and RICE (2008) simulated the generation of off-fault plastic strain during propagation of dynamic ruptures on frictional faults surrounded by a solid governed by Coulomb plastic yielding. While these studies provide important insights for various topics, they do not account for the evolution of elastic properties that accompanies the generation of cracking and inelastic strain, and they are also typically done within 2-D "plane strain" frameworks.

In the present work we attempt to understand some general aspects of the evolution of large strike-slip fault zone structures. The study is based on three-dimensional numerical simulations with a regional lithospheric model consisting of a seismogenic crust governed by damage rheology over a viscoelastic substrate (BEN-ZION and LYAKHOVSKY, 2006). Using this framework with parameters constrained by laboratory and geophysical observations, we examine the evolving geometrical and elastic properties of fault zones and the associated deformation patterns.

In the next section we summarize observational results on fault zone structures that are relevant to our study. In Section 3 we review the damage rheology framework and key aspects of the numerical model employed in this work. In Section 4 we use geophysical observations of strength degradation and recovery within active fault zones to narrow the range of admissible damage rheology parameters. Section 5 contains the results of our parameter-space study on structural evolution of large strike-slip fault zones. Our simulations produce for ranges of realistic conditions flower structures with depth and secondary faulting within stepovers comparable to those documented in geological and seismic studies. The results support the view that fault zones display highly localized slip embedded within a wider shallow damage zone. The implications of the results to plate boundary dynamics and suggestions for continuing studies are discussed in Section 6.

2. *Geological and Geophysical Observations of Fault Zone Structure*

The geometrical properties of fault structures and earthquake slip zones have been documented in many geological and geophysical studies. In general, strike-slip fault zones display a nested hierarchy of damage zones and slip surfaces that form "flower structures" with depth (e.g., WILCOX *et al.*, 1973; SYLVESTER and SMITH, 1976; SYLVESTER, 1988). In a typical fault structure, the principal slip zone is surrounded by gouge and embedded within a tabular or wedge-shaped damage zone (BEN-ZION and SAMMIS, 2003, and references therein). The extent of the damage zone may be defined as the region in which the density of deformation features exceeds the average regional level of deformation in the surrounding host rock (CHESTER, 1995).

Studies of earthquake slip in exhumed faults and paleoseismic trenches indicate that within the top few kilometers of the crust the majority of coseismic slip is accommodated along very narrow slip zones (e.g., SIBSON, 2003; ROCKWELL and BEN-ZION, 2007). The highly localized slip zone and surrounding ultracataclasite layer are referred to as the "core" of the fault zone. This fault core is typically parallel to the macroscopic slip vector and is surrounded by a cataclasite layer which is a few meters thick (e.g., CHESTER and CHESTER, 1998; SCHULZ and EVANS, 2000). The damage zone (DZ) around the fault core typically consists of a zone of intense damage, and possibly pulverized rocks, with a width of a few hundred meters (DOR *et al.,* 2006, 2008), which is surrounded by a broader, several kilometers wide zone, of distributed damage. The latter is probably a relic structure of the progressive coalescence and localization of the active fault zone over time (AMBRASEYS, 1970; KIM *et al.*, 2004; SIBSON, 2003).

The near-surface observations of distributed DZ are supported and complemented by a variety of geophysical studies that associate the DZ with a negative gravity anomaly and low seismic velocities (e.g., STIERMAN, 1984; MOONEY and GINZBURG, 1986), along with anisotropic seismic waves (e.g., COCHRAN *et al.*, 2003; LIU *et al.* 2004; PENG and BEN-ZION, 2004, 2005) and elevated seismic scattering (e.g., REVENAUGH, 2000). Within the fault zone, seismic waves may be trapped in a narrow zone of intense coherent damage that is significantly distinct from the wider distributed damage zone (e.g., BEN-ZION and AKI,

1990; LI *et al.*, 1990). Such fault zone trapped waves have been observed along large faults of the North Anatolian Fault System (NAFS), the Eastern California Shear Zone (ECSZ) and the San Andreas Fault System (e.g., LI *et al.*, 1994; BEN-ZION *et al.*, 2003). Systematic inversions of trapped waves indicate ~100 m wide tabular zones that extend typically to ~3–4 km depth and are characterized by strong attenuation and ~30–50% shear-wave velocity (V_S) reduction relative to their surroundings (PENG *et al.*, 2003; KORNEEV *et al.*, 2003; LEWIS *et al.*, 2005).

Damage zones also show up in geodetic measurements that detect amplification of deformation signals along fault zones. BEN-ZION *et al.* (1990) observed amplified strain and water-level signals at several locations along the Mojave segment of the San Andreas fault. FIALKO *et al.* (2002) and HAMIEL and FIALKO (2007) interpreted measurements of Interferometeric Synthetic Aperture Radar (InSAR) along several large strike-slip faults in terms of compliant zones that are 1–2 kilometers wide, 3 km to over 10 km deep, and which have rigidity (μ) reduction of 50%–70% relative to the host rock. The discrepancies in dimensions (particularly, width) of the geodetically determined "compliant zones" and seismically determined "trapping structures" probably reflect differences between the broader long-term, quasi-passive, damage structure and the narrower active zone associated with recent earthquake ruptures. Other effects may feed into these discrepancies. For example, distributed microfractures affect the geodetically observed static strength more than they affect the seismically observed dynamic strength of rock (e.g., IDE, 1936; EISSA and KAZI, 1988). Other seismic observations such as seismic anisotropy (e.g., COCHRAN *et al.*, 2003; LIU *et al.* 2004; PENG and BEN-ZION, 2004, 2005) and elevated scattering (e.g., REVENAUGH, 2000) near large faults conform with the geodetically determined wide damage zones (1–2 km wide at the top few kilometers), indicating perhaps that the "trapping structures" are much smaller than the entire damage zone.

To date, the best evidence of high localization of seismic slip at depths larger than 3–5 km is associated with the general tendency of seismicity to localize along relatively-straight fault segments to zones with width that is comparable to the smallest dimension that is resolvable by the data analysis. In places with good network coverage and relocated seismicity, the width of such zones is only a few tens of meters (e.g., POUPINET *et al.*, 1984; NADEAU *et al.*, 1994; SCHAFF *et al.*, 2002; MCGUIRE and BEN-ZION, 2005; THURBER *et al.*, 2006).

A significant deviation from the relatively simple DZ structure described above occurs at fault stepover zones. Fault zones generally display higher geometrical complexity and broader damage zones within stepovers than along relatively straight segments. While the major fault segments reflect a positive feedback of strain weakening and strain localization along the fault cores, persisting geometrical features such as fault offsets, kinks, and bends, can produce strain hardening that leads to local complexity and secondary fractures at different scales (BEN-ZION and SAMMIS, 2003). Many studies have characterized macroscopic structural features within enlarged damage zones at geometrical irregularities (e.g., SEGALL and POLLARD, 1980; KIM *et al.*, 2004). Our study attempts to clarify the evolution of structural properties of fault zones along relatively straight segments as well as near large persisting stepovers.

3. Damage Rheology Framework

3.1. Theoretical Background

In the past decade continuum damage mechanics models have been successfully applied (e.g., BERCOVICI and RICARD, 2003; TURCOTTE and GLASSCOE, 2004) in various studies of long-term tectonic deformation. LYAKHOVSKY *et al.* (1997a,b), HAMIEL *et al.* (2004) and references therein developed a thermodynamically-based continuum damage model for evolving elastic properties of rocks sustaining irreversible brittle deformation. The employed damage rheology is applicable to volumes with a sufficiently large number of cracks that allow quantitative description through properties of the crack distribution rather than those of the individual cracks (LYAKHOVSKY and MYASNIKOV, 1984, 1985). The model generalizes the strain energy function of a solid to account for first-order macroscopic effects of distributed cracks (i.e., damage), and makes the elastic moduli functions of an evolving damage state variable α representing the local crack density. An undamaged material with $\alpha = 0$ is the ideal solid governed by 3-D linear elasticity, while a material with $\alpha = 1$ is completely destroyed. Using the balance equations of energy and entropy, the damage model quantifies the effective elastic behavior of a cracked solid for all intermediate values of the damage variable ($0 < \alpha < 1$). Detailed reviews and recent developments of the model can be found in BEN-ZION and LYAKHOVSKY (2006) and LYAKHOVSKY and BEN-ZION (2008). Here we only summarize the main ingredients of the model that are relevant for our work.

The effects of distributed cracks (i.e., existing damage) on the elastic properties of a solid are accounted for in the damage model by generalizing the strain energy function to the form:

$$U = \frac{1}{\rho}\left(\frac{\lambda}{2}I_1^2 + \mu I_2 - \gamma I_1\sqrt{I_2}\right) \tag{1}$$

where $I_1 = \varepsilon_{kk}$ and $I_2 = \varepsilon_{ij}\varepsilon_{ij}$ are the first and second invariants of the elastic strain tensor ε_{ij}, ρ is the mass density, λ and μ are the Lamé parameters, and γ is a third modulus of a damaged solid. The first two terms of equation (1) give the classical strain potential of linear elasticity (e.g., MALVERN, 1969). The third term may be derived using the effective medium theory of BUDIANSKY and O'CONNELL (1976) for non-interacting cracks that dilate and contract in response to tension and compression (LYAKHOVSKY et al., 1997b), or by expanding the strain energy potential as a general second-order function of I_1 and I_2 and eliminating non-physical terms (BEN-ZION and LYAKHOVSKY, 2006).

The kinetic aspects of the damage rheology involve making the elastic moduli functions of the damage state variable, and deriving an equation for the evolution of damage. LYAKHOVSKY *et al.* (1997a) showed that the leading term of the damage evolution equation, satisfying energy conservation and nonnegative entropy production, can be written as

$$\frac{d\alpha}{dt} = \begin{cases} C_d I_2(\xi - \xi_0), & \text{for } \xi \geq \xi_0 \\ C_1 \cdot \exp\left(\frac{\alpha}{C_2}\right) I_2(\xi - \xi_0), & \text{for } \xi < \xi_0 \end{cases} \tag{2}$$

where $\xi = I_1/\sqrt{I_2}$ is referred to as the strain-invariants ratio, and C_d, C_1, C_2 are damage rate parameters further described in Sections 3.2 and 4.1. The parameter ξ_0 is a yielding threshold separating states of deformation involving material degradation ($d\alpha/dt > 0$) when $\xi > \xi_0$, and material healing ($d\alpha/dt < 0$) when $\xi < \xi_0$. AGNON and LYAKHOVSKY (1995) and LYAKHOVSKY *et al.* (1997a) related this parameter to the angle of internal friction by considering the critical shear stress for Mohr-Coulomb sliding. They obtained $\xi_0 = -0.8$ for rock with internal friction coefficient of $f = 0.6$ and Poisson's ratio $v = 0.25$, and noted that this value varies only slightly (-0.7 to -0.9) for rocks with Poisson's ratio between 0.2 and 0.3.

Equation (2) was derived assuming for simplicity

$$\begin{aligned} \lambda &= \lambda_0 = \text{constant}; \\ \mu &= \mu_0 + \alpha \gamma_m \xi_0; \\ \gamma &= \alpha \gamma_m; \end{aligned} \tag{3}$$

where γ_m is the maximum value of the third elastic modulus defined by normalization of the damage variable. The dependencies of elastic moduli on the damage variable produce the following changes during loading: As the damage variable α increases, the shear modulus μ decreases, Poisson's ratio v increases, and the modulus γ increases from 0 (damage free) to γ_m. Following the onset of positive damage evolution above the elastic limit at $\xi = \xi_0$ and before the final macroscopic failure, the model incorporates a gradual accumulation of inelastic strain, ε_{ij}^v, described in Appendix 1. When the damage variable reaches a critical value α_{cr}, there is brittle instability leading to rapid conversion of deviatoric elastic strain to permanent plastic strain. The reduced deviatoric stress at the end of the brittle failure episode typically leads to a state $\xi < \xi_0$ that is associated (equation (2)) with healing. The exponential dependency of the damage recovery (healing) is motivated by the logarithmic healing with time that is observed for rocks and other materials (e.g., DIETERICH, 1978; 1979). LYAKHOVSKY *et al.* (2005) showed that the above damage model reproduces the main observed features of rate- and state-dependent friction, and constrained the healing parameters C_1, C_2 by comparing model calculations with lab Frictional data.

The main components of the numerical procedure, utilizing the Fast Lagrangian Analysis of Continua (FLAC) algorithm (e.g., CUNDALL and BOARD, 1988; POLIAKOV *et al.*, 1993), is presented in Appendix 1. To simulate long-term deformation processes with appropriate boundary conditions at the edges of our model domain, we use boundary conditions with variable forces (LYAKHOVSKY and BEN-ZION, 2008). These boundary conditions account for the stress buildup and abrupt drop during each seismic cycle, and for the evolution of elastic properties and cumulative plastic strain within the model domain. Appendix 2 presents an overview of the variable-force boundary conditions, and Appendix 3 applies the conditions in a test study verifying the viscoelastic component of our code. Additional details on the employed damage model and comparisons of results with laboratory fracture and friction data are given by LYAKHOVSKY *et al.* (1997a,b, 2005), HAMIEL *et al.* (2004, 2006), BEN-ZION and LYAKHOVSKY (2006) and LYAKHOVSKY and BEN-ZION (2008).

3.2. Damage Model Parameters

A fundamental set of results of previous damage-based models is that rheological damage parameters have significant impact on the evolving geometrical properties of fault zones, seismicity patterns and spatial distribution of deformation. LYAKHOVSKY *et al.* (2001) and BEN-ZION and LYAKHOVSKY (2006) suggested that damage zone structure is primarily controlled by (1) the ratio between loading rate and healing rate, (2) the overall degree of "brittleness" of crustal deformation which may be parameterized by the seismic coupling coefficient χ, and (3) the susceptibility to propagation of rupture associated with dynamic weakening and related dynamic time scale τ_r. Below we review the main material parameters and outline their effect on fault zone evolution. Table 1 presents a synthesis of the plausible range of values for each parameter, and justifications for these values.

As discussed in the context of equation (2), the employed damage rheology includes three material parameters that affect the rate of damage evolution with time: C_1 and C_2 are both healing rate parameters, and C_d is the damage accumulation rate parameter. In our simulations, the loading rate is closely linked to the specified tectonic strain rate and the value of C_d is fixed. Therefore the ratio of loading rate to healing rate is governed by the healing rate parameters C_1 and C_2. Simulations with high healing rates compared to the loading rate result in rapid near-complete healing of fault damage. In such settings, ruptured fault zones quickly regain their strength, enabling larger interseismic stress accumulation and coseismic stress drops, and the evolving fault zones have more complex geometries than cases with low healing rates.

The degree to which crustal deformation is brittle is controlled by the material parameter C_v, the coefficient of damage-related inelastic deformation (Appendix 1). This material parameter determines the ratio of aseismic to seismic components of deformation, or the seismic coupling coefficient, given (BEN-ZION and LYAKHOVSKY, 2006) by $\chi = 1/(1 + R)$ through the nondimensional R value with $R = \mu_0 C_v$ and μ_0 being the initial rigidity of the material. LYAKHOVSKY and BEN-ZION (2008) demonstrated that higher crustal C_v values induce larger components of aseismic deformation in the seismogenic zone, and therefore lower coseismic stress-drops. The susceptibility to rupture propagation determines the degree of dynamic weakening and dynamic time scale τ_r during the occurrence of brittle instability. LYAKHOVSKY *et al.* (2001) found that higher

Table 1

Damage rheology parameters and their constraints.

Parameter	Preferred range	Justification for value, and references	Comments
C_d Damage accumulation rate	0.5–5 s^{-1}	Lyakhovsky *et al.* (1997a): Fracture experiments with granite at relatively high confining pressure (100 MPa)	Lyakhovsky *et al.* (2005) show that at shallow depths ($z < -5$ km) C_d increases with the decrease in confining pressure ($C_d^{surface} > 10\ s^{-1}$)
C_1, C_2 Healing rate parameters	$C_1 = 10^{-24}$–$10^{-4}\ s^{-1}$ $C_2 = 0.1$–0.01	Lyakhovsky *et al.* (2005): Analysis of 1-D damage and comparison with rate and state dependent friction parameters	See discussion and new constraints in Section 4
ξ_o Critical strain invariant ratio (generalized internal friction)	−0.7 to −1	Agnon and Lyakhovsky (1995): Analytic analysis indicates that a friction coefficient of $f \sim 0.6$–0.7 corresponds to $\xi_0 = -0.8$ Lyakhovsky *et al.* (1997a): 3-D faulting experiments yield $\xi_0 = -0.7$ to -1	Lyakhovsky *et al.* (1997a) also indicate a week dependency of ξ_0 on Poisson's ratio, ν, but concluded that $\xi_0 = -0.8$ for various rocks with $\nu = 0.2$–0.3
C_v Damage-related inelastic strain accumulation	10^{-4}–$5 \cdot 10^{-6}\ MPa^{-1}$	Yang and Ben-Zion (2009): Analysis of aftershock sequence in southern California, and comparison to damage rheology predictions	1. $C_v \cdot d\alpha/dt$ is the damage related compliance 2. Based on these C_v values, the fraction χ of elastic strain released during a seismic cycle as brittle deformation is estimated to be 30%-85%
τ_r Characteristic time scale for seismic wave damping	$3 \cdot 10^2$–$3 \cdot 10^4$ s	Ben-Zion and Lyakhovsky (2006): Largest aftershock magnitude analysis (comparing numeric results and expected log-linear relation)	

τ_r values induce larger ruptures and lead to relatively simple failure histories consisting of system-sized events occurring in a single fault zone.

The above material parameters are currently constrained mainly by analytical considerations and by fracture and friction experiments (Table 1). In the next section, we use geophysical data to narrow the admissible range of healing rate parameters, and to better relate these parameters to natural deformation processes observed along active fault systems.

4. Geophysical Constraints on Healing Rate Parameters

4.1. Parameters C_1 and C_2, and Healing as a Function of Time

As follows from equation (2), the healing rate depends on the strain magnitude (I_2), strain invariant ratio (ξ), level of material damage (α), and material properties including critical strain invariant ratio (ξ_0) and two healing rate parameters C_1 and C_2. The critical strain invariant ratio is well constrained and fairly constant ($\xi_0 \approx -0.8$). The strain invariants ratio (ξ) varies from $-\sqrt{3}$ to $+\sqrt{3}$. The post failure shear strain (I_2) is mostly controlled by the lithostatic pressure and may vary with depth by two orders of magnitude or less. Therefore, the rate and the overall effectiveness of the healing process are primarily determined by the rate parameters C_1 and C_2. While C_2 determines the rate dependence on the damage state α and varies within one order of magnitude, C_1 may vary by many orders of magnitude (LYAKHOVSKY *et al.*, 2005). Depending on the combination of these parameters the healing process may be fast or slow, and may yield insignificant or near-complete healing of the damaged material over long timescales. Equation (2) indicates that very large C_1 corresponds essentially to zero memory, in which case damaged material heals rapidly and almost completely. If damage accumulation is also rapid (C_d is very large) the model will display ideal elastoplastic behavior. Very small C_1 yields insignificant healing except for the special case of $C_2 \approx 0$ in which healing is near instantaneous and complete. Extremely large C_2 values lead to a healing rate proportional to C_1 (i.e., $d\alpha/dt = C_1 I_2(\xi - \xi_0)$). In such cases, for $C_1 < 1 \cdot 10^{-10}$ s^{-1} the healing is slow and insignificant while for $C_1 > 1 \cdot 10^{-05}$ s^{-1} the healing is fast and almost complete (Fig. 1).

To better understand the healing process and constrain the parameters C_1 and C_2, we define a time scale for healing (τ_h) during which the relative change of the elastic moduli (μ, γ) is above an arbitrarily chosen rate of 0.1% yr^{-1} ($d\alpha/dt = 3 \cdot 10^{-11}$sec^{-1}). Simplifying equation (2) for a uniform strain invariant ratio suitable for healing, assuming ($\xi - \xi_0$) to be of the order of one, and setting the healing rate to this chosen threshold, the expected damage level is given by

$$\alpha_f = C_2 \ln\left[3 \cdot 10^{-11}/C_1 I_2\right], \quad (4)$$

where α_f is the damage level as the healing becomes slower than 0.1% yr^{-1}. This value is referred to hereafter as the final damage level, even though slow although possibly

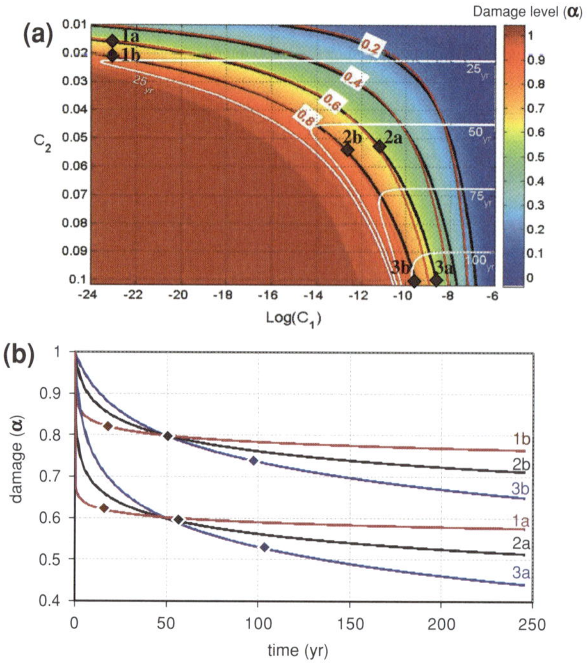

Figure 1

(a) Damage levels of typical crustal material at 10 km depth after 50 years of healing, for the damage model parameter space suggested by LYAKHOVSKY *et al.* (2005). Black and red lines show contours of equal damage and final damage (α_f), respectively. White lines are contours of equal healing time scale (τ_h). Black diamonds indicate sets of healing parameters referred to in Figure 1b. **(b)** Damage as a function of time after failure for six sets of C_1 and C_2 parameters (indicated in Fig. 1a). The healing time scale (τ_h) and final damage level (α_f) of each healing process are indicated by diamonds (red – 1a,b; black- 2a,b; blue – 3a,b). Note that some healing occurs after α_f is attained (see Section 4.1).

significant healing continues after α_f is reached (at geological time scales). Substituting α_f in equation (8) of LYAKHOVSKY *et al.* (2005) for the damage level as a function of time, the time scale for healing (i.e., the duration required to reduce the damage level from $\alpha = 1$ to α_f) can be estimated by

$$\tau_h = \frac{\exp\left(\frac{1-\alpha_f}{C_2}\right) - 1}{I_2 \frac{C_1}{C_2} \exp\left(\frac{1}{C_2}\right)}. \tag{5}$$

Within the healing parameter space that represents materials that undergo significant healing (i.e., damaged materials with $\alpha_f < 0.8$), the time scale for healing (τ_h) falls in the range 10–110 yr (Fig. 1a). The time scale for healing indicates how fast a damaged material heals to a near constant damage level, however it does not reveal the final damage level. Therefore, materials with similar τ_h may exhibit a wide range of final damage levels (Fig. 1). Figure 1b displays six healing processes determined by six sets of

C_1 and C_2 parameters (marked and labeled in Fig. 1a). Healing processes with short time scales (e.g., 1a and 1b in Fig. 1b) display higher initial healing rate and faster decay of the healing rate compared to processes with longer time scales (e.g., 3a and 3b in Fig. 1b). As evident from Figure 1b, in order to fully determine the healing parameters of a material one would either need to establish the damage level at two distinct times after failure or to determine the current damage level, the duration since failure and the healing time scale relevant to the specific healing process.

LYAKHOVSKY *et al.* (2005) suggested that the parameter C_2 is closely related to the parameter b of rate and state friction ($b \sim 10^{-1}$; $C_2 \sim 10^{-2}$–10^{-1}), and obtained the following relation between parameters C_1 and C_2:

$$C_1 \approx BC_2 \exp\left(-\frac{\alpha_0}{C_2}\right) \Big/ \varepsilon^2_{cmp}, \tag{6}$$

where B ($\sim$1–2 s^{-1}) is a laboratory-determined time scale for the evolution of static friction with hold time (DIETERICH, 1972, 1978) and ε_{cmp} is the compaction strain estimated by the ratio between lithostatic stress and the bulk modulus (K). LYAKHOVSKY *et al.* (2005) estimated $\varepsilon_{cmp} \sim 10^{-2}$ for crustal rocks at seismogenic depths, but noted that this strain level may vary significantly for various lithologies and depths. Since the lithospheric stress within the seismogenic zone (depth 1–20 km) ranges between 20 MPa and 400 MPa, and the bulk modulus of typical crustal rocks varies by an order of magnitude (CHRISTENSEN and MOONEY, 1995), it is safe to assume that the compaction strain could vary by 2–3 orders of magnitude. Given that the parameter B may differ from the well-constrained lab-based values and that compaction strain may vary significantly, we consider (Fig. 2) a wide range of C_1 values per C_2 value (six orders of magnitude).

4.2. *In Situ Geophysical Constraints for Healing Parameters C_1 and C_2*

We use data from seismic surveys along large fault zones with significant fault-related damage to better constrain the healing parameters C_1 and C_2. Simplifying equation (3) for uniform shear deformation $\mu = \mu_0(1-\alpha)$, and using the relation between rigidity, density and shear wave velocity ($\mu = \rho V_s^2$), we convert reported seismic velocity and rigidity reductions (Section 2) into damage level estimates. The seismic and geodetic studies indicate that major strike-slip fault zones rapidly heal in the top few km to $\alpha \approx 0.75$ during the early postseismic stage (BEN-ZION *et al.*, 2003; PENG *et al.*, 2003; LEWIS *et al.*, 2005), and thereafter display damage levels of $\alpha > 0.5$ (HAMIEL and FIALKO, 2007; FIALKO *et al.*, 2002; FIALKO, 2004). While the healing at greater depth is expected to be higher, these observations may indicate that damage zones of large active faults do not completely heal over time scales of typical earthquake cycles. This argument is supported by the abundance of ancient fault zones that remain weaker than the surrounding rock (TCHALENKO, 1970; SENGOR *et al.*, 2005; ARMIJO *et al.*, 1996; POWELL and WELDON, 1992; EVANS *et al.*, 2000). Additional support comes from previous numerical models (LYAKHOVSKY *et al.*, 2001; FINZI *et al.*, 2006) and experimental work (TENTHOREY *et al.*,

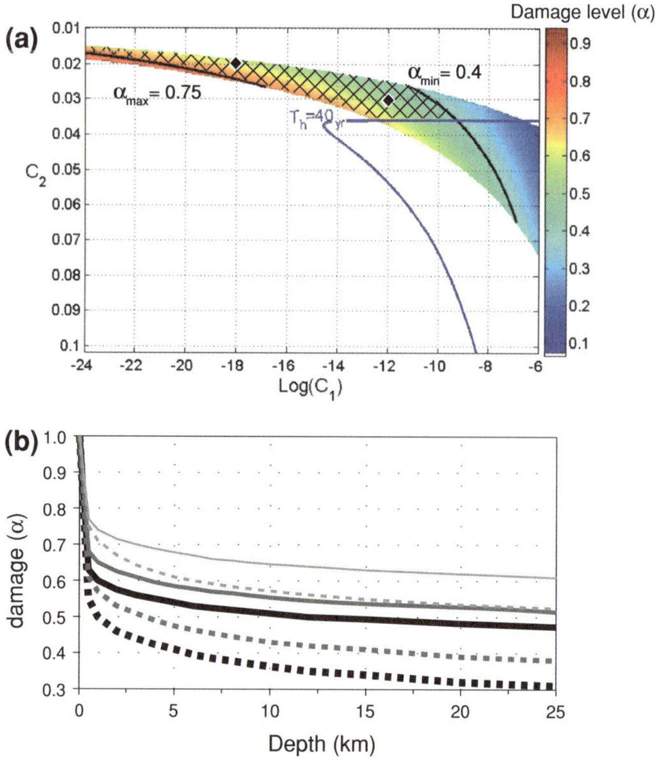

Figure 2

(**a**) Geophysical, analytical and laboratory-based constraints on healing parameters. Damage levels at a depth of 1 km after 50 years of healing are shown for C_1 values within three orders of magnitude of their expected values (Eq. (6)). The hatched region bracketed by the maximum healing time scale (τ_h = 40 yr) and the minimum and maximum admissible damage levels ($0.75 > \alpha_f > 0.4$) represents the healing parameters relevant to models of natural damage zones (see text). Diamond symbols indicate two healing parameter sets plotted in Figure 2b for illustration. (**b**) Estimated damage versus depth after 0.1 yr (gray lines), 10 yr (dark gray lines) and 100 years of healing (thick black lines) under lithospheric stress conditions, for healing parameters representative of natural fault zones (solid lines: $C_1 = 1\cdot10^{-18}\ s^{-1}$, $C_2 = 0.02$; dashed lines: $C_1 = 1\cdot10^{-12}\ s^{-1}$, $C_2 = 0.03$).

2003) showing that damage zones do not heal completely during the earthquake cycle. The above is expected to be valid for low-porosity crystalline rocks. In contrast, deformation bands in sandstones and other high porosity rocks are frequently denser and stronger than their host rock (Aydin and Johnson, 1983; Shipton and Cowie, 2003). The limited healing argument implies that long-term interseismic healing in low porosity rocks is typically minor and that damage generation and healing in such rocks occurs predominantly in the seismogenic crust during the co- and early postseismic interval (e.g., over weeks to months). This is supported by various postseismic healing rate studies (e.g., Li *et al.*, 2006; Karabulut and Bouchon, 2007; Peng and Ben-Zion, 2006; Schaff and

BEROZA, 2004; RUBENSTEIN and BEROZA, 2004; BAISCH and BOKELMANN, 2001; WU *et al.*, 2009).

We note that the similar interseismic damage levels for the different fault zones mentioned above may reflect resolution limitations of the seismic and geodetic methods (i.e., perhaps materials with $\alpha < 0.5$ are not reliably detected by these techniques). If this is the case, the argument that long-term healing is minor may not be valid. Furthermore, the healing parameters may be pressure- and/or temperature-dependent (as the damage accumulation rate parameter C_d). Further work should be done to better constrain the healing parameters at seismogenic depths. Finally, the healing computations in our parameter space study do not take into account ongoing deformation (and damage accumulation) due to continuous tectonic loading or nearby earthquakes, and therefore they may underestimate damage levels in natural systems.

Based on the above considerations we suggest two general constraints for the healing parameters suitable for models of natural processes: (a) the minimum damage level expected in shallow crustal fault zones during the interseismic stage should be above $\alpha \approx 0.4$, and (b) the healing time scale representative of natural damage zones should be shorter than $\tau_h \approx 40$ yrs (yielding healing rates of 5–10% yr^{-1} after 4–5 months of healing, and a very low rate of approximately 0.1% yr^{-1} after 40 years of healing). The resulting healing parameter space is outlined in Figure 2 by a hatched pattern. The admissible values of C_1 and C_2 in that subspace are $10^{-24}\ \text{s}^{-1}$ to $10^{-10}\ \text{s}^{-1}$ and 0.015 to 0.035, respectively.

A final analytical constraint for healing parameters can be derived from the convexity condition for macroscopic failure used in our damage rheology framework (LYAKHOVSKY *et al.*, 1997a). This stability condition indicates that near the surface, where normal stress is low compared to shear stress and the strain invariants ratio is approximately zero, the maximum sustainable (stable) damage level is approximately $\alpha = 0.75$. Therefore, healing parameters that yield lasting damage levels greater than $\alpha = 0.75$ at shallow depths ($z < 3$ km) are assumed to be nonrealistic (Fig. 2).

5. *Damage and Strain Distribution across Active Strike-slip Faults*

To investigate the structure of damage zones associated with evolving strike-slip fault systems, we use several realizations of a three-dimensional model of transform plate boundary incorporating damage rheology in the upper crust.

5.1. Model Setup

A typical model setup (Fig. 3) consists of a layered seismogenic crust governed by damage rheology, underlain by viscoelastic lower crust and upper mantle layers. The modeled region is 100–250 km in the along-strike direction, 100 km wide and 50 km deep. A detailed description of such a model setup is given by BEN-ZION and LYAKHOVSKY

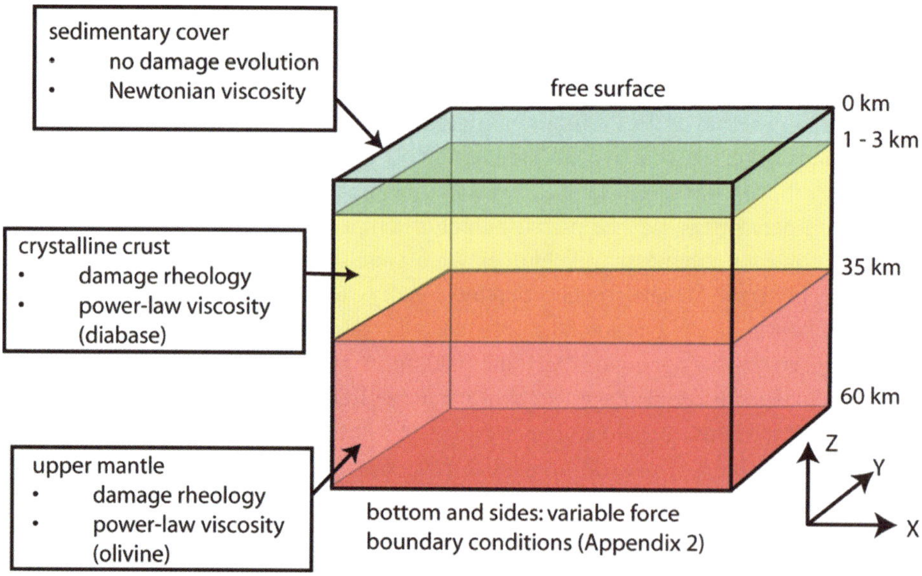

Figure 3
A block diagram of typical 3-D lithospheric structure used in the numerical simulations. The fault parallel extent of the model domain varies from 100 to 250 km in different simulations. Imposed damage (not shown in Fig. 3) is applied only in a few simulations of long-term fault stepover evolution (i.e., initial conditions of high-resolution fault stepover models included damage zones representing a segmented fault).

(2006). Here we only summarize the main ingredients. A diabase flow law is used to represent the rheology of the crystalline crust (Carter and Tsenn, 1987) and a dunite/olivine flow law is used for the upper mantle (Kirby and Kronenberg 1987). We assume a geothermal gradient of 20 C km^{-1}. The assumed flow laws and geotherm are kept fixed in our simulations. A range of damage model parameters, chosen from the values given in Table 1, is used in the models (Table 2). Since we are not trying to characterize surface damage structures, and as frequent failure of surface elements due to low confining stress would be computationally time consuming, we suppress damage accumulation in the simulated surface layer (typically top 3 km of the crust) by setting there $C_d = 0$. However, based on test models with $C_d > 0$ within the surface layer (not presented here), we expect that the surface damage zone is slightly wider and consists of locally higher damage levels than the underlying (simulated) damage zone.

A variable force boundary condition (Lyakhovsky and Ben-Zion, 2008, 2009) is applied to the sides and bottom of the model domain, simulating a constant far-field fault-parallel velocity with relative rate of 32 mm/yr (corresponding to the San Andreas Fault). These boundary conditions are further discussed in Appendix 2. The boundary driving forces are set to represent imposed fault zones outside the model domain, and they induce faulting near the centers of the north and south edges of the model (see for illustration Fig. A1). The top model boundary is stress free.

Table 2

Model parameters used in our fault evolution study (omitting models repeated with near-identical parameters)

Model name	Grid spacing (km)	Healing C_1 (s^{-1})	Healing C_2	Dynamic weakening τ_r	Seismic ratio χ	Initial damage heterogeneity	Initial τ_r heterogeneity
NB_0	3.2	6.0E-11	0.07	1.0E + 4	80%	0%	30%
NB_2	3.2	1.0E-13	0.02	1.0E + 4	80%	0%	30%
NB_6	2.2	1.0E-13	0.02	5.0E + 3	61%	0%	30%
NB_7	2.2	1.0E-13	0.02	5.0E + 3	99%	0%	30%
Nb_1_lap	3.2	1.0E-13	0.02	1.0E + 4	80%	0%	30%
Nb_3_lap	3	1.0E-13	0.02	1.0E + 4	80%	0%	30%
Nb_5_lap	2.2	1.0E-13	0.02	1.0E + 4	80%	0%	30%
Nb_7_lap	2.2	1.0E-13	0.02	3.0E + 3	67%	0%	30%
Nb_8_lap	2.2	1.0E-13	0.02	3.0E + 3	97%	0%	30%
Nb_9_lap	2.2	1.0E-13	0.02	5.0E + 3	57%	0%	30%
Nb_11_lap	2.2	5.0E-11	0.05	4.0E + 3	61%	0%	30%
Nb_geos_4	1.6	1.0E-13	0.02	1.0E + 4	80%	30%	30%
Nb_geos_6	1.6	1.0E-13	0.02	2.0E + 3	67%	30%	30%
Prop_2	3.2	1.0E-13	0.03	8.0E + 3	80%	25%	30%
Prop_7	3.2	2.0E-09	0.15	1.0E + 4	80%	15%	30%
Prop_9	3.2	6.0E-11	0.07	1.0E + 4	80%	25%	30%
Prop_lap_2	3.2	1.0E-13	0.03	9.0E + 3	80%	25%	30%
Prop_lap_6	3.2	1.0E-13	0.03	1.0E + 4	80%	10%	30%
Prop_run_a9	2.2	1.0E-12	0.03	5.0E + 3	61%	10%	30%
Prop_geos_5	2.2	1.0E-12	0.03	7.0E + 3	72%	10%	30%
Prop_lin_4.5	4.5	1.0E-13	0.03	1.0E + 4	80%	0%	30%
Stepover_8	0.25	1.0E-10	0.03	3.0E + 1	40%	100%	30%
Stepover_9	0.25	1.0E-10	0.03	1.0E + 2	40%	100%	30%
Stepover_10	0.25	1.0E-10	0.025	4.0E + 2	40%	50%	30%
Stepover_11	0.25	1.0E-10	0.025	6.0E + 2	40%	100%	30%
long-term_1	0.6	1.0E-12	0.03	6.0E + 4	80%	25%	30%
long-term_5	0.6	1.0E-20	0.02	6.0E + 4	80%	25%	30%

Other material parameters were set to represent crustal materials and were not varied in our models (these parameters include: ξ_0 = -0.8. $C_d = 5\ s^{-1}$; Sedimentary layer: density $\rho = 2.4 \cdot 10^{-3}$ Kg m^{-3}, Newtonian viscosity $\eta = 10^{19}$ Pa S; Crustal rheology: Young's modulus E = 80 GPa, Poisson's ratio $v = 0.3$, $\rho = 2.8 \cdot 10^{-3}$ Kg m^{-3}, viscosity flow law coefficients: $A = 6.3 \cdot 10^{-20}$ Pa^{-n} S^{-1}, n = 3.05, Q = 276 Kj mol^{-1}; Mantle rheology: E = 150 GPa, $v = 0.3$, $\rho = 3.3 \cdot 10^{-3}$ Kg m^{-3}, $A = 7.0 \cdot 10^{-14}$ Pa^{-n} S^{-1}, n = 3.0, Q = 520 Kj mol^{-1})

5.2. Model Output: Examples and Interpretation

The model outputs include the level of damage α and strain ε. We calculate surface velocities, rigidity μ and other related quantities from these variables. Since α, ε and μ are computed throughout the model domain, both plan views (including depth slices) and cross-sectional views of these parameters may be plotted at any time step. Figures 4 and 5 show examples of model outputs, and illustrate features which correspond to observed geological structures such as fault segments, stepovers, and flower structures.

Contiguous sets of elements that fail repeatedly, resulting in a higher level of α (and a lower μ) than their surroundings, are interpreted as fault segments (Figs. 4, 5). Because of their relative weakness, these fault segments are also the centers of high velocity gradients

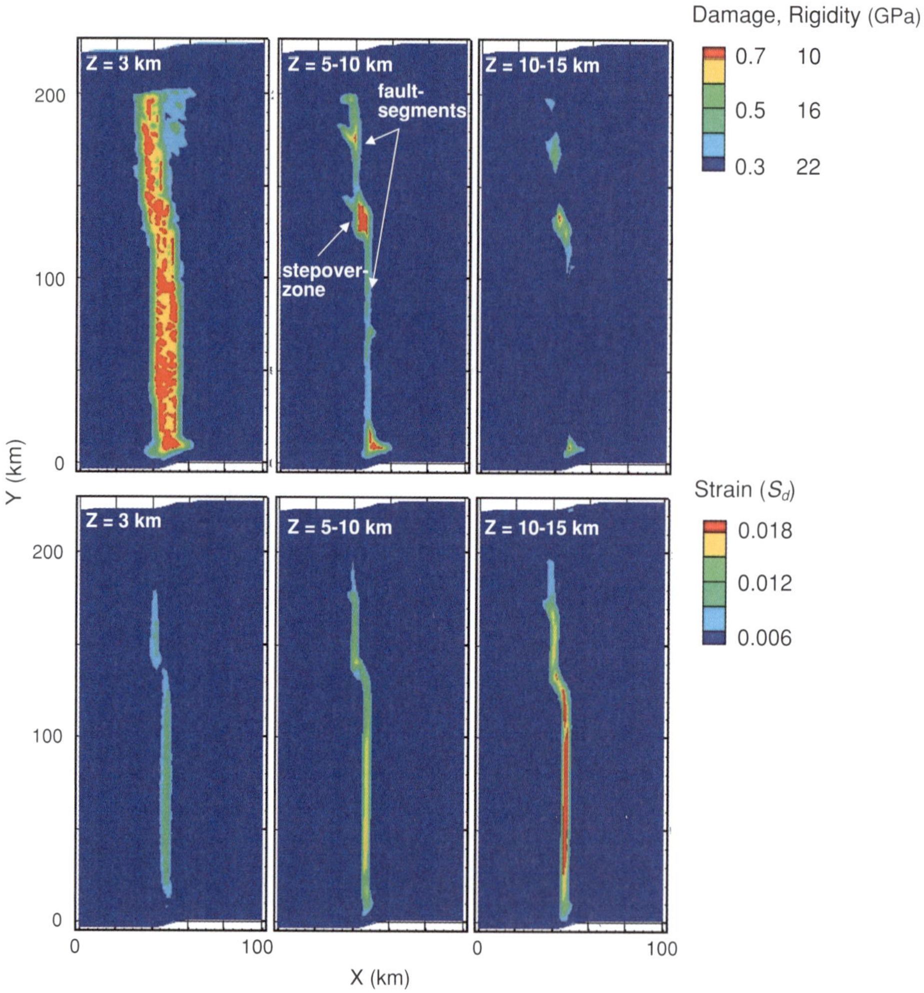

Figure 4
Plan views of a segmented strike slip fault at several depths, showing damage level (α), rigidity (μ) and the second invariant of the deviatoric strain, S_d (S_d = sqrt($e_{ij} \cdot e_{ij}$) where $e_{ij} = \varepsilon_{ij}$-$\delta_{ij}$ ε_{kk} /3 and δ_{ij} is the Kronecker delta). Shallow (z = 3 km) damage (top left panel) is distributed within the stepover and around the fault segments. At depth (z = 5–10 km) damage is highly localized along the fault segments and is distributed within the fault stepover. At the lower part of the seismogenic crust, damage within stepover may persist long after the localized damage along fault segments heals.

and high strain rate (Figs. 4, 5). Cross-sectional profiles through modeled fault segments (Fig. 5) display "flower structures" with depth, which comprise localized damage along the active fault core with a superimposed, broader zone of distributed damage in the top 3–10 kilometers of the crust (Figs. 4, 5). Based on these observations, we define two damage subzones that are distinct in their evolution patterns, damage level, and spatial

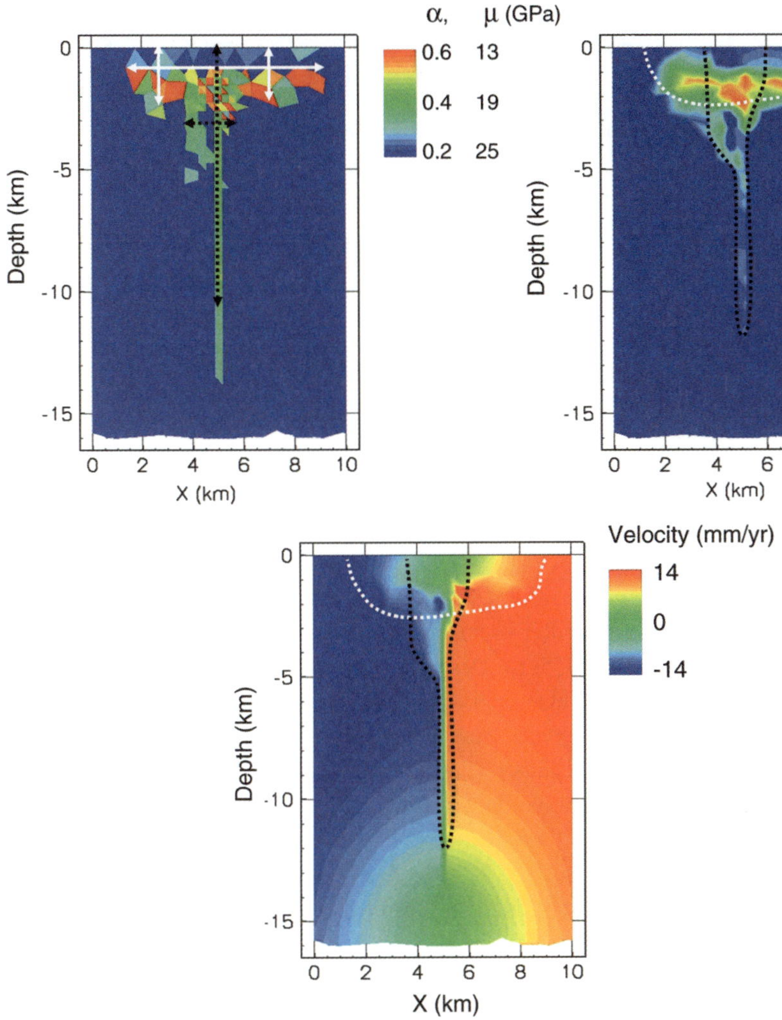

Figure 5
Cross-sectional views of a typical "flower type" damage zone along a strike-slip fault, displaying damage levels and rigidity (**A** and **B**), and fault-parallel velocity (**C**). Annotations in (**A**) show the dimensions of the Localized Active Fault (LAF) damage (black lines) and Distributed Off Fault (DOF) damage (white lines) as they were measured in this study. Dashed black and white lines show the location of the LAF and DOF damage zones (respectively) in the smoothed contour plots (**B**, **C**).

distribution: (1) Localized Active-Fault (LAF) damage which represents the highly localized damage along the active fault cores (Figs. 4, 5). The LAF damage is coseismically very high along the primary slip zone, but it rapidly heals at depth. (2) Distributed Off-Fault (DOF) damage which is sustained cumulative damage resulting from many earthquakes. The DOF damage develops during the early stages of

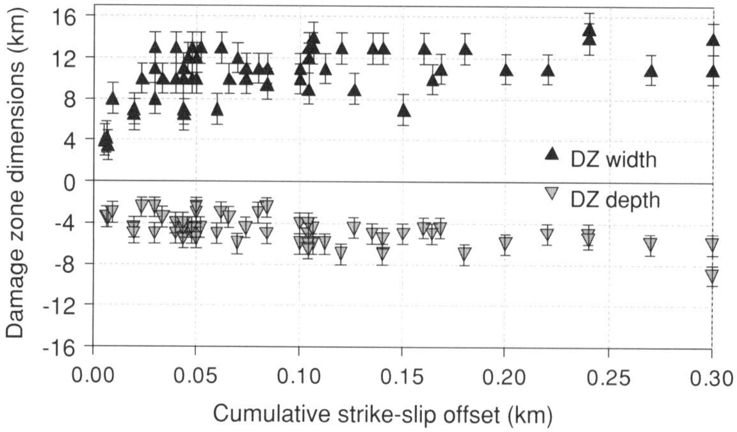

Figure 6
DOF damage zone width and depth plotted against cumulative strike-slip offset. After an initial stage with relatively fast damage zone growth, the DOF damage zone dimensions remain fairly constant (at offsets exceeding 0.05 km).

fault-system evolution, and thereafter its spatial extent is stable and the degree of damage within it evolves locally (Figs. 4, 5, 6). Descriptive analyses of damage structures along simulated strike-slip fault segments are given in Section 5.3.

Plan views of the model domain show several examples of stepover zones where segments are offset from one another (Fig. 4). Our models produce just extensional stepovers; that is, during propagation, new offset faults form in areas where end-effects from existing faults contribute a tensional mean normal stress. Stepovers are characterized by a wide DOF damage zone, and by high levels of damage, high strains, and low rigidity within the stepover. Descriptive analyses of damage structures within simulated stepover-zones are given in Section 5.4.

Before we discuss the results of our models, we need to (1) introduce observable quantities which may be systematically measured and then used to compare between model results, (2) define a threshold criterion for the maturity of our modeled faults to ensure that we base our analysis only on models that were run long enough to form mature damage zones, and (3) confirm that the model results presented here are fairly insensitive to numerical element dimensions. This is necessary because the models shown on Table 2 were run for different durations, and for a range of element dimensions.

To systematically describe the spatial extent of damage zones around faults we have chosen a threshold damage level of $\alpha = 0.35$ (presumably above any expected background damage level). The four quantities used to define the extent of damage are the widths and depths of both the LAF and DOF damage zones. Measurements of these quantities are performed on plan-view and cross-section plots of the simulated fault zones (without smoothing). The width of the DOF damage represents the maximal spatial extent of the damage zone, and its depth is the average depth extent of the shallow distributed damage away from the active fault core (Fig. 5a). The width of the LAF damage represents the maximum width of the localized damage along the fault core, and its depth is the maximum depth extent of the damage zone (Fig. 5a). To facilitate comparison with geodetic studies we define also the Fault Compliant Zone (FCZ) in our models as the volume in which the average material rigidity is reduced by 50% relative to the host rock, consistent with the compliant zones of FIALKO *et al.* (2002). The simulated FCZ typically consists of the entire LAF damage and most of the DOF damage.

Strike-slip fault systems evolve over time, first becoming complex and then gradually simplifying to a more continuous configuration with fewer fault segments (e.g., BEN-ZION and SAMMIS, 2003 and references therein; LYAKHOVSKY and BEN-ZION, 2009). This complicates directly comparing model runs that may have been cut off at different evolutionary stages. Figure 6 illustrates that the width and depth of the DOF damage initially grows rapidly, starting to stabilize after a total relative displacement of approximately 0.05 km. This simulated stage of rapid DOF damage growth represents the initial stage of fault growth and complexity increase after which the fault configuration starts to stabilize and the strain localizes along the fault. Hence to analyze mature damage zone structures we can only use fault simulations with a total displacement larger than 0.05 km.

All results displayed on subsequent plots are for models with mature damage zones in which at least 0.05 km of displacement has accrued (corresponding to a modeled time interval of about 1600 years).

Since brittle failure in our simulations is associated with an abrupt transition from initial (static) to final (dynamic) stress levels, the numerical models are inherently discrete (e.g., RICE and BEN-ZION, 1996) and some aspects of the results are expected to be grid-dependent. An analysis of our entire set of damage zone simulations indicates that while the spatial extent of damage is somewhat grid-size dependent, the average level of damage within voluminous damage zones (e.g., stepover zones) is probably not grid-size dependent (Fig. 7). Furthermore, the spatial extent of damage zones in our models with element dimensions between 0.6 and 4 km (Fig. 7) are not significantly sensitive to element size. Results of models within this range of element dimensions will be shown together on subsequent plots. Models with finer and coarser elements were also run. For models with 4.5 km elements, results were smeared forming apparently wider and deeper damage structures. Models with 0.25 km elements ran very slowly and were numerically unstable. These simulations were terminated after 10–20 days of CPU time, during which the simulated damage zones did not reach a stable width. Based on Figure 6 we interpret the narrow and shallow damage zones that formed in these simulations as being immature, and we therefore do not incorporate them in our analysis.

5.3. Damage, Rigidity and Strain Distribution across Strike-slip Fault Segments

Simulated damage zones along strike-slip faults form flower structures consisting of a shallow DOF damage zone 6 to 14 kilometers wide and up to 7 kilometers deep, and a

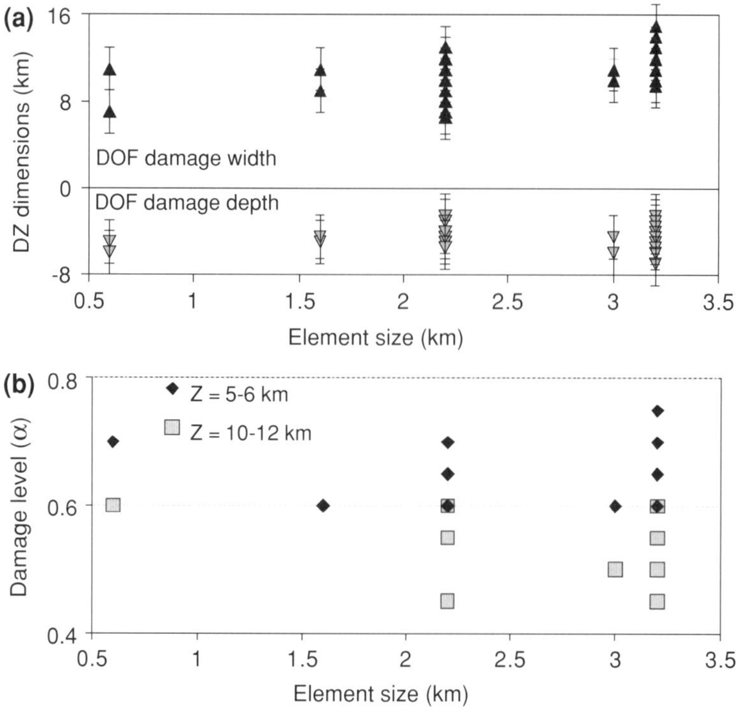

Figure 7
Sensitivity of model results to element size. (**a**) For models with element dimensions between 0.6 and 3.2 km, DOF damage zone dimensions are fairly insensitive to grid size. (**b**) Damage levels in fault stepovers at depth of 5–6 km and 10–12 km. These damage levels are insensitive to element size within the range we model. The scatter in results on both panels is due to different values of damage model parameters and variations in cumulative slip.

more intense LAF damage zone around the fault core. The DOF damage is the result of cumulative damage from past earthquakes (Figs. 5, 6). The LAF damage is narrow and steeply dipping, and it may extend to the bottom of the seismogenic zone (Figs. 4, 5) coseismically and in the early postseismic interval. The deeper parts of the LAF damage zone are only one model element wide and appear discontinuous (Figs. 4, 5). This probably indicates a tendency to evolve to a much narrower extent than our model element dimensions, in agreement with observed extreme localization of active slip zones (e.g., CHESTER and CHESTER, 1998; SIBSON, 2003; ROCKWELL and BEN-ZION, 2007). The highest damage levels along fault segments are usually found at the shallowest part of the LAF damage zone. The parts of the LAF damage in the shallow crust that are more than one element thick may correspond to the observed distributed DZ along the surface trace of large strike-slip faults.

The flower structure in our simulations is a robust feature that shows little dependence on damage parameters (C_1, C_2, χ, τ_r), and limited sensitivity to the presence of various

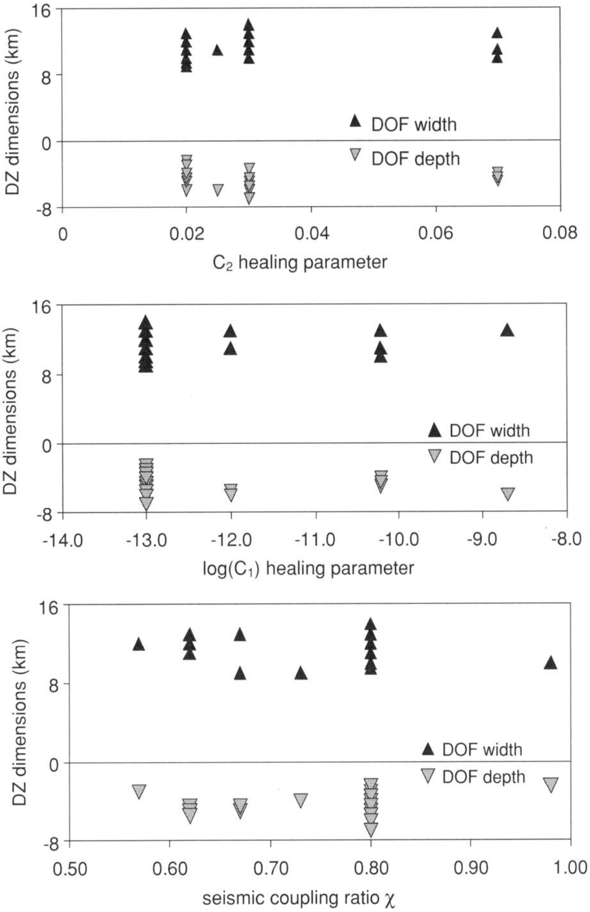

Figure 8
Width and depth of the distributed part of the damage zone (DOF damage) for models with a wide range of material parameters. DOF damage zone dimensions are insensitive to the healing parameters and the seismic coupling ratio.

degrees of material heterogeneities (Table 2). The insensitivity of the DOF damage to healing parameters (Fig. 8) may imply that the shallow crust is readily damaged and it experiences limited healing regardless of material parameters. This apparent insensitivity to the healing parameters may also indicate that the current range of modeled healing parameters is insufficient and further analysis is needed to understand the role of healing in damage zone evolution. The deeper sections of the LAF damage are more sensitive to the healing rate parameters (Fig. 9a). This reflects the fact that healing processes are sensitive functions of the normal stress. Our results indicate that the depth extent of the LAF damage (after the early postseismic interval) ranges from the entire seismogenic zone for materials with extremely slow healing ($C_2 > 0.05$; $\tau_h > 55$ yrs; Fig. 9b) to a few

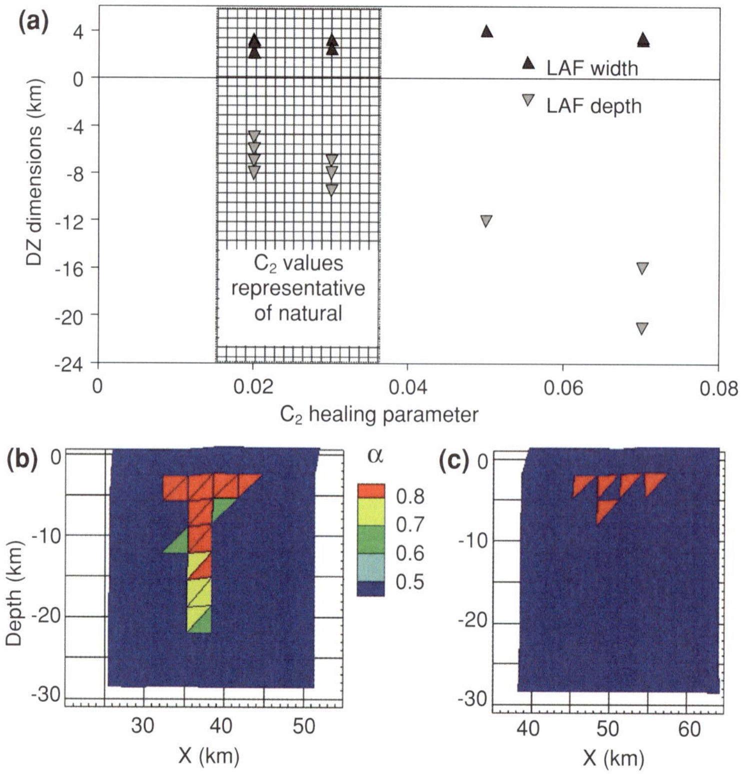

Figure 9

(**a**) Width and depth of the LAF damage zone as a function of healing parameter C_2 (width measured at 5–8 km depth, just below the DOF damage). The hatched region indicates admissible values of C_2 for modeling natural processes (see section 4). (**b, c**) Examples of deep and shallow fault core damage zones in models with long (**b**) and short (**c**) healing time scales. The geometry of these zones is stable through most of the interseismic interval.

kilometers for the faster healing materials ($C_2 < 0.03$; $\tau_h > 25$ yrs; Fig. 9c). Model realizations with healing parameters representative of natural processes (see Section 4) indicate that during most of the seismic cycle the contiguous well-developed LAF damage is limited to the top section of the crust (e.g., Figs. 9c, 10a). These results are consistent with numerical simulations of plastic strain generation during dynamic rupture (Ben-Zion and Shi, 2005), and analyses of large seismic data sets recorded around active faults (e.g., Ben-Zion *et al.*, 2003; Peng *et al.*, 2003; Korneev *et al.*, 2003; Cochran *et al.*, 2003; Lewis e*t al.*, 2005; Graymer *et al.*, 2007).

The elastic strength of simulated fault zones is reduced as the damage level increases. The relation between rigidity and damage level (equation (2)) implies that the effective rigidity is primarily a function of α. However, the effective rigidity is further reduced near the surface and along the fault core because the strain invariants ratio ξ is higher at these

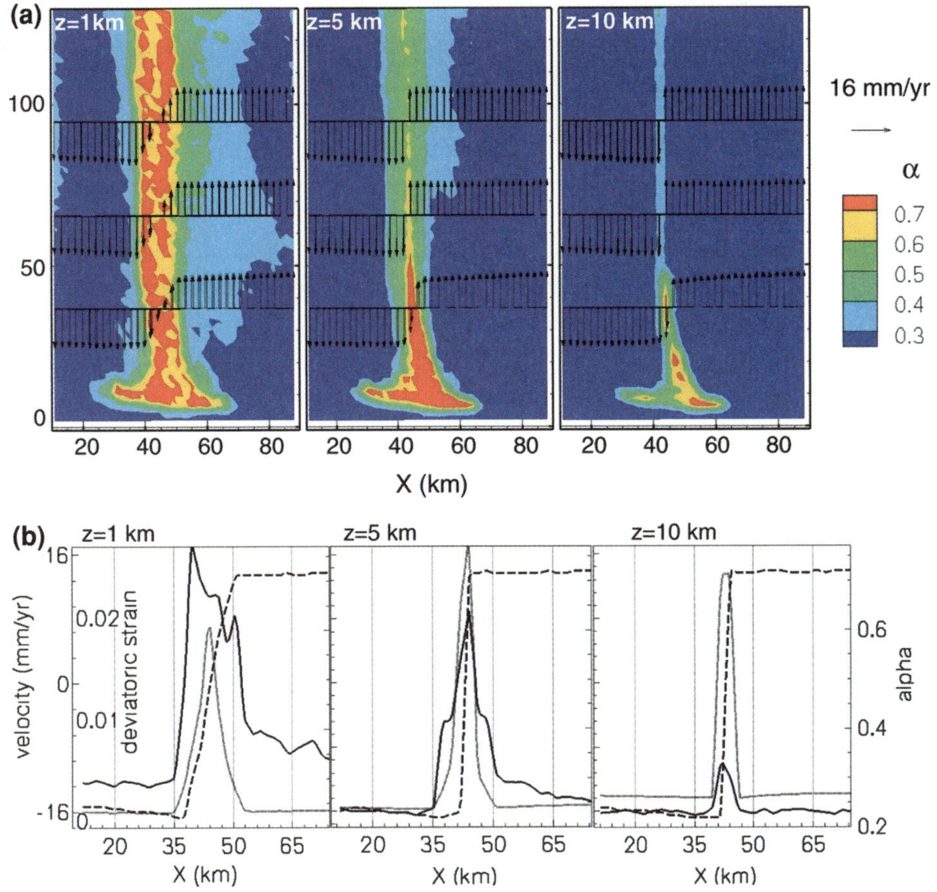

Figure 10
Correlation between damage and strain across simulated fault-zones. (**a**) Damage at depths 1, 5, 10 km and superimposed fault-parallel velocity. (**b**) Plots of damage level (solid curve, right ordinate), long-term (multi-cycle) fault parallel velocity (dashed curve, outer-left ordinate) and deviatoric strain (gray dotted curve, inner-left ordinate) for the same depths, extracted along the central section shown in (**a**) (at y = 65 km).

localities. Our damage zone models indicate that significant rigidity reduction (50–70% reduction) occurs within the shallow DOF damage and along the upper part of the fault-core (e.g., top 5 km of the LAF damage). Along the LAF at depths exceeding 12 km healing is rapid and the long-lasting (interseismic) reduction on μ is relatively small (Figs. 4, 5).

Our fault evolution models indicate that the bulk of inelastic strain is concentrated in the highly damaged cores of the fault zones (i.e., along the LAF damage). However, our models also exhibit low-gradient strain beyond the fault cores in the uppermost crust (top 3–5 km). The modeled off-fault strain is typically confined to a shallow layer approximately 10–15 km wide that exhibits significant DOF damage ($\alpha > 0.4$: see

Fig. 10). Where the shallow DOF damage is of lower intensity ($\alpha < 0.4$) our models do not display significant strain, suggesting that relatively thin layers of slightly damaged rock may not modify regional surface deformation patterns and may not be easily detectable by means of geodesy. At distances greater than 10 km from fully formed fault-zones, the total deviatoric strain is negligible and the long-term average strain rate is uniform, indicating that the undamaged upper crust behaves kinematically as a rigid block. The correlation between damage and strain distribution confirms recent interpretations of observed surface deformation patterns above fault-related compliant zones (e.g., FIALKO *et al.*, 2002; FIALKO, 2004). This correlation indicates that geodetically observed compliant zones are related to relatively high damage levels ($\alpha > 0.4$), and suggests that parts of the damage zones ($\alpha < 0.4$) may not be geodetically observable.

5.4. Fault Stepovers

While major fault segments display a positive feedback of strain-weakening and localization along highly damaged fault cores, persisting geometrical features such as fault offsets, kinks, and bends, may display strain-hardening and produce local complexity and new fractures at different scales (BEN-ZION and SAMMIS, 2003). In our models, segmented fault zones display continuous distributed seismic and aseismic deformation within fault stepover zones. This consists of aseismic strain and small earthquakes; however, moderate earthquakes ($M_L < 5.5$) also occur occasionally. While much of the damage along fault segments heals during the interseismic stage, the damage level at stepovers remains persistently high (Fig. 11a, compare with Fig. 9c). Fault stepovers and segment termination zones undergo significant damage accumulation during the interseismic stage, typically displaying average α between 0.5 and 0.75 (measured at a depth of 5–10 km), depending on the healing rate parameters. The damage within stepovers extends to greater depths than along fault segments, and in many simulations it reaches the bottom of the seismogenic zone (Figs. 11, 4). Models with realistic healing parameters typically exhibit significant damage ($\alpha > 0.5$) to depths of 10–15 km (Fig. 11). An important implication of the permanently elevated damage level within fault stepovers is that these regions of reduced μ (Fig. 4) affect rupture propagation and strong ground motion patterns.

5.5. Fault System Complexity as a Function of Time

According to our results, the DOF damage dimensions depend on the maturity of the fault system. Simulations with a wide range of material properties and numerical characteristics (e.g., element dimensions and boundary conditions) indicate that the damage zone grows until the fault accumulates an offset of about 0.05–0.1 km (Fig. 6). During this early evolutionary stage, the fault system's complexity increases as additional segments nucleate and propagate, forming new damage zones. This stage culminates as deformation localizes along narrow slip zones. As mentioned, an important exception to

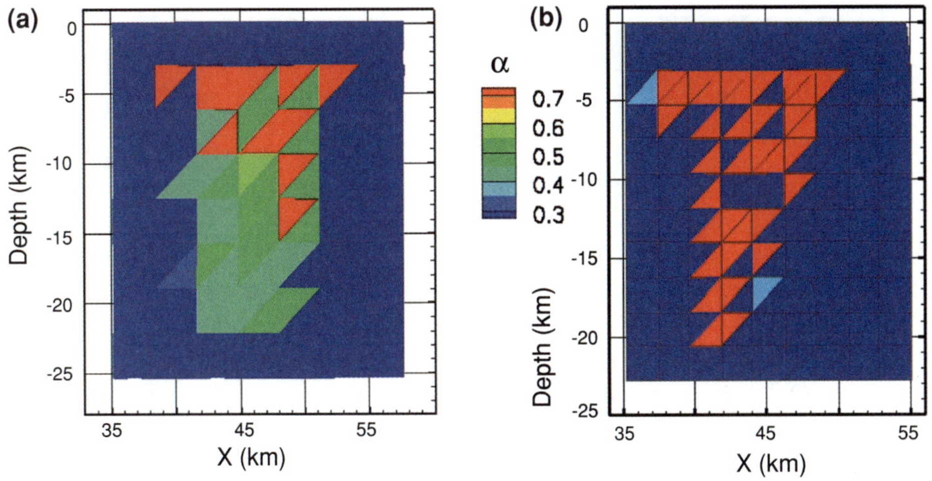

Figure 11
Cross-sections of damage patterns within fault stepovers. These zones of fault complexity are interseismically active, retaining high damage levels throughout the entire seismic cycle. Significant differences in damage level at depth (i.e., at Z > 10 km) arise from varying healing parameters: Models with fast and efficient healing (**a**) yield lower damage levels than those with slow and inefficient healing (**b**). Based on our healing rate analysis (Section 4), the faster healing models (e.g., **a**, and Fig. 9c) are better representatives of natural DZ processes. The healing rates in these models are identical to those in Figures 9b and 9c.

this evolution stage is the widening of the DZ at bends and stepovers and other local fault complexities (Figs. 12a-c, 13). The fault system evolution continues as through-going faults bridge sites of fault complexity such as stepovers (Figs. 12d, 13). When through-going faults fully form and bridge the entire depth of the stepovers, these structures may become inactive. After this stage the width and depth of distributed damage remain fairly constant, until the existing fault configuration cannot accommodate the evolving regional stress, at which point new faults form and migration of faulting may occur. Parallel faults may sustain simultaneous damage accumulation (Fig. 13) or exhibit alternating deformation periods before the initial faults become inactive and the new fault configuration stabilizes. These results are consistent with existing multi-disciplinary knowledge of fault systems (BEN-ZION and SAMMIS, 2003 and references therein; DOLAN *et al.*, 2007) and the related numerical results of LYAKHOVSKY and BEN-ZION (2009).

6. *Discussion and Conclusions*

We perform a large numerical parameter-space study relating to general aspects of structural evolution of large strike-slip faults and related deformation fields, using a layered lithospheric model with an upper crust governed by a continuum damage rheology. One barrier to the widespread incorporation of damage rheology in crustal

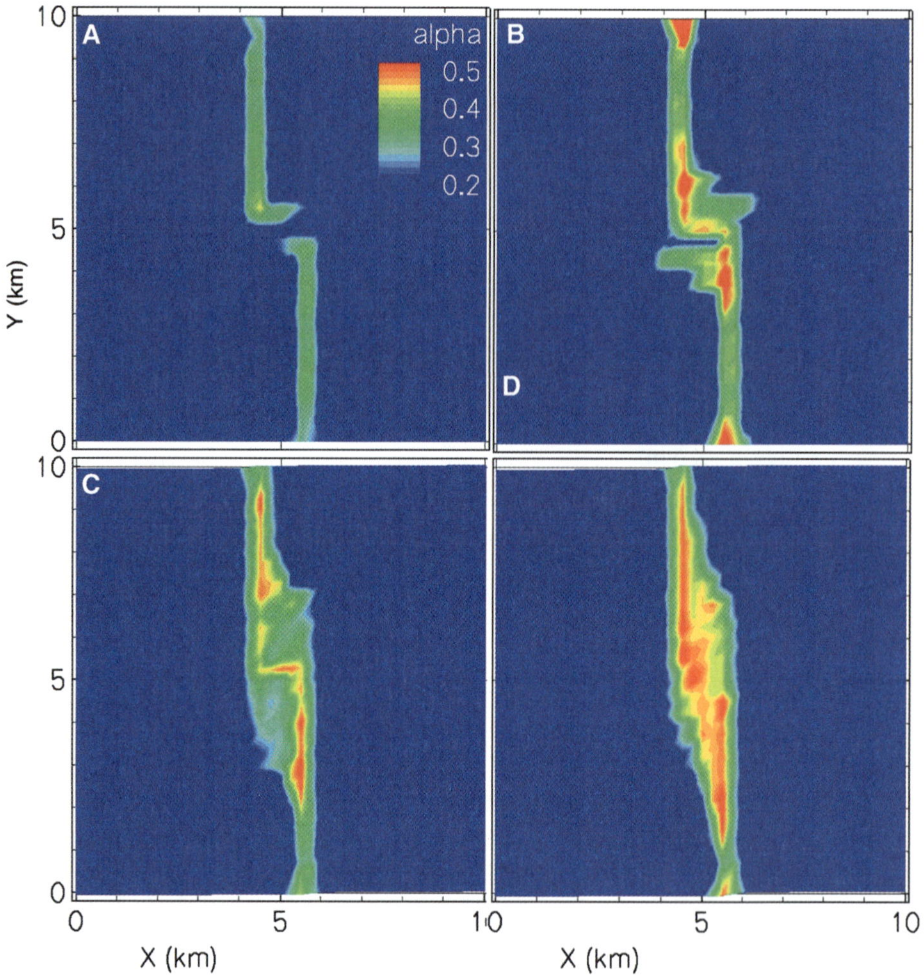

Figure 12
Fault stepover evolutionary stages displayed in four snap-shots of damage levels around a stepover zone (at 3 km depth). (**a**) Segmented fault. (**b**, **c**) Extensive damage accumulation within the stepover zone. In (**c**) distinct lateral (subsidiary) "faults" (regions of high damage) link between the two fault segments. (**d**) Formation of a through-going fault through the entire stepover zone. These results are from high-resolution, small-domain models focusing on stepovers.

deformation models has been the numerous damage model parameters, whose relationships with observable phenomena are sometimes unclear. Considerable work in recent years has gone into relating these parameters to results of brittle deformation experiments and thermodynamic theory (e.g., LYAKHOVSKY *et al.*, 2001, 2005; HAMIEL *et al.*, 2004, 2006). We take this further by using the observed shear modulus reduction in damaged fault zones to constrain the ranges of likely values for healing parameters C_1 and C_2. We find that admissible values of C_1 and C_2 are 10^{-24} to 10^{-10} s^{-1} and 0.015

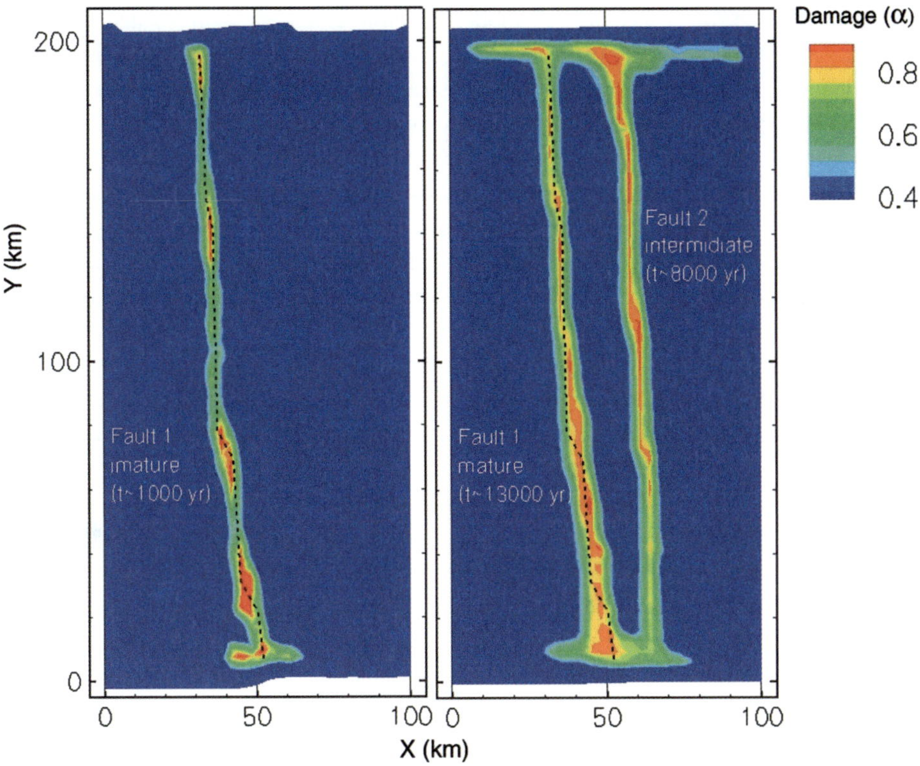

Figure 13

Fault evolution snap-shots showing damage levels at 10 km depth. The originally segmented fault (left panel; notice large stepovers) is smooth with time (right, mature fault 1; note that some of the apparent small stepovers are actually numerical artifacts formed because the fault is not parallel to the grid). An additional fault formed after approximately 5000 yrs to better accommodate regional stress (right, fault 2). Annotations of times in the images indicate the age of each fault in the simulation. Finely dashed line long fault 1 in both panels indicates its original segmented outline (as shown in left panel).

to 0.035, respectively (Fig. 2). This range is significantly reduced relative to previous studies (e.g., LYAKHOVSKY *et al.*, 2005).

Our models with reasonable damage and viscoelastic parameters yield general deformation patterns that are comparable to those seen in natural strike-slip fault systems. Flower structures, stepovers, localized strain around fault segments and permanent damage in the shallow upper crust and within stepovers are all reproduced. Due to model simplifications (e.g., no damage accumulation in the surface layer) and element size limitations, our models cannot conclusively predict the details of surface damage patterns, the width of the fault core, or the geometry of small faults and fractures within flower structures and stepover regions. Our simulations would probably yield narrower damage zones if we incorporated depth-dependent damage-rate parameters as suggested by LYAKHOVSKY *et al.* (2005).

The fault stepover zones in our models exhibit extensive damage and elasticity degradation sustained during many earthquake cycles. The simulated tensional stepovers show damage patterns consistent with intense tensile fracturing and dilation, and therefore are expected to exhibit long-lived enhanced permeability. Such damage patterns are consistent with recent structural evolution models for dilational stepovers (De Paola *et al.*, 2007), and with mineral exploration studies that relate hydrothermal ore deposits to long-lasting extensive damage and increased permeability within fault stepovers (Micklethwaite and Cox, 2004; Sheldon and Micklethwaite, 2007). The permanent damage zones our models predict should be detectable with detailed seismic and geodetic imaging studies.

An important implication of the predicted damage zones at stepovers concerns the interaction between damaged material and propagating earthquakes. During the interseismic stage, weakened stepover zones experience continuous earthquakes and proportionally more inelastic strain than the surrounding crust. This reduces the interseismic stress accumulation in the stepover region, which could aid earthquake rupture arrest. Various studies address rupture propagation across stepovers using quasi-static models (e.g., Segall and Pollard, 1980), dynamic models (Harris and Day, 1999; Harris *et al.*, 1991) and field observations (e.g., Wesnousky, 2006). Such studies show that it is easier for a rupture to jump across dilational stepovers (such as those our model produces) than compressional stepovers, because dilational normal stress brings nearby faults closer to failure. Sibson (1985) and Hamiel *et al.* (2005) argue, however, that during an earthquake, the normal stress change in a dilational stepover could lead to a sudden opening of fluid-filled cracks, reducing pore pressure and causing material hardening. An analogous effect is also seen in some of the models of Harris and Day (1993), where fluid within a dilational stepover inhibits the ability of rupture to propagate across it. As long-term damage accumulation and coseismic pore pressure decrease have competing effects on the elastic strength of a stepover, further study will be required to clarify the ramifications of damage for rupture propagation and arrest. We also note that the extensive rock damage near stepovers should produce amplified ground motion, and hence higher seismic hazard, at those regions.

While the presented model simulations have several limitations, the following features appear to be robust:

- Distributed fault zone damage develops early in the evolution of a fault system (approaching steady-state damage zone dimensions after .05 km of total slip in our models).
- The strain generally localizes to narrow segments which are wider near the surface than at depth (representing flower structures).
- Along fault segments, the damage heals postseismically at depths exceeding 5–10 km and is permanent at shallower depths.
- The off-fault strain distribution correlates with the permanent shallow damage exhibited along fault-zones.

- Stepover regions may be permanently damaged to mid-crustal depths, and such damage zones should be detectable with focused seismic and geodetic studies.

Our numerical simulations of fault zone evolution suggest that the overall aspects of fault zone deformation along large faults at seismogenic depths can be modeled effectively over time intervals of several large earthquake cycles by a collection of narrow segmented zones or planar surfaces (i.e., without incorporating off-fault damage evolution). However, the evolving fault zone structures and off-fault damage play important roles in the evolution of geometrical properties of fault sections, the formation of fault systems, and in the deformation patterns along plate boundaries. Although regional scale static stress transfer in the crust is probably only moderately affected by the presence of shallow weakened damage zones along fault segments, long-lived volumes of extensively damaged material within fault stepovers and near other persistent geometrical irregularities can significantly affect earthquake propagation, strong ground motion, (locally) crustal stress state and surface deformation patterns.

Acknowledgments

We thank Yann Klinger and Harsha Bhat for useful comments that enhanced the readability of the paper. The study was supported by the Southern California Earthquake Center (based on NSF Cooperative Agreement EAR-0106924 and USGS Cooperative Agreement 02HQAG0008) and the US–Israel Binational Science Foundation, Jerusalem Israel (2004046).

Appendix 1: Numerical Procedure

The numerical code we use for our modeling utilizes the Fast Lagrangian Analysis of Continua (FLAC) method (e.g., CUNDALL and BOARD, 1988; POLIAKOV *et al.*, 1993). This formulation is explicit in time and it continuously updates the shape functions of tetrahedral elements allowing large deformations to be simulated. The general procedure involves solving the equations of motion to determine nodal velocities and to calculate the elastic component of the element strains. The total strain tensor is calculated by summing the elastic strain component, a viscous strain component and a component representing the damage-related inelastic strain ε_{ij}^{v}. HAMIEL *et al.* (2004) introduced the gradual inelastic strain, ε_{ij}^{v}, whose accumulation rate is proportional to the rate of damage accumulation:

$$\frac{d\varepsilon_{ij}^{v}}{dt} = \begin{cases} C_v \dfrac{d\alpha}{dt}\tau_{ij} & \text{for } \dfrac{d\alpha}{dt} > 0 \\ 0 & \text{for } \dfrac{d\alpha}{dt} \le 0 \end{cases} \qquad \text{(A1)}$$

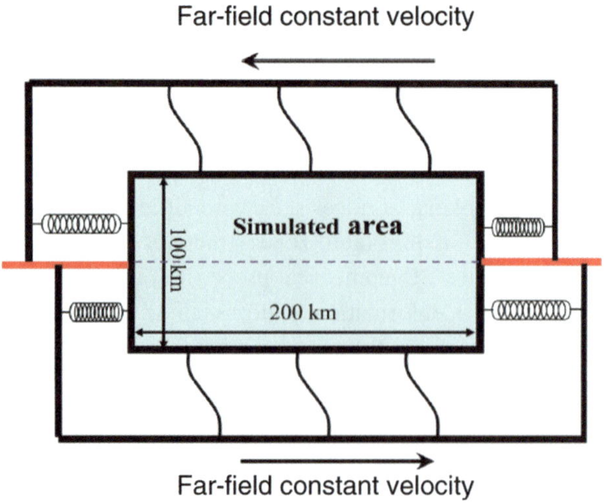

Figure A1
Schematic illustration of the variable force boundary condition. The model domain is coupled to a constant, fault-parallel velocity in the far-field, with the degree of coupling depending on a stiffness parameter (springs). Red lines indicate shear localization imposed by the boundary conditions as they were applied in our models (localized shear is also applied on the bottom boundary of our models, not illustrated here). Tectonic velocities and domain dimensions are for illustration and may be varied.

C_v is a material constant and $\tau_{ij} = \sigma_{ij} - \sigma_{kk}\delta_{ij}/3$ is the deviatoric stress tensor. The compliance, or inverse of viscosity ($C_v \cdot d\alpha/dt$), relates the deviatoric stress to the rate of irreversible strain accumulation. Ben-Zion and Lyakhovsky (2006) connected the rate of irreversible strain accumulation with partitioning between seismic and aseismic deformation in the seismogenic zone, and showed that the fraction of elastic strain released seismically, referred to as the seismic coupling coefficient χ, can be estimated as:

$$\chi = \frac{1}{1+R} \tag{A2}$$

The damage rheology constitutive and kinetic relations provide element stresses and local values of damage, which are used to update material properties (moduli μ and γ) and to calculate the nodal force balances applied in the next time step. This procedure continues through many time steps until damage in any element reaches the critical level $\alpha = \alpha_{cr}$ associated with brittle instability. The brittle failure at any element may lead to rupture propagation. This is simulated with a quasi-dynamic procedure for calculating stress drop and plastic strain in regions sustaining brittle instabilities (Lyakhovsky and Ben-Zion, 2008). Their formulation connects the damage rheology parameters with dynamic friction of simpler frameworks, and the plastic strain accumulation is governed by a procedure that is equivalent to Drucker-Prager plasticity.

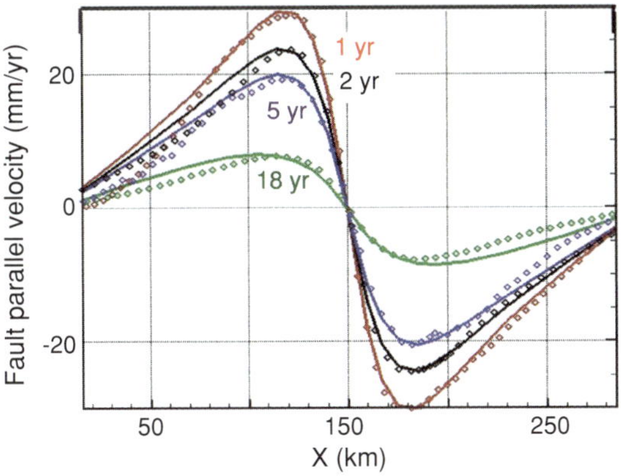

Figure A2
Surface velocities from our damage-disabled model (points) compared to a finite element solution of viscoelastic postseismic deformation (solid lines). The various lines and points represent velocities after 1 year (red), 2 years (black), 5 years (purple) and 18 years (green) of postseismic deformation. Maxwell time T_m= 110 years, other model detail are given in Appendix 2.

Appendix 2: Boundary Conditions

Numerical models of evolving fault systems typically incorporate either constant velocity or constant stress boundary conditions. LYAKHOVSKY and BEN-ZION (2008) show that these boundary conditions may generate very different velocity distributions during the interseismic period, and they conclude that neither of these boundary conditions is appropriate for simulations of fault system evolution over many earthquake cycles. To perform such simulations, LYAKHOVSKY and BEN-ZION (2008) introduced a modified boundary condition in which forces are proportional to a stiffness of virtual springs multiplied by the mismatch (slip-deficit) between the far-field plate motion and displacement of the boundary nodes (Fig. A1). Unlike the constant-velocity or constant-stress boundary conditions, this boundary condition accounts for stress accumulation and strain-rate decrease during the interseismic periods, as well as abrupt stress reduction at the model boundary during seismic events. The variable force condition realistically adjusts the forces applied on the model domain boundaries according to the evolution of elastic properties, stress state, and seismic events within the simulated domain (LYAKHOVSKY and BEN-ZION, 2008).

We constructed a series of test models to demonstrate the performance of this boundary condition and to study the spring stiffness parameter. Using these models we compared postseismic deformation in large domain models with fixed fault parallel boundaries, to results obtained from narrow-domain models with the variable force boundary condition. In the large models the domain width was set to be twenty times larger than the prescribed rupture depth (so that minor deformation is expected to occur near the fault parallel boundaries), and the fault-parallel boundaries were fixed. In the narrow-domain models, the fault perpendicular width was 140 km (i.e., fault parallel boundaries were 70 km from the fault). In both the narrow- and wide- domain models the rheology was set to be viscoelastic without damage evolution, and with realistic material properties ($\mu = 30$ GPa, $\nu = 0.25$, elastic plate thickness 20 km, and Newtonian substrate viscosity of 10^{19} Pa s). The analysis of stiffness values and their effect on boundary forces and velocities shows that using a very large spring stiffness ($> 5\cdot10^6$ MPa) results in a constant velocity on the boundary nodes. Free boundary conditions are approximated with the use of very low ($< 5\cdot10^2$ MPa) stiffness values. Using intermediate stiffness values, the force applied on boundary nodes varies as a response to deformation processes within the model domain, except for rare instances in which a constant force is applied on boundary nodes that experience a very large mismatch between the plate motion and the boundary displacement. To best represent constant velocities far from the fault zone (i.e., at the boundaries of the very large models) the spring stiffness in the narrow-domain models was set to $5\cdot10^3$ to 10^4 MPa. This is similar to the value used in the models presented in this paper (10^4 MPa).

Appendix 3: Code Verification: Viscoelastic Component

To verify that the numerical procedure properly models viscoelastic deformation, we simulated postseismic surface deformation following a kinematically imposed earthquake, using a model in which no damage evolution was allowed. We compared our results with those from a viscoelastic finite-element model (GAEA, SAUCIER and HUMPHREYS, 1992). This extends previous comparisons which yielded interseismic velocities comparable to analytic solutions (LYAKHOVSKY *et al.*, 2001).

The two models were set up with identical domain size (300 × 250 × 100 km), fault depth (20 km), coseismic stress drop (20 MPa), lithospheric structure, and material properties ($\mu = 30$ GPa, $\nu = 0.25$, elastic plate thickness 20 km, and Newtonian substrate viscosity of 10^{19} Pa s). In both models the fault-parallel side and bottom boundaries were fixed, and the top boundary was stress-free. To simulate stress-free fault-perpendicular end boundaries our damage-disabled model incorporated the variable force boundary condition with a low stiffness, which is approximately equivalent to the stress-free boundary condition used in the FEM. Figure A2 shows surface velocities calculated with both models, for several time epochs after the modeled earthquake (1, 2, 5, and 18 years). The minor differences between the models are due in part to applying the variable force boundary condition with spring stiffness slightly too high (10^3 MPa) and therefore not simulating completely stress-free end boundaries. This slightly suppressed the velocities away from the fault.

REFERENCES

AGNON, A. and LYAKHOVSKY, V. *Damage distribution and localization during dyke intrusion.* In (G. Baer and A. Heimann, eds), *The Physics and Chemistry of Dykes* (Rotterdam, Balkema, 1995) pp. 65–78.

AMBRASEYS, N. N. (1970), *Some characteristic features of the North Anatolian fault zone*, Tectonophysics *9*, 143–165.

AMPUERO, J.-P. and BEN-ZION, Y. (2008), *Cracks, pulses and macroscopic asymmetry of dynamic rupture on a bimaterial interface with velocity-weakening friction*, Geophys. J. Int. *173*, 674–692, doi: 10.1111/j.1365-246X.2008.03736.x.

ANDREWS, D. J. (2005), *Rupture dynamics with energy loss outside the slip zone*, J. Geophys. Res. *110*, B01307, doi:10.1029/2004JB003191.

AYDIN, A. and JOHNSON, A.M. (1983), *Analysis of faulting in porous sandstones*, J. Struct. Geol. *5*, 19– 31.

ARMIJO, R., MEYER, B., KING, G. C. P., RIGO, A., and PAPANASTASSIOU, D. (1996), *Quaternary evolution of the Corinth Rift and its implications for the Late Cenozoic evolution of the Aegean*, Geophys. J. Int. *126*, 11–53.

BAISCH. S., and BOKELMANN, G. H. R. (2001), *Seismic waveform attributes before and after the Loma Prieta earthquake: Scattering change near the earthquake and temporal recovery*, J. Geophys. Res. *106*, 16,323-16,337.

BEN-ZION, Y. (1996), *Stress, slip and earthquakes in models of complex single-fault systems incorporating brittle and creep deformations*, J. Geophys. Res. *101*, 5677–5706.

BEN-ZION, Y. and AKI, K. (1990), *Seismic radiation from an SH line source in a laterally heterogeneous planar fault zone*, Bull. Seismol Soc. Am. *80*, 971–994.

BEN-ZION, Y. and ANDREWS, D. J. (1998), *Properties and implications of dynamic rupture along a material interface*, Bull. Seismol. Soc. Am. *88*(4), 1085–1094.

BEN-ZION, Y., HENYEY, T., LEARY, P. and LUND, S. (1990), *Observations and implications of water well and creepmeter anomalies in the Mojave segment of the San Andreas fault zone*, Bull. Seismol. Soc. Am. *80*, 1661–1676.

BEN-ZION, Y. and LYAKHOVSKY, V. (2006), *Analysis of aftershocks in a lithospheric model with seismogenic zone governed by damage rheology*, Geophys. J. Int. *165*, 197–210.

BEN-ZION, Y., PENG, Z., OKAYA, D., SEEBER, L., ARMBRUSTER, J. G., OZER, N., MICHAEL, A. J., BARIS, S. and AKTAR, M. (2003), *A shallow fault zone structure illuminated by trapped waves in the Karadere-Duzce branch of the North Anatolian Fault, western Turkey*, Geophys. J. Int. *152*, 699–717.

BEN-ZION, Y. and SAMMIS, C. (2003), *Characterization of Fault Zones*, Pure Appl. Geophys. *160*, 677–715.

BEN-ZION, Y. and SHI, Z. (2005), *Dynamic rupture on a material interface with spontaneous generation of plastic strain in the bulk*, Earth Planet. Sci. Lett. *236*, 486–496, doi: 10.1016/j.epsl.2005.03.025.

BERCOVICI, D. and RICARD, Y. (2003), *Energetics of a two-phase model of lithospheric damage, shear localization and plate-boundary formation*, Geophys. J. Int. *152* (3), 581–596 doi:10.1046/j.1365-246X.2003.01854.x

BUDIANSKY, B. and O'CONNELL, R. J. (1976), *Elastic moduli of a cracked solid*, Int. J. Sol. Struct. *12*, 81–97.

CARTER, N. L. and TSENN, M. C. (1987), *Flow properties of continental lithosphere*, Tectonophysics *136*, 27–63.

CHESTER, F. M. and CHESTER, J. S. (1998), *Ultracataclasite structure and friction processes of the Punchbowl Fault, San Andreas System, California*, Tectonophysics *295*, 199–221.

CHESTER, F. M. (1995), *Geologic studies of deeply exhumed faults of the San Andreas System, Southern California*: Collaborative research with Saint Louis University and Utah State University: NEHRP annual project summary, award No. 94G2457, v. 37.

CHESTER, F. M., EVANS, J. P., and BIEGEL, R. L. (1993), *Internal structure and weakening mechanisms of the San Andreas Fault*, J. Geophys. Res. *98*, 771–786.

CHRISTENSEN, N. I., and MOONEY, W. D. (1995), *Seismic velocity structure and composition of the continental crust: A global view*, J. Geophys. Res. *100* (B6), 9761–9788.

COCHRAN, E. S., VIDALE, J. E., and LI, Y. G. (2003), *Near-fault anisotropy following the Hector Mine earthquake*, J. Geophys. Res. *108*(B9), 2436, doi:10.1029/2002JB002352.

COWIE, P. A., VANNESTE, C., and SORNETTE, D. (1993), *Statistical Physics Model for the Spatiotemporal Evolution of Faults*, J. Geophys. Res. *98*(B12), 21,809–21,821.

CUNDALL, P. A. and BOARD, M. *A microcomputer program for modeling large-strain plasticity problems*. In *Numerical Methods in Geomechanics*, Proc. 6th Int. Conf. Numerical Methods in Geomechanics, Innsbruck, (ed. Swoboda), (C., Rotterdam, Balkema, 1988) pp. 2101–2108.

DE PAOLA, N., HOLDSWORTH, R. E., COLLETTINI, C., MCCAFFREY, K. J. W., and BARCHI, M. R. (2007), *The structural evolution of dilational step-overs in regional transtensional zones*. (Cunningham W. D. and Mann,

P., eds), *Tectonics of Strike-Slip Restraining and Releasing Bends* (Geolog. Soc., London. Special Publications), *290*, pp. 433–445.

DIETERICH, J. H. (1972), *Time-dependent friction in rocks*, J. Geophys. Res. *77*, 3690–3697.

DIETERICH, J. H. (1978), *Time-dependent friction and the mechanics of stick-slip*, Pure Appl. Geophys. *116*, 790–805.

DIETERICH, J. H. (1979), *Modeling of rock friction 1. Experimental results and constitutive equations*, J. Geophys. Res. *84*, 2161–2168.

DOLAN, J. F., BOWMAN, D. D., and SAMMIS, C. G. (2007), *Long-range and long-term fault interactions in Southern California, Geology*, *35*, 855–858.

DOR, O., ROCKWELL, T. K. and BEN-ZION, Y. (2006), *Geological observations of damage asymmetry in the structure of the San Jacinto, San Andreas and Punchbowl Faults in Southern California: A possible indicator for preferred rupture propagation direction*, Pure Appl. Geophys. *163*, 301–349, doi:10.1007/s00024-005-0023-9.

DOR, O., YILDIRIM, C., ROCKWELL, T.K., BEN-ZION, Y., EMRE, O., SISK, M., DUMAN, T. Y. (2008), *Geologic and geomorphologic asymmetry across the rupture zones of the 1943 and 1944 earthquakes on the North Anatolian Fault: Possible signals for preferred earthquake propagation direction*, Geophys. J. Int., doi: 10.1111/j.1365-246X.2008.03709.x.

DUNHAM, E. M. and RICE, J. R. (2008), *Earthquake slip between dissimilar poroelastic materials*, J. Geophys. Res., in press.

EISSA, E. A. and KAZI, A. (1988), *Relation between static and dynamic Young's Moduli of Rocks*, Int. J. Rock Mech. Min. Sci. Geomech. Abstr. *25* (6), 479–482.

EVANS, J. P., SHIPTON, Z. K., PACHELL, M. A., LIM, S. J., and ROBESON, K. *The structure and composition of exhumed faults, and their implication for seismic processes.* In Proc. of the 3rd Confer. on *Tecto. problems of the San Andreas system*, (Stanford University 2000).

FIALKO, Y. (2004), *Probing the mechanical properties of seismically active crust with space geodesy: Study of the coseismic deformation due to the 1992 M_w7.3 Landers (southern California) earthquake*, J. Geophys Res. *109*, B03307, doi:10.1029/2003JB002756.

FIALKO, Y., SANDWELL, D., AGNEW, D., SIMONS, M., SHEARER, P., and MINSTER, B. (2002), *Deformation on nearby faults induced by the 1999 Hector Mine earthquake*, Science *297*, 1858–1862.

FINZI, Y., HEARN, E. H., LYAKHOVSKY, V., and BEN-ZION., Y. (2006), *3-D viscoelastic damage rheology models of strike-slip fault systems and their associated surface deformation*, EOS Trans. AGU, *87*(52), Fall Meet. Suppl., Abstract T21C-0425.

GRAYMER, R.W., LANGENHEIM, V.E., SIMPSON, R.W., JACHENS, R.C., and PONCE, D.A. (2007), *Relatively simple through-going fault planes at large-earthquake depth may be concealed by the surface complexity of strike-slip faults.* In (Cunningham, W.D., and Mann, Paul, eds.) *Tectonics of Strike-Slip Restraining and Releasing Bends*, (Geological Society of London Special Publication 2007), vol. 290, pp. 189–201, doi: 10.1144/SO290.5 0305-8719/07.

HAMIEL, Y., LYAKHOVSKY, V., and AGNON, A. (2005), *Rock dilation, nonlinear deformation, and pore pressure change under shear*, Earth Planet. Sci. Lett. *237*, 577–589.

HAMIEL, Y., KATZ, O., LYAKHOVSKY., RECHES, Z. and FIALKO, Y. (2006), *Stable and unstable damage evolution in rocks with implications to fracturing of granite*, Geophys. J. Int. *167*, 1005–1016.

HAMIEL, Y., LIU, Y., LYAKHOVSKY, V., BEN-ZION, Y., and LOCKNER, D. (2004), *A visco-elastic damage model with applications to stable and unstable fracturing*, Geophys. J. Int. *159*, 1155–1165.

HAMIEL, Y. and FIALKO, Y. (2007), *Structure and mechanical properties of faults in the North Anatolian Fault system from InSAR observations of coseismic deformation due to the 1999 Izmit (Turkey) earthquake*, J. Geophys Res. 112, B07412, doi:10.1029/2006JB004777.

HARRIS, R. A. and DAY, S. M. (1999), *Dynamic 3-D simulations of earthquakes on en echelon faults*, Geophys. Res. Lett. *26*, 2089–2092.

HARRIS, R. A. and DAY, S. M. (1993), *Dynamics of fault interaction: Parallel strike-slip faults*, J. Geophys. Res. *98*, 4461–4472.

HARRIS, R. A., ARCHULETA, R. J., and DAY, S. M. (1991), *Fault steps and the dynamic rupture process: 2-D numerical simulations of a spontaneously propagating shear fracture*, Geophys. Res. Lett. *18*, 893–896.

HEARN, E. H. and FIALKO, Y. (2009), *Coseismic deformation of Mojave compliant zones and crustal stresses*, J. Geophys. Res., in press.

HICKMAN, S., SIBSON, R.H., and BRUHN, R. (1995), *Introduction to a special section, mechanical involvement of fluids in faulting*, J. Geophys. Res. *100*, 12,831–12,840.

IDE, J. M. (1936), *Comparison of statically and dynamically determined Young's modulus of rocks*, Proc. Nat. Acad. Sci., U.S.A. *22*, 81–92.

KARABULUT, H. and BOUCHON, M. (2007), *Spatial variability and non-linearity of strong ground motion near a fault*, Geoph. J. Int. *170*, 1, 262–274.

KIM, Y. S., PEACOCK, D. C. P., and SANDERSON, D. J. (2004), *Fault damage zones*, J. Struct. Geology *26*, 503–517.

KING, G. (1986), *Speculations on the geometry of the initiation and termination processes of earthquake rupture and its relation to morphology and geological structure*, Pure Appl. Geophys. *124*, 567–585.

KIRBY, S. H. and KRONENBERG, A. K. (1987), *Rheology of the lithosphere: Selected topics*, Rev. Geophys. *25*, 1219–1244.

KORNEEV, V.A., NADEAU, R.M., and MCEVILLY, T.V. (2003), *Seismological studies at Parkfield IX: Fault-one imaging using guided wave attenuation*, Bull. Seismol. Soc. Am. *93*, 1415–1426.

LEWIS, M. A., PENG, Z., BEN-ZION, Y., and VERNON, F. (2005), *Shallow seismic trapping structure in the San Jacinto fault zone*, Geophys. J. Int. *162*, 867–881, doi:10.1111/j.1365-246X.2005.02684.x.

LI, Y. G., LEARY, P., AKI, K., and MALIN, P. (1990), *Seismic trapped modes in the Oroville and San Andreas fault zones*, Science, *249*, 763–766.

LI, Y.-G., AKI, K., ADAMS, D., HASEMI, A., and LEE, W. H. K. (1994), *Seismic guided waves trapped in the fault zone of the Landers, California, earthquake of 1992*, J. Geophys. Res. *99*(B6), 11,705–11,722.

LI, Y.-G., CHEN, P., COCHRAN, E. S., VIDALE, J. E., and BURDETTE, T. (2006), *Seismic evidence for rock damage and healing on the San Andreas fault associated with the 2004 M 6.0 Parkfield Earthquake*, Bull. Seismol. Soc. Am. *96*, 4B, S349–S363.

LIU, Y., TENG T. L., and BEN-ZION, Y. (2004), *Systematic analysis of shear-wave splitting in the aftershock region of the 1999 Chi-Chi earthquake: Evidence for shallow anisotropic structure and lack of systematic temporal variations*, Bull. Seismol. Soc. Am. *94*, 2330–2347.

LYAKHOVSKY, V. and MYASNIKOV, V. P. (1984), *On the behavior of elastic cracked solid*, Phys. Solid Earth *10*, 71–75.

LYAKHOVSKY, V. and MYASNIKOV, V. P. (1985), *On the behavior of visco-elastic cracked solid*, Phys. Solid Earth *4*, 28–35.

LYAKHOVSKY, V., BEN-ZION, Y., and AGNON, A. (1997a), *Distributed damage, faulting, and friction*, J. Geophys. Res. *102*, 27,635–27,649.

LYAKHOVSKY, V., RECHES, Z., WEINBERGER, R., and SCOTT, T.E. (1997b), *Non-linear elastic behavior of damaged rocks*, Geophys. J. Int. *130*, 157–166.

LYAKHOVSKY, V., BEN-ZION, Y., and AGNON, A. (2001), *Earthquake cycle, faults, and seismicity patterns in rheologically layered lithosphere*, J. Geophys. Res. *106*, 4103–4120.

LYAKHOVSKY, V., BEN-ZION, Y., and AGNON, A. (2005), *A viscoelastic damage rheology and rate- and state-dependent friction*, Geophys. J. Int. *161*, 179–190.

LYAKHOVSKY, V. and BEN-ZION, Y. (2008), *Scaling relations of earthquakes and aseismic deformation in a damage rheology model*, Geophys. J. Int. *172*, 651–662, doi: 10.1111/j.1365-246X.2007.03652.x.

LYAKHOVSKY, V., and BEN-ZION, Y. (2009), *Evolving fault zone structures in a damage rheology model*, Geochemistry, Geophysics, Geosystems, in review.

MALVERN, L.E. *Introduction to the Mechanics of a Continuum Medium* (New Jersey, Prentice-Hall, Inc., 1969), 713 pp.

MCGUIRE, J. and BEN-ZION, Y. (2005), *High-resolution imaging of the Bear Valley section of the San Andreas Fault at seismogenic depths with fault-zone head waves and relocated seismicity*, Geophys. J. Int. *163*, 152–164, doi: 10.1111/j.1365-246X.2005.02703.x.

MICKLETHWAITE, S. and COX, S. F. (2004), *Fault-segment rupture, aftershock-zone fluid flow and mineralization*, Geology *32*, 813–816.

MOONEY, W. D. and GINZBURG, A. (1986), *Seismic measurements of the internal properties of fault zones*, Pure Appl. Geophys. *124*, 141–157.

NADEAU, R., ANTOLIK, M., JOHNSON, P., FOXALL, W., and MCEVILLY, T. V. (1994), *Seismological studies at Parkfield III: Microearthquake clusters in the study of fault-zone dynamics*, Bull. Seismol. Soc. Am. *83*, 247–263.

Oglesby, D. D., Day, S. M., Li, Y.-G., and Vidale, J. E. (2003), *The 1999 Hector Mine Earthquake: The dynamics of a branched fault system*, Bull. Seismol. Soc. Am. *93*, 6, 2459–2476.

Olson, J., and Pollard, D. D. (1989), *Inferring paleostresses from natural fracture patterns: A new method*, Geology *17*, 4, 345–348.

Peng, Z. and Ben-Zion, Y. (2004), *Systematic analysis of crustal anisotropy along the Karadere- Duzce branch of the north Anatolian fault*, Geophys. J. Int. *159*, 253-274, doi:10.1111/j.1365-46X.2004.02379.x.

Peng, Z. and Ben-Zion, Y. (2005), Spatio-temporal variations of crustal anisotropy from similar events in aftershocks of the 1999 M7.4 İzmit and M7.1 Düzce, Turkey, earthquake sequences, Geophys. J. Int. *160*(3), 1027–1043, doi: 10.1111/j.1365-246X.2005.02569.x.

Peng, Z. and Ben-Zion, Y. (2006), *Temporal changes of shallow seismic velocity around the Karadere-Duzce branch of the north Anatolian fault and strong ground motion*, Pure Appl. Geophys. *163*, 567–600, doi: 10.1007/s00024-005-0034-6.

Peng, Z., Ben-Zion, Y., Michael, A. J., and Zhu, L. (2003), *Quantitative analysis of seismic trapped waves in the rupture zone of the 1992 Landers, California earthquake: Evidence for a shallow trapping structure*, Geophys. J. Int. *155*, 1021–1041.

Poliakov, A., Cundall, P., Podladchikov, Y., and Lyakhovsky, V. *An explicit inertial method for the simulation of viscoelastic flow: an evaluation of elastic effects on diapiric flow in two- and three-layers model.* In Proc. NATO Advanced Study Institute on *Dynamic Modeling and Flow in the Earth and Planets*, (Runcorn, K.E. and Stone, D., eds) (Dordrecht, Kluwer, 1993) pp. 175–195.

Poupinet, G., Ellsworth, W. L., and Frechet, J. (1984), *Monitoring velocity variations in the crust using earthquake doublets: An application to the Calaveras Fault, California*, J. Geophys. Res. *89*(B7), 5719–5731.

Powell, R. E. and Weldon, R. J. (1992), *Evolution of the San Andreas Fault*, Annu. Rev. Earth Planet. Sci. *20*, 431–468.

Revenaugh, J. (2000), *The relation of crustal scattering to seismicity in southern California*, J. Geophys. Res. *105*(B11), 25,403–25,422.

Rice, J.R. and Ben-Zion, Y. (1996), *Slip complexity in earthquake fault models*, Proc. Natl. Acad. Sci. U.S.A. *93*, 3811–1818.

Rockwell, T. K. and Ben-Zion, Y. (2007), *High localization of primary slip zones in large earthquakes from paleoseismic trenches: Observations and implications for earthquake physics*, J. Geophys. Res. *112*, B10304, doi:10.1029/2006JB004764.

Rubinstein, J.L. and Beroza, G.C. (2004), *Evidence for widespread strong ground motion in the M_w 6.9 Loma Prieta eartquake*, Bull. Seismol. Soc. Am. *94*, 1595–1608.

Rudnicki, J. W. and Rice, J. R. (2006), *Effective normal stress alteration due to pore pressure changes induced by dynamic slip propagation on a plane between dissimilar materials*, J. Geophys. Res. *111*, B10308, doi:10.1029/2006JB004396.

Saucier, F. and Humphreys, E.D. (1993), *Horizontal crustal deformation in Southern California from joint models of geologic and very long baseline interferometry measurements.* In *Contributions of Space Geodesy to Geodynamics* (D.E. Smith and D.L. Turcotte, eds.), pp. 139–176, (AGU Geodyn. Ser. Vol. *23*, Washington D.C. 1993).

Schaff, D. P., Bokelmann, G. H. R., Beroza, G. C., Waldhauser, F., and Ellsworth, W. L. (2002), *High-resolution image of Calaveras Fault seismicity*, J. Geophys. Res. 107(B9), 2186, doi:10.1029/2001JB000633.

Schaff, D. P. and Beroza, G. C. (2004), *Coseismic and postseismic velocity changes measured by repeating earthquakes*, J. Geophys. Res. *109*, B10302, doi:10.1029/2004JB003011.

Schulz, S. E. and Evans, J. P. (2000), *Mesoscopic structure of the Punchbowl Fault, Southern California and the geologic and geophysical structure of active strike-slip faults*, J. of Struct. Geol. *22*, 913–930.

Segall, P. and Pollard, D. D. (1980), *Mechanics of discontinuous faults*, J. Geophys. Res. *85*, 4337–4350, 1980.

Sengor, A. M. C., Tuysz, O., Imren, C., Sakinc, M., Eyidogan, H., Gorur, N., Le Pichon, X., and Rangin, C. (2005), *The North Anatolian Fault: A new look*, Annu. Rev. Earth Planet. Sci. *33*, 37–112.

Sheldon, H. A., and Micklethwaite, S. (2007), *Damage and permeability around faults: Implications for mineralization*, Geology *35*, 10, 903–906.

Shipton, Z. K., and Cowie, P. A. (2003), *A conceptual model for the origin of fault damage zone structures in high-porosity sandstone*, J. Struct. Geol. *25*, 3, 333–344

Sibson, R.H. (1985), *Stopping of earthquake ruptures at dilational fault jogs*, Nature *316*, 248–251.

SIBSON, R. H. (2003), *Thickness of the seismic slip zone*, Bull. Seismol. Soc. Am. *93*, 3, 1169–1178.
STIRLING, M. W., WESNOUSKY, S. G., and SHIMAZAKI, K. (1996), *Fault trace complexity, cumulative slip, and the shape of the magnitude-frequency distribution for strikeslip faults: a global survey*, Geophys. J. Int. *124*, 833–868.
STIERMAN, D. J. (1984), *Geophysical and geological evidence for fracturing, water circulation and chemical alteration in granitic rocks adjacant to major strike-slip faults*, J. Geophys. Res. *89*, B7, 5849–5857.
SYLVESTER, A. G. (1988), *Strike-slip faults*, Geol. Soc. Am. *100*, 1666–1703.
SYLVESTER, A. G. and SMITH, R., (1976), *Tectonic transpretions and basement controlled deformation in the San Andreas fault zone, Salton trough, California*, AAPG Bull. *60*, 2081–2102.
TCHALENKO, J. S. (1970), *Similarities between shear zones of different magnitudes*, Geolog. Soc. Am. Bull. *81*, 1625–1640.
TEMPLETON, E. L. and RICE, J. R. (2008), *Off-fault plasticity and earthquake rupture dynamics, 1. Dry materials or neglect of fluid pressure changes*, J. Geophys. Res.
TENTHOREY, E., COX, S. F., and TODD, H. F. (2003), *Evolution of strength recovery and permeability during fluid-rock reaction in experimental fault zones*, Earth Planet. Sci. Lett. *206*(1–2), 161–172.
THURBER, C., ZHANG, H., WALDHAUSER, F., HARDEBECK, J., MICHAEL, A., and EBERHART-PHILLIPS, D. (2006), *Three-dimensional compressional wavespeed model, earthquake relocations, and focal mechanisms for the Parkfield, California, Region*, Bull. Seismol. Soc. of Am. *96*, 4B, S38–S49, doi: 10.1785/0120050825.
TURCOTTE, D. L. and GLASSCOE, M.T. (2004), *A damage model for the continuum rheology of the upper continental crust*, Tectonophysics *383*, 71–80.
WESNOUSKY, S. G. (2006), *Predicting the endpoints of earthquake ruptures*, Nature *444*, 358–360.
WESNOUSKY, S. (1994), *The Gutenberg-Richter or characteristic earthquake distribution, which is it?*, Bull. Seismol. Soc. Am. *84*, 1940–1959.
WILCOX, R. E., HARDING, T. P., and SEELY, D. R. (1973), *Basic wrench tectonics*, AAPG Bull. *57*, 74–96.
WU, C., PENG, Z. and BEN-ZION, Y. (2009), *Non-linearity and temporal changes of fault zone site response associated with strong ground motion*, Geophys. J. Int. *176*, 265–278, doi: 10.1111/j.1365-246x.2008.04005.x.
YAMASHITA, T. (2007), *Postseismic quasi-static fault slip due to pore pressure change on a bimaterial interface*, J. Geophys. Res. *112*, B05304, doi:10.1029/2006JB004667.
YANG, W. and BEN-ZION, Y. (2009), *Observational analysis of correlations between aftershock prodectivities and regional conditions in the context of a damage rheology model*, Geophys. J. Int., *177*, 481–499 doi: 10.1111/j.1365-246x.2009.0414s.

(Received May 7, 2008, revised September 30, 2008)
Published Online First: June 30, 2009

To access this journal online:
www.birkhauser.ch/pageoph

Pure appl. geophys. 166 (2009) 1575–1594
0033–4553/09/101575–20
DOI 10.1007/s00024-009-0511-4

© Birkhäuser Verlag, Basel, 2009

Pure and Applied Geophysics

Segmentation along Strike-Slip Faults Revisited

GHISLAIN DE JOUSSINEAU[1,2] and ATILLA AYDIN[1]

Abstract—Fault segmentation and fault steps and their evolution are relevant to the dynamics and size of earthquake ruptures, the distribution of fault damage zones and the capacity of fault seal. Furthermore, segment interactions and coalescence are the fundamental processes for fault growth. To contribute to this end, we investigated the architecture of strike-slip faults by combining field observations in the Valley of Fire State Park, Nevada, and the published data sets.

First, we studied the trace complexity for 49 faults with offsets ranging from 12 m to 460 km. We established that the number of fault steps (hence fault segments) per unit length is correlated to the maximum fault offset by a negative power law. The faults have longer segments and fewer steps when their offsets increase, indicating the progressive growth, smoothening and simplification of the fault traces as a function of the offset, as proposed by previous investigators.

Second, we studied the dimensions of the segments and steps composing $\sim$20 of the previous fault systems. The mean segment length, mean step length and mean step width are all correlated to the maximum fault offset by positive power laws over four orders of magnitude of the offset. In addition, the segment length distributions of four of the faults with offsets ranging from 80 m to 100 km are all lognormal, with most of the segment lengths falling in the range of one to five times the maximum offset of the faults. Finally, the fault steps have an approximately constant length-to-width ratio indicating that, regardless of their environment, strike-slip faults have a remarkable self-similar architecture probably due to the mechanical processes responsible for fault growth. Our data sets can be used as tools to better predict the geometrical attributes of strike-slip fault systems with important consequences for earthquake ruptures, the distribution and properties of fault damage zones, and fault sealing potential.

Key words: Strike-slip faults, fault trace complexity, fault step, fault segment length, fault step length, fault step width, hierarchical self-similar fault architecture.

1. *Introduction*

Strike-slip faults typically have a complex architecture with numerous segments of various lengths separated by steps (relay zones) of a broad range of sizes. This segmented character of strike-slip fault traces has a direct impact on the dynamics and size of earthquake ruptures, which are believed to initiate generally at small steps (asperities, i.e., features temporarily opposing slip) and terminate at large steps (barriers, i.e., features

[1] Rock Fracture Project, Department of Geological and Environmental Sciences, Stanford University, 450 Serra Mall, Building 320, Stanford, CA 94305, USA.
[2] Beicip-Franlab, 232 Avenue Napoléon Bonaparte, 92502 Rueil-Malmaison Cedex, France.
E-mail: ghislain.dejoussineau@beicip.com

permanently resisting slip) (Segall and Pollard, 1980; Sibson, 1985, 1986; Barka and Kadinsky-Cade, 1988; Aki, 1989; Harris *et al.*, 1991; Harris and Day, 1999; Otsuki and Dilov, 2005; Wesnousky, 2006; Shaw and Dieterich, 2007). Specifically, it has been proposed that the probability of earthquake ruptures to jump from one fault segment to the next decreases exponentially with stepover distance (Shaw and Dieterich, 2007) and that, in the absence of any linking structure between fault segments, strike-slip fault steps wider than 3–5 km systematically arrest seismic ruptures (Harris *et al.*, 1991; Wesnousky, 2006). However, further improvement in the comprehension of dynamic rupture along, and fluid flow across, faults requires a sound knowledge of fault discontinuities and their spatial distribution as a function of fault size or slip.

The segmentation of strike-slip fault systems also plays a critical role in controlling the location, orientation and length of fault splays (Segall and Pollard, 1980; de Joussineau *et al.*, 2007). These splays, in turn, impact the distribution of damage and the geometrical and statistical properties of secondary fault networks around strike-slip faults (Flodin and Aydin, 2004; Myers and Aydin, 2004; de Joussineau and Aydin, 2007). Moreover, the fault architecture affects the repartition of slip and stress along fault systems, with important consequences for damage zone distribution (Cartwright *et al.*, 1995; Knott *et al.*, 1996; de Joussineau and Aydin, 2007; Finzi *et al.*, 2009). Finally, fault segmentation controls the development of fault core that forms preferentially in stepovers because of high strain levels therein. This is critical for fluid flow in the subsurface since damage zones typically have higher permeability than the parent rock whereas fault cores usually have lower permeability (Caine *et al.*, 1996; Jourde *et al.*, 2002; Aydin, 2000; Odling *et al.*, 2004).

Notwithstanding its importance, the architecture of strike-slip faults in the subsurface cannot easily be detected accurately by seismic techniques which typically identify faults as line elements and fail to separate closely-spaced segments. Consequently, attempts to better predict the geometrical features of strike-slip faults have been made. Based on the analysis of strike-slip fault traces, Wesnousky (1988) and Stirling *et al.* (1996) found that the number of fault steps (hence fault segments) per unit length is a decreasing function of the maximum fault offset. This structural evolution by which strike-slip fault traces smoothen and straighten progressively with increasing offset is expected to impact the seismological behavior of faults (Wesnousky, 1988; Scholz, 2002; Ben-Zion and Sammis, 2003) as well as the distribution of fault damage zones (de Joussineau and Aydin, 2007). However, due to the lack of data for the entire spectrum of fault size, it is still unclear whether the relationship between strike-slip fault trace complexity and fault offset hold true in all situations.

From a different angle, Aydin and Nur (1982) analyzed the shape of fault steps along major strike-slip faults and showed that the length to width ratio of these steps was more-or-less constant. Based on this finding, they proposed conceptual models for the structural evolution of strike-slip fault systems, in which the coalescence of neighboring basins and/or the formation of new fault strands parallel to the existing ones account for the progressive enlargement of fault steps with increasing offset.

From an experimental point of view, OTSUKI and DILOV (2005) determined that strike-slip faults developing in rock samples subjected to cyclic loading had a hierarchical self-similar architecture. At each stage of the fault evolution, the segments were composed of smaller segments and steps related to earlier stages of fault evolution, and the fault length was correlated to the mean segment length, mean step length and mean step width by positive power laws. This hierarchical self-similar architecture of strike-slip faults is expected to strongly impact the nucleation and termination of earthquake ruptures. However, to our knowledge, no comparable data set is available from natural faults, thus making it uncertain whether the results of these experiments are applicable to geological situations.

In this paper, we combine field studies of strike-slip fault zones, published data sets and analyses of published fault maps to develop a basis for a better understanding and forecasting of the architecture of strike-slip faults. First, we confirm the existence of a negative power law relationship between the number of steps per unit length and the maximum fault offset over five orders of magnitude of the offset. Second, we show that the mean segment length is correlated to the maximum fault offset by a positive power law. We also analyze the segment length distribution in a series of faults and observe common features between the distributions independently of the fault offset. Finally, we establish that there is a positive power law relationship between the mean step length and mean step width and the maximum fault offset, and confirm that the length to width ratio of strike-slip fault steps falls into a narrow range.

This work provides additional data for the remarkable self-similarity of the architecture of strike-slip faults, and forms a basis for an improved understanding of strike-slip fault characteristics. In turn, these results have critical implications for the dynamics of seismic ruptures, fault growth processes, the distribution and properties of fault damage zones and fault sealing capacity.

2. *Methods*

In this paper, we follow AYDIN and NUR (1982) and WESNOUSKY (1988) and use the term ‘step’ to refer to the relay zones between the segments along strike-slip faults (regardless of their extensional or compressive nature). We present data on the architecture of a fairly large population of strike-slip faults obtained from our own field study, data sets published by other workers and analyses of fault maps in the literature. The new field data were collected in the Valley of Fire State Park, located approximately 60 miles NE of Las Vegas in southeastern Nevada (Fig. 1a). In this area, excellent outcrops of aeolian Jurassic Aztec Sandstone allow to study of well-exposed strike-slip fault networks at centimeters to kilometers scale. The mechanisms of formation and evolution of these fault networks and their geometrical and statistical properties were previously documented by MYERS and AYDIN (2004), FLODIN and AYDIN (2004) and DE JOUSSINEAU and AYDIN (2007).

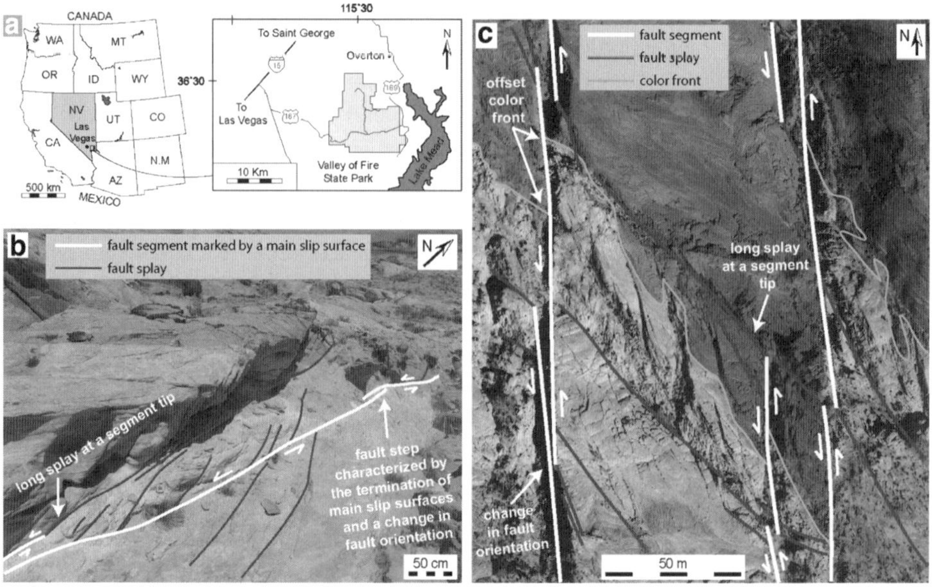

Figure 1
(a) Geographic location of the Valley of Fire State Park, SE Nevada (modified from FLODIN and AYDIN, 2004); (b) Ground photograph showing the architecture of a small fault system (slip ~ 10 m); (c) Aerial photograph showing the architecture of two large fault systems (slip ~ 100 m or more). The tips of fault segments are often associated with large splays (b, c). Fault steps are usually characterized by the termination of main slip surfaces (b) and by changes in the general orientation of the fault zone (b, c). Offset color fronts provide an estimate of the amount of lateral slip for large faults (c).

The size and number of segments and steps along a series of faults in the Valley of Fire were first determined on high-quality aerial photographs and were then validated in the field, with a final (field-based) resolution of a few tens of meters for the largest faults studied (offset $\geq$ 600 m), a few meters for intermediate faults (14 m < offset < 600 m) and a few tens of centimeters for the smallest faults studied (offset $\leq$ 14 m). Criteria used to decipher the fault segmentation in the field were the morphology and orientation of the fault zones, the continuity of slip surfaces, the continuity and thickness variations of fault core and fault damage zone and the location of main splays along the faults that may indicate the tips of large segments (DE JOUSSINEAU *et al.*, 2007) (Figs. 1b and 1c). The maximum separation on map view was considered as the maximum fault offset, that is faults are assumed to be pure strike-slip. For small faults (with ~10 m of offset), the maximum fault offset was determined in the field based on the displacement of markers such as dune boundaries or color bands inherited from diagenetic processes (EICHHUBL *et al.*, 2004) or other structural markers like older shear bands and compaction bands. For intermediate or large faults (with ~ 100 m of offset or more), the maximum fault offset was determined on aerial photographs based on the displacement of major chemical reaction fronts marking the limit between different diagenetic units appearing in various colors in the Aztec Sandstone (EICHHUBL *et al.*, 2004; Fig. 1c). These offsets were also field-checked afterwards.

We used the published data from WESNOUSKY (1988) and STIRLING *et al.* (1996) together with new data obtained from analyses of published fault maps. The fault examples selected in the literature were those for which detailed maps were available and the maximum fault offsets were known. These examples come from sedimentary, metamorphic and volcanic environments, making the results of the study applicable to a wide range of settings.

For each fault example, the dimensions of segments and steps were measured directly on the fault maps. The number of steps per unit length (km) was obtained by dividing the number of steps by the length (in km) over which the faults were analyzed. The mean values of the segment length, step width and step length are arithmetic average values. For the steps, the configurations of both overlapping and underlapping steps were considered. The imprecision in the data is shown in the plots by error bars where applicable. It is related either to determination of the maximum fault offset, the number of steps per kilometer (i.e., the number of segments composing the fault systems) or both, and was estimated by the authors of the original studies.

Finally, certain studies provided intervals of values instead of exact values for the maximum fault offset or the number of steps per km, making it difficult to plot the data. In this case, we calculated an arithmetic average value for the parameter considered, assigned it to the fault studied, and considered the imprecision in the estimation of the parameter to be the difference between this average value and the lower and upper boundaries of the interval. For example, if a fault offset was reported to be in the interval 1–5 km, we considered that the fault offset was 3 ± 2 km.

3. Results

3.1. Fault Trace Characterization

Here we examine the trace complexity of a series of strike-slip faults in the field in the Valley of Fire State Park and in the literature to complement the data sets in WESNOUSKY (1988) and STIRLING *et al.* (1996). Figure 2 shows plots of the number of steps per km as a function of the maximum strike-slip offset for 49 faults with offsets ranging from 12 m to 460 km, with the source of the data indicated. Data in Figure 2 (also given in Table 1) are best fit by a negative power law relationship: $y = 0.26x^{-0.81}/R^2 = 0.74$, where the number of steps and the maximum offset are plotted on y and x axis, respectively. In addition, best fits to the data from the Valley of Fire, data of STIRLING *et al.* (1996) and data of WESNOUSKY (1988) are given and discussed in Section 4.1. Confidence intervals calculated for the mean and single values of the number of steps with a confidence threshold of 95% are also represented, as in all following comparable figures. The best-fit relationship determined over a large range of data in Figure 2 indicates that the strike-slip faults considered here for

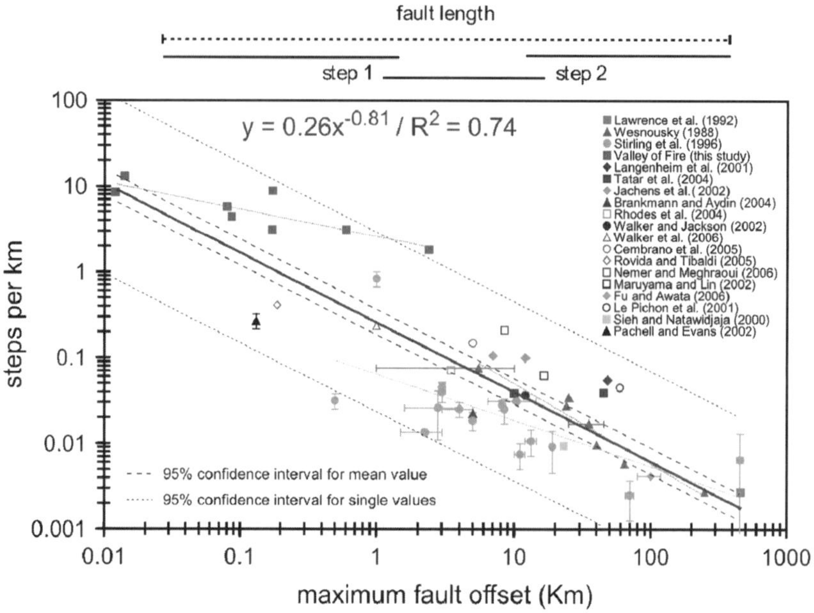

Figure 2

Relationships between the maximum fault offset and the number of steps per km for the strike-slip faults from this study and the published literature as referenced on the figure. The best (power) fits to the general data (solid blue line) and to individual data (Valley of Fire: dotted red line; STIRLING *et al.*, 1996: dotted green line; WESNOUSKY, 1988: dotted pink line) are given. The (95%) confidence intervals for the mean (dashed black lines) and single (dashed blue lines) values of the number of steps per km are also shown.

the statistical analyses have longer segments and fewer steps with increasing slip magnitudes, which is related to the fundamental process of fault growth by segment coalescence (CARTWRIGHT *et al.*, 1995; DE JOUSSINEAU and AYDIN, 2007). These results extend and confirm those from previous studies (WESNOUSKY, 1988; STIRLING *et al.*, 1996).

3.2. *Segment Length*

Here we consider the segment length in ~ 20 of the faults in Figure 2, for which segments and their tips and steps could be identified so that adequate mean segment lengths, mean step lengths and mean step widths (see Section 3.3) could be measured. Figure 3 shows a clear positive power law relationship between the mean segment length and the maximum fault offset: $y = 2.16x^{0.89} / R^2 = 0.87$, where the mean segment length and the maximum offset are plotted on y and x axis, respectively. Results in Figure 3 are consistent with those in the experimental study by OTSUKI and DILOV (2005). In their study, the data were plotted against fault length instead of fault offset as in our work. However, as these two fault parameters are thought to be correlated (COWIE and SCHOLTZ,

Table 1

*Data plotted in Figure 2. *Note that also a value of 65 km was proposed in the literature*

Source study	Fault name	Scale	Max. fault offset (km)	Number of steps per km
This study	Classic	1/3800	0.175	8.9
	12 m	1/1000	0.012	8.5
	86 m	1/3800	0.086	4.4
	172 m	1/3800	0.172	3.1
	Lonewolf	1/3800	0.08	5.78
	14 m	1/1000	0.014	13.1
	Baseline Mesa	1/37000	2.4	1.81
	600 m	1/37000	0.6	3.09
Wesnousky (1988)	Newport-Ingelwood	1/1050000	5.5 ± 4.5	0.0758
	Calaveras-Green Valley	1/345000	24	0.0275
	San Jacinto	1/323000	25	0.0345
	Whittier-Elsinore	1/4000000	40	0.00965
	Garlock	1/3030000	64	0.00585
	San Andreas	1/2940000	250	0.00275
	North Anatolian	1/4550000 to 1/2850000	35 ± 10*	0.017
Stirling *et al.* (1996)	Helendale	1/530000	3	0.045 ± 0.007
	Calico-Mesquite	1/870000	8.2	0.028 ± 0.004
	Pisgah	1/500000	10.4 ± 4	0.031
	Camp Rock	1/570000	2.8 ± 1.2	0.026 ± 0.015
	Lenwood	1/480000	2.25 ± 0.75	0.0135 ± 0.0005
	San Miguel-Vallecitos	1/1090000	0.5	0.0315 ± 0.0065
	MTL (Shikoku island)	1/1330000	5	0.0185 ± 0.0045
	Neodani	1/670000	4 ± 1	0.025 ± 0.005
	Atera	1/600000	8.5 ± 1.5	0.025 ± 0.008
	Atotsugawa	1/590000	3	0.040 ± 0.01
	Tanna	1/310000	1	0.835 ± 0.165
	Wairau	1/550000	455 ± 25	0.0065 ± 0.0065
	Hope	1/1560000	19	0.00925 ± 0.00473
	Wellington	1/1820000	11 ± 1	0.0075 ± 0.0025
	Altun	1/5000000	70 ± 5	0.0025 ± 0.00125
	Haiyuan	1/1670000	13.25 ± 1.25	0.01072 ± 0.00358
Lawrence *et al.* (1992)	Chaman Transform Zone	1/2500000	460	0.00273
Langenheim *et al.* (2001)	Las Vegas Valley Shear Zone (North trend)	1/1030000	48	0.0549
Tatar *et al.* (2004)	Amanos	1/800000	45	0.039
	East Hatay	1/800000	10	0.03825
Jachens *et al.* (2002)	Ludlow	1/1750000	12	0.1
	Blackwater	1/1750000	7	0.10638
Brankman and Aydin (2004)	Mattinata	1/510000	5	0.02234
Rhodes *et al.* (2004)	Mae Kuang	1/360000	3.5	0.07143
Walker and Jackson (2002)	Gowk	1/9500000	12	0.03659
Walker *et al.* (2006)	Jid	1/230000	1	0.23684
Cembrano *et al.* (2005)	Coloso	1/180000	5	0.14865
Rovida and Tibaldi (2005)	Buesaco	1/77000	0.188	0.4116
Nemer and Meghraoui (2006)	Roum	1/180000	8.5	0.211

Table 1

contd.

Source study	Fault name	Scale	Max. fault offset (km)	Number of steps per km
Maruyama and Lin (2002)	Arima-Takatsuki Tectonic Line	1/440000	16.5	0.0625
Fu and Awata (2006)	Kunlun	1/10000000	100 ± 20	0.00422
Le Pichon *et al.* (2001)	Marmara	1/1300000	59	0.045
Sieh and Natawidjaja (2000)	Great Sumatran	1/4620000	23	0.00947
Pachell and Evans (2002)	Gemini	1/62500	0.131	0.2688 ± 0.0538

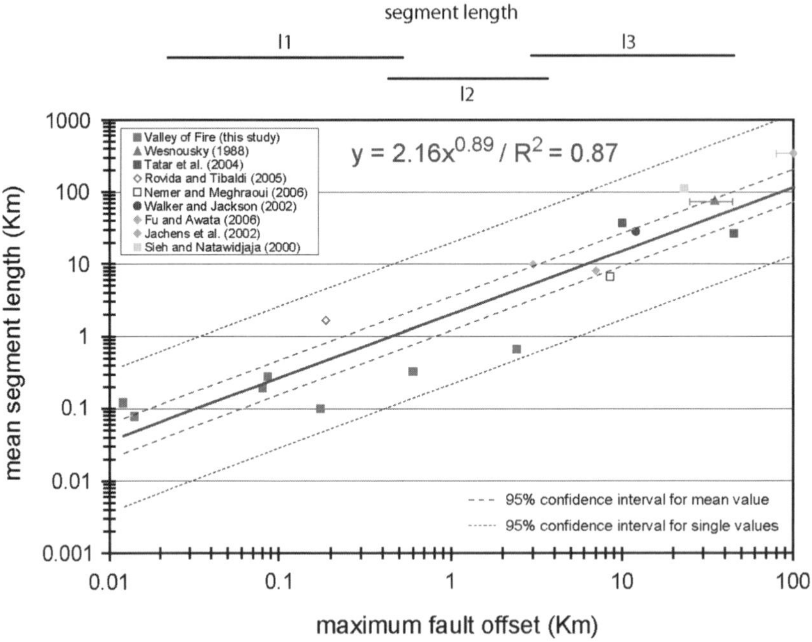

Figure 3

Relationships between the maximum fault offset and the mean segment length for strike-slip faults from this study and the published literature as referenced on the figure. The best (power) fit to the data and the (95%) confidence intervals for the mean and single values of the mean segment length are shown.

1992; Dawers *et al.*, 1993; Schlische *et al.*, 1996), direct comparisons could be made between our results and those in Otsuki and Dilov (2005).

Figure 4 provides additional constraints on strike-slip fault segment lengths by showing segment length distributions for four of the faults in Figure 3 with offsets ranging from 80 m to 100 km. All of these distributions are lognormal. Moreover, for each fault, all of the segments have a length falling in the range of zero to ten times the maximum fault offset 'd', with most of the segment lengths in the interval 1d–5d (see data in Table 2).

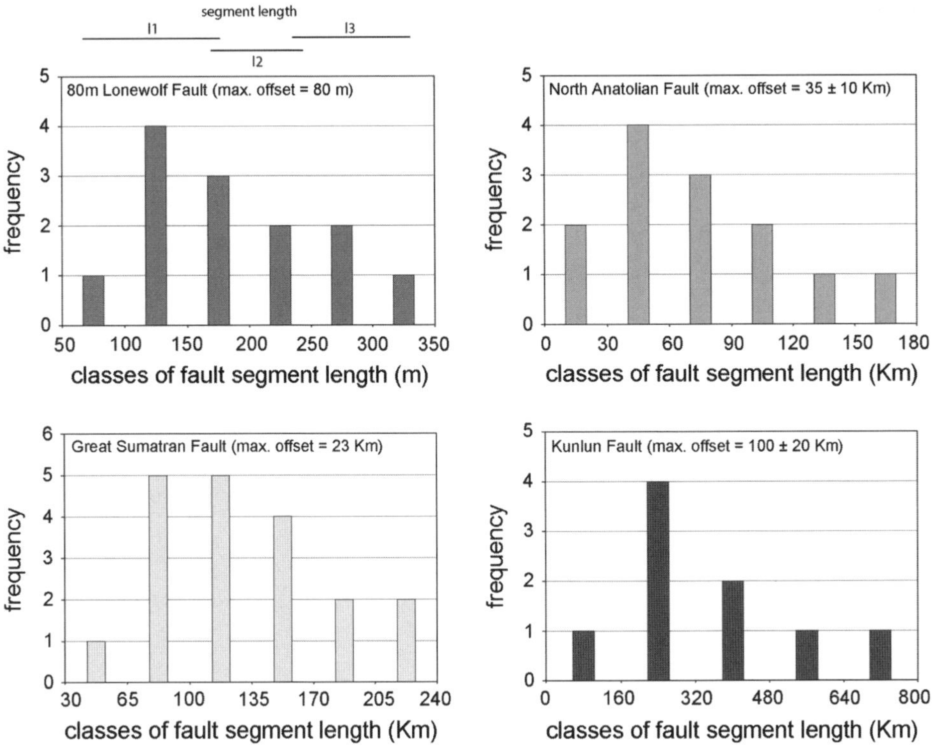

Figure 4
Segment-length frequency distributions for four strike-slip faults. Fault names and offsets are indicated.

3.3. Step Dimensions

We also investigated the mean step length and mean step width in the same faults as in Figure 3, augmented by two new fault examples where enough steps could be measured but insufficient segment length data were available. The data set in Figure 5 presents a clear positive power law relationship between the mean step length and the maximum fault offset: $y = 0.30x^{0.77}/R^2 = 0.84$, where the mean step length and the maximum offset are plotted on y and x axis, respectively. Figure 6 illustrates the same kind of relationship between the mean step width and the maximum fault offset: $y = 0.12x^{0.80}/R^2 = 0.81$, where the mean step width and the maximum offset are plotted on y and x axis, respectively. Again, these results are consistent with those in OTSUKI and DILOV (2005).

Finally, Figure 7 shows the relationships between step width and step length for the faults analyzed in this study and for the faults compiled in AYDIN and NUR (1982). The data of the two sets align very well and are best fit by a quasi-linear positive power law relationship: $y = 2.69x^{0.97}/R^2 = 0.93$, where the step length and step width are plotted on y and x axis, respectively. This relationship is very similar to that in AYDIN and NUR (1982) and indicates that the length-to-width ratio of fault steps falls into a narrow range with an average value of slightly less than 3 ($\sim$2.7 for this study).

Table 2

*Data plotted in Figure 4. *Note that also a value of 65 km was proposed in the literature*

Source study	Fault name	Scale	Max. fault offset, d (km)	Segment length (km)	Segment length (d)
This study	Lonewolf	1/3800	0.08	0.079	0.99
				0.102	1.28
				0.129	1.61
				0.139	1.74
				0.149	1.86
				0.188	2.35
				0.191	2.39
				0.191	2.39
				0.201	2.51
				0.234	2.93
				0.271	3.39
				0.274	3.43
				0.347	4.34
Sieh and Natawidjaja (2000)	Great Sumatran	1/4620000	23	35	1.52
				60	2.61
				60	2.61
				65	2.83
				70	3.04
				70	3.04
				85	3.7
				85	3.7
				90	3.91
				95	4.13
				95	4.13
				120	5.22
				125	5.43
				150	6.52
				150	6.52
				160	6.96
				180	7.83
				200	8.7
				220	9.57
Wesnousky (1988)	North Anatolian	1/4550000 to 1/2850000	35 ± 10 (*)	23	0.65
				27	0.78
				46	1.3
				47	1.35
				59	1.67
				59	1.69
				70	1.99
				77	2.2
				78	2.24
				96	2.73
				102	2.92
				127	3.63
				179	5.12

Table 2

contd.

Source study	Fault name	Scale	Max. fault offset, d (km)	Segment length (km)	Segment length (d)
Fu and Awata (2006)	Kunlun	1/10000000	100 ± 20	135	1.35
				230	2.3
				270	2.7
				300	3
				300	3
				330	3.3
				340	3.4
				480	4.8
				680	6.8

4. *Discussion and Implications*

4.1. Scale- and Resolution-related Biases

Because this paper aims to analyze the architecture of strike-slip faults over a wide range of magnitude of the fault offset, the data presented in Figures 2 to 7 were obtained from faults of very different sizes that were mapped at different scales, i.e., with different resolutions. These variations in resolution of the mapping, inherent to any compilation work, could possibly introduce a bias in the data sets. In particular, the trends observed in Figures 2, 3, and 5–7 may be affected by the mapping resolution, with the larger faults being mapped with a lesser degree of detail than the smaller ones, i.e., tending to have longer segments and fewer steps. However, there are strong arguments supporting the existence of the trends in the data sets. First, the concept that faults become smoother as they mature is well accepted. In particular, previous studies by Wesnousky (1988) and Stirling *et al.* (1996) have established clear relationships between maximum strike-slip fault offset and strike-slip fault trace complexity over up to three orders of magnitude of the fault offset (offsets ranging from 500 m to 455 km). Our study complements those results and expands them to smaller fault scales (offsets down to 12 m). In addition, the same trends are observed in the data corresponding to the faults from the Valley of Fire. These faults have been mapped with great care and at resolutions that are nearly uniform in a range of scales, so that scale and resolution biases were avoided to the extent possible. The fact that clear relationships between maximum fault offset and number of steps per km, mean segment length, mean step length and mean step width are observed in these data suggests that the trends in the general data sets in Figures 2, 3 and 5-7 have geological and mechanical origins which are likely related to fault growth processes, as discussed in the next section.

Examination of the Valley of Fire data in Figures 2–3 and 5–6 may suggest the existence of trends that are different from the trends in other data sets and the general trends of the combined data. For example, in Figure 2, data from the Valley of Fire may

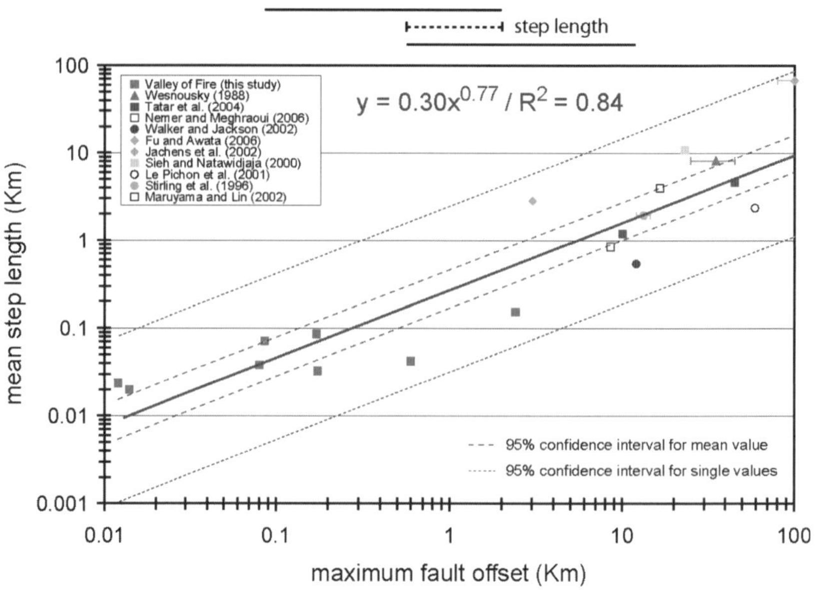

Figure 5
Relationships between the maximum fault offset and the mean step length for strike-slip faults from this study and the published literature as referenced on the figure. The best (power) fit to the data and the (95%) confidence intervals for the mean and single values of the mean step length are shown.

be fit by a power law with an exponent of ~0.32, whereas the power trend for the combined data has an exponent of ~0.81, and power trends fitting the data of Stirling *et al.* (1996) and Wesnousky (1988) have exponents of ~0.52 and ~0.94, respectively. This is a very common observation in any combined data set. For example, Schlische *et al.* (1996) provided a combined data set for the relationships between fault length and maximum fault slip. It is clear in their data set that each particular population of data can be characterized by its own trend that is different from the general trend of the combined data. Similar observations can be made in the data set in Dawers *et al.* (1993). Such differences in the trends of the different individual data sets were generally attributed, among others, to different rock types and geological contexts or different mechanical behaviors of faulted rocks (Dawers *et al.*, 1993; Schlische *et al.*, 1996). They appear to reflect the actual variability that exists in nature.

However, it is also possible that the scaling parameters may change with the size of the structural features and the scale of observation, i.e., that the data could be characterized by different trends at different scale intervals. For the relationships between fault length and maximum fault slip, it has been proposed that there is a characteristic dimension at which scaling parameters change in relation with either a change in the geometry of fault growth or different material properties encountered by faults that propagate into deeper parts of the crust as their sizes increase (Dawers *et al.*, 1993).

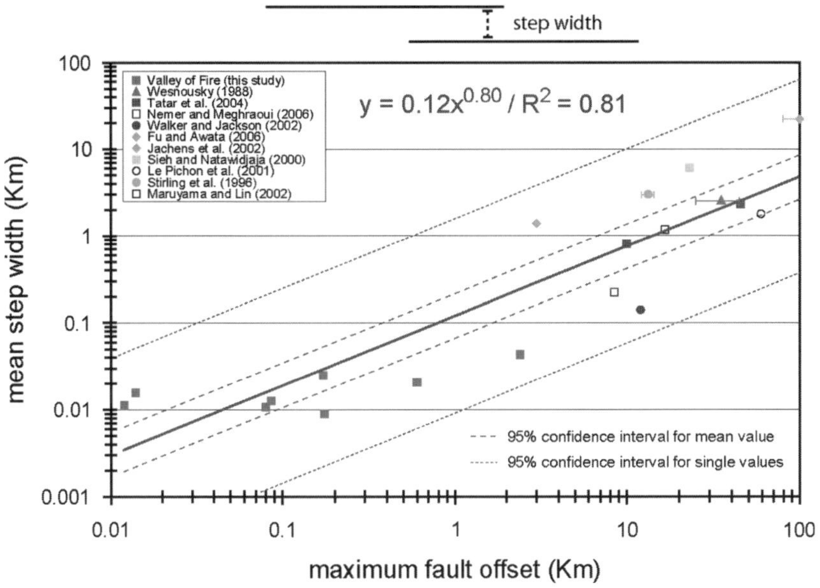

Figure 6

Relationships between the maximum fault offset and the mean step width for strike-slip faults from this study and the published literature as referenced on the figure. The best (power) fit to the data and the (95%) confidence intervals for the mean and single values of the mean step width are shown.

In our study, the different trends in Figure 2 may partly owe their existence to scale differences. The data obtained from faults mapped at a relatively small scale in the Valley of Fire may be skewed towards high numbers of steps per unit length, hence exhibit lower power law exponents than the data obtained from faults mapped at larger scales. Also, the different trends in Figure 2 may indicate a change in scaling parameters with the size of features, in relationship to changes in fault growth processes. However, several observations suggest that this may not be the case. First, no clear change in the scaling parameters for a given value of the fault offset is observed in the data. There is no clear break in the data in Figure 2 but a relative scatter of data at all scales. For instance, data based on the studies by STIRLING *et al.* (1996), PACHELL and EVANS (2002), ROVIDA and TIBALDI (2005) or WALKER *et al.* (2006) plot below the Valley of Fire data for similar strike-slip fault offsets. Second, the trend fitting the data of STIRLING *et al.* (1996) has an exponent that is closer to the one in the Valley of Fire data than to the one in the data of WESNOUSKY (1988), although STIRLING *et al.* (1996) and WESNOUSKY (1988) both studied large-scale faults. This suggests that the different trends in Figure 2 are more related to different contexts and lithologies or different mechanical behaviors of rocks than to changes in scaling relationships with the size of features. It follows that the global trends in Figures 2, 3, 5 and 7 may be more-or-less valid, even if the existence of possible biases due to scale variability in the original fault maps cannot be completely ruled out.

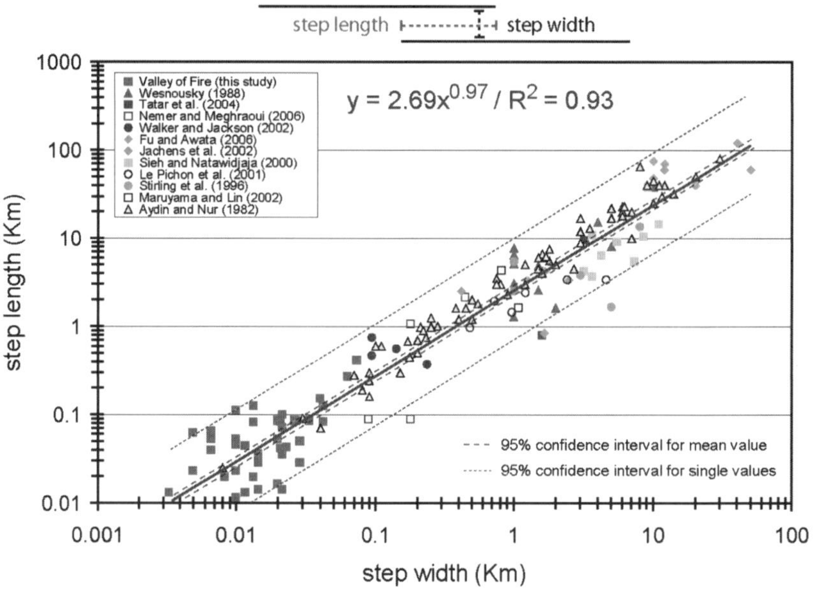

Figure 7
Relationships between step width and step length for strike-slip faults from this study and the published literature as referenced on the figure. The best (nearly-linear power) fit to the data and the (95%) confidence intervals for the mean and single values of the step length are shown.

4.2. Fault Growth and Evolution

It is now widely accepted that brittle faults start from a series of short echelon arrays and grow by linkage of the neighboring segments (SEGALL and POLLARD, 1980; CARTWRIGHT *et al.*, 1995; MYERS and AYDIN, 2004; KIM *et al.*, 2004). It then follows that steps are destructed by through-going planar surfaces to produce more continuous and smoother fault segment geometry at larger scales while new segments achieve a greater capability to interact with each other at a greater distance (AYDIN and SCHULTZ, 1990; SCHOLZ, 2002). BEN-ZION and SAMMIS (2003) characterized this process of fault growth and maturation by summing up that fault structures tend to evolve toward geometric simplicity at all scales. The data related to the size and density distribution of fault steps and the evolution of fault segment length we presented in this paper are consistent with these notions.

It appears that the underlying mechanical principle for many of these phenomena is rather simple. First, for a mode-II fracture, the stress components at a point in the regions away from the fracture tips decay as $(a/r)^2$ for 2-D (POLLARD and SEGALL, 1987) and $(a/r)^3$ for 3-D (BEN-ZION and SAMMIS, 2003), where “a” is the half fault length and “r” is the radial distance to the center of the fault. Second, fault interaction is an important factor in the final fault geometry. AYDIN and SCHULTZ (1990) investigated the influence of fault geometry as well as the sense of step using a 2-D mode-II fracture interaction model.

They concluded that the relative locations of the neighboring fracture tips strongly impact echelon mode-II fracture geometry although this influence is more-or-less independent of the sense of echelon step. They also determined that fault propagation energy first increases as the fault tips approach each other and then decreases sharply as the fracture tips pass by each other. The data on the step size and distribution are consistent with these notions.

Finally, our analyses of strike-slip fault steps complement those in AYDIN and NUR (1982) and confirm that the length-to-width ratio of fault steps is more-or-less constant. This supports the conceptual models of evolution of strike-slip fault systems in which the size of fault steps progressively increases through the coalescence of neighboring relay zones and/or the formation of new fault strands parallel to the existing ones during fault growth.

4.3. Seismic Rupture

Our results indicate that the trace complexity of strike-slip faults is a smoothly decreasing function of the fault offset, implying that larger faults have fewer and larger steps and longer segments than smaller faults. This may have important implications for fault growth processes and earthquake ruptures. First, it is explicit evidence for the universality of fault segment linkage and coalescence or the straightening of fault traces by obliteration and/or elimination of steps at certain stages of fault development. Second, strength properties of fault zones partly depend on the presence of fault steps which oppose the slip and act as asperities or barriers for rupture events along the faults (SIBSON, 1986; AKI, 1989). As the distribution of steps along strike-slip faults is a function of the fault offset, it follows than the strength properties of faults may also be partly correlated with the offset. This supports the proposition by WESNOUSKY (1988) that strike-slip faults, together with their structural evolution, may undergo a seismological evolution whereby the size and frequency distribution of slip events are a function of the fault offset.

Our results also provide a basis for applying the experimental results of OTSUKI and DILOV (2005) to natural faults. These authors observed that strike-slip faults developing sequentially in rock samples had a hierarchical self-similar architecture wherein the fault segments of a higher hierarchical rank (i.e., newly developed) were composed of smaller segments and steps of a lower hierarchical rank (i.e., developed during previous stages of fault evolution). In addition, the length of the fault systems was correlated to the mean segment length, mean step length and mean step width by positive power laws. Based on these observations, they derived the Gutenberg-Richter law and the empirical relationship of the length of the seismic nucleation zone to the seismic moment and proposed that the seismic rupture mimics the hierarchical geometry of fault zones, by nucleating at a small step of a lower hierarchical rank and terminating at a large step of higher hierarchical rank. Our analyses of natural strike-slip fault zones yielded results that compare well with those of OTSUKI and DILOV (2005). This suggests that natural strike-slip faults may have the same hierarchical self-similar architecture as the faults in their experiments. Thus, their reasoning in terms of the impact of fault architecture on the dynamics of seismic ruptures could also apply to natural geological faults. This idea is also supported by recent conceptual models for the development of strike-slip fault zones (MYERS and AYDIN, 2004; FLODIN and AYDIN, 2004; DE JOUSSINEAU and AYDIN, 2007), which proposed that strike-slip faults have a hierarchical pattern where, at each stage of the fault evolution, the fault segments can be broken into a series of smaller segments and steps related to earlier phases of fault development. It is likely that new, short segments are also generated as faults grow. However, the rate of generation of such short segments along a given mature fault is probably small compared with the rate of development of large segments by coalescence of earlier, shorter segments.

We should, however, note that predicting the distribution and size (length and width) of steps along faults may not be sufficient for forecasting the nucleation site and extent of seismic ruptures. Other important parameters have an effect on fault strength properties and the dynamics of earthquake ruptures. Particularly, ruptures may be influenced by the orientation of fault segments with respect to the regional tectonic stress field. Examples of a rupture jumping across a major fault step because the orientation of the fault segment that was being ruptured was becoming unfavorable have been documented, as in the 1992 Landers earthquake (BOUCHON *et al.*, 1998). However, even if the mechanics of earthquake rupture is complex and its size and location cannot entirely be forecasted by knowing or estimating the pattern and architecture of the related geological faults, this knowledge still provides valuable information regarding one of the important parameters influencing the nucleation and termination of earthquake ruptures.

4.4. Prediction of Subsurface Fault Attributes and Fluid Flow

This study, by establishing the general self-similar architecture of strike-slip faults and the relationships between the maximum fault offset and the dimensions of fault segments and steps, may allow one to better forecast the geometrical attributes of strike-slip faults. This has important implications for the internal structure of fault zones, the distribution of damage zones and the properties of secondary fault networks associated with strike-slip fault systems.

First, it may be possible to estimate the width of steps along a given fault of known length or offset (both being correlated). Steps are locations of the faults where fault rocks and damage zones develop preferentially by progressive fracturing and crushing of the host rock because of high stress concentration therein. Thus, determining the dimensions of fault steps could provide the size of the largest pockets of initially fractured rocks and later fault core along the mature fault zones. This is important in terms of fluid flow because initially highly fractured steps are important conduits whereas the fault core developed later on could act as a barrier for fluids when characterized by very low cross-fault permeability (AYDIN, 2000; FLODIN *et al.*, 2005). Second, being able to predict the architecture of large strike-slip faults with resolvable slip magnitude could provide insights into the distribution and characteristics of associated secondary faults, which may not be detected by seismic techniques if they accommodate small magnitudes of slip. These secondary faults could develop as splays associated with the tip regions of the main fault segments (FLODIN and AYDIN, 2004; DE JOUSSINEAU and AYDIN, 2007; Fig. 1c), and may have length and angular properties correlated with the length of and spacing between segments composing the strike-slip fault systems (DE JOUSSINEAU *et al.*, 2007). Thus evaluating the number of segments composing the main strike-slip fault along with their length and spacing (given by the step width) could provide ways of estimating the location, frequency, and length and angular properties of the associated secondary faults. The prediction of such secondary faults is critical because they strongly influence the geometrical and statistical properties of damage zones (DE JOUSSINEAU and AYDIN, 2007). They could also enhance fracture connectivity by linking distant faults (as in Fig. 1c), and may increase the compartmentalization of the medium when they have a continuous core with low permeability.

5. *Conclusions*

We have provided new data for the remarkable self-similar architecture of strike-slip fault systems from the Valley of Fire State Park (SE Nevada), in addition to published data sets and maps. The new data set collected from the same region and rock type complements the published data and interpretations and lends support for the previous results related to the general trends in fault scaling. First, we have determined that the number of fault steps (hence fault segments) per unit length is correlated to the maximum fault offset by a negative power law over five orders of magnitude of the offset, suggesting that fault traces become progressively smoother with increasing offset. Second, we have determined that the mean segment length, mean step length and mean step width are all correlated to the maximum fault offset by positive power laws. In addition, we have analyzed the segment length distributions in four of the faults with offsets ranging from 80 m to 100 km. These distributions are similar in such that they are all lognormal, with most of the fault segments with a length in the interval of one to five times the fault offset. Finally, the steps composing the strike-slip fault systems have an approximately constant length-to-width ratio. These results are consistent with the commonly accepted concepts that strike-slip fault trace geometries become smoother and simpler as their offsets increase.

These fault scaling relationships usually have slightly different trends for the data sets from the Valley of Fire State Park, Nevada, our detailed study area, and the data sets from other geographic and geological environments. These differences, as proposed by the earlier investigators, could be due to the types and rheology of rocks, geological environments, and the fault growth processes. It is also possible that the scale of observation may cause artifacts in the fault scaling relationships.

These conclusions shed light onto the processes involved in the growth of natural faults over a broad range of scales. The data sets may be used as tools to better characterize the architecture of subsurface strike-slip fault systems and to estimate useful parameters for subsurface applications, with critical implications for earthquake ruptures and fluid flow in the Earth's crust.

Acknowledgements

We acknowledge support from the Department of Energy, Basic Energy Sciences, Geosciences Research Program (Grant # DE-FG03-94ER14462 to Atilla Aydin and David D. Pollard). We thank the Valley of Fire State Park Rangers for permission to work in some areas of the Park where regular visitors are not allowed, and Ann-Laure Moreau for providing assistance in the field. Finally, we acknowledge the comments and suggestions by Yehuda Ben Zion, Mike Oskin and an anonymous reviewer.

REFERENCES

AKI, K. (1989), *Geometric features of a fault zone related to the nucleation and termination of an earthquake rupture*. In Proc. Conf. XLV *Fault Segmentation and Controls of Rupture Initiation and Termination*, U.S. Geological Survey Open-File Report 89–315, 1–9.

AYDIN, A. (2000), *Fractures, faults, and hydrocarbon entrapment, migration and flow*, Marine Petrol. Geol. *17*, 797–814.

AYDIN, A. and SCHULTZ, R.A. (1990), *Effect of mechanical interaction on the development of strike-slip faults with echelon patterns*, J. Struct. Geol. *12*, 123–129.

AYDIN, A. and NUR, A. (1982), *Evolution of pull-apart basins and their scale independence*, Tectonics *1*, 91–105.

BARKA, A. and KADINSKY-CADE, K. (1988), *Strike-slip fault geometry in Turkey and its influence on earthquake activity*, Tectonics *7*, 663–684.

BEN-ZION, Y. and SAMMIS, C.G. (2003), *Characterization of fault zones*, Pure Appl. Geophys. *160*, 677–715.

BOUCHON, M., CAMPILLO, M., and COTTON, F. (1998), *Stress field associated with the rupture of the 1992 Landers, California, earthquake and its implications concerning the fault strength at the onset of the earthquake*, J. Geophys. Res. *103*, 21091–21097.

BRANKMAN, C.M. and AYDIN, A. (2004), *Uplift and contractional deformation along a segmented strike-slip fault system: The Gargano Promontory, southern Italy*. J. Struct. Geol. *26*, 807–824.

CAINE, J.S., EVANS, J.P., and FORSTER, C.B. (1996), *Fault zone architecture and permeability structure*, Geology *24*, 1025–1028.

CARTWRIGHT, J.A., TRUDGILL, B.D., and MANSFIELD, C.S. (1995), *Fault growth by segment linkage: An explanation for scatter in maximum displacement and trace length data from the Canyonlands Grabens of SE Utah*, J. Struct. Geol. *17*, 1319–1326.

CEMBRANO, J., GONZÁLEZ, G., ARANCIBIA, G., AHUMADA, I., OLIVARES, V., and HERRERA, V. (2005), *Fault zone development and strain partitioning in an extensional strike-slip duplex: A case study from the Mesozoic Atacama fault system, Northern Chile*, Tectonophysics *400*, 105–125.

COWIE, P.A. and SCHOLZ, C.H. (1992a), *Displacement-length scaling relationship for faults: Data synthesis and discussion*, J. Struct. Geol. *14*, 1149–1156.

DAWERS, N.H., ANDERS, M.H., and SCHOLZ, C.H. (1993), *Growth of normal faults: Displacement-length scaling*, Geology *21*, 1107–1110.

DE JOUSSINEAU, G. and AYDIN, A. (2007), *The evolution of the damage zone with fault growth and its multiscale characterization*, J. Geophys. Res. *112*, B12401, doi:10.1029/2006JB004711.

DE JOUSSINEAU, G., MUTLU, O., AYDIN, A., and POLLARD, D.D. (2007), *Characterization of strike-slip fault-splay relationships in sandstone*, J. Struct. Geol. *29*, 1831–1842.

EICHHUBL, P., TAYLOR, W.L., POLLARD, D.D., and AYDIN, A. (2004), *Paleo-fluid flow and deformation in the Aztec Sandstone at the Valley of Fire, Nevada—Evidence for the coupling of hydrogeologic, diagenetic and tectonic processes*, Geol. Soc. Am. Bull. *116*, 1120–1136.

FINZI, Y., HEARN, E.H., BEN-ZION, Y., and LYAKHOVSKY, V. (2009), *Structural properties and deformation patterns of evolving strike-slip faults: Numerical simulations incorporating damage rheology*, Pure Appl. Geophys., in press.

FLODIN, E., and AYDIN, A. (2004), *Evolution of a strike-slip fault network, Valley of Fire State Park, southern Nevada*, Geol. Soc. Am. Bull. *116*, 42–59.

FLODIN, E.A., GERDES, M., AYDIN, A., and WIGGINS, W. D. (2005), *Petrophysical properties of cataclastic fault rock in sandstone*. In *Faults, Fluid Flow, and Petroleum Traps* (eds. Sorkhabi, R., and Tsuji, Y.) American Association of Petroleum Geologists Memoir *85*, 197–227.

FU, B. and AWATA, Y. (2006), *Displacement and timing of left-lateral faulting in the Kunlun Fault Zone, northern Tibet, inferred from geologic and geomorphic features*, J. Asian Earth Sci. 29, 253–265.

HARRIS, R.A., ARCHULETA, R.J., and DAY, S. M. (1991), *Fault steps and the dynamic rupture process: 2-d simulation of a spontaneously propagating shear fractures*, Geophys. Res. Lett. *18*, 893–896.

HARRIS, R.A. and DAY, S.M. (1999), *Dynamic 3D simulation of earthquakes on en echelon faults*, Geophys. Res. Lett. *26*, 2089–2092.

JACHENS, R.C., LANGENHEIM, V.E., and MATTI, J.C. (2002), *Relationship of the 1999 Hector Mine and 1992 Landers fault ruptures to offsets on Neogene faults and distribution of late Cenozoic basins in the Eastern California Shear Zone*, Bull. Seismol. Soc. Am. *92*, 1592–1605.

JOURDE, H., FLODIN, E.A., AYDIN, A., and DURLOFSKY, L.J. (2002), *Computing permeability of fault zones in eolian sandstone from outcrops measurements*, Am. Assoc. Petrol. Geologists Bull. *86*, 1187–1200.

KIM, Y., PEACOCK, D.C.P., and SANDERSON, D.J. (2004), *Fault damage zones*, J. Struct. Geol. *26*, 503–517.

KNOTT, S.D., BEACH, A., BROCKBANK, P.J., LAWSON BROWN, J., MCCALLUM, J.E., and WELBON, A.I. (1996), *Spatial and mechanical controls on normal fault populations*, J. Struct. Geol. *18*, 359–2372.

LANGENHEIM, V.E., GROW, J.A., JACHENS, R.C., DIXON, G.L., and MILLER, J.J. (2001), *Geophysical constraints on the location and geometry of the Las Vegas Valley Shear Zone, Nevada*, Tectonics *20*, 189–209.

LAWRENCE, R.D., HASAN KHAN, S., and NAKATA, T. (1992), *Chaman Fault, Pakistan-Afghanistan*, Annales Tectonicae *6*, 196–223.

LE PICHON, X., ŞENGÖR, A.M.C., DEMIRBAĞ, E., RANGIN, C., İMREN, C., ARMIJO, R., GÖRÜR, N., ÇAĞATAY, N., MERCIER DE LEPINAY, B., MEYER, B., SAATÇILAR, R., and TOK, B. (2001), *The active Main Marmara Fault*, Earth Planet. Sci. Lett. *192*, 595–616.

MARUYAMA, T. and LIN, A. (2002), *Active strike-slip faulting history inferred from offsets of topographic features and basement rocks: A case study of the Arima-Takatsuki Tectonic line, southwest Japan*, Tectonophysics *344*, 81–101.

MYERS, R. and AYDIN, A. (2004), *The evolution of faults formed by shearing across joint zones in sandstone*, J. Struct. Geol. *26*, 947–966.

NEMER, T. and MEGHRAOUI, M. (2006), *Evidence of coseismic ruptures along the Roum fault (Lebanon): A possible source for the AD 1837 earthquake*, J. Struct. Geol. *28*, 1483–1495.

ODLING, N.E., HARRIS, S.D., and KNIPE, R.J. (2004), *Permeability scaling properties of fault damage zones in siliclastic rocks*, J. Struct. Geol. *26*, 1727–1747.

OTSUKI, K. and DILOV, T. (2005), *Evolution of self-similar geometry of experimental fault zones; implications for seismic nucleation and earthquake size*, J. Geophys. Res. *110*, B03303, doi:10.1029/2004JB003359.

PACHELL, M.A. and EVANS, J.P. (2002), *Growth, linkage, and termination processes of a 10-km-long strike-slip fault in jointed granite: The Gemini fault zone, Sierra Nevada, California*, J. Struct. Geol. *24*, 1903–1924.

POLLARD, D.D. and SEGALL, P., *Theoretical displacement and stresses near fractures in rock: With applications to faults, joints, veins, dikes, and solution surfaces*. In *Fracture Mechanics of Rock* (ed. Atkinson, B.K.) (Academic Press 1987) pp 277–349.

RHODES, B.P., PEREZ, R., LAMJUAN, A., and KOSUWAN, S. (2004), *Kinematics and tectonic implications of the Mae Kuang Fault, northern Thailand*, J. Asian Earth Sci. *24*, 79–89.

ROVIDA, A., and TIBALDI, A. (2005), *Propagation of strike-slip faults across Holocene volcano-sedimentary deposits, Pasto, Colombia*, J. Struct. Geol. *27*, 1838–1855.

Schlische, R.W., Young, S.S., Ackermann, R.V., and Gupta, A. (1996), *Geometry and scaling relations of a population of very small rift-related normal faults*, Geology *24*, 683–686.

Scholz, C.H., *The Mechanics of Earthquakes and Faulting*, 2nd revised edition (Cambridge University Press 2002).

Segall, P. and Pollard, D.D. (1980), *Mechanics of discontinuous faults*, J. Geophys. Res. *85*, 4337–4350.

Shaw, B.E. and Dieterich, J.H. (2007), *Probabilities for jumping fault segment stepovers*, Geophys. Res. Lett. *34*, L01307, doi:10.1029/2006GL027980.

Sibson, R.H. (1985), *Stopping of earthquake ruptures at dilational fault jogs*, Nature *316*, 248–251.

Sibson, R.H. (1986), *Rupture interaction with fault jogs*. In *Earthquake Source Mechanics* (eds. Das, S., Boatwright, J., and Scholz, C.H.) (Am. Geophys. Union 1986) pp. 157–167.

Sieh, K. and Natawidjaja, D. (2000), *Neotectonics of the Sumatran fault, Indonesia*, J. Geophys. Res. *105*, 28295–28326.

Stirling, M.W., Wesnousky, S.G., and Shimazaki, K. (1996), *Fault trace complexity, cumulative slip, and the shape of the magnitude-frequency distribution for strike-slip faults: A global survey*, Geophys. J. Internat. *124*, 833–868.

Tatar, O., Piper, J.D.A., Gürsoy, H., Heimann, A., and Koçbulut, F. (2004), *Neotectonic deformation in the transition zone between the Dead Sea Transform and the East Anatolian fault Zone, Southern Turkey: A palaeomagnetic study of the Karasu Rift Volcanism*, Tectonophysics *385*, 17–43.

Walker, R. and Jackson, J. (2002), *Offset and evolution of the Gowk fault, S.E. Iran: a major intra-continental strike-slip system*, J. Struct. Geol. *24*, 1677–1698.

Walker, R.T., Bayasgalan, A., Carson, R., Hazlett, R., McCarthy, L., Mischler, J., Molor, E., Sarantsetseg, P., Smith, L., Tsogtbadrakh, B., and Thompson, G. (2006), *Geomorphology and structure of the Jid right-lateral strike-slip fault in the Mongolian Altay mountains*, J. Struct. Geol. *28*, 1607–1622.

Wesnousky, S.G. (1988), *Seismological and structural evolution of strike-slip faults*, Nature *335*, 340–342.

Wesnousky, S.G. (2006), *Predicting the endpoints of earthquake ruptures*, Nature *444*, 358–360.

(Received August 26, 2008, accepted March 30, 2009)
Published Online First: July 29, 2009

To access this journal online:
www.birkhauser.ch/pageoph

Pure appl. geophys. 166 (2009) 1595–1627
0033–4553/09/101595–33
DOI 10.1007/s00024-009-0519-9

© Birkhäuser Verlag, Basel, 2009

Pure and Applied Geophysics

Influence of Outcrop Scale Fractures on the Effective Stiffness of Fault Damage Zone Rocks

W. Ashley Griffith,[1,3] Pablo F. Sanz,[2,4] and David D. Pollard[1]

Abstract—We combine detailed mapping and microstructural analyses of small fault zones in granodiorite with numerical mechanical models to estimate the effect of mesoscopic (outcrop-scale) damage zone fractures on the effective stiffness of the fault zone rocks. The Bear Creek fault zones were active at depths between 4 and 15 km and localize mesoscopic off-fault damage into tabular zones between two subparallel boundary faults, producing a fracture-induced material contrast across the boundary faults with softer rocks between the boundary faults and intact granodiorite outside the boundary faults. Using digitized fault zone fracture maps as the modeled fault geometries, we conduct nonlinear uniaxial compression tests using a novel finite-element method code as the experimental "laboratory" apparatus. Map measurements show that the fault zones have high nondimensional facture densities (>1), and damage zone fractures anastamose and intersect, making existing analytical effective medium models inadequate for estimation of the effective elastic properties. Numerical experiments show that the damage zone is strongly anisotropic and the bulk response of the fault zone is strain-weakening. Normal strains as small as 2% can induce a reduction of the overall stiffness of up to 75%. Fracture-induced effective stiffness changes are large enough to locally be greater than intact modulus changes across the fault due to juxtaposition of rocks of different lithologies; therefore mesoscopic fracturing is as important as rock type when considering material or bimaterial effects on earthquake mechanics. These results have important implications for earthquake rupture mechanics models, because mesoscopic damage zone fractures can cause a material contrast across the faults as large as any lithology-based material contrast at seismogenic depths, and the effective moduli can be highly variable during a single rupture event.

Key words: Damage zone, effective medium, Young's modulus, finite-element method, bimaterial.

1. *Introduction*

Fault zones are heterogeneous structures. In some cases, this heterogeneity can be inherited and may be related either to rock lithology or preexisting structural elements (e.g., Bergbauer and Martel, 1999; Davatzes and Aydin, 2003; Myers and Aydin, 2004;

[1] Department of Geological and Environmental Sciences, Stanford University, Stanford, CA, 94305 U.S.A. E-mail: dpollard@stanford.edu
[2] Civil and Environmental Engineering, Stanford University, Stanford, CA, 94305 U.S.A. E-mail: pablo.f.sanz@exxonmobil.com
[3] Istituto Nazionale di Geofisica e Vulcanologia, Via di Vigna Murata, 605, 00143-Roma, Italia. E-mail: griffith@ingv.it
[4] ExxonMobil Upstream Research Company.

SEGALL and POLLARD, 1983a, 1983b). Also, heterogeneity is commonly related to the processes responsible for fault growth, slip, and evolution: Rocks adjacent to faults are typically fractured due to brittle deformation associated with repeated slip events on the main fault surface(s).

Together, the primary fault surface and the adjacent, fractured rocks comprise two structural domains of the most commonly acknowledged idealization of a fault zone (e.g., CHESTER *et al.*, 1993): (1) the *fault core* and (2) *the damage zone*. The fault core typically consists of highly comminuted and altered rock and hosts most of the shear deformation along one or more principal slip zones (e.g., CHESTER and CHESTER, 1998). The damage zone can be symmetric or asymmetric about the fault core, and typically consists of variably fractured rock where microscopic and macroscopic fracture density increases with proximity to the fault core (e.g., CHESTER *et al.*, 2004; SCHULTZ and EVANS, 2000).

The structure of fault zones in the Mt. Abbot quadrangle (Fig. 1) of the central Sierra Nevada (e.g., MARTEL, 1990; MARTEL and BOGER, 1998; MARTEL and POLLARD, 1989; MARTEL *et al.*, 1988; PACHELL and EVANS, 2002) deviate from this idealized fault zone model. Individual small faults typically consist of a thin fault core, <1–2 cm in thickness, surrounded by granodiorites largely undamaged except for wing cracks concentrated around the fault tips (SEGALL and POLLARD, 1983a). In more mature fault zones, the subject of this study, total offset is typically on the order of several meters to >100 meters, and damage in the form of outcrop-scale fractures is concentrated between paired faults, and rocks immediately outside of the paired faults are largely undamaged (Fig. 1).

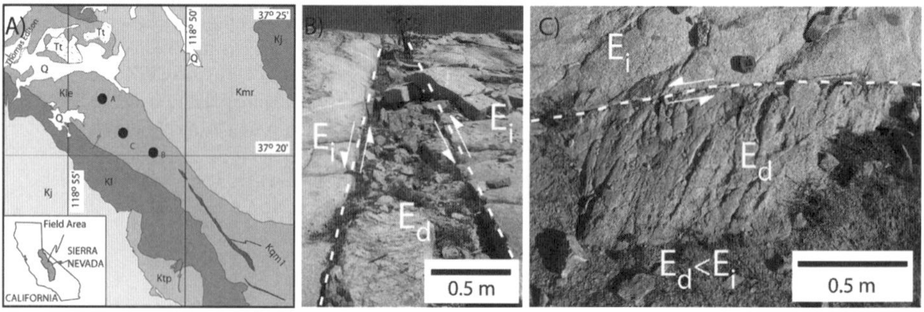

Figure 1

(A) Simplified geological map of a portion of the Bear Creek drainage (modified from BERGBAUER and MARTEL, 1999; GRIFFITH *et al.*, 2008; and LOCKWOOD and LYDON, 1975). The main geologic units shown are, from oldest to youngest, unspecified Cretaceous and Jurassic plutons (Kj), the Lamarck Granodiorite (Kl), the Mount Givens Granodiorite (Kmo), the Lake Edison Granodiorite (Kle), Quartz monzonite of Mono Recesses (Kmr), Cretaceous quartz monzonite and granite (Kqm1), Granitic rock of uncertain affinity (Ktp), Tertiary olivine trachybasalt (Tt), and Quaternary deposits (Q). Black circles and letter designations A-C refer to the locations of top, middle, and bottom fault zone maps illustrated in Figure 3. (B) and (C) Fault zones consist of discrete subvertical left-lateral boundary faults (1–2 cm thick) with damage in the form of outcrop-scale cross fractures localized between. Extensive damage is confined to the region between the boundary faults, and rocks outside of the boundary faults are typically intact. The fault zone in (A) is approximately 1 meter wide. Only one boundary fault is exposed in the fault zone pictured in (B). The stiffness E_d of the damaged granodiorite is expected to be less than the stiffness E_i of the intact granodiorite.

Increased fracture density leads to a decrease in the effective uniaxial stiffness (FAULKNER *et al.*, 2006; WALSH, 1965) and strength (LOCKNER *et al.*, 2000) of the damage zone rocks. The effective stiffnesses, or moduli, describe the stress-strain responses of fault zone materials. Changes in the effective moduli during fault zone evolution have important implications for the mechanics of slip events that occur on them. The effective elastic moduli of rocks have been shown theoretically and experimentally to influence a number of earthquake phenomena including, but not limited to:

(i) Velocity of shear ruptures. The shear modulus of fault rocks limits the rupture velocity of earthquakes: Off-fault damage has been shown to slow earthquake rupture significantly in laboratory experiments (e.g., BIEGEL *et al.*, 2008). In undamaged materials the limiting velocity for 'subshear' ruptures is $\sim$92% of the shear-wave velocity (v_s) where $v_s = \sqrt{G/\rho}$ and G is the shear modulus and ρ is the mass density.

(ii) Directivity of earthquake ruptures on faults that juxtapose dissimilar materials. Juxtaposition of rocks with different elastic moduli across a fault interface may influence the directivity of pulse-like ruptures if the contrast is great enough: ANDREWS and BEN-ZION (1997) showed that for a two-dimensional plain strain mode II rupture on a bimaterial interface with a constant coefficient of friction, slip pulses propagate in the direction of slip in the more compliant medium when the wave-speed contrast between the media is 20% or greater. While some doubt has been cast on the importance of this mechanism for rupture directivity in nature, due to effects related to time-weakening friction and three-dimensional rupture (ANDREWS and HARRIS, 2005; HARRIS and DAY, 2005; XIA *et al.*, 2005), the bimaterial mechanism might be relevant for primarily mode II ruptures (AMPUERO and BEN-ZION, 2008; BEN-ZION, 2006a, 2006b). DOR *et al.* (2006) examined damage distributions around several large faults and concluded that asymmetrically distributed damage, where more damage occurs on the stiffer side of the fault, was consistent with repeated unilateral ruptures over time with the same directivity. However, damage created by repeated unilateral ruptures may alter the effective modulus of the originally stiffer rocks.

(iii) Balance between heat production and thermal pressurization of pore fluids during faulting. Damage zone fracturing can work against thermal pressurization during earthquake ruptures by increasing the effective compressibility of the fault zone rocks (e.g., LACHENBRUCH, 1980; MASE and SMITH, 1987; NODA and SHIMAMOTO, 2005; RICE, 2006; SIBSON, 1973). Recent studies have used laboratory measurements of cylindrical samples taken from fault zones to constrain model values of compressibility (e.g., RICE, 2006; WIBBERLEY, 2002). If fault wall rocks are heavily fractured, however, their bulk compressibility is expected to differ relative to that of undamaged rocks (e.g., WALSH, 1965).

In mechanical studies, fault zones are commonly idealized as a discrete discontinuity or a thin shearing layer between identical or dissimilar isotropic, homogeneous materials, where the material, properties are assigned values believed to be representative of the entire continuum (e.g., ANDREWS and BEN-ZION, 1997; BEN-ZION and SAMMIS, 2003; POLLARD and SEGALL, 1987). Considerable attention has been given to measuring representative material parameters, including effective elastic moduli, of fault core and damage zone materials in natural fault zones (e.g., FAULKNER *et al.*, 2006; LOCKNER *et al.*, 2000; WIBBERLEY, 2002; WIBBERLEY and SHIMAMOTO, 2003) in an attempt to constrain the appropriate model material properties. These studies have confirmed the correlation between microscopic damage and alteration in fault zones and material properties such as strength and compressibility.

In addition, extensive literature has been developed regarding the analytical estimation of the properties of damaged rocks (e.g., BUDIANSKY and O'CONNELL, 1976; HUDSON, 1980; HUDSON *et al.*, 1992; KACHANOV, 1987, 1992; LYAKHOVSKY *et al.*, 1997; WALSH, 1965). However, most of these studies focus on (1) small deformations, and (2) sparse populations of small, microscopic cracks. On the other hand, slight attention has been given to the role of mesoscopic, intersecting fractures on the effective moduli of rocks in fault zones.

The three-dimensional extent of damage zones around large, crustal scale faults has been elucidated by seismological analysis of trapped waves (BEN-ZION *et al.*, 2003; LI *et al.*, 1994) and by geodesy (FIALKO, 2004). BEN-ZION *et al.* (2003) and FIALKO (2004) inferred structures that reduced the shear-wave velocity by a factor of two in damage zones on the order of 100 m and 2 km thick, respectively. In both cases the inferred damage zone exists primarily over the top few km, suggesting that extensive damage zone development is confined to the uppermost part of the crust, above the seismogenic zone (5–15 km). BEN-ZION and SHI (2005) used two-dimensional simulations of dynamic rupture on a material interface separating dissimilar materials to illustrate that increasing confining pressure tends to decrease off-fault plastic strain associated with dynamic rupture significantly at depths below about 3 km. Extensive damage zone development may be limited further at seismogenic depths by increased healing rates along faults and fractures relative to shallower depths (e.g., FINZI *et al.*, 2009, and references therein).

The goal of this study is to quantify the effect of mesoscopic fractures in small ($\sim$0.5–3 m thick) damage zones along faults exhumed from seismogenic depths (4–15 km) on the effective stiffness of the damage zone rocks. We mapped and sampled faults and damage zone fractures in granodiorites in the Bear Creek drainage of the Sierra Nevada (e.g., MARTEL *et al.*, 1988; SEGALL and POLLARD, 1983). These faults and fractures contain both hydrothermal mineral deposits and pseudotachylytes (e.g., GRIFFITH *et al.*, 2008), suggesting that they slipped under variably saturated conditions and underwent seismic deformation. We calculate the contribution of the damage zone fractures to the bulk elastic properties of the fault zones using numerical mechanical models. The models accurately represent the mapped geometry of the damage zone fractures and treat the fractures as frictional interfaces, and boundary conditions are chosen to represent expected conditions at seismogenic depths at which the faults were active.

2. *Study Area*

Fracturing in the Sierra Nevada batholith has been documented in the Bear Creek drainage, in the southern half of the Mount Abbot quadrangle (LOCKWOOD and LYDON, 1975; MARTEL *et al.*, 1988; SEGALL and POLLARD, 1983a,). These studies have concentrated on the Lake Edison Granodiorite ($\sim$88 Ma). The ages are based on K-Ar analyses of hornblende (BERGBAUER and MARTEL, 1999). The depth of emplacement of the plutons is estimated at 4–15 km based on amphibole geobarometry pressure estimates of 100–400 MPa (AGUE and BRIMHALL, 1988). Syn-to-post-magmatic structures in Bear Creek occur in the form of nearly vertical dikes, shear zones, joints, and faults. Radiometric age constraints tie the fracturing to the period immediately after plutonic intrusion during cooling and, perhaps, emplacement of adjacent plutons (BERGBAUER and MARTEL, 1999; SEGALL *et al.*, 1990; STERN *et al.*, 1981).

The faults of Bear Creek strike parallel to the joints and slipped left-laterally. Three macroscopic stages of fracturing/faulting are recognized. The first stage (Fig. 2I) is characterized by a single set of pre-faulting joints that strike predominantly ENE and are mineralized principally by quartz, epidote, and chlorite (MARTEL *et al.*, 1988; SEGALL and POLLARD, 1983a). The traces of the ENE-striking joints are as long as several tens of meters, they have opened as much as a few centimeters, and they are spaced centimeters to several meters apart. The second stage (Fig. 2II) is characterized by "small faults" (MARTEL *et al.*, 1988; SEGALL and POLLARD, 1983b), with dimensions similar to those of the joints and as much as 2 m of slip. The small faults also commonly display oblique opening mode fractures (wing cracks) near their ends (SEGALL and POLLARD, 1983a). Microstructurally, small faults record a diverse history of deformation mechanisms ranging from crystal plastic shear to brittle fracture and comminution, and frictional melting (Fig. 2A–C; GRIFFITH *et al.*, 2008). The third stage (Fig. 2III) is characterized by "simple fault zones" (MARTEL *et al.*, 1988), tabular volumes of variably fractured rock typically 0.5–3 m thick which are the subject of this study. These zones are bounded by faults $\leq$1–2 centimeters thick that, in turn, are linked by obliquely striking fractures. The simple fault zones are as long as one kilometer, but have segments tens of meters long, similar to the lengths of the longest joints. The simple fault zones accommodate as much as 20 meters of slip, with the slip being concentrated on the boundary faults. Simple fault zones rarely coalesce into "compound fault zones" (MARTEL, 1990). These have lengths as great as several kilometers, widths as great as several meters, and accommodate up to 140 meters of slip (PACHELL and EVANS, 2002). This commonly observed hierarchy implies that the width of damage zone development along the Bear Creek faults is limited by pre-existing anisotropy, as the faults grew from subparallel cooling joints, and the damage zones are confined to tabular zones between the subparallel faults.

The damaged rock within the fault zones is variably fractured, ranging from a few discrete cross-fractures to pulverized epidote and chlorite-rich gouge (Fig. 1B, C). In the more densely fractured cases the fault zones can be identified geomorphically in the field by the presence of narrow troughs, the walls of which approximate the boundary fault

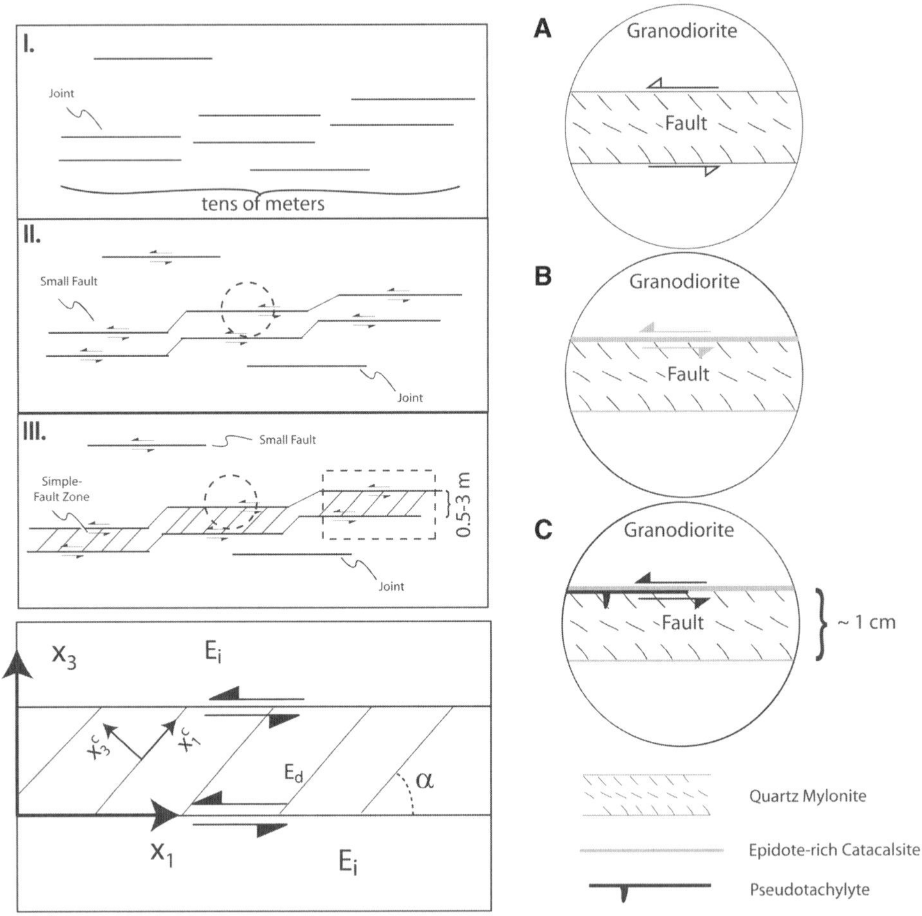

Figure 2
Conceptual model of macroscopic and microscopic evolution of faults in Bear Creek. I-III correspond to the macroscopic evolution of faults suggested by Martel *et al.* (1988), where in stage I joints form in response to thermal stresses; stage II is marked by slip nucleation along joints and subsequent linkage to form small faults; and in stage III simple fault zones composed of paired small faults linked by secondary fractures form due to the interaction of neighbouring faults. Dashed circles in II and III indicate the location of A, B, and C. A-C represent the microstructural evolution of faults from crystal plastic shear on quartz mylonites to cataclastic overprint and pseudotachylyte generation. Bottom left illustrates the fault zone configuration with fracture-induced effective modulus contrast between the interior damage zone rocks (E_d) and intact granodiorites (E_i) outside of the boundary faults and defines the global and crack-local reference frames.

surfaces (Fig. 1B). In rare cases, fault zones are exposed in outcrop along limited distances parallel to the strike of the boundary faults (<10 m). Martel *et al.* (1988) mapped and described some of these exposures. For this study we mapped and sampled two additional fault zones in the Bear Creek area, and we include a simplified version of Martel's fault zone map from the "Waterfall Site" (Fig. 3). The fault zones always consist of two paired boundary faults, each with a left-lateral sense of shear. In less mature fault zones, the trend of cross-fractures typically makes small (30–50°) angles with the boundary faults; however in more mature faults additional cross fractures at higher angles tend to infill the fault zones (Fig. 3A). The boundary faults are typically less than 1–2 cm in thickness (Fig. 4A), and the cross fractures are typically much less than 1 cm in thickness (Fig. 4B, C). Measured offset on cross fractures is always less

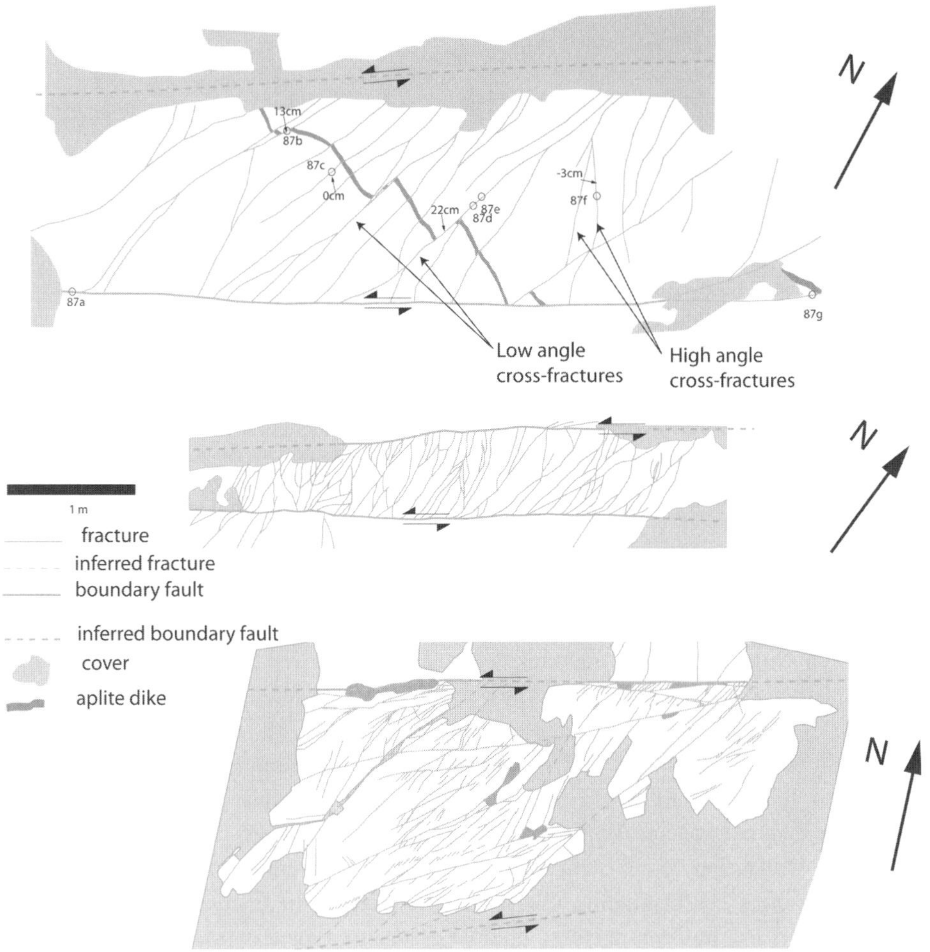

Figure 3

Three Bear Creek fault zones mapped and sampled for this study. In all cases the fault zones consist of a tabular zone with cross fractures bounded by left-lateral boundary faults. Fault zones typically accommodate up to 20 m of total offset. Top fault zone is the Big Juniper Fault Zone (BJFZ). Note offset aplite dike cutting through the entire fault zone. Middle fault zone is from the Trail Junction Meadows site of SEGAL and POLLARD (1983). Bottom fault zone is modified from the map of MARTEL *et al.* (1988) from the Waterfall site. Scale bar applies for all three mapped fault zones.

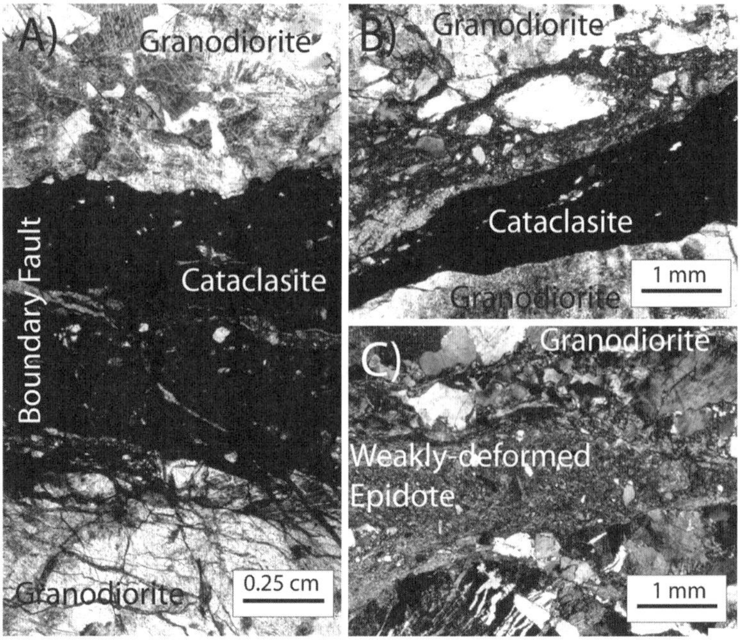

Figure 4
Microstructures of boundary faults and cross fractures. (A) Boundary fault from the BJFZ under plane polarized light. Fault consists of highly comminuted fine-grained cataclasite. This fault accommodated several meters of slip. (B) Cross fracture with 22 cm slip. Lighter colored coarse cataclasite is cut by finer grained cataclasite. (C) Cross fracture with less than 3 cm slip. Entire cataclasite zone is dominated by coarse grained, euhedral epidote and chlorite.

than that on boundary faults and can vary from zero to several tens of centimeters. Offset, if any, along low angle cross fractures is left-lateral, however fractures oriented at high angles relative to the boundary faults can show small ($\leq$a few centimeters) right-lateral offset.

In thin section, boundary faults contain variably comminuted grains consisting of quartz and feldspar clasts from the host rock, precursory quartz mylonites, epidote, chlorite, and minor sphene. While quartz and feldspar clasts are sourced from the host rock and precursory mylonites, epidote and chlorite were formed by fluid-rock interaction after the onset of brittle faulting (GRIFFITH *et al.*, 2008). In fractures with little or no apparent offset (<a few centimeters), epidote grains are typically prismatic in shape and show very little comminution (Fig. 4C). Qualitative analysis of thin sections suggests that slip along these faults is typically accompanied by grain size reduction and rounding of clasts (e.g., Fig. 4). In boundary faults, the core material typically consists of very fine grained cataclasite and ultracataclasite. Boundary faults and cross fractures with large measured (tens of centimeters to meters) offsets commonly show evidence of multiple principal slip zones. In the case of the cross-fracture shown in Figure 4B, an early

cataclastic fabric has been overprinted by a fine-grained ultracataclasite which is dark to opaque in plane polarized light.

Conceptual models proposed for the formation of these paired fault zones attribute the formation and localization of damage to slip on the boundary faults. MARTEL and POLLARD (1989) hypothesized that the cross fractures formed as opening mode wing cracks formed at the tips of small slip patches on the boundary faults. MARTEL and BOGER (1998) attributed the cross fractures to three-dimensional interaction between the boundary faults at overlaps near the fault's upper and lower (mode III) tips. Because of limited exposure, it is difficult to assess the along-strike continuity of fault zones, and whether damage exists between boundary faults along their entire lengths, or just near extensional step-overs along boundary faults. Regardless of the formation mechanism, a few observations are clear based on field mapping and microstructural analysis:

(i) Damage in the form of mesoscopic fractures is consistently concentrated between thin (<1–2 cm) boundary faults, and the host granodiorite shows little mesoscopic fracturing outside of the boundary faults; and
(ii) The cross fractures form as opening mode fractures filled variably by secondary epidote, chlorite, and clasts of the host rock. The cross fractures slipped in association with slip events on the boundary faults, resulting in comminution of fracture mineral fillings. Continued slip events along boundary faults were apparently accompanied by reactivation of existing cross fractures and by increased damage by formation of new fractures.

Continued slip on boundary faults after localization of damage on the interior of the paired fault zones results in an increasingly strong material contrast across each of the paired boundary faults, where damage within the fault zone induces an effectively softer material on the inside of the fault zone relative to the unfractured granodiorite outside of the fault zone.

3. *Fault Zone Idealization*

We calculate effective values for fault zone stiffness in a global reference frame where x_1 is parallel to the boundary faults and the outcrop face is parallel to the x_1-x_3 plane (Fig. 5). For analytical calculations (Appendix A), an additional fracture-local coordinate system is centered on each fracture, with the x_3 axis orthogonal to the cross-fractures, at an angle α to the positive x_1 axis in the global reference frame (Fig. 5).

We confine the numerical investigation to the Big Juniper Fault Zone (BJFZ) shown in Figure 3A. This fault zone was chosen because abundant offset markers (including an aplite dike that transects the entire fault zone, and an offset zenolith on the antithetic fracture at sample location 87f, Fig. 3A) provide constraints on slip on several cross fractures and boundary faults. This allows us to confirm that the model reproduces the observed slip distributions under approximate loading conditions operating when the faults were active. This confirmation is discussed in more detail in the section *Model Geometry Confirmation*. The fractured damage zone is tabular in shape and bounded by two subparallel faults. To standardize application of external boundary conditions we created a fault mesh that is rectangular in shape and represents the average width of the fault damage zone (Fig. 5). This simplification required extension of some of the cross fractures partially covered by sediment to the model boundary. A comparison of the mapped fault geometry and the model mesh is shown in Figure 6.

Here we are interested in estimating the effective elastic stiffness of fractured damage zone rocks in the reference frame defined above. The numerical simulations focus on calculating the effective apparent Young's modulus of the fault zone as a representation of the overall stiffness. In the isotropic elastic compliance tensor Young's modulus ($E = 1/s_{11} = 1/s_{22} = 1/s_{33}$) defines the relationship between an applied normal stress and the parallel normal strain under *uniaxial* stress conditions:

$$\sigma_{33} = E\varepsilon_{33}, \quad \sigma_{11} = \sigma_{22} = \sigma_{12} = \sigma_{13} = \sigma_{23} = 0. \tag{1}$$

In this case Young's modulus is defined with uniaxial loading in the x_3 direction. However, because the fault zones presented in this study are extensively fractured by cross fractures that anastamose, but are approximately subparallel, the fault zone should behave as an anisotropic body as a whole. In the case that fractures within a damaged rock consist of a single set of parallel fractures, the rock body is expected to be *transversely anisotropic*. In the crack-local reference frame (Fig. 2) where the normals to the fractures are parallel to the x_3^c axis, the compliance tensor is completely defined by five independent constants:

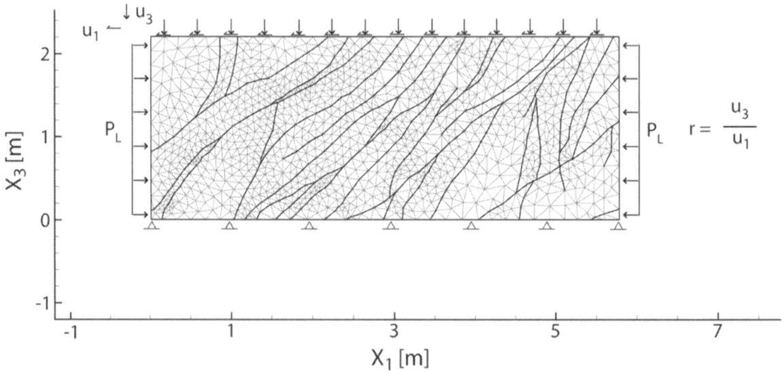

Figure 5
Discretized fault zone (BJFZ from Fig. 3) with external boundary conditions defined for the model geometry validation.

$$s_{IJ} = \begin{bmatrix} s_{11} & s_{12} & s_{13} & 0 & 0 & 0 \\ s_{12} & s_{11} & s_{13} & 0 & 0 & 0 \\ s_{13} & s_{13} & s_{33} & 0 & 0 & 0 \\ 0 & 0 & 0 & s_{44} & 0 & 0 \\ 0 & 0 & 0 & 0 & s_{44} & 0 \\ 0 & 0 & 0 & 0 & 0 & s_{66} \end{bmatrix}, \quad s_{66} = \frac{s_{11} - s_{22}}{2}. \tag{2}$$

As the compliance matrix is written above, x_3^c is the axis of symmetry. Here, the Young's modulus in the x_3^c direction is equivalent to $1/s_{33}$. If the compliance matrix is transformed by a rotation about α to the global reference frame, compliances s_{45}, s_{54}, s_{16}, s_{61}, s_{26}, and s_{62} become nonzero. In the transformed compliance tensor, S_{ij}', $1/s_{33}'$ then represents the *apparent* Young's modulus in the x_3 direction, as principal stresses are not necessarily parallel to principal strains in this case.

4. Numerical Model

We use a finite-element method code based on continuum and contact mechanics (SANZ, 2008; SANZ *et al.*, 2007, 2008) to simulate deformation on the mapped fault geometry of the Big Juniper Fault Zone (BJFZ, Fig. 3A). The mechanical models are

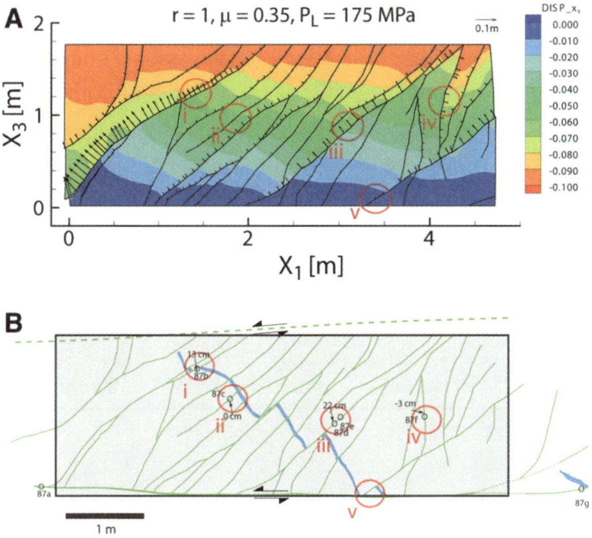

Figure 6

Comparison of numerically computed slip distributions with geological observations. (A) Slip distributions (arrows) and x_3-parallel displacements (contours) for experiment with r = ΔV/ΔH = 1 and μ = 0.35. Arrows normal to faults indicate left-lateral slip if pointing to the upper left and right-lateral slip if pointing to the lower right. Length of arrows corresponds to computed slip magnitude. (B) Mapped fault geometry with offsets along cross fractures measured in the field.

two-dimensional with plane strain assumed, and the fault zone fractures are represented as an array of material interfaces with the mapped geometry, separating isotropic, homogeneous material with constitutive properties representative of undamaged granodiorite: $E = 60$ GPa (BIRCH, 1966) and $\nu = 0.25$ (GERCEK, 2007). Constants and boundary conditions used in models discussed in the text are summarized in Table 1. Contact mechanics (Appendix B) enables us to model slip and/or opening on the fractures. Such relative displacements of the fracture surfaces requires frictional contact constraints along these interfaces.

A regularized Coulomb friction law simulates the mechanical response of the interfaces (Appendix B; LAURSEN and SIMO, 1993; WRIGGERS, 2002). Each contact surface has three mechanical parameters: A normal elastic stiffness k_N, a tangential elastic stiffness k_T, and a sliding friction coefficient μ. Frictional slip is path-dependent in nature and its evolution requires the integration of the constitutive law in finite steps. The integration is performed by a return-mapping algorithm similar to a non-associative plasticity model (WRIGGERS, 2002). We discretize the interfaces with node-to-segment contact elements (LAURSEN and SIMO, 1993; WRIGGERS, 2002). A more detailed discussion of the mechanical framework and of the numerical implementation can be found in Appendix B and in SANZ (2008) and SANZ *et al.* (2007).

4.1. Model Geometry Confirmation

In order to confirm that the overall model response is representative of the natural fracture system, we use geologic observations to assign boundary conditions that approximate conditions in which the faults were active and compare the resulting slip distribution to the distribution mapped in Figure 3. Because none of the internal fault zone fractures cut or offset the boundary faults in the field, we prescribe displacement

Table 1

Constants and boundary conditions imposed on numerical simulations discussed in text

Test Purpose	E, Intact (MPa)	ν, Intact	Total u_1 (m)	Total u_3 (m)	σ_1 (MPa)	σ_3 (MPa)	μ	k_T (GPa/m)	k_N (GPa/m)
Geometry Confirmation r = 1	60	0.25	0.25*	0.25	$P_L = 175$	-	0.35	10,000	10,000
E_1 ($\mu = 0.3$)	60	0.25	0.1	-	-	$P_L = 175$	0.3	10,000	10,000
E_1 ($\mu = 0.5$)	60	0.25	0.1	-	-	$P_L = 175$	0.5	10,000	10,000
E_1 ($\mu = 0.7$)	60	0.25	0.1	-	-	$P_L = 175$	0.7	10,000	10,000
E_3 ($\mu = 0.3$)	60	0.25	-	0.26	$P_L = 175$	-	0.3	10,000	10,000
E_3 ($\mu = 0.5$)	60	0.25	-	0.26	$P_L = 175$	-	0.5	10,000	10,000
E_3 ($\mu = 0.7$)	60	0.25	-	0.26	$P_L = 175$	-	0.7	10,000	10,000

*indicates displacements were applied tangent to the element edges on the upper half of the model. In all other cases displacements are applied normal to the model edges

boundary conditions on the upper and lower model boundaries parallel to the boundary faults (i.e., parallel to x_1) with the assumption that any displacement normal to the boundary faults is continuous along the entire boundary. The lower boundary is fixed (zero displacement boundary conditions) while displacements u_1 and u_3 are prescribed tangential and normal to the upper boundary, respectively (Fig. 5). Because the fault zone extends laterally in the x_1 direction beyond the map, and some cross fractures intersect the x_3-parallel model boundaries (meaning displacements are expected to be nonuniform along x_3), we avoid prescribing displacements along these boundaries. Instead we prescribe constant normal tractions equivalent to a lithostatic load (P_L) for a burial depth of 7 km, within the range of amphibole geobarometry estimates for emplacement depths of 4–15 km (AGUE and BRIMHALL, 1988). Assuming a gradient of 25 MPa/km and a depth of 7 km we have $P_L = 175$ MPa (Table 1).

In all cases external boundary conditions are applied in two steps. In the first loading stage, we prescribe an isotropic stress state where $\sigma_{11} = \sigma_{33} = P_{\mathrm{L}}$ (Table 1) and no slip occurs anywhere in the model. In the second loading stage we subject the modeled fault zone to a range of combinations of displacements u_1 and u_3 defined by the ratio $r = u_3/u_1$. The choice of r depends on the goal of each set of simulations. For example, in determination of the effective Young's modulus parallel to x_3, $r = \infty$. For the model confirmation we use $r = 1$. The observation of left- and right-lateral slip on cross fractures suggests that the fault zone has been sheared parallel to and shortened normal to the boundary faults. In each simulation, the frictional strength of the fracture interfaces is characterized by a single constant coefficient of friction μ. We consider a range of possible values for this parameter, and run each loading configuration for $\mu = 0.3$–0.7. Fracture stiffnesses k_N and k_T are assigned constant values of 10,000 GPa/m in all simulations (Table 1). The loading conditions that produce a slip sense and relative magnitudes of slip on model cross fractures similar to geological observations are interpreted to approximate loading conditions on the fault zone *in situ*. Because we cannot constrain the finite deformation path of the fault zone from its initial configuration to the final configuration, we are only concerned with approximating the finite, relative slip magnitudes on the cross fractures.

Figure 6 shows the modeled slip distributions on the cross fractures for $r = 1$ and $\mu = 0.35$ (Table 1). Other values of μ show similar slip distributions with different absolute magnitudes. Under the loading configuration $r = 1$, slip magnitudes are seen to reach a maximum along the faults with the largest measured offsets in the field at locations (i) and (iii). At location (ii) the modeled slip is zero, agreeing with the offset measured in the field at the corresponding location. At location (iv) the modeled slip is right-lateral, consistent with field measurements along the high-angle, antithetic cross fractures. Only at location (v) do the model results disagree significantly with the measured offset in the field, as the offset shown on the map is significantly greater than the modeled slip. This discrepancy is likely due to the fixed lower model boundary. Away from the boundaries, however, the model appears to confirm the response of the fault zone to remotely applied loads.

4.2. Apparent Young's Moduli

We perform uniaxial compression tests on the model fault zone in two configurations to calculate the apparent Young's moduli $E_1 = 1/s_{11}$' and $E_3 = 1/s_{33}$', where the prime denotes the component of the transformed compliance tensor rotated from the crack-local coordinate system to the global coordinate system. The contraction directions are parallel to x_1 (fault parallel) and x_3 (fault normal), respectively. These experimental configurations are illustrated in Figure 7. For E_1 (Fig. 7A), the right model boundary is displaced to the left, in the negative x_1 direction. The left boundary is frictionless, but is fixed (zero displacement) normal to the boundary. As in the model geometry confirmation experiments, the Young's modulus experiments are initiated in an isotropic stress state, where $\sigma_{11} = \sigma_{33} = P_L$ (Table 1). For the remainder of the experiment, the upper and lower boundaries are subjected to a constant normal traction equal to P_L (Table 1). Such constant stress boundary conditions have been shown to limit internal stress relaxation leading to an accelerated failure process (e.g., BEN-ZION, 2008; LYAKHOVSKY et al., 2001). Because we are not attempting to simulate an entire earthquake cycle, but instead

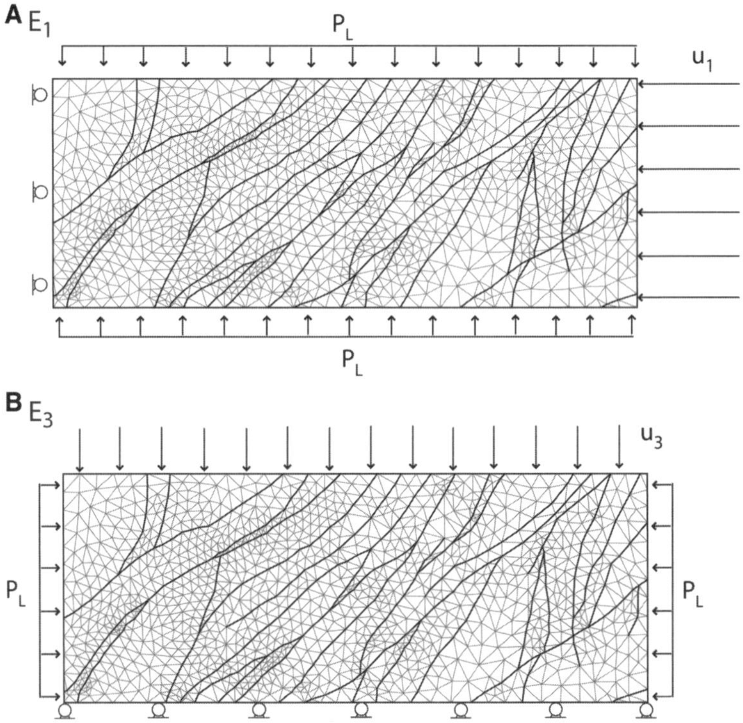

Figure 7
External boundary conditions for apparent Young's modulus uniaxial compression tests. See text for discussion.

mimicking deformation experiments in the laboratory, these boundary conditions should be sufficient. Stresses are calculated from forces recorded at the nodes on the x_1-parallel boundaries at each loading step, and the average values of these stresses are used in the Young's modulus calculations. The model configuration for the calculation of E_3 (Table 1) is identical to the model configuration for E_1 rotated by 90° (Fig. 7B).

Figure 8 shows the stress-strain curves and apparent effective Young's moduli, E_1 and E_3 ($1/s_{11}$' and $1/s_{33}$') calculated for simulations with three different μ values assigned to the damage zone fractures ($\mu = 0.3$, 0.5, and 0.7; Table 1). At each loading step, the apparent Young's modulus is calculated as the discrete tangent of the corresponding stress-strain curve. Because the FEM model assumes plane strain deformation, and the plane strain Young's modulus, E_{ps}, is related to the actual Young's modulus, E, by $E = E_{PS}/(1 - \upsilon^2)$, we multiply the model results by $(1-\upsilon^2)$, where υ is the background isotropic value of Poisson's ratio ($\upsilon = 0.25$). Note that in both cases, E_1 and E_3 initially

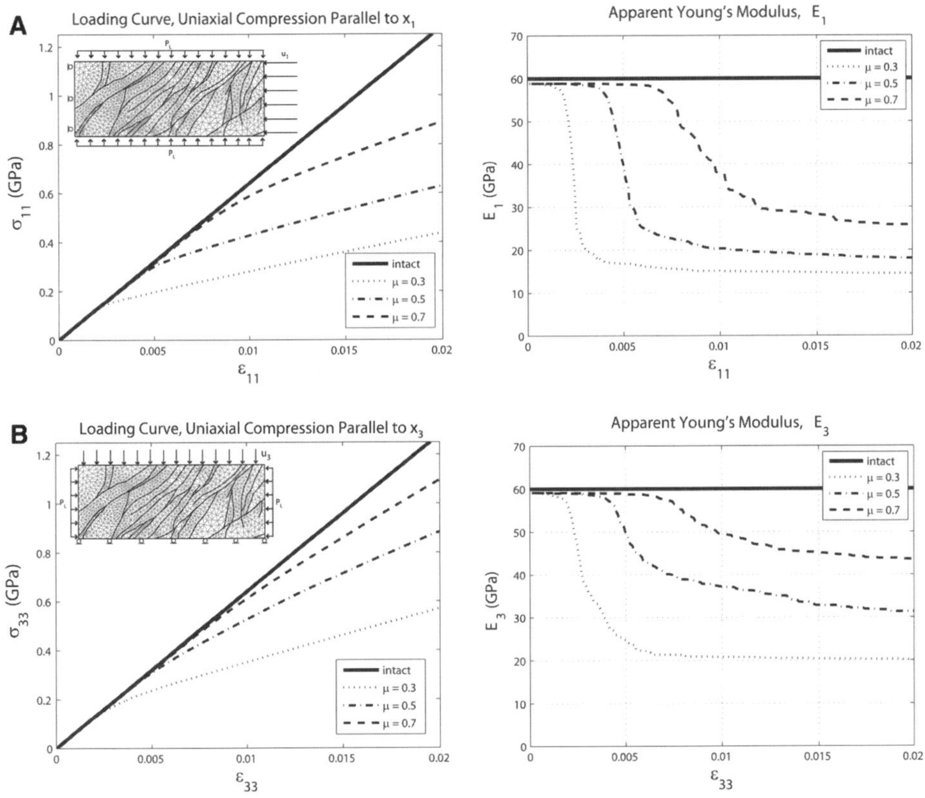

Figure 8

x_3-parallel Young's modulus (E_3) and x_1-parallel Young's Modulus (E_1) for three different friction coefficients compared to the background modulus of the intact rock. Upper and lower model boundaries are modelled as frictionless plates. Left plot shows the loading curve and right plot shows the Young's modulus values varying with the corresponding normal strain.

remain constant with increasing strain at a level that is slightly ($\sim$1 GPa) less than Young's modulus for the intact granodiorite, and then progressively decreases, approaching a residual constant value at ε_{11} or $\varepsilon_{33} = 0.02$, respectively. The constant Young's modulus at the beginning of each experiment is slightly lower than the intact value due to the finite stiffness assigned at the fracture contacts ($k_N = k_T = 10{,}000$ GPa/m). In all cases, the apparent Young's modulus follows a strain-softening behavior, and achieves the maximum reduction as the normal strain in the shortening direction reaches ε_{11} or $\varepsilon_{33} = 0.02$. Residual values of E_1 (Fig. 9A) at $\varepsilon_{11} = 0.02$ are 15 GPa for $\mu = 0.3$, 19 GPa for $\mu = 0.5$, and 26 GPa for $\mu = 0.7$. These apparent Young's moduli represent reductions from the intact value of 75%, 68%, and 57%, respectively. Residual values of E_3 (Fig. 9B) at $\varepsilon_{33} = 0.02$ are 21 GPa for $\mu = 0.3$, for 33 GPa for $\mu = 0.5$, and 46 GPa for $\mu = 0.7$. These apparent Young's moduli represent reductions from the intact value of 65%, 45%, and 23% respectively.

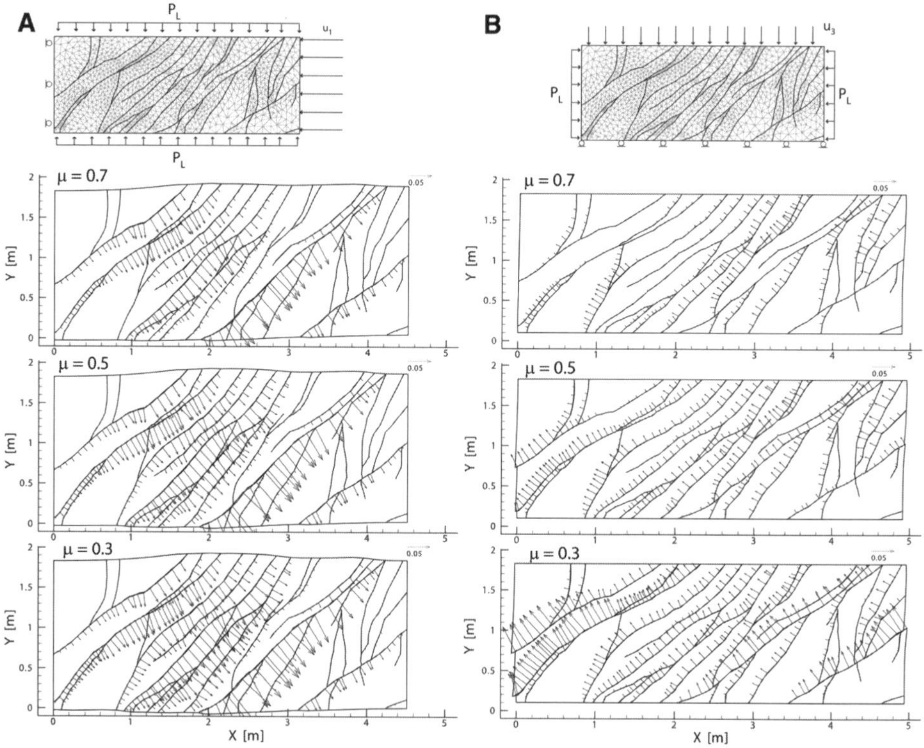

Figure 9
Cumulative slip distribution on cross fractures for each experiment plotted in Figure 8. Slip distributions represent (A) results from the x_1-parallel apparent Young's modulus (E_1) tests, and (B) results of the x_3-parallel apparent Young's modulus (E_3) tests. Top, $\mu = 0.7$; middle $\mu = 0.5$; bottom, $\mu = 0.3$. Length of arrows represents slip magnitude. Arrows with negative x component represent left-lateral slip. Larger amounts of slip result in larger reductions in the apparent Young's modulus.

Slip distributions for each simulation are shown in Figure 9, where the lengths of the arrows represent the magnitude of slip, and arrows pointing to the western side of the fractures indicate left-lateral slip. The strain-dependent reduction of effective moduli of fault zone rocks is related to slip on the cross fractures: As a slip patch nucleates and spreads along a fracture the fault zone becomes more compliant (Fig. 9). This is consistent with greater modulus reductions at lesser μ. As strain progresses, however, all of the fractures optimally oriented for slip become active, and the modulus reduces to a constant value. Continued modulus reduction would require the formation of additional fractures. Note that the maximum slip on any of the fractures is on the order of 10 cm.

The fault zone is more compliant under x_1-parallel loading, and the influence of frictional properties on the bulk compliance is less for this case as seen in the smaller spread between maximum apparent Young's modulus reduction (Fig. 8) and the larger and more extensive slip distributions relative to the case of x_3-parallel loading (Fig. 9). This difference highlights the anisotropic behavior induced by the damage zone fractures.

4.3. *v*, *G* and *K*

The numerical simulations focus on the calculation of the effective Young's modulus because the loading configuration is straightforward to implement. The test for the shear modulus, G (defined as $\sigma_{ij} = 2G\varepsilon_{ij}$, $i \neq j$), and the bulk modulus, K (defined as $1/3\sigma_{\alpha\alpha} = K\varepsilon_{\alpha\alpha}$), is more complicated. The calculation of G requires that the fault zone be loaded in pure shear; however, because the model fracture interfaces have zero tensile strength, any prescribed symmetric shear tractions or displacements along the model boundaries, in the absence of an increasing normal load, would cause the interfaces to separate. The test for K is not difficult to implement, however an isotropic stress state in the model would result in zero frictional slip along the faults (i.e., $K_{\text{damaged}} = K_{\text{intact}}$). This is precisely what happens during the initial loading phase of all of the experiments: loading the fault zone isotropically to P_L returns the intact bulk modulus.

Instead of calculating these moduli directly, we calculate the effective Poisson's ratio for the uniaxial compression tests, and then approximate the corresponding isotropic G and K (Fig. 10). Poisson's ratio is found by calculating normal strain from the difference between the average node displacements on either side of the model in the direction orthogonal to the applied displacement, and substituting these values into $\upsilon = -\varepsilon_{11}/\varepsilon_{33}$. The results are shown in Figure 10A for the two compression tests with $\mu = 0.5$. The solid black line is the background isotropic value of $v = 0.25$. In the initial loading stages, v is considerably lower than the isotropic elastic value, again due to stiffness effects of the fractures. As frictional slip commences on the fractures, v steadily increases to a value of $v = 0.3$, at ε_{11} or $\varepsilon_{33} = 0.004$, at which point the values diverge and that for compression parallel to x_1 approaches $v = 0.5$, the value corresponding to an incompressible elastic material.

Substituting the modeled apparent E and v values into equations (6) and (7), we calculate the effective isotropic G and K for $\mu = 0.5$ (Figs. 10B, C). Because the overall

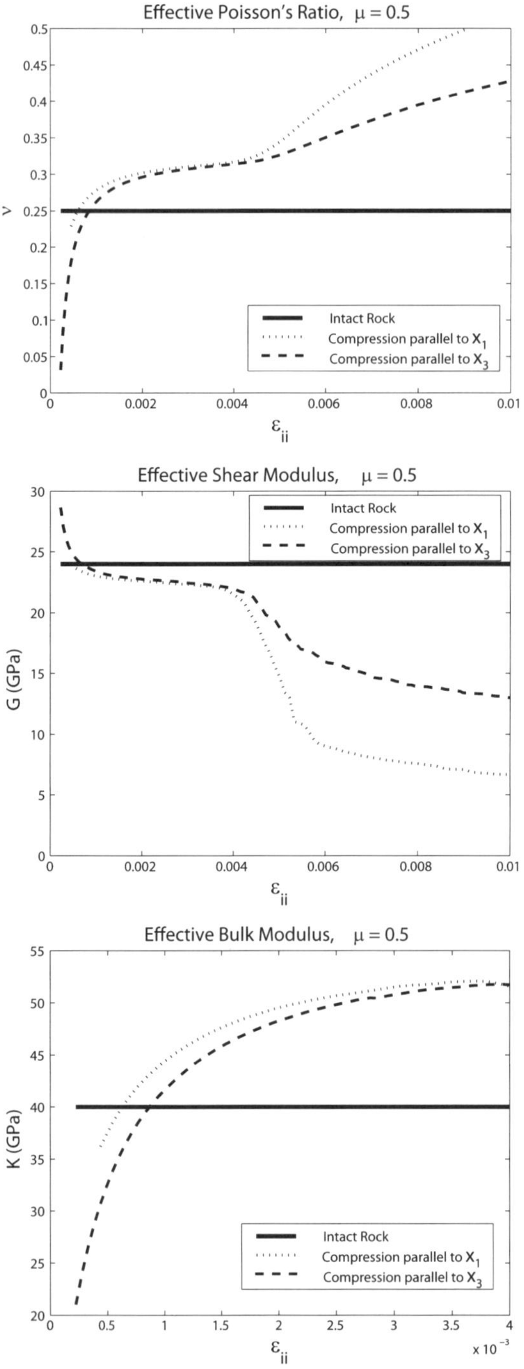

Figure 10

ν, G, and K calculated from uniaxial compression tests with $\mu = 0.5$.

fault zone response is highly anisotropic and strain-softening (i.e., inelastic), these values are rough approximations, but illustrate the influence of the outcrop-scale fractures on the compliance of the fault zones. Apparent G decreases from an initial state larger than the background isotropic G, and steadily degrades along a trend roughly inverse that of ν. Because the fault zone becomes incompressible at strains of $\varepsilon_{ii} \approx 0.01$, we only show G for values below this threshold. The maximum reduction of $\sim$75% in apparent G (Fig. 10B) relative to G_{intact} (solid line) occurs for the x_1-parallel compression case (dotted line).

Apparent K is plotted only up to normal strains of ε_{11} or $\varepsilon_{33} \approx 0.004$ (Fig. 10C). Beyond this value apparent K approaches infinity as ν approaches 0.5 (i.e., the fault zone becomes incompressible). Presumably the formation of additional fractures would be encouraged as this incompressible state is reached during deformation of the fault zone *in situ*. Unlike E, ν, and G, the calculated K converges for the two loading conditions with increasing normal strain, as both the fault zone approaches incompressible.

5. *Discussion*

Both sets of numerical uniaxial compression tests (Fig. 8) show large reductions in effective elastic moduli within the fault zone for normal strains up to 0.02 and total slip on cross-fractures of up to 10 cm. These simulations show that frictional slip on outcrop-scale fractures can result in maximum reductions in the apparent effective Young's modulus as great as 75%. This is on the order of contrasts imposed by the juxtaposition of rocks of different lithologies. For example, the Young's modulus of several common rock types is tabulated in Table 2, from Birch (1966). This implies that mesoscopic damage zone fracturing can impose a significant material contrast. With more slip and associated damage, this material contrast could become even greater, although hydrothermal mineral precipitation observed in these fault zones (Martel *et al.*, 1988; Pachell and Evans, 2002) suggests a concurrent healing process that would act to stiffen the fault zone.

This is a significant reduction in effective stiffness for fault zones which were active at depths greater than 4 km (Griffith *et al.*, 2008; Martel *et al.*, 1988; Segall *et al.*,

Table 2

Some Young's moduli reported for common rocks, tabulated from Birch (1966)

Young's modulus, E (GPa):		
Rock Type	**Dry Modulus**	**Wet Modulus**
Granodiorite, Weston Mass.	60	
Granite, Westerly	40, 71	
Limestone, Solenhofen	55	55
Shale, siltstone	44	19
Halite, Upper Zechstein series	28	

1990). While multiple lines of geophysical (Ben-Zion *et al.*, 2003; Fialko, 2004) and theoretical (Ben-Zion and Shi, 2005) evidence suggest that significant damage development should be suppressed at such depths, it is clear from field and microscopic evidence that (a) the fault and damage zone fractures remained weaker than the surrounding rocks despite hydrothermal mineral precipitation as evidenced by overprinting fault fabrics and (b) that macroscopic damage zone fractures continued to form throughout the fault zone evolution, as evidenced by the cross-fracture hierarchy and range of fracture filling textures, ranging from undisturbed joint filling to highly comminuted cataclasite. This observation implies that either (i) such localized damage zones might be common even at seismogenic depths, or (ii) this is a special case unique to the Bear Creek faults. In either case, it should be stressed that these damage zones are very thin (0.5–3 m) relative to flower structures (up to several km) and pulverization zones common at shallower depths along crustal scale faults. It is evident that the width of the Bear Creek damage zones was limited by the pre-existing anisotropy, as damage is consistently confined to the region between two subparallel boundary faults. Certainly, the formation of damage could have been encouraged by fault interaction based on the quasi-static analyses of Martel and Pollard (1989) or Martel and Boger (1998). On the other hand, stress drops along these faults have been inferred to be anomalously high, on the order of 100 MPa (Griffith *et al.*, 2009). Such large stress drops could potentially combat the retarding effects of dynamic damage formation due to increasing confining pressure at growing depths (Andrews, 2005). In any case, the results of this study suggest that the existence of damage zones of meter-scale thickness should not be precluded along active faults at seismic depths, and that these damage zones may significantly reduce the rock stiffness adjacent to the principal slip zone.

The numerical simulations conducted in this study also show that the frictional slip on outcrop-scale factures leads to a bulk nonlinear strain-softening response of the damage zone. This implies that the effective stiffness could vary dramatically throughout a single rupture event, and throughout the evolution of the fault zone, even in the short term (i.e., at strains much less than 2%). While it is clear that the cross fractures formed as opening-mode features and many of the cross fractures slipped on multiple occasions during the evolution of the fault zone, the timing and rate of deformation are not known. The cross fractures are compositionally and microstructurally similar to the boundary faults, however this does not provide sufficient information to ascertain whether or not they slipped simultaneously with rupture events on the boundary faults. Also, because the cross fractures do not contain pseudotachylyte, it is difficult to assess the rate at which these fractures slipped.

These numerical simulations are instructive because they allow for the relaxation of some of the key assumptions of the analytical models. Compared to the micromechanical models (Appendix A), there is no maximum fracture density requirement, and compared to the stiff parallel joint model of Amadei and Savage (1993) the deformation of the fractures need not be entirely elastic. Assumptions common to all of the analytical models are that the fractures are parallel, do not intersect, and are uniformly distributed throughout the entire fractured rock body. Clearly none of these are valid assumptions based on mapping of the fault zone fractures. The stiff parallel joint model may be the most appropriate model for a mesoscopically fractured damage zone, however this model is difficult to constrain for rocks *in situ*. Even if it were reasonable to assume that the fractures deformed elastically, deformability of the fractures in the stiff parallel joint model is based on the fracture stiffness. Because fracture stiffness is strongly dependent on normal load (e.g., YOSHIOKA and SCHOLZ, 1989a, 1989b) and may vary by several orders of magnitude both spatially and temporally, the fracture stiffness may change appreciably during a transient slip event and laterally along each fracture depending on the fracture filling and roughness of the fracture faces. Instead, the numerical approach accurately represents the mapped geometry of the fault zone and faults and fractures interact and slip when their Coulomb strength is exceeded. The contact problem is more easily constrained for the numerical experiments, as friction coefficients for fault rocks are independent of the normal load.

There are a number of limitations to the current study, some of which could be improved with future work: (1) First, we assume that the frictional properties of the faults are constant throughout a deformation cycle. This is a simplification of the real system as frictional strength is expected to change during and between slip events. (2) We do not include effects of fluid pressure and transport. It is apparent that hydrous conditions persisted throughout the fault activity. The presence of fluids along these faults would add contrasting effects of weakening the fractures by lowering the effective normal stress and increasing the peak strength of the fractures by enhancing the healing process between slip events. (3) The rocks between the modeled fractures are assumed to deform elastically, which fails to represent the observed pattern of increasing fracture density with fault zone evolution. A more realistic approach for future studies would be to represent the faults with a plastic rheology, or the nonlinear damage rheology of LYAKHOVSKY *et al.* (1997). (4) Finally, these results are scale-sensitive. While the bulk response of the fault zone to external loads is significantly different than for the intact case, the fault zone experiences large internal deformation gradients. Therefore the expected local stiffness is expected to vary considerably throughout the fault zone.

6. Conclusions

We combine detailed mapping and microstructural analysis of fault zones in the Bear Creek drainage of the central Sierra Nevada with numerical Finite-Element methods to estimate the contribution of outcrop-scale fractures to the effective stiffness of the fault rocks. The Bear Creek fault zones were active at seismogenic (4–15 km) depths and localize mesoscopic off-fault damage into thin (0.5–3 m) tabular zones between two subparallel boundary faults, producing a fracture-induced material contrast across the boundary faults. Frictional slip on the mesoscopic damage zone fractures results in a bulk strain-softening response of the fault damage zone, thus the effective moduli are expected to vary during individual slip events and during the longer term evolution of the fault zones. The material within the boundary faults can be as much as 75% softer than the adjacent undamaged granodiorite, even within juvenile fault zones (~8–10 m total offset). For larger, more mature fault zones this material contrast could be significantly greater, even at seismogenic depths.

Acknowledgements

Thoughtful reviews by Y. Ben-Zion, R. Weinberger, and an anonymous reviewer greatly improved this manuscript. G. Di Toro, G. Pennacchioni, and S. Nielsen assisted in field work, and the National Forest Service graciously provided sampling permission in the John Muir National Wilderness. This work was supported financially by the Rock Fracture Project, Stanford University, National Science Foundation Grant No.CMG-0417521 (Collaborations in Mathematical Geosciences), and the U.S. Geological Survey (USGS), Department of Interior, under USGS award number 08HQGR0010. The views and conclusions contained in this document are those of the authors and should not be interpreted as necessarily representing the official policies, either expressed or implied, of the U.S. Government.

Appendix A: Analytical Models

Analytical effective medium models present an efficient means for calculating effective elastic properties of fractured rock masses given an estimate of the nondimensional spatial density of fractures, and can be used to invert for fracture density given seismic velocities for the fractured rock. A more recent formulation replaces the fracture density parameter with a damage variable which allows the modeled cracks to contract or dilate during deformation (LYAKHOVSKY *et al.*, 1997). An attractive feature of these models is that given some basic assumptions about the anisotropy of the fractured medium (e.g., that it is transversely anisotropic), one can define the complete anisotropic compliance or stiffness tensor. Here we evaluate the applicability of several analytical effective medium models for describing the bulk response of the BJFZ.

Most effective medium theories for cracked solids are based on micromechanics (e.g., WALSH, 1965a, b; KACHANOV, 1992; JAEGER *et al.*, 2007 and references therein), and consist of several classes, including (i) zero interaction; (ii) self-consistent; and (iii) wave scattering. The zero interaction models treat each fracture as if it were isolated as a single fracture in an otherwise undamaged material, and then sums the effect of all of the cracks within the body. In the self-consistent models (e.g., O'CONNELL and BUDIANSKY, 1974), each crack is treated similar to the zero interaction models, except that the effect of each crack is added to the overall moduli of the cracked material in an iterative process, so that the effect of each successive fracture is calculated against an increasingly lower background modulus. The first two general formulations, then, provide lower and upper bounds on the effects of the cracks on the effective elastic moduli, respectively. A third class of effective medium theories (e.g., HUDSON, 1980) approaches the problem using scattering theory analysis of the mean field in an isotropic elastic solid embedded with thin cracks. Like the self-consistent method, this approach also takes into account interaction between cracks.

A common feature of most micromechanical models described above is that the fractures are represented by a single spatial parameter, the fracture density, Γ, represented by (e.g. WALSH, 1965):

$$\Gamma_{3D} = \frac{1}{V}\sum_{i=1}^{N} a_i^3, \quad \text{and} \quad \Gamma_{2D} = \frac{1}{A}\sum_{i=1}^{N} a_i^2, \tag{A1}$$

where N is the number of fractures, a is the fracture half-radius or half-length, V is the total rock volume, A is the total rock area, and subscripts 3-D and 2-D refer to the three- and two-dimensional forms of the fracture density, respectively. All of the models yield similar results for fracture densities less than $\sim\Gamma_{3D} \approx 0.2$ but disagree at higher densities (KACHANOV, 1992; JAEGER *et al.*, 2007). Whereas these micromechanical models are sufficient for describing overall effects of microcracks in sparsely fractured rocks, their limitations make them inadequate for descriptions of many natural mesoscopic and macroscopic fracture systems, which typically have fracture densities on the order of $\Gamma_{2D} \approx 1$ to 2 (RENSHAW, 1997). Using equation (A1) above, the digitized BJFZ fracture map yields a fracture density of $\Gamma_{2D} = 1.36$ (Fig. A1); therefore we must select an effective medium model which allows higher fracture densities.

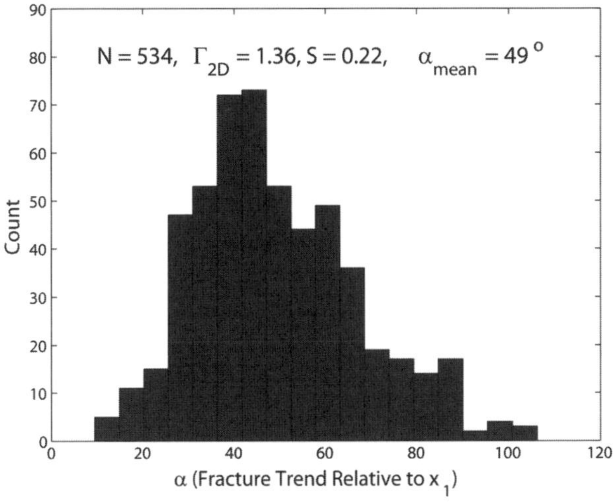

Figure A1

Fracture orientation (α) relative to the positive x_1 axis. N is the number of measured segments, S is the calculated spacing in meters, α_{mean} is the mean fracture orientation, and Γ_{2D} is the two-dimensional fracture density.

A different approach accounts for nonlinear elastic behavior of rocks due to opening and closing of cracks (e.g., ZOBACK and BYERLEE, 1975; WEINBERGER *et al.*, 1994) by adding a second-order term in the elastic potential of a Hookean solid (LYAKHOVSKY *et al.*, 1997b). Damage mechanics has been shown to successfully match experimental rock deformation for arbitrary loading paths (e.g., HAMIEL *et al.*, 2006). This approach does not suffer from exactly the same limitation as the methods described above, as it uses a scalar damage variable which depends on the change in the elastic strain energy potential during deformation (e.g., LYAKHOVSKY *et al.*, 1997a). However, this approach is difficult to apply to the present case of estimating the overall response of macroscopic frictional fractures, as it estimates stiffness changes independent of any previous structural features. Therefore there is no statistical way to represent the macroscopic damage zone fractures in the model.

Still another approach to estimating the elastic moduli of fractured rocks is to treat the fractures as evenly distributed and parallel with characteristic normal and shear stiffness (AMADEI and SAVAGE, 1993). In this approach, the spatial distribution is characterized by the fracture spacing, S, rather than the fracture density, Γ. Fractures are idealized with negligible thickness and infinite length, and their mechanical behavior is characterized by a normal stiffness k_n and shear stiffness k_s. The method is independent of fracture density and subscripts n and s are chosen here to differentiate these stiffnesses from the stiffnesses k_N and k_T used as regularization parameters in the numerical models. For one set of fractures perpendicular to the positive x_3^c axis the compliances are:

$$\begin{aligned} s_{11} &= s_{22} = \frac{1}{E}, \\ s_{33} &= \frac{1}{E} + \frac{1}{k_n S}, \\ s_{44} &= s_{55} = \frac{1}{G} + \frac{1}{k_s S}, \\ s_{66} &= \frac{1}{G}, \\ c_{12} &= c_{13} = -\frac{\upsilon}{E}. \end{aligned} \tag{A2}$$

We idealize the fractures of the BJFZ (Fig. 3) as a set of parallel joints with spacing S and orientation α relative to the positive x_1 direction. In order to estimate values of these parameters, we take advantage of the digitized maps produced during the finite-element mesh generation. The fractures are divided into short subequal length segments, where the length of each segment equals the corresponding segment lengths in the Finite-Element grid (Fig. 5). The orientations of all of the fracture segments relative to the positive x_1 axis are plotted in Figure A1. There is one dominant fracture orientation at $\alpha \approx 49^{\circ}$ with a spread of one standard deviation $= 16^{\circ}$. We calculate the fracture spacing using the area method described by WU and POLLARD (1995), yielding $S \approx 0.22$ m. Using these parameters, a large range of effective moduli can be calculated depending on the choice of k_n and k_s. Because fracture stiffness may vary over several orders of magnitude and is highly sensitive to the normal traction acting on the fracture (e.g., WORTHINGTON and LUBBE, 2007; YOSHIOKA and SCHOLZ, 1989a, b) we attempt to use the numerical results to constrain these parameters for the BJFZ.

Using the effective medium model, the compliances are calculated in the crack-local reference frame (Fig. 2) using eqns. (2) and (A2). We calculate the apparent Young's moduli parallel to the x_1 and x_3 axes in the global reference frame in the numerical experiments. These correspond to $E_3 = 1/s_{33}'$ and $E_1 = 1/s_{11}'$ in the compliance matrix transformed into global reference frame (**S**') by rotating around the mean fracture orientation angle $\alpha = 49^{\circ}$ using the Bond transformation matrix N (e.g., MAVKO *et al.*, 1998):

$$S'(k_n, k_s) = NS(k_n, k_s)N^T. \tag{A3}$$

In equation (A2), s_{33} is a function of E, S, and k_n, while s_{11} is only a function of E. However, through the rotational transformation (equation (A3)) from crack-local to global reference frame s_{11}' and s_{33}' both become functions of k_n and k_s as follows. By defining the two strain and stress vectors as:

$$\begin{aligned} \varepsilon_1 &= S'(k_n, k_s)\sigma_1; \quad \sigma_1 = [1 \quad 0 \quad 0 \quad 0 \quad 0 \quad 0]^T, \\ \varepsilon_3 &= S'(k_n, k_s)\sigma_3; \quad \sigma_3 = [0 \quad 0 \quad 1 \quad 0 \quad 0 \quad 0]^T, \end{aligned} \tag{A4}$$

one can calculate the x_1 and x_3-parallel apparent Young's moduli. The problem of solving for the crack stiffnesses can be reduced to two equations in two unknowns:

$$E_1 = \frac{\varepsilon_1}{\sigma_1} = \frac{1}{s_{11}'(k_n, k_s)} \quad \text{and} \quad E_3 = \frac{\varepsilon_3}{\sigma_3} = \frac{1}{s_{33}'(k_n, k_s)}. \tag{A5}$$

The expressions for s_{11}' and s_{33}' are:

$$\begin{aligned} s_{11}' &= \frac{1}{E} + \frac{\sin^4 \alpha}{k_n S} + \frac{\sin^2 2\alpha}{4k_s S}, \\ s_{33}' &= \frac{1}{E} + \frac{\cos^4 \alpha}{k_n S} + \frac{\sin^2 2\alpha}{4k_s S}. \end{aligned} \tag{A6}$$

The other components of the transformed compliance matrix can be found using equations (A2) and (A3) and are written down in AMADEI and SAVAGE (1993). Finally, solving these equations for k_n and k_s yields:

$$\begin{aligned} k_n &= \frac{-1 + 2\cos^2 \alpha}{S(s_{33}' - s_{11}')}, \\ k_s &= -\frac{E \sin^2 2\alpha(\sin^4 \alpha - \cos^4 \alpha)}{4S([1 - Es_{33}'] \sin^4 \alpha - [1 - Es_{11}'] \cos^4 \alpha)}. \end{aligned} \tag{A7}$$

Substituting these values of k_n, and k_s constrained by the numerical calculations of the apparent Young's moduli into equation (A2), one can define the complete transversely anisotropic compliance matrix. As an example, substituting values of $E = 60$ GPa and $\alpha = 49^{\circ}$, as well as $E_1 = 20$ GPa and $E_3 = 37$ GPa from the uniaxial compression tests for $\mu = 0.5$ at ε_{11} or $\varepsilon_{33} = 0.01$ into equation (A7) yields crack stiffnesses of $k_n = 28$ GPa/m and $k_s = -55$ GPa/m. Clearly, negative crack stiffness is not valid. Although this procedure is valid for a very small range of results at extremely small strains (ε_{11} or $\varepsilon_{33} << 0.005$), this analytical model and the numerical experiments are not compatible.

There are several sources of this incompatibility. The most obvious are the constitutive assumptions built into the models. The analytical model assumes that the fractures deform only elastically based on k_n and k_s. Under this assumption, the largest reduction in the apparent Young's modulus occurs when the fractures are orthogonal to the direction of maximum compression. Under this configuration the classical Coulomb friction model predicts zero sliding, thus no reduction in the Young's modulus. A second source of incompatibility is the scale of the problem. The BJFZ, and, for that matter, all of the fault zones in the Bear Creek area are tabular in shape and much longer (in the x_1 direction) than they are wide (in the x_3 direction). This and a nonuniform fracture distribution lead to large stress and strain variations across them during the numerical uniaxial tests. An assumption of the analytical model is that the rock mass is large enough to contain numerous fractures, yet small enough to make stress and strain gradients within it negligible. Clearly this criterion is not satisfied. Finally, while the cross fractures of the BJFZ have a preferred orientation (Fig. A1), they anastamose and intersect such that there is a large spread in fracture orientations across the fault zone, and the fractures are not infinite in length. Thus the assumption of an array of long parallel fractures is an insufficient idealization for the fault zone.

Appendix B: Numerical Model

Rock masses are disordered discontinuous media in which the main hydraulic and mechanical features are the discontinuities (Bieniawski, 1978). Consequently, a very important aspect of modeling rock masses is the appropriate representation of these discontinuities, either as individual entities (explicitly through interfaces) or as a homogenized system (implicitly as diffuse discontinuities, i.e., continuum approach). The suitability of these models depends on the problem scale relative to the fracture density and on the configuration of the discontinuities. Continuum approaches should be used for rock masses with no discontinuities (intact rock) or with many of them (equivalent continuum rock mass), where the equivalent properties of the latter can be established through a homogenization process or through empirical relations (e.g., Bieniawski, 1978; Hoek and Diederichs, 2006; Sonmez *et al.*, 2006).

To adequately capture the mechanics of the fault damage zone, we implement a model that explicitly captures discontinuous displacement fields. Here we develop a model based upon continuum mechanics and frictional contacts in which the coupling between the deformation of both the rock matrix and the rock discontinuities is realistically accounted for. A key goal of this study is to model the fracture evolution accounting for frictional sliding. A constitutive model for geologic interfaces considering unilateral contact and inelastic relative motion is presented and implemented into a finite-element code following SANZ (2008).

With contact mechanics it is possible to capture a normal discontinuity (gap), a tangential discontinuity (slip), or no discontinuity (stick or bonded) along a pre-defined interface. A critical aspect of the implementation is the choice of method to convert the variational inequality of a contact constraint into an equality suitable for the finite element framework. This implementation applies the penalty method to impose the contact constraints and a regularized friction law to simulate the mechanical response of the interfaces (LAURSEN, 2002; WRIGGERS, 2002). The numerical integration of these path-dependent interface models is performed by a return mapping algorithm similar to a non-associative plasticity model (WRIGGERS, 2002).

Contact Kinematics and Traction Measures

The contact condition is defined in the initial (i.e., undeformed) configuration via a minimum distance problem or closest point projection (e.g., LAURSEN, 2002; WRIGGERS, 2002) such that every point $\mathbf{X}^1$ on a contact surface can be related to a point $\mathbf{X}^2$ on the other contact surface. The relative displacement between these two contact points is defined as

$$\mathbf{g} = \mathbf{u}(\mathbf{X}^2) - \mathbf{u}(\mathbf{X}^1) = \mathbf{u}^2 - \mathbf{u}^1, \tag{B1}$$

where $\mathbf{u}^1$ and $\mathbf{u}^2$ are the displacements of the material points $\mathbf{X}^1$ and $\mathbf{X}^2$, respectively.

The relative displacement $\mathbf{g}$ can be additively decomposed into a normal part $\mathbf{g}_N$ and a tangential part $\mathbf{g}_T$ as follows

$$\mathbf{g} = \mathbf{g}_N + \mathbf{g}_T, \tag{B2}$$

where

$$\mathbf{g}_N = (\mathbf{n}_N \otimes \mathbf{n}_N) \cdot \mathbf{g}; \quad \mathbf{g}_T = \mathbf{g} - \mathbf{g}_N = (\mathbf{1} - \mathbf{n}_N \otimes \mathbf{n}_N) \cdot \mathbf{g}, \tag{B3}$$

where $\mathbf{n}_N$ is the unit outward normal to the contact surface, $\mathbf{1}$ is the second-order identity tensor, and the symbol $\otimes$ denotes a juxtaposition, e.g., $(\mathbf{a} \otimes \mathbf{b})_{ij} = a_i b_j$.

Let the contact traction $\mathbf{t}$ acting on the interface be decomposed into a normal traction $\mathbf{t}_N$ and a tangential traction $\mathbf{t}_T$ as follows

$$\mathbf{t} = \mathbf{t}_N + \mathbf{t}_T = t_N \mathbf{n}_N + t_T \mathbf{n}_T = \sum_{A=1}^{2} t_A \mathbf{n}^{(A)}, \tag{B4}$$

where $t_N = \mathbf{t} \cdot \mathbf{n}_N$ is the contact pressure (negative for compression), $t_T = \|\mathbf{t}_T\|$ and $\mathbf{n}_T = \mathbf{t}_T / \|\mathbf{t}_T\|$ is a unit vector tangent to the surface in the direction of $\mathbf{t}_T = \mathbf{t} \cdot (\mathbf{1} - \mathbf{n}_N \otimes \mathbf{n}_N)$. The component $A = 1$ corresponds to the direction $\mathbf{n}_N$ perpendicular to the interface, and the component $A = 2$ corresponds to the tangential direction $\mathbf{n}_T$.

Interface Elastic Model

Analogous to plasticity theory, the total relative displacement **g** and relative velocity $\dot{\mathbf{g}}$ are decomposed into an elastic part and a plastic part

$$\mathbf{g} = \mathbf{g}^e + \mathbf{g}^p \Leftrightarrow \dot{\mathbf{g}} = \dot{\mathbf{g}}^e + \dot{\mathbf{g}}^p. \tag{B5}$$

The relationship between the traction **t** and the elastic relative displacement **g** follows Hooke's law and can be written in rate form as

$$\dot{\mathbf{t}} = \mathbf{C}^e \cdot \dot{\mathbf{g}}^e \Leftrightarrow \begin{bmatrix} \dot{t}_N \\ \dot{t}_T \end{bmatrix} = \begin{bmatrix} k_n & 0 \\ 0 & k_T \end{bmatrix} \begin{bmatrix} \dot{g}^e_N \\ \dot{g}^e_T \end{bmatrix}, \tag{B6}$$

where $\dot{g}^e_N$ and $\dot{g}^e_T$ are the normal and tangential elastic components of the relative velocity between the contacting surfaces, and

$$\mathbf{C}^e = k_N(\mathbf{n}_N \otimes \mathbf{n}_N) + k_T(\mathbf{n}_T \otimes \mathbf{n}_T), \tag{B7}$$

a second-order tensor, is the elastic stiffness of the interface. In this work, it is assumed that Eq. (B7) is linear, and the components of the elastic stiffness $\mathbf{C}^e_{AB} = \partial t_A / \partial g^e_B$ are constants. However, there may be applications in which the use of stress dependent $\mathbf{C}^e_{AB}$ are desirable such as in problems involving rock joints subject to large variations of compressive stress in which it is known that k_N and k_T are monotonically increasing functions of increasing compressive stress (e.g., YOSHIOKA and SCHOLZ, 1989a, b). To provide symmetric normal and tangential stress changes for positive and negative changes in the relative normal and tangential displacement, respectively, the off diagonal components of $\mathbf{C}^e$ are zero. The tangential and normal elastic stiffness, $k_T = \partial t_T / \partial g^e_T$ and $k_N = \partial t_N / \partial g^e_N$, are measurable and numerous values have been reported in the literature (e.g., MAHTAB, 1969; GOODMAN et al., 1972; ROSSO, 1976; BELYTSCHKO et al., 1984). In the numerical simulations large values for the tangential stiffness used to approximate the rigid-plastic behavior of the classical Coulomb friction law, and a high value of the normal elastic stiffness were used to avoid unrealistic interpenetration between contacting surfaces (Table 1).

Interface Frictional Model

The frictional behavior of the interface is represented by the Coulomb law. Similarly to a yield function in elastoplasticity the function $\mathbf{\Phi}(\mathbf{t}_N, t_T) = 0$ (Fig. B1), a conical surface in the $(\mathbf{t}_N, t_T)$ stress space, defines the onset of irreversible relative motion as

$$\mathbf{\Phi}(\mathbf{t}_N, t_T) = \|\mathbf{t}_T\| - (c - \mu t_T) \leq 0, \tag{B8}$$

where the shear strength of the interface is defined by the coefficient of friction μ .

The classical Coulomb friction law has an indeterminate, or multi-valued, structure when viewed as a law relating tangential stresses to slip. To avoid the non-differentiability of the friction law at the onset of sliding (i.e., zero relative tangential displacement), we use a regularized version of the stick-slip behavior (Fig. B2).

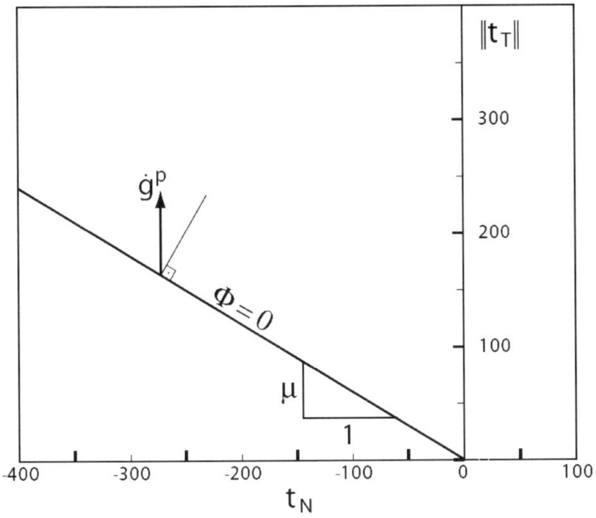

Figure B1
Coulomb slip criterion plotted on graph of tangential versus normal traction. The tangential evolution of the inelastic relative motion (parallel to the vertical axis) illustrates the non-associative nature of the classical Coulomb friction law.

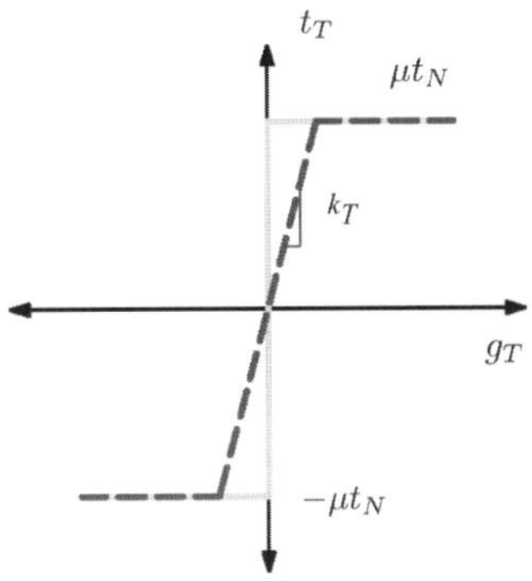

Figure B2
Schematic representation of the regularization of the Coulomb friction law. Dashed line represents the regularization of the exact Coulomb rigid-slip formulation (represented by a gray solid line).

The evolution of the inelastic component of the relative displacement $\dot{\mathbf{g}}^p$ is described through a non-associative flow rule as

$$\dot{\mathbf{g}}^p = \dot{\lambda}\mathbf{n}_T, \tag{B9}$$

where as in plasticity $\dot{\lambda}$ is the non-negative plastic multiplier.

When an interface undergoes inelastic relative motion (i.e., frictional slip) the second-order interface tangent stiffness takes the following form

$$\mathbf{C}^{ep} = k_N(\mathbf{n}_N \otimes \mathbf{n}_N) - \mu k_N(\mathbf{n}_T \otimes \mathbf{n}_N). \tag{B10}$$

The derivation of the interface tangent stiffness and details regarding the numerical implementation can be found in SANZ (2008).

REFERENCES

AGUE, J. J. and BRIMHALL, G. H. (1988), *Magmatic arc asymmetry and distribution of anomalous plutonic belts in the batholiths of California: Effects of assimilation, crustal thickness, and depth of crystallization*. Geol. Soc. Am. Bull. *100*, 912–927.

AMADEI, B. and SAVAGE, W. Z. *Effects of joints on rock mass strength and deformability*. In *Comprehensive Rock Engineering*, Vol. I (ed. J. A. HUDSON), pp. 331–365 (Pergamon Press 1993).

AMPUERO, J.-P. and BEN-ZION, Y. (2008) *Cracks, pulses and macroscopic asymmetry of dynamic rupture on a bimaterial interface with velocity-weakening friction*, Geophys. J. Int. *173*, 674–692, doi: 10.1111/j.1365-246X.2008.03736.x.

ANDREWS, D. J. and BEN-ZION, Y. (1997), *Wrinkle-like slip pulse on a fault between different materials*, J. Geophys. Res.*102*(B1), 553–571.

ANDREWS, D. J. (2005), *Rupture dynamics with energy loss outside the slip zone*, J. Geophys. Res. *110*, doi:10.1029/2004JB003191.

ANDREWS, D. J. and HARRIS, R. A. (2005), *The wrinkle-like slip pulse is not important in earthquake dynamics*, Geophys. Res. Lett. *32*, doi:10.1029/2005GL023996.

BELYTSCHKO, T., PLESHA, M. E., and DOWDING, C. H. (1984) *A computer method for stability analysis of caverns in jointed rock*, Int. J. Numer. Anal. Methods Geomech. *8*, 473–492.

BEN-ZION, Y. (2006a), *Comment on "The wrinkle-like slip pulse is not important in earthquake dynamics"* by D. J. ANDREWS and R. A. HARRIS, Geophys. Res. Lett. *33*, L06310, doi:10.1029/2005GL025372.

BEN-ZION, Y., (2006b), *A comment on "Material contrast does not predict earthquake rupture propagation direction"* by R. A. HARRIS and S. M. DAY, Geophys. Res. Lett. *33*, L13310, doi:10.1029/2005GL025652.

BEN-ZION, Y. (2008), *Collective behavior of earthquakes and faults: Continuum-discrete transitions, progressive evolutionary changes and different dynamic regimes*, Rev. Geophys. *46*, doi:10.1029/2008rg000260.

BEN-ZION, Y., PENG, Z., OKAYA, D, SEEBER, L., ARMBRUSTER, J. G., OZER, N., MICHAEL, A. J., BARIS, S. and AKTAR, M. (2003), *A shallow fault zone structure illuminated by trapped waves in the Karadere-Duzce branch of the North Anatolian Fault, western Turkey*, Geophys. J. Int. *152*, 699–717.

BEN-ZION, Y. and SAMMIS, C. G. (2003), *Characterization of fault zones*, Pure Appl. Geophys. *160*, 677–715.

BEN-ZION, Y. and SHI, Z. (2005), *Dynamic rupture on a material interface with spontaneous generation of plastic strain in the bulk*, Earth Planet. Sci. Lett. *236*, 486–496, doi: 10.1016/j.epsl.2005.03.025.

BERGBAUER, S. and MARTEL, S. J. (1999), *Formation of joints in cooling plutons*, J. Struct. Geol. *21*(7), 821–835.

BIEGEL, R. L., SAMMIS, C. G., and ROSAKIS, A. J. (2008), *An experimental study of the effect of off-fault damage on the velocity of a slip pulse*, J. Geophys. Res. *113*, doi:10.1029/2007JB005234.

BIENIAWSKI, Z. T. (1978), *Determining rock mass deformability: Experience from case histories*, Int. J. Rock Mech. and Min. Sci. and Geomech. Abs. *15*, 237–248.

BIRCH, F. (1966), *Compressibility: Elastic constants*. In *Handbook of Physical Constants*, Vol. 97 (ed. S. P. CLARK), pp. 97–173. Geological Society of America, Memoir.

BORJA, R. I., SAMA, K. M., and SANZ, P. F. (2003), *On the numerical integration of three-invariant elastoplastic constitutive models*, Comp. Meth. Appl. Mech. Eng. *192*, 1227–1258.

CHESTER, F. M., EVANS, J. P., and BIEGEL, R. L. (1993), *Internal structure and weakening mechanisms of the San Andreas fault*, J. Geophys. Res. *98*(B1), 771–786.

CHESTER, F. M. and CHESTER, J. S. (1998), *Ultracataclasite structure and friction processes of the San Andreas fault*, Tectonophysics *295*, 195–221.

CHESTER, F. M., CHESTER, J. S., KIRSCHNER, D. L., SCHULTZ, S. E., and EVANS, J. P. *Structure of large-displacement, strike-slip fault zones in the brittle continental crust*. In *Rheology and Deformation in the*

Lithosphere at Continental Margins (ed. G. D. KARNER, B. TAYLOR, N. W. DRISCOLL, and D. L. KOHLSTEDT) (Columbia University Press 2004).

CROOK, A. J. L., OWEN, D. R. L., WILSON, S. M., and YU, J. G. (2006), *Benchmarks for the evolution of shear localisation with large relative sliding in frictional materials*, Comp. Meth. Appl. Mech. Eng. *195*, 4991–5010.

CROOK, A. J. L., WILSON, S. M., YU, J. G., and OWEN, D. R. L. (2006), *Predictive modeling of structure evolution in sandbox experiments,* J. Struct. Geol. *17*, 409–421.

DAVATZES, N. C. and AYDIN, A. (2003), *Overprinting faulting mechanisms in high porosity sandstones of SE Utah*, J. Struct. Geol. *25*, 1795–1813.

DOR, O., ROCKWELL, T. K., and BEN-ZION, Y. (2006), *Geological observations of damage asymmetry in the structure of the San Jacinto, San Andreas and Punchbowl faults in southern California: A possible indicator for preferred rupture propagation direction*. Pure Appl. Geophys. *163*, 301–349.

FAULKNER, D. R., MITCHELL, T. M., HEALY, D., and HEAP, M. J. (2006), *Slip on "weak" faults by the rotation of regional stress in the fracture damage zone*, Nature *444*, doi:10.1038/nature05353.

FIALKO, Y. (2004), *Probing the mechanical properties of seismically active crust with space geodesy: Study of the co-seismic deformation due to the 1992 M_w7.3 Landers (southern California) earthquake*, J. Geophys Res. *109*, doi:10.1029/2003JB002756.

FINZI, Y., HEARN, E. H., BEN-ZION, Y., and LYAKHOVSKY, V. (2009), *Structural properties and deformation patterns of evolving strike-slip faults: Numerical simulations incorporating damage rheology*, Pure Appl. Geophys. *166*.

GERCEK, H. (2007), *Poisson's ratio values for rock*, Int. J. Rock Mech. Min. Sci. *44*, 1–13.

GOODMAN, R. E., HEUZE, F. E., and OHNISHI, Y. (1972), *Research on strength-deformability water pressure relationship for faults in direct shear*. Final Report, ARPA Contract H0210020, Univ of Cal., Berkeley.

GRIFFITH, W. A., DI TORO, G., PENNACCIHIONI, G., and POLLARD, D. D. (2008), *Thin pseudotachylytes in faults of the Mt. Abbot quadrangle, Sierra Nevada: Physical Constraints for Small Seismic Slip Events*, J. Struct. Geol. *30*, 1086–1094.

GRIFFITH, W. A., DI TORO, G., PENNACCHIONI, G., POLLARD, D. D., and NIELSEN, S. (2009), *Static stress drop associated with brittle slip events on exhumed faults*, J. Geophys. Res. *114*, B02402, doi:10.1029/2008JB005879.

HAMIEL, Y., KATZ, O., LYAKHOVSKY, V., RECHES, Z., and FIALKO, Y. (2006), *Stable and unstable damage evolution in rocks with implications to fracturing of granite*, Geophys. J. Int., doi: 10.1111/j.1365-246X.2006.03126.

HAMIEL, Y., LYAKOVSKY, V., STANCHITS, S., DRESEN, G., and BEN-ZION, Y. (2008), *Brittle deformation and damage-induced seismic wave anisotropy in rocks*, Geophys. J. Int., in review, 2008.

HARRIS, R. A. and DAY, S. M. (2005), *Material contrast does not predict earthquake rupture propagation direction*, Geophys. Res. Lett. *32*, doi:10.1029/2005GL023941.

HOEK, E. and DIEDERICHS, M. S. (2006), *Empirical estimation of rock mass modulus*, Int. J. Rock Mech. Min. Sci. *43*, 203–215.

HUDSON, J. A. (1980), *Overall properties of a cracked solid*. Cambridge Phil. Soc., Math. Proc. *88*, 371–384.

HUDSON, J. A., LIU, R., and CRAMPIN, S. (1996), *The mechanical properties of materials with interconnected cracks and pores*, Geophys. J. Int. *124*, 105–112.

JAEGER, J. C., COOK, N. G. W., and ZIMMERMANN, R. W. (2007), *Fundamentals of Rock Mechanics* (Blackwell Publishing, Malden, MA, U.S.A., 2007).

KACHANOV, M. (1992), *Effective elastic properties of cracked solids: Critical review of some basic concepts*, Appl. Mech. Rev. *45*, 304–335.

LACHENBRUCH, A. H. (1980), *Frictional heating, fluid pressure, and the resistance to fault motion*, J. Geophys. Res. *85*, 6,097–6,112.

LAURSEN, T. A. *Computational contact and impact mechanics* (Springer-Verlag, Berlin and Heidelberg 2002).

LAURSEN, T. A. and SIMO, J. C. (1993), *A continuum-based finite element formulation for the implicit solution of multibody large deformation frictional contact problems*, Int. J. Num. Meth. Eng. *36*, 3451–3485.

LI, Y.-G., AKI, K., ADAMS, D., HASEMI, A., and LEE, W. H. K. (1994), *Seismic guided waves trapped in the fault zone of the Landers, California, earthquake of* 1992, J. Geophys. Res. *99*, 11,705–11,722.

LOCKNER, D., TANAKA, H., ITO, H., and IKEDA, R. (2000), *Permeability and strength of core samples from the Nojima fault of the 1995 Kobe earthquake*. In *Proc. Internat. Workshop on the Nojima Fault Core and Borehole Analysis*, U.S. Geological Survey Open-file Report 00-129.

LOCKWOOD, J. P. and LYDON, P. A. (1975), *Geologic map of the Mount Abbot quadrangle, California.* Geological Survey Geologic Quadrangle *GQ-1155* scale 1:62,500.

LYAKHOVSKY, V. and BEN-ZION, Y. (2008), *Scaling relations of earthquakes, aseismic deformation and evolving fault structures in a damage rheology model*, Geophys. J. Int., *172*, 651–662, doi: 10.1111/j.1365-246X.2007.03652.x.

LYAKHOVSKY, V., BEN-ZION, Y., and AGNON, A. (1997a), *Distributed damage, faulting, and friction*, J. Geophys. Res. *102*, 27635–27649.

LYAKHOVSKY, V., RECHES, Z., WEINBERGER, R., and SCOTT, T. E. (1997b), *Non-linear elastic behaviour of damaged rocks*, Geophys. J. Int., 157–166.

MAHTAB, M. A. (1969), Three-dimensional finite element analysis of joint and rock slopes. Ph.D. Thesis, Univ. of California, Berkeley.

MARTEL, S. J., POLLARD, D. D., and SEGALL, P. (1988), *Development of simple strike-slip fault zones, Mount Abbot Quadrangle, Sierra Nevada, California*. Geol. Soc. Am. Bull. *100*, 1,451–1,465.

MARTEL, S. J. and POLLARD, D. D. (1989), *Mechanics of slip and fracture along small faults and simple strike-slip fault zones in granitic rock,* J. Geophys. Res. *94*, 9,417–9,428.

MARTEL, S. J. (1990), *Formation of compound strike-slip fault zones, Mount Abbot quadrangle, California*, J. Struct. Geol. *12*, 869–882.

MARTEL, S. J. and BOGER, W. A. (1998), *Geometry and mechanics of secondary fracturing around small three-dimensional faults in granitic rock*. J. Geophys. Res.*103*, 21,299–21,314.

MASE, C. W. and SMITH, L. (1987), *Effects of frictional heating on the thermal, hydrologic, and mechanical response of a fault*, J. Geophys. Res. *92*, 6249–6272.

MAVKO, G., MUKERJI, T., and DVORKIN, J. *The Rock Physics Handbook; Tools for Seismic Analysis in Porous Media* (CambridgeUniversity Press 1998).

MYERS, R. and AYDIN, A. A. (2004), *The evolution of fault zones formed by shearing across joint zones in sandstone*, J. Struct. Geol. *26*, 947–956.

NODA, H. and SHIMAMOTO, T. (2005), *Thermal pressurization and slip-weakening distance of a fault: An example of the Hanaore Fault, southwest Japan*, Bull. Seismol. Soc. Am. *95*(4), 1224–1233.

O'CONNELL, R. J. and BUDIANSKI, B. (1974), *Seismic velocities in dry and wet cracked solids*, J. Geophys. Res. *79*, 5412–5426.

PACHELL, M. A. and EVANS, J. P. (2002), *Growth, linkage, and termination processes of a 10-km-long, strike-slip fault in jointed granite: The Gemini fault zone, Sierra Nevada, California.* J. Struct. Geol. *24*, 1903–1924.

POLLARD, D. D. and SEGALL, P. *Theoretical displacements and stresses near fractures in rock: With applications to faults, joints, veins, dikes, and solution surfaces.* In Fracture Mechanics of Rock (ed. B. K. Atkinson), pp. 277–349 (Academic Press Inc. 1987).

RENSHAW, C. E. (1997), *Mechanical controls on the spatial density of opening-mode fracture networks*, Geology *25*(10), 923–926.

RICE, J. R. (2006), *Heating and weakening of faults during earthquake slip*, J. Geophys. Res. *111*, doi:10.1029/2005JB004006.

ROSSO, R. (1976), *A comparison of joint stiffness measurements in direct shear, triaxial compression and in situ*, Int. J. Rock Mech. Min. Sci. Geomech Abst. *13*, 167–172.

SANZ, P. F. (2008), *Modeling rock folding with large deformation frictional contact mechanics*, Ph.D. Thesis, Stanford University, California, USA.

SANZ, P. F., BORJA, R. I., and POLLARD, D. D. (2007), *Mechanical aspects of thrust faulting driven by far-field compression and their implications to fold geometry*, Acta Geotech. *2*, 17–31.

SANZ, P. F., POLLARD, D. D., ALLWARDT, P. F., and BORJA, R. I. (2008), *Mechanical models of fracture reactivation and slip on bedding surfaces during folding of the asymmetric anticline at Sheep Mountain, Wyoming*, J. Struct. Geol. *30*, 1177–1191, doi:10.1016/j.jsg.2008.06.002.

SCHULTZ, S. E. and EVANS, J. P. (2000), *Mesoscopic structure of the Punchbowl Fault, Southern California and the geologic and geophysical structure of active strike-slip faults*, J. Struct. Geol. *22*, 913–930.

SEGALL, P. and POLLARD, D. D. (1983a), *Nucleation and growth of strike slip faults in granite*, J. Geophys. Res. *88*, 555–568.

SEGALL, P. and POLLARD, D. D. (1983b), *Joint formation in granitic rock of the Sierra Nevada*, Geol. Soc. Am. Bull. *94*, 563–575.

SEGALL, P., MCKEE, E. H., MARTEL, S. J., and TURRIN, B. D. (1990), *Late Cretaceous age of fractures in the Sierra Nevada batholith, California*. Geology *18*, 1,248–1,251.
SIBSON, R. H. (1973), *Interactions between temperature and pore-fluid pressure during earthquake faulting and a mechanism for partial or total stress relief*, Nature *243*, 66–68.
SIMO, J. C. and TAYLOR, R. L. (1985), *Consistent tangent operators for rate-independent elastoplasticity*, Computer Methods in Applied Mechanics and Engineering *48*, 101–118.
SONMEZ, H., GOKCEOGLU, C., NEFESLIOGLU, H. A., KAYABASI, A. (2006), *Estimation of rock modulus: For intact rocks with an artificial neural network and for rock masses with a new empirical equation*, Int. J. Rock Mech. Min. Sci. *43*, 224–35.
STERN, T. W., BATEMAN, P. C., MORGAN, B. A., NEWALL, M. F., and PECK, D. L. (1981), *Isotopic U-Pb ages of zircon from the granitoids of the central Sierra Nevada, California*, U.S. Geol. Surv. Prof. Paper *1185*, 17.
WALSH, J. B. (1965), *The effect of cracks on the uniaxial elastic compression of rocks*, J. Geophys. Res. *70*, 399–411.
WEINBERGER, R., RECHES, Z., EIDELMAN, A., and SCOTT, T. S. (1994), Tensile properties of rocks in four-point beam tests under confining pressure. In Proc. the First North American Rock Mechanics Symp., Austin, Texas, (eds. P. Nelson and S.E. Laubach), pp. 435–442.
WIBBERLEY, C. A. J. (2002), *Hydraulic diffusivity of fault gouge zones and implications for thermal pressurization during seismic slip*, Earth Planets Space *54*, 1153–1171.
WIBBERLEY, C. A. J. and SHIMAMOTO, T. (2003), *Internal structure and weakening effects of major strike-slip fault zones: The Median Tectonic Line in Mie Prefecture, Southwest Japan*, J. Struct.Geol. *25*, 59–78.
WORTHINGTON, M. H. and LUBBE, R. (2007), *The scaling of fracture compliance,* Geol. Soc. Lond. Spec. Pub. *207*, 73–82.
WRIGGERS, P. (1995), *Finite-element algorithms for contact problems*, Arch. Comput. Meth. Eng. *4*, 1–49.
WRIGGERS, P., *Computational Contact Mechanics* (Wiley and Sons 2002).
WU, H. and POLLARD, D. D. (1995), *An experimental study of the relationship between joint spacing and layer thickness*, J. Struct. Geol. *17*(6), 887–905.
YOSHIOKA, N. and SCHOLZ, C. H. (1989), *Elastic properties of contacting surfaces under normal and shear loads 1. Theory,* J. Geophys. Res. *94*, 17681–17690.
YOSHIOKA, N. and SCHOLZ, C. H. (1989), *Elastic properties of contacting surfaces under normal and shear loads 2. Comparison of Theory with Experiment,* J. Geophys. Res. *94*, 17691–17700.
ZOBACK, M. D. and BYERLEE, J. D. (1975), *The effect of microcrack dilatancy on the permeability of Westerly granite*, J. Geophys. *Res. 80*, 752–755.

(Received August 14, 2008, revised January 26, 2009)
Published Online First: June 27, 2009

To access this journal online:
www.birkhauser.ch/pageoph

Pure appl. geophys. 166 (2009) 1629–1648
0033–4553/09/101629–20
DOI 10.1007/s00024-009-0512-3

© Birkhäuser Verlag, Basel, 2009

Pure and Applied Geophysics

Effects of Off-fault Damage on Earthquake Rupture Propagation: Experimental Studies

Charles G. Sammis,[1] Ares J. Rosakis,[2] and Harsha S. Bhat[1,2]

Abstract—We review the results of a recent series of papers in which the interaction between a dynamic mode II fracture on a fault plane and off-fault damage has been studied using high-speed photography. In these experiments, fracture damage was created in photoelastic Homalite plates by thermal shock in liquid nitrogen and rupture velocities were measured by imaging fringes at the tips. In this paper we review these experiments and discuss how they might be scaled from lab to field using a recent theoretical model for dynamic rupture propagation. Three experimental configurations were investigated: An interface between two damaged Homalite plates, an interface between damaged and undamaged Homalite plates, and the interface between damaged Homalite and undamaged polycarbonate plates. In each case, the velocity was compared with that on a fault between the equivalent undamaged plates at the same load. Ruptures on the interface between two damaged Homalite plates travel at sub-Rayleigh velocities indicating that sliding on off-fault fractures dissipates energy, even though no new damage is created. Propagation on the interface between damaged and undamaged Homalite is asymmetric. Ruptures propagating in the direction for which the compressional lobe of their crack-tip stress field is in the damage (which we term the 'C' direction) are unaffected by the damage. In the opposite 'T' direction, the rupture velocity is significantly slower than the velocity in undamaged plates at the same load. Specifically, transitions to supershear observed using undamaged plates are not observed in the 'T' direction. Propagation on the interface between damaged Homalite and undamaged polycarbonate exhibits the same asymmetry, even though the elastically "favored" '+' direction coincides with the 'T' direction in this case. The scaling properties of the interaction between the crack-tip field and pre-existing off-fault damage (i.e., no new damage is created) are explored using an analytic model for a nonsingular slip-weakening shear slip-pulse and verified using the velocity history of a slip pulse measured in the laboratory and a direct laboratory measurement of the interaction range using damage zones of various widths adjacent to the fault.

Key words: dynamic rupture, fracture damage, supershear rupture, asymmetric propagation, fault zone, slip pulse.

1. *Introduction*

Although earthquakes are commonly modeled as frictional instabilities on planar faults, real faults, many of which have been exhumed from seismogenic depths, have a more complex structure shown schematically in Figure 1. Most slip occurs in a highly

[1] Department of Earth Sciences, University of Southern California, Los Angeles, CA 90089-0740, U.S.A.
[2] Graduate Aeronautical Laboratories, 1200 E. California Blvd., Pasadena, CA 91125, U.S.A.

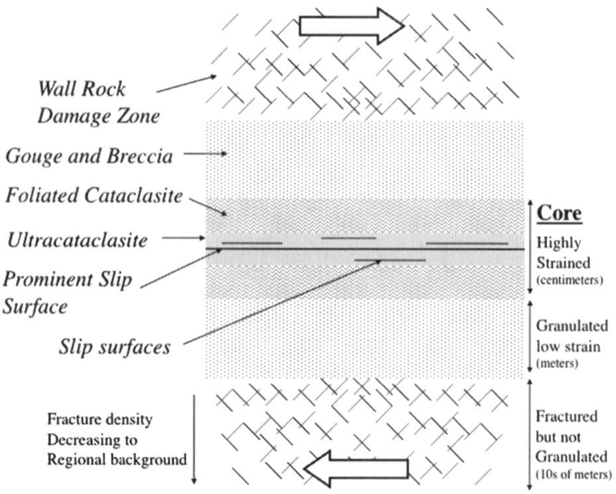

Figure 1
Schematic cross section of an idealized fault at seismogenic depth. The nested layered structure is described in the text. The widths of the layers vary from fault to fault, and the structure is often less symmetric than shown here.

sheared "core," typically a few centimeters thick, and composed of extremely fine-grained granulated rock (known as cataclasite) which is commonly altered to clay mineralogy, particularly at shallow depths. Slip within the core is often localized onto principal slip surfaces a few mm thick and composed of still finer-grained ultracataclasites. The core is bordered by layers of coarser granulated rock commonly termed gouge or fault breccia. These layers are typically meters thick and appear to have accommodated little or no shear strain. For some large displacement strike-slip faults, "pulverized rocks" in which individual grains are shattered but the rock fabric is undisturbed have been observed to distances of 100–300 meters from the fault plane in formations that were at or near the surface during faulting (Dor *et al.*, 2006). The granular layers are bordered by highly fractured (but not granulated) wall rock within which the fracture density decreases to the regional background value over a distance of a few hundred meters. More detailed descriptions of fault zone structure, deviations from the ideal symmetry in Figure 1, and discussions as to how it might have formed are given by Ben-Zion and Sammis (2003) and Biegel and Sammis (2004).

The focus of this paper is on how fault zone structure affects earthquake rupture propagation, with special emphasis on a series of recent laboratory measurements of rupture velocity on faults in damaged photoelastic plates by Biegel *et al.* (2008, 2009) and Bhat *et al.* (2009). Biegel *et al.* (2008) observed a reduction in rupture velocity caused by a symmetric distribution of damage about a premachined fault, and measured the spatial extent of the interaction between the rupture tip and the off-fault damage. Biegel *et al.* (2009) and Bhat *et al.* (2009) studied propagation asymmetries caused by slip on the interface between damaged and undamaged materials, as would be the case if an earthquake rupture propagated along one side of a fault zone. A key question here is: How wide does the fault zone have to be in order to affect rupture propagation? We address this question at the end of the paper by using a recent dynamic slip pulse model (RICE *et al.*, 2005) as guidance in scaling observations in the laboratory to natural earthquakes.

Consider first the case of symmetric off-fault damage with no material contrast. A first-order effect of fracture damage on rupture propagation is to lower the elastic stiffness of the material. By reducing the shear-wave speed c_S, the effect on rupture velocity is to lower the limiting Rayleigh speed, which is 0.92 c_S for Mode II ruptures. However, BIEGEL *et al.* (2008) found that damage reduced the rupture velocity below that expected, based solely on the lower shear-wave speed. They ascribed this additional reduction to a further dynamic reduction in modulus and to anelastic losses associated with frictional slip on the myriad of small off-fault fractures that comprise the damage.

BIEGEL *et al.* (2009) and BHAT *et al.* (2009) extended these measurements to ruptures that propagate on the interface between damaged and undamaged materials. In these experiments, the off-fault damage produced additional asymmetries in the propagation of ruptures beyond those expected from the associated contrast in elastic stiffness. Propagation asymmetries ascribed to the damage were observed to be stronger than those due to elastic contrasts.

2. *Experimental Apparatus*

All experiments described here used the apparatus shown in Figure 2. Square plates of the photoelastic polymers Homalite and polycarbonate were prepared with a premachined fault at an angle α to the edge as in Figure 3. The plates were loaded with a uniaxial stress P and a bilateral rupture was nucleated using a high-voltage pulse to explode a wire in a small hole that crossed the center of the fault plane. The explosion reduced the normal stress on an $\sim$1-cm long patch of the fault plane, allowing shear slip that nucleated the rupture. The voltage pulse was also used to trigger a pair of high-speed digital cameras that recorded a series of images of the shear stress field revealed as fringes in polarized laser light.

The upper panel in Figure 4 shows four frames taken at the times indicated during one of the Homalite experiments. Fringes associated with stress concentration at the rupture fronts and the S wave generated by the nucleation event can be identified. The lower panel in Figure 4 shows the corresponding instantaneous velocity as a function of time. Instantaneous velocity was found by differentiating an interpolated cubic spline fit to the displacements using a MATLAB® utility (BIEGEL *et al.*, 2009; BHAT *et al.*, 2009). Note that propagation is symmetric and both rupture tips accelerate to the limiting (Rayleigh) rupture velocity $c_R = 0.92c_S$ for mode II propagation and then transition to supershear velocities approaching the P-wave speed (note, $c_P/c_S = 2.1$ for Homalite).

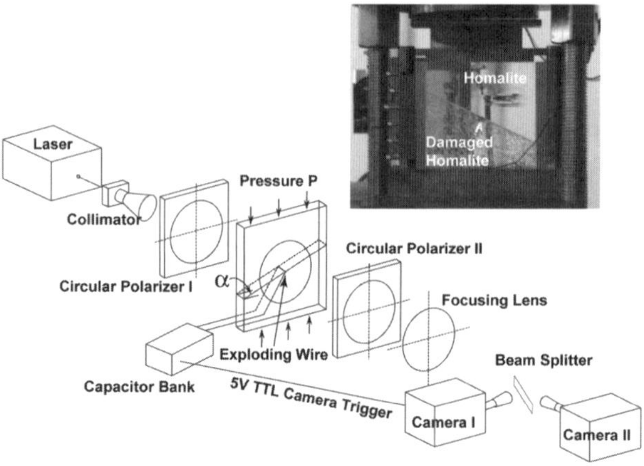

Figure 2
Schematic diagram of apparatus used to take a series of high-speed photographs of dynamic ruptures on premachined faults in photoelastic Homalite and polycarbonate plates. The inset shows a sample in the loading frame used to apply uniaxial stress *P*.

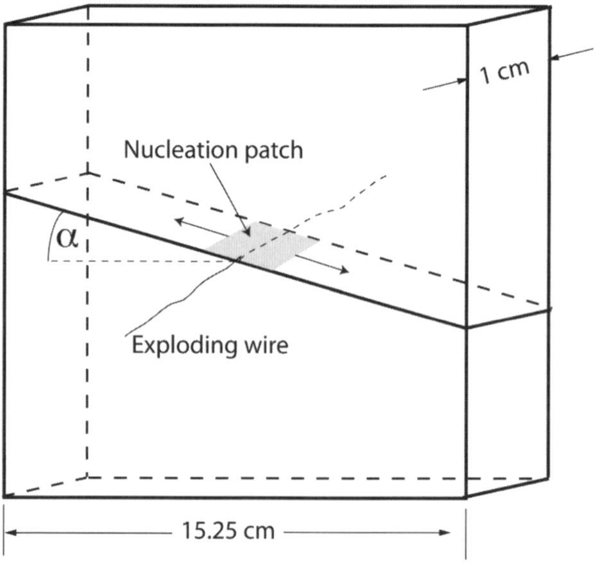

Figure 3
Sample geometry showing premachined fault at angle α. The exploding wire is used to create a patch of low effective normal stress that nucleates a bilateral rupture at the center of the fault plane.

3. Effect of Symmetric Off-Fault Damage on Rupture Velocity

BIEGEL *et al.* (2008) introduced fracture damage into Homalite plates by using a razor knife to score a mesh of scratches on both sides of the plate spaced about 2 mm apart and

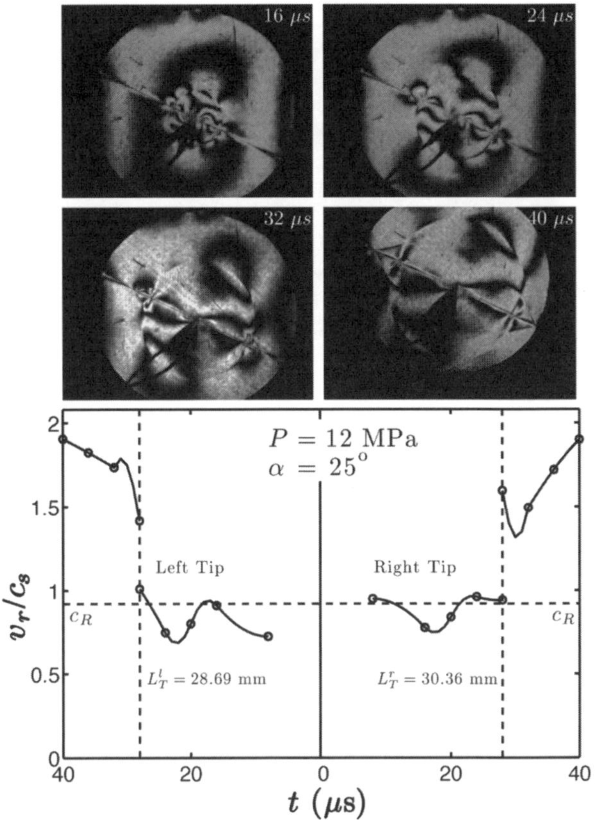

Figure 4

Dynamic rupture on the interface between two Homalite plates. The upper four panels show frames from the sequence of high-speed photographs taken at the times indicated. Note the butterfly-like stress concentrations at the rupture front in the frame taken at 24 μs. The shear stress concentration associated with the S wave generated at nucleation is most evident at 2 and 7 o'clock in the frames taken at 32 and 40 μs. The Mach cone produced by supershear propagation is most evident in the frame taken at 40 μs. The lower panel shows the instantaneous velocity of the rupture tips as a function of time. Note that propagation is symmetric. Both tips propagate near the Rayleigh velocity $c_R = 0.92c_S$ for about 29 μs before transitioning to supershear velocities approaching c_P.

oriented at +/− 45° to the loading axis, followed by immersion in liquid nitrogen for 45 seconds. The thermal shock produced a network of fractures with a mean spacing of about 1 cm as shown in Figure 5. Travel-time curves measured for both rupture fronts and for the S-wave are plotted in Figure 6 where they are compared with those for the undamaged Homalite. Note that rupture velocity and S-wave speed in the damaged plates are lower than those in the undamaged plates. More importantly, the ratio v_r/c_S is significantly smaller in the damaged material where it is less that the Rayleigh speed ($v_r < 0.92c_S^{DH}$, where c_S^{DH} is the shear speed in damaged Homalite). The implication is that the rupture velocity is further reduced by anelastic loss due to frictional sliding on the

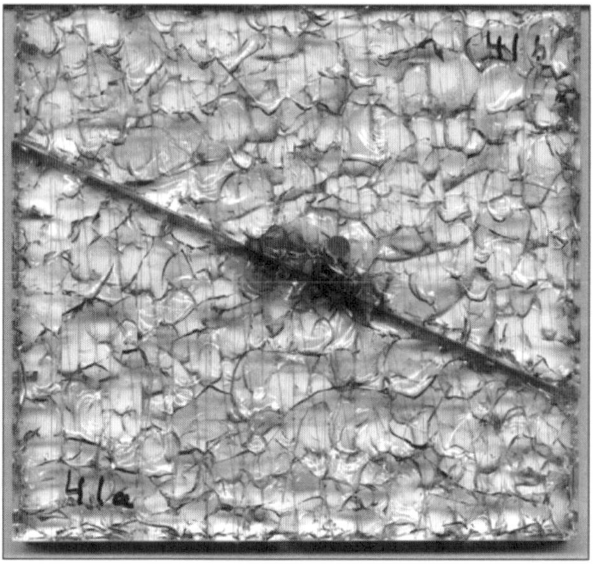

Figure 5
Fracture-damaged sample produced by first scoring both sides of the Homalite plates with a razor knife and then submersing in liquid nitrogen for 45 seconds. The center of the fault plane has been darkened by the explosion that nucleated rupture.

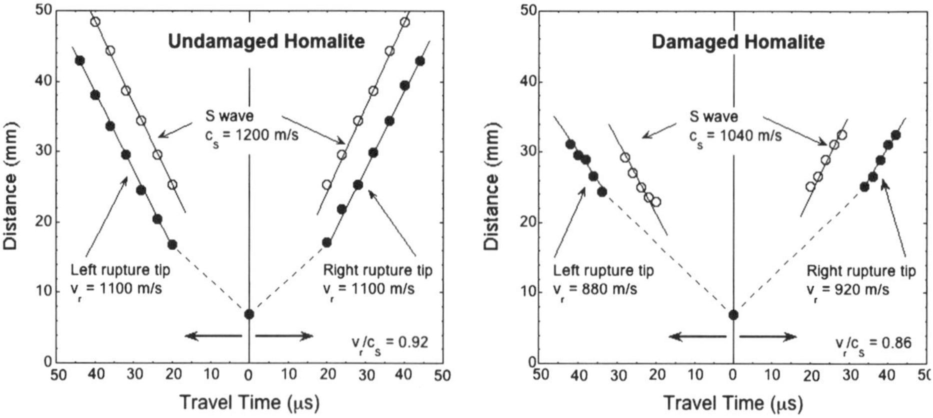

Figure 6
Travel time curves for a rupture propagating on the interface between two undamaged Homalite plates on the left, and between two fracture-damaged Homalite plates on the right. S-wave travel times are also plotted. $P = 12$ MPa and $\alpha = 25°$ in both tests. Note that rupture in undamaged Homalite did not transition to supershear as in Figure 4. Since the load was the same, the difference was probably due to the slightly rougher sliding surface in this case (see Rosakis *et al.*, 2008).

off-fault fractures. No additional off-fault damage was created by the rupture in these experiments.

4. *Effect of Asymmetric Off-Fault Damage on Rupture Velocity and Directionality*

BIEGEL *et al.* (2009) used the apparatus in Figure 2 to measure the velocity of ruptures on the interface between damaged and undamaged Homalite. As illustrated in Figure 7, the symmetry in these experiments is broken in two ways: elastically and anelastically. The elastic asymmetry is caused by the change in elastic modulus across the fault plane, and results in different propagation velocities in the '+' direction (the direction in which the lower velocity damaged Homalite moves) and in the opposite '−' direction. The physical cause of this elastic asymmetry is tension across the fault plane at the tip of the rupture propagating in the '+' direction (WEERTMAN, 1980; HARRIS and DAY, 1997; COCHARD and RICE, 2000; RANJITH and RICE, 2001; BEN-ZION, 2001; XIA *et al.*, 2005b; SHI and BEN-ZION, 2006; RUBIN and AMPUERO, 2007; AMPUERO and BEN-ZION, 2008).

Prior experimental studies in such elastic bi-materials by XIA *et al.* (2005b) found that ruptures in the '+' direction propagate at the generalized Rayleigh wave speed while those in the opposite '−' direction transition to super shear velocities that approach $P_{\text{slow}} \equiv c_P^{\text{slow}}$, the P-wave speed in the material having slower wave speeds. Theoretical studies have shown that, depending on the friction law, loading and nucleation conditions, a transition to supershear propagation is also possible in the '+' direction with velocity approaching $P_{\text{fast}} \equiv c_P^{\text{fast}}$, the P-wave speed in the material having faster wave

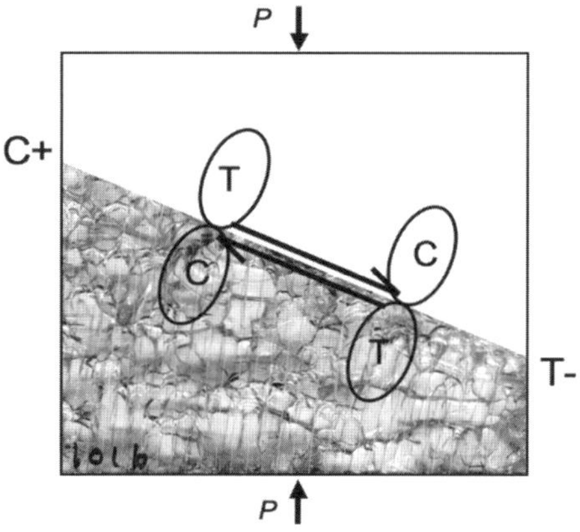

Figure 7
Asymmetries in an undamaged Homalite plate in contact with a damaged Homalite plate. Since the damaged Homalite is slightly less stiff than the undamaged Homalite, the '+' direction of propagation is to the left, which by convention is the direction of motion of the less stiff material. The anelastic asymmetry is denoted by the 'C' propagation direction for which the compressional lobe of the crack-tip stress concentration travels through the damage and the 'T' direction for which the crack tip places the damage in tension.

speeds (Cochard and Rice, 2000; Ranjith and Rice, 2001, Shi and Ben-Zion, 2006). Bhat *et al.* (2009) observed simultaneous supershear propagation in the '+' direction at P_{fast} and in the '−' direction at P_{slow}. Shi and Ben-Zion (2006) observed such simultaneous supershear propagation velocities in simulations with a nucleation procedure that imposed supershear rupture in the nucleation zone.

We hypothesize that the anelastic asymmetry arises because one fracture tip has the compressive lobe of its stress concentration in the damage (which we term the 'C' direction) while the other tip has its tensile lobe in the damage (which we term the 'T' direction). For example, in Figure 7 the rupture tip moving to the left is labeled 'C+' because it is moving in the '+' direction (the direction of motion of the more compliant damaged Homalite) and 'C' because the compressive stress concentration lies in the damage. Following this convention, the tip moving to the right is labeled 'T−'. We propose that the physical cause of the anelastic asymmetry is that tension in the 'T' direction enhances sliding on the off-fault cracks that comprise the damage while compression in the 'C' direction suppresses such sliding. This asymmetry is evident in Figure 8, which shows the velocities measured by Biegel *et al.* (2009) for a rupture on the interface between damaged and undamaged Homalite. Note that the rupture running in the 'C+' direction moves at P_{fast}, the expected velocity in the '+' direction for an elastic bi-material. Off-fault damage appears to have little additional effect in the 'C' direction, presumably because sliding is suppressed by the crack-tip compression. However, note that the rupture in the 'T−'direction stops. The interpretation is that tension enables energy dissipation by frictional sliding in the off-fault damage near the crack tip that completely suppresses propagation. The full set of rupture velocities measured by Biegel *et al.* (2009) on interfaces between Homalite and damaged Homalite is summarized in Figure 9 where they are compared with measurements at the same loads for the undamaged Homalite system. Note that ruptures in the 'C+' direction are only slightly affected by the damage while those in the 'T−' direction are severely slowed or even stopped, especially at the highest loads.

The rupture velocity in the all-damaged Homalite sample at $P = 12$ MPa (Figs. 5 and 6) is also plotted in Figure 9 where it is seen to be the same as the velocity in the 'T−' direction on the interface between Homalite and damaged Homalite (also at 12 MPa). This is the expected result based on our hypothesis that the compressive side of the rupture does not see the damage.

Bhat *et al.* (2009) extended this work by measuring rupture velocities on the interface between damaged Homalite and undamaged polycarbonate (which has a slightly lower shear-wave speed than does damaged Homalite). However, for comparison, they first measured the velocity of ruptures on the interface between undamaged Homalite and undamaged polycarbonate. The results of a typical experiment are shown in Figure 10 where four frames of the high-speed sequence (taken at the times indicated) are shown in the upper panel and the velocities are shown in the lower panel. Note that left rupture propagating in the '+' direction transitions to supershear velocities approaching P_{fast} while the right rupture propagating in the '−' direction approaches P_{slow}. This bimaterial

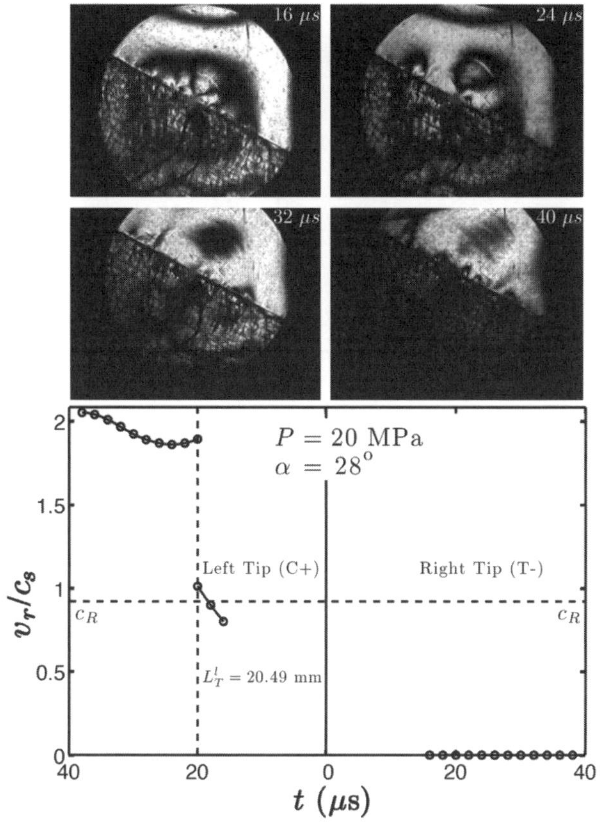

Figure 8
Dynamic rupture on the interface between damaged and undamaged Homalite plates. The upper four panels show frames at the times indicated selected from the sequence of high-speed photographs. The lower panel shows the instantaneous velocity of the rupture tips as a function of time. Note that the left tip propagating in the 'C+' direction transitions to supershear as in Figure 4 but that rupture propagation in the 'T−' direction is completely suppressed, presumably by energy loss on the off-fault damage activated by the tensile lobe of the stress field.

system was explored in more detail by Xia *et al.* (2005b), although they never observed supershear in both directions as in Figure 10, probably because the propagation distance to the supershear transition in their experiments was longer than the radius of the observable circle due to their rougher sliding surfaces (see Rosakis *et al.*, 2008 for a discussion of the supershear transition length).

The elastic and anelastic asymmetries that arise when damaged Homalite is in sliding contact with undamaged polycarbonate are illustrated in Figure 11. Since undamaged polycarbonate has a lower elastic stiffness than does damaged Homalite, propagation directions in these experiments are 'T+' to the left and 'C−' to the right (see Table 1 for a summary of the elastic properties of Homalite, damaged Homalite, and polycarbonate).

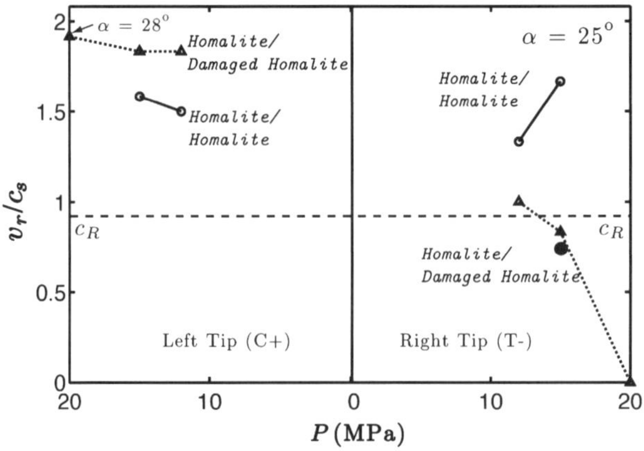

Figure 9
Summary of experimental results comparing rupture velocity in damaged bimaterial samples with those in undamaged Homalite plates. Note that the off-fault damage has little effect on ruptures traveling in the 'C+' direction but a major effect on those propagating in the 'T−' direction. Bimaterial ruptures which have the tensile lobe in the damage did not transition to supershear and, at large uniaxial loads, did not propagate at all. The solid circle is for damaged Homalite/damaged Homalite at $P = 12$ MPa from Figure 6 (from Biegel *et al.*, 2009).

Figure 12 shows the rupture velocities on the interface between damaged Homalite and polycarbonate for $P = 15$ MPa and $\alpha = 25°$. Note that in the 'T+' direction the rupture velocity approaches the generalized Rayleigh speed while in the 'C−' direction it slowly increases above the generalized Rayleigh speed, possibly toward P_{slow}, the P-wave speed in polycarbonate. The full set of rupture velocities measured by Bhat *et al.* (2009) on interfaces between damaged Homalite and polycarbonate is summarized in Figure 13 where they are compared with the measured velocity of ruptures at the same loads on the interface between undamaged Homalite and polycarbonate. As in the previous case of Homalite in contact with damaged Homalite, ruptures in the 'C' direction are only slightly affected by the damage. Like rupture velocities in the undamaged system, they appear to be increasing toward P_{slow}, but only at higher loads. Ruptures in the 'T' direction are more severely affected by the damage. The transition to supershear, which takes place in the undamaged system, is suppressed in the damaged system where ruptures propagated at the generalized Rayleigh speed for applied loads up to 15 MPa.

5. *Scaling Laboratory Studies to Natural Fault Zones*

Since the anelastic propagation asymmetry is caused by the interaction of the crack tip stress concentration with the off-fault damage, it is important to know the spatial extent of this crack tip field and how it scales from the ruptures in photoelastic polymer plates in the laboratory to earthquakes on natural faults in rock. The answer can be found

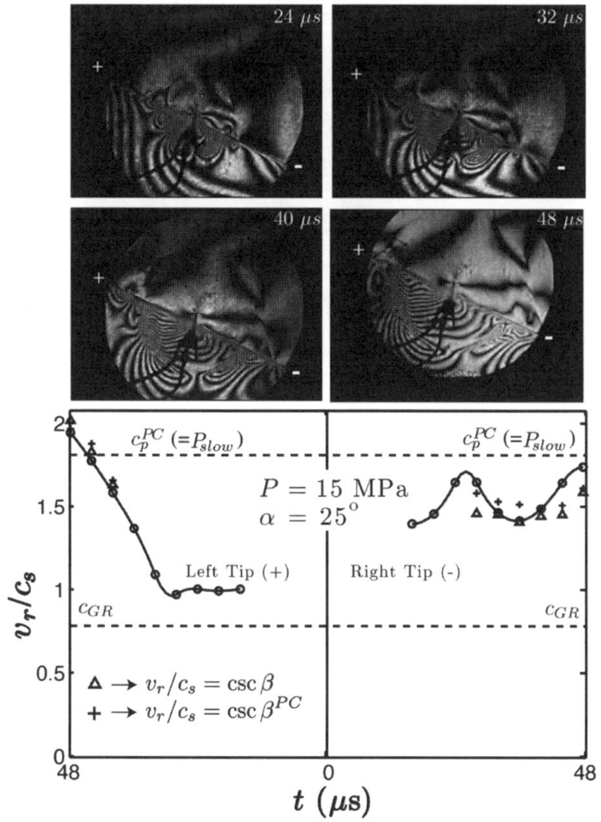

Figure 10

Dynamic rupture on the interface between undamaged Homalite and undamaged polycarbonate plates. The upper four panels show frames at the times indicated selected from the sequence of high-speed photographs. The lower panel shows the instantaneous velocity of the rupture tips as a function of time. Note that the left tip propagating in the '+' direction transitions to supershear velocities approaching P_{fast}. The right tip propagating in the '−' direction also transitions to supershear velocities approaching P_{slow}. The + and Δ symbols indicate propagation velocities determined for the angle β of the Mach cone in Homalite and in polycarbonate respectively (from BHAT *et al.*, 2009).

using an analytic model for a dynamic slip pulse developed by RICE *et al.* (2005) (hereafter referred to as RSP). The geometry and parameters of the model are illustrated in Figure 14. The spatial extent of the off-fault stress field depends on four nondimensional parameters: the ratio of components of the prestress $\sigma_{xx}^o/\sigma_{yy}^o$, the ratio of the rupture velocity to the shear-wave speed v_r/c_S, the ratio of the residual to peak stress τ_r/τ_p, and the length of the slip-weakening zone relative to the length of the slip pulse R/L. As illustrated in Figure 15, RSP found that the spatial extent of the stress concentration is larger for smaller values of $\sigma_{xx}^o/\sigma_{yy}^o$ (when the stress vector makes a

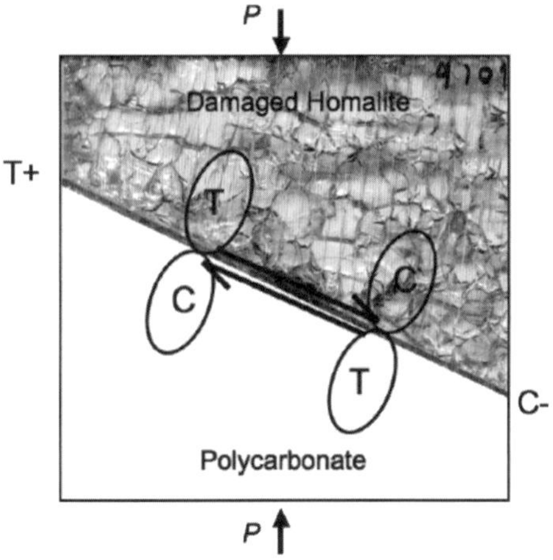

Figure 11
Asymmetries in an undamaged polycarbonate plate in contact with a damaged Homalite plate. Since the polycarbonate is slightly less stiff than the damaged Homalite, the '+' direction of propagation is to the left. The anelastic asymmetry is denoted by the 'C' propagation direction for which the compressional lobe of the crack tip stress concentration travels through the damaged Homalite and the 'T' direction for which the crack tip places the damage in tension. Note that the directions are now 'T+' and 'C−' and are different from the 'C+' and 'T−' directions for Homalite in contact with damaged Homalite in Figure 7.

Table 1

Elastic properties of sample materials

	c_P (m/s)	c_S (m/s)	ν
Homalite	2498[1]	1200[1]	0.35[1]
Damaged Homalite	2200[3]	1040[2]	0.25[3]
Polycarbonate	2182[1]	960[1]	0.38[1]

[1] Rosakis *et al.* (2008)
[2] Biegel *et al.* (2008)
[3] O'Connell and Budiansky (1974)

larger angle with the fault plane) and for larger values of v_r/c_S (where v_r is limited to sub-Rayleigh propagation). Note in this figure that all spatial variables are scaled by R_o^*, the length of the slip-weakening zone in the limit of $L \to \infty$and $v_r \to 0^+$. Note also in Figure 15 that the region of Coulomb failure extends to a distance of about R_o^* at high rupture velocities and when the pre-stress vector is at a large angle to the fault plane.

By fitting their model to Heaton's (1990) estimates of L, δ, and v_r for seven earthquakes, RSP were able to estimate the characteristic displacement δ_1, the dynamic stress drop $(\sigma_{yx}^o - \tau_r)$, the fracture energy G, and R_o^* for each earthquake evaluated at the centroid depth of the rupture (typically about 7 km). Although the seismic parameters estimated by Heaton may be crude, they yield fracture energies that are consistent with

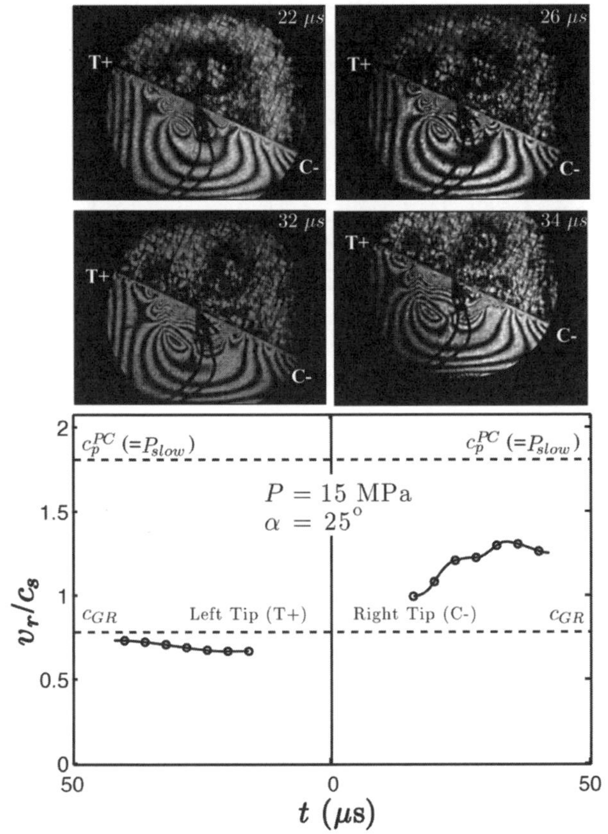

Figure 12

Dynamic rupture on the interface between damaged Homalite and undamaged polycarbonate plates. The upper four panels show frames at the times indicated selected from the sequence of high-speed photographs. The lower panel shows the instantaneous velocity of the rupture tips as a function of time. Note that the left tip propagating in the 'C−' direction transitions to supershear as in Figure 10 but that rupture propagation in the 'T+' direction does not transition to supershear, presumably due to energy loss on the off-fault damage activated by the tensile lobe of the stress field (from BHAT *et al.*, 2009).

other estimates both in magnitude and in their dependence on δ (see, e.g., RICE, 2006). RSP found that R_o^* is on the order of 1–40 meters, which is comparable to the width of natural fault zones at seismogenic depth inferred by LI and MALIN (2008), It should be noted that the width of natural fault zones at depth is still controversial and may be significantly narrower (BEN-ZION *et al.*, 2003; PENG *et al.*, 2003; LEWIS *et al.*, 2005; FINZI *et al.*, 2009). On the other hand, the spatial distribution of near-fault seismicity before and after large earthquakes has been interpreted to indicate a damage zone 10–100 meters wide at seismogenic depth (POWERS and JORDAN, 2009) or even wider (HAUKSSON, 2009).

We can similarly fit the RSP model to a slip pulse documented by LU *et al.* (2007) in a Homalite experiment shown in Figure 16. The propagation velocity of the pulse was

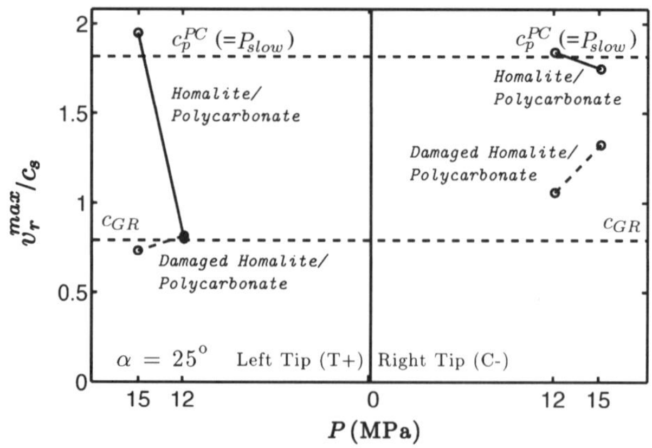

Figure 13
Summary of experimental results comparing rupture velocity in damaged and undamaged bimaterial plates. Note that the off-fault damage has little effect on ruptures traveling in the 'C−' direction but a significant effect on those propagating in the 'T+' direction. Bimaterial ruptures which have the tensile lobe in the damage did not transition to supershear (from BHAT *et al.*, 2009).

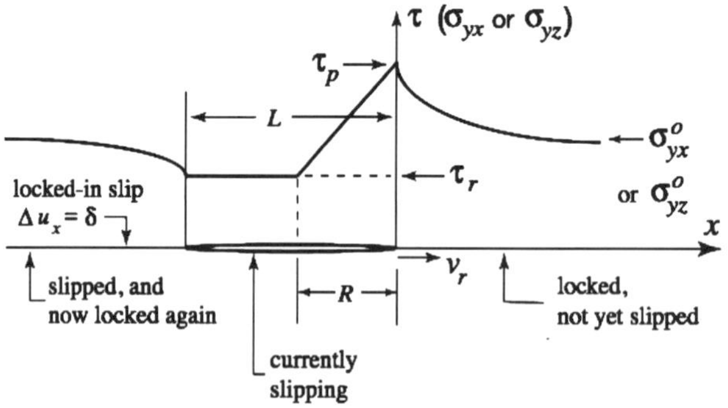

Figure 14
Geometry and parameters of the nonsingular slip-weakening slip-pulse model from RICE *et al.* (2005).

$v_r = 1045$ m/s. Its duration was about 10.2 μs with a rise-time of about 4 μs. This gives $L = 10.7$ mm, $R = 4.2$ mm and, assuming R/L is approximately the ratio of the rise time to the duration, $R/L \sim 0.4$. Integrating the area under the curve gives $\delta = 28$ μm. Since $c_S = 1200$ m/s for Homalite, we have $v_r/c_S = 0.87$. The density of Homalite is 1262 kg/m3 and the shear modulus is $\mu = 1.82$ GPa.

The fracture energy G can be calculated using the graph of $G/(\mu\delta^2/\pi L)$ vs v_r/c_S with R/L as a parameter given as Figure 15 in RSP. Using the values of v_r/c_S, R/L, μ, δ, and L from the previous paragraph gives $G = 21.2$ J/m^2. It is interesting that $G/\delta \approx 1$ MJ/m^3, close to the value estimated for earthquakes by RICE (2006).

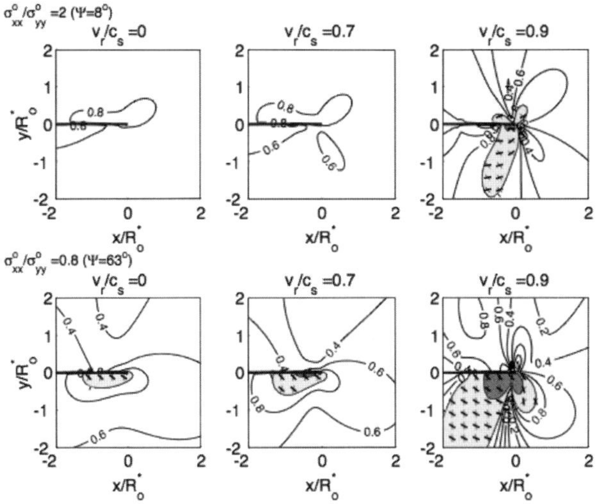

Figure 15
Contour plot of $\tau_{max}/\tau_{Coulomb}$ as a function of position surrounding the tip of a propagating slip pulse when ratio of the weakening distance to the slip length of the pulse is $R/L = 0.1$. Within the shaded areas $\tau_{max}/\sigma_{Coulomb} > 1$ and slip on pre-existing fractures is possible. The short heavy lines within the shaded areas indicate the orientation of optimal planes for right-lateral slip, the short lighter lines for left-lateral slip. The three columns of plots are for different values of the scaled propagation velocity $v_r/c_S = 0$, 0.7 and 0.9. In the top row of plots $\sigma^o_{xx}/\sigma^o_{yy} = 2$, in the bottom row $\sigma^o_{xx}/\sigma^o_{yy} = 0.8$. All calculations in this figure assume $\tau_r/\tau_p = 0.2$. From RICE *et al.* (2005).

This value of G can then be used to find the dynamic stress drop according to Eq. (15) in RSP: $(\sigma^o_{yx} - \tau_r) = G/\delta = 0.76$ MPa. For the applied uniaxial load of 10 MPa and the fault angle of 20°, the shear and normal stress on the fault plane are $\sigma^o_{yx} = 3.21$ MPa and $\sigma_n = 8.83$ MPa, respectively. The residual stress on the fault plane can be calculated as $\tau_r = \sigma^o_{yx} - (\sigma^o_{yx} - \tau_r) = 2.45$ MPa, which gives a dynamic coefficient of friction of $f_d = \tau_r/\sigma_n = 0.28$. The characteristic slip can then be found as $\delta_1 = \delta(\sigma^o_{yx} - \tau_r)/(\tau_p - \tau_r) = 7.5$ μm. Finally, the slip-weakening distance can be estimated using Eq. (13) in RSP, where we take $(\tau_p - \tau_r) = \sigma_n(0.6 - 0.28)$

$$R^*_o = \frac{9\pi}{16(1-\nu)} \frac{\mu G}{(\tau_p - \tau_r)^2} = 11\text{mm}. \tag{1}$$

BIEGEL *et al.* (2008) tested the prediction that the interaction between the crack tip stress field and the damage extends to a distance of about R^*_o. They fabricated a set of samples with a damage band of half width w surrounding the fault plane (Figure 17, left panel) and another set that was totally damaged except for a band of undamaged Homalite of half width w (Fig. 17, right panel). The resultant rupture velocities as a function of w for both damaged and undamaged bands are summarized in Figure 18. Note

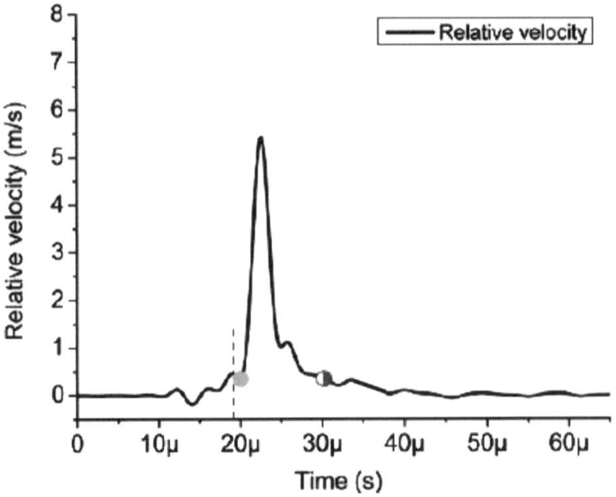

Figure 16
A narrow sub-Rayleigh pulse on the interface between two Homalite plates with $\alpha = 20°$ and $P = 10$ MPa showing the history of relative velocity across the fault plane recorded at 20 mm away from the nucleation point. The dashed line indicates the shear-wave arrival. The solid dot indicates the estimated initiation time of interface sliding. The half-filled dot indicates an estimate of interface locking time (from Lu *et al.*, 2007).

that all rupture velocities were sub-Rayleigh and that most of the change in velocity occurs over a distance $0 < w < 1$ cm, which is about the value of R_o^* estimated above.

6. *Discussion*

An asymmetric distribution of off-fault damage in a fault zone has been shown to produce strongly directional propagation on the fault plane. The physical source of this directionality is hypothesized to be the interaction between the crack-tip stress field and the off-fault fractures. Ruptures travel more slowly in the ‘T’ direction for which the tensile lobe of the stress concentration travels through the damage. By comparison, ruptures that travel in the ‘C’ direction for which the compressive lobe is in the damage propagate as if there were no off-fault damage. Our physical interpretation is that crack-tip compression immobilizes off-fault cracks while crack tip tension enhances frictional slip. It is surprising that the directionality observed in laboratory experiments is so strong (see summaries in Figs. 9 and 13), especially since there was no evidence that new damage was created. The lack of new damage in the experiments is not surprising since the scale-length of the existing damage is about 1 cm—the same size as Ro*. For real earthquakes, Rice *et al.* (2005) estimated Ro* at seismogenic depth to be on the order of meters. Earthquakes should be capable of generating new damage to distances of meters

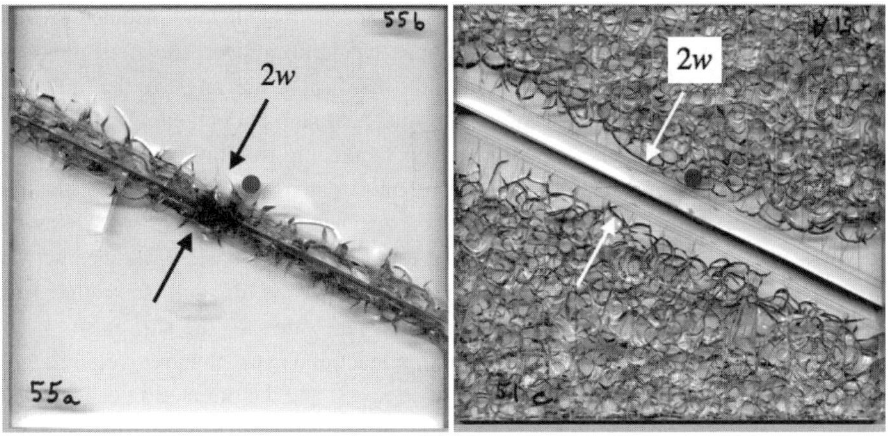

Figure 17
The Homalite plate on the left has a damage band of width $2w$ centered on the fault. The plate on the right has an undamaged band of width $2w$ centered on the fault.

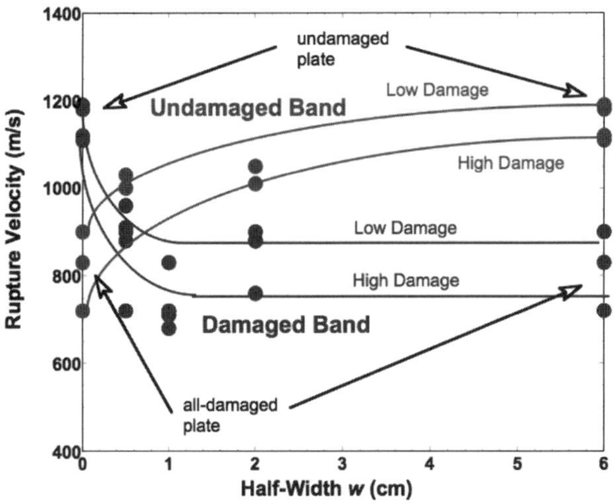

Figure 18
Rupture velocity as a function of w for both damaged and undamaged bands shown in Figure 17. Note that the majority of the decrease in velocity for the damage bands and increase in velocity for the undamaged bands occurs for bands with w less than about 1 cm (from BIEGEL *et al.*, 2008).

from the fault plane, and the generation of new damage is expected to make directional propagation even stronger.

We now return to the central question of how wide a fault zone must be if it is to affect the velocity and symmetry of an earthquake rupture. The simple answer is that the fault zone must be as wide or wider than Ro^*, the approximate spatial distance to which stress concentration at the rupture tip produces Coulomb slip on the most favorably oriented fractures in the case when the tectonic stress is at a high angle to the fault plane and the rupture velocity is near the Rayleigh limit. Although the preceding analysis used the RICE *et al.* (2005) slip pulse model with slip-weakening friction, other models based on crack-like ruptures and which represent off-fault damage using Coulomb plasticity also find that the nonlinear interaction extends a distance on the order of Ro^* (ANDREWS, 2005; BEN-ZION and SHI, 2005; TEMPLETON and RICE, 2008). The basic difference between slip-pulse and crack-like models is that the spatial extent of the off-fault interaction grows with the propagation of a crack, but reaches a constant value for the slip pulse.

In all these models, the spatial extent of the interaction is also quantitatively related to the ratio of the rupture velocity to the shear-wave speed and the orientation of the remote "tectonic" stress field relative to the fault plane. Additional factors which may affect the interaction between a rupture and off-fault damage but which are not discussed here include the effects of water in the fault zone (see, eg., RICE, 2006), the influence of the nucleation process on the ensuing rupture (SHI and BEN-ZION, 2006; AMPUERO and BEN-ZION, 2008; SHI *et al.*, 2008), and the case of supershear rupture. The experimental results reviewed here found that off-fault damage suppresses supershear propagation in all cases in which the tensile lobe of the rupture tip stress field encounters off-fault damage. This may be one reason why supershear earthquake ruptures are relatively rare.

Acknowledgements

The authors wish to thank Yehuda Ben-Zion and an anonymous reviewer for many helpful suggestions. The authors acknowledge support of the National Science Foundation collaborative grant EAR-0711171 to the University of Southern California and the California Institute of Technology. This research was supported by the Southern California Earthquake Center. SCEC is funded by NSF Cooperative Agreement EAR-0106924 and USGS Cooperative Agreement 02HQAG0008. The SCEC contribution number for this paper is 1227.

REFERENCES

AMPUERO, J.-P. and BEN-ZION, Y. (2008), *Cracks, pulses and macroscopic asymmetry of dynamic rupture on a bimaterial interface with velocity-weakening friction*, Geophys. J. Int. *173*, 674–692, doi: 10.1111/j.1365-246X.2008.03736.x.

ANDREWS, D.J. (2005), *Rupture dynamics with energy loss outside the slip zone*, J. Geophys. Res. *110*, B01307, doi:10.1029/2004JB003191.

BEN-ZION, Y. (2001), *Dynamic ruptures in recent models of earthquake faults*, J. Mech. Phys. Sol. *49*, 2209–2244.

BEN-ZION, Y. and SAMMIS, C.G. (2003), *Characterization of Fault Zones*, Pure Appl. Geophys. *160*(3), 677–715.

BEN-ZION, Y. and SHI, Z. (2005), *Dynamic rupture on a material interface with spontaneous generation of plastic strain in the bulk*, Earth Planet. Sci. Lett. *236*, 486–496, DOI, 10.1016/j.epsl.2005.03.025.

BEN-ZION, Y., PENG, Z., OKAYA, D., SEEBER, L., ARMBRUSTER, J.G., OZER, N., MICHAEL, A.J., BARIS, S., and AKTAR, M. (2003), *A shallow fault zone structure illuminated by trapped waves in the Karadere-Duzce branch of the North Anatolian Fault, western Turkey*, Geophys. J. Int. *152*, 699–717.

BHAT, H.S., BIEGEL, R.L., ROSAKIS, A.J., and SAMMIS, C.G. (2009), *The effect of asymmetric damage on dynamic shear rupture propagation II: With mismatch in bulk elasticity*, J. Geophs. Res., in press.

BIEGEL, R.L. and SAMMIS, C.G. (2004), *Relating fault mechanics to fault zone structure*, Adv. Geophys. *47*, 65–111.

BIEGEL, R.L., SAMMIS, C.G., and ROSAKIS, A.J. (2008), *An experimental study of the effect of off-fault damage on the velocity of a slip pulse*, J. Geophys. Res. *113*, B04302, doi:10.1029/2007JB005234.

BIEGEL, R.L., BHAT, H.S., SAMMIS, C.G., and ROSAKIS, A.J. (2009), *The effect of asymmetric damage on dynamic shear rupture propagation I: No mismatch in bulk elasticity*, J. Geophys. Res., in press.

COCHARD, A. and RICE, J.R. (2000), *Fault rupture between dissimilar materials- Ill-posedness, regularization, and slip-pulse response*, J. Geophy. Res. *105* (B11), 25,891–25,907.

DOR, O., BEN-ZION, Y., ROCKWELL, T.K., and BRUNE, J. (2006), *Pulverized rocks in the Mojave section of the San Andreas Fault Zone*, Earth Planet. Sci. Lett. *245*, 642–654.

FINZI, Y., HEARN, E.H., BEN-ZION, Y., and LYAKHOVSKY, V. (2009), *Structural properties and deformation patterns of evolving strike-slip faults: Numerical simulations incorporating damage rheology*, Pure Appl. Geophys., in press.

HARRIS, R.A. and DAY, S.M. (1997), *Effects of a low-velocity zone on a dynamic rupture*, Bull. Seismol. Soc. Am. *87*(5), 1267–1280.

HAUKSSON, E. (2008), *Spatial separation of large earthquakes, aftershocks, and background seismicity: Analyisi of interseismic and coseismic seismicity patterns in southern California*, Pure Appl. Geophys., in press.

HEATON, T.H. (1990), *Evidence for and implications of self-healing pulses of slip in earthquake rupture*, Phys. Earth Planet. Int. *64*(1), 1–20.

LEWIS, M.A., PENG, Z., BEN-ZION, Y., and VERNON, F. (2005), *Shallow seismic trapping structure in the San Jacinto fault zone*, Geophys. J. Int. *162*, 867–881, doi:10.1111/j.1365-246X.2005.02684.x.

LI, Y.G. and MALIN, P.E. (2008), *San Andreas Fault damage at SAFOD viewed with fault-guided waves*, Geophys. Res. Lett. *35*(8), L08304, doi:10.1029/2007GL032924.

LU, X., LAPUSTA, N., and ROSAKIS, A.J. (2007), *Pulse-like and crack-like ruptures in experiments mimicking crustal earthquakes*, Proc. Natl. Acad. Sci. USA *104*(48), 18,931–18,936, doi:10.1073/pnas.0704268104.

O'CONNELL, R.J. and BUDIANSKY, B. (1974), *Seismic velocities in dry and saturated cracked solids*, J. Geophys. Res. *79*(35), 5412–5426.

PENG, Z., BEN-ZION, Y., MICHAEL, A.J., and ZHU, L. (2003), *Quantitative analysis of seismic trapped waves in the rupture zone of the 1992 Landers, California earthquake: Evidence for a shallow trapping structure*, Geophys. J. Int. *155*, 1021–1041.

POWERS, P.M. and JORDAN, T.H. (2009), *Distribution of seismicity across strike-slip faults in California*, J. Geophys. Res., in press.

RANJITH, K. and RICE, J.R. (2001), *Slip dynamics at an interface between dissimilar materials*, J. Mech. Phys. Solids *49*, 341–361.

RICE, J.R. (2006), *Heating and weakening of faults during earthquake slip*, J. Geophys. Res. *111*, B05311, doi:10.1029/2005JB004006,

RICE, J.R., SAMMIS, C.G., and PARSONS, R. (2005), *Off-fault secondary failure induced by a dynamic slip pulse*, Bull. Seismol. Soc. Amer. *95*(1), 109–134.

ROSAKIS, A.J., XIA, K., LYKOTRATIS, G., and KANAMORI, H. (2008), *Dynamic shear rupture in frictional interfaces: Speeds, Directionality and Modes*. In *Treatise in Geophysics* (ed. H. KANAMORI), Vol. 4, 153–192.

RUBIN, A.M. and AMPUERO, J.-P. (2007), *Aftershock asymmetry on a bimaterial interface*, J. Geophys. Res. *112*, B05307, doi:10.1029/2006JB004337.

SHI, Z.Q. and BEN-ZION, Y. (2006), *Dynamic rupture on a bimaterial interface governed by slip-weakening friction*, Geophys. J. Int. *165*, 469–484.

SHI, Z.Q., BEN-ZION, Y., and NEEDLEMAN, A. (2008), *Properties of dynamic rupture and energy partition in a two-dimensional elastic solid with a frictional interface*, J. Mech. Phys. Solids *56*, 5–24, doi:10.1016/j.jmps.2007.04.006.

TEMPLETON, E.L. and RICE, J.R. (2008), *Off-fault plasticity and earthquake rupture dynamics, 1. Dry materials or neglect of fluid pressure changes*, J. Geophys. Res. *113*, B09306, doi:10.1029/2007JB005529.

WEERTMAN, J. (1980), *Unstable slippage across a fault that separates elastic media of different elastic constants*, J. Geophys. Res. *85*, 1455–1461.

XIA, K., ROSAKIS, A.J., and KANAMORI, H. (2005a), *Supershear and sub-Rayleigh to supershear transition observed in laboratory earthquake experiments*, Exp. Tech. *29*, 63–66.

XIA, K., ROSAKIS, A.J., KANAMORI, H., and RICE, J.R. (2005b), *Laboratory earthquakes along inhomogeneous faults: Directionality and supershear*, Science *308*, 681–684.

(Received October 6, 2008, accepted March 23, 2009)
Published Online First: July 29, 2009

To access this journal online:
www.birkhauser.ch/pageoph

Pure appl. geophys. 166 (2009) 1649–1667
0033–4553/09/101649–19
DOI 10.1007/s00024-009-0515-0

© Birkhäuser Verlag, Basel, 2009

Pure and Applied Geophysics

Geometry of the Nojima Fault at Nojima-Hirabayashi, Japan – I. A Simple Damage Structure Inferred from Borehole Core Permeability

David A. Lockner,[1] Hidemi Tanaka,[2] Hisao Ito,[3] Ryuji Ikeda,[4] Kentaro Omura,[5] and Hisanobu Naka[6]

Abstract—The 1995 Kobe (Hyogo-ken Nanbu) earthquake, M = 7.2, ruptured the Nojima fault in southwest Japan. We have studied core samples taken from two scientific drillholes that crossed the fault zone SW of the epicentral region on Awaji Island. The shallower hole, drilled by the Geological Survey of Japan (GSJ), was started 75 m to the SE of the surface trace of the Nojima fault and crossed the fault at a depth of 624 m. A deeper hole, drilled by the National Research Institute for Earth Science and Disaster Prevention (NIED) was started 302 m to the SE of the fault and crossed fault strands below a depth of 1140 m. We have measured strength and matrix permeability of core samples taken from these two drillholes. We find a strong correlation between permeability and proximity to the fault zone shear axes. The half-width of the high permeability zone (approximately 15 to 25 m) is in good agreement with the fault zone width inferred from trapped seismic wave analysis and other evidence. The fault zone core or shear axis contains clays with permeabilities of approximately 0.1 to 1 microdarcy at 50 MPa effective confining pressure (10 to 30 microdarcy at *in situ* pressures). Within a few meters of the fault zone core, the rock is highly fractured but has sustained little net shear. Matrix permeability of this zone is approximately 30 to 60 microdarcy at 50 MPa effective confining pressure (300 to 1000 microdarcy at *in situ* pressures). Outside this damage zone, matrix permeability drops below 0.01 microdarcy. The clay-rich core material has the lowest strength with a coefficient of friction of approximately 0.55. Shear strength increases with distance from the shear axis. These permeability and strength observations reveal a simple fault zone structure with a relatively weak fine-grained core surrounded by a damage zone of fractured rock. In this case, the damage zone will act as a high-permeability conduit for vertical and horizontal flow in the plane of the fault. The fine-grained core region, however, will impede fluid flow across the fault.

Key words: Nojima fault, fault structure, permeability, strength, Kobe earthquake.

[1] US Geological Survey, 345 Middlefield Rd MS977, Menlo Park, CA 94025, USA. E-mail: dlockner@usgs.gov
[2] Department of Earth and Planetary Sciences, University of Tokyo, Tokyo, Japan.
[3] Center for Deep Earth Exploration, Japan Agency for Marine-Earth Science and Technology, Yokohama, Japan.
[4] Department of Earth and Planetary Sciences, Hokkaido University, Hokkaido, Japan.
[5] National Research Institute for Earth Science and Disaster Prevention, Tsukuba, Japan.
[6] Ehime University, Matsuyama, Japan.

1. Introduction

The scientific drillholes crossing the Nojima fault at depth in the epicentral region of the 1995 Kobe earthquake, M = 7.2, provide a unique opportunity to study the mechanical and fluid transport properties of an active fault immediately after a major rupture event. Most first-hand evidence of the properties of active faults comes from examination of surface exposures that have typically undergone long and complicated histories of uplift and alteration. Examination of fault rock associated with rupture nucleation or significant energy release on strike-slip faults presents a particular problem since there is little vertical slip to bring deeper rocks to the surface. Seismic, gravity, electromagnetic and other remote sensing techniques can provide information about the deep structure of active faults, however a complete understanding of fault zone properties in the hypocentral regions of damaging earthquakes requires direct observation by the drilling of deep scientific boreholes. Recent examples of these include the boreholes crossing the Nojima fault, the Taiwan Continental Drilling Program TCDP (crossing at 1.1 km depth the Chelungpu fault that slipped in the 1999 Chi-Chi earthquake) (Ma *et al.*, 2006) and the San Andreas Fault Observatory at Depth SAFOD (crossing at 3 km depth the San Andreas fault near Parkfield, California (Hickman *et al.*, 2004).

Some recent studies of exhumed fault zones, especially mature zones that have sustained large total offset (Faulkner *et al.*, 2003; Wibberley and Shimamoto 2003) reveal broad, complex structures with multiple slip surfaces and highly variable permeability. Logging data of the SAFOD borehole show a similar complex structure with at least two active strands separated by broad zones of material with fluctuations in seismic velocity, resistivity and density (Hickman *et al.*, 2007). While the active traces of the San Andreas at SAFOD are centimeters in width, they are embedded in zones of heterogeneous material that has undergone shearing, rotation and mixing over the history of the fault.

At the other end of the spectrum of fault zone complexity are numerous laboratory and field observations of much simpler fault zone structures. Probably the simplest form, as suggested by studies of exhumed faults (Caine *et al.*, 1996; Caine and Forster, 1999; Chester *et al.*, 1993; Chester and Logan, 1986), consists of a narrow clay-rich or fine-grained core surrounded by a damage zone of highly fractured rock. The bulk of the shear deformation associated with the fault occurs in the narrow core zone, implying that this is the weakest portion of the fault structure. However, due to the fine grain size of the clay or crushed material in the fault core, it is expected to have a relatively low permeability. In addition to mesoscopic fractures, the surrounding damage zone contains a high density of microcracks, both within and between grains, although it is likely to have sustained relatively little total shear strain. Similar structures have been observed in laboratory samples that have been loaded to failure in shear (Moore and Lockner, 1995). If the fault occurs in low-porosity crystalline rock (such as the granodiorite of the Nojima fault), the high crack density in the damage zone results in significant pore volume increase and an increase in transport properties such as fluid permeability (Caine *et al.*, 1996; Caine and

FORSTER, 1999) and electrical conductivity (LOCKNER and BYERLEE, 1986). Furthermore, the fact that shear has been captured repeatedly by the narrow fault core implies that this material should have lower strength than the surrounding host rock and damage zone material. The contrast in material properties between damage zones associated with active faults and the surrounding country rock can often be observed at macroscopic to near crustal scales with remote geophysical techniques such as magnetotelluric and seismic profiling. Most recently, trapped seismic waves, in which the fault zone acts as a waveguide, have been observed in active faults (BEN-ZION *et al.*, 2003; LI and LEARY, 1990), and in particular on the Nojima fault (LI *et al.*, 1998).

The M7.2 Kobe earthquake nucleated beneath the Akashi Strait on January 17, 1995 (Fig. 1). The surface rupture on Awaji Island (to the SW of the epicenter) was about 10 km in length with 1 to 2 meters of lateral offset along the Nojima fault (LIN *et al.*, 1995; LIN and UDA, 1996; NAKATA *et al.*, 1995). Surface rupture progressively diminished for about 13 km to the SW of the epicenter (LIN and UDA, 1996). Two scientific drillholes, crossing the Nojima fault 7 km SW of the epicenter at Nojima-Hirabayashi, were initiated about one year after the earthquake by the Geological Survey of Japan (GSJ) and the National Research Institute for Earth Science and Disaster Prevention (NIED). The GSJ borehole site was 75 m to the SE of the fault trace and crossed the fault at an oblique angle at a depth of 624 meters (Fig. 2). Fault core material at this depth contained smectite and other weak clay but also significant amounts of fine-grained quartz, feldspar, calcite and relatively strong alteration minerals (MIZOGUCHI *et al.*, 2008; MOORE *et al.*, 2008; OHTANI *et al.*, 2000; TANAKA *et al.*, 2001, 2007). The NIED hole was started 302 m to the SE of the fault trace. This drillhole intersected two clay-rich shear zones at depths of 1140 m and 1312 m. A more diffuse shear zone was also crossed at about 1800 meters (see Fig. 2). These three zones are referred to in this paper as the 1140, 1312, and 1800 m NIED shear zones. Strength and permeability measurements reported here, as well as petrographic observations presented by (MOORE *et al.*, 2008), all indicate that unlike the other shear zone crossings, the 1800 m NIED shear zone was not activated by the 1995 Kobe earthquake. TANAKA *et al.* (2007) concluded that the main slip of the Kobe earthquake occurred at the 1140 m shear zone. In their interpretation, based on petrographic analysis, geochemistry and well-logging data, the Nojima Fault Zone (NFZ) spanned a minimum depth range in the NIED borehole from 1054.0 to 1189.6 m (Fig. 2).

In the present study, we report on laboratory tests of 22 core samples taken from the NIED and GSJ drillholes following the Kobe earthquake. Rock strength and matrix permeability measurements were carried out to provide fault zone properties for profiles across the fault at two depths. Due to the limited number of samples and the difficulties associated with sample preparation, this study was intended to be exploratory in nature. In fact, the permeability and strength measurements are in excellent agreement with the idealized fault structural model described above. Additional data regarding petrography and microcrack characteristics are provided in a companion paper (MOORE *et al.*, 2008).

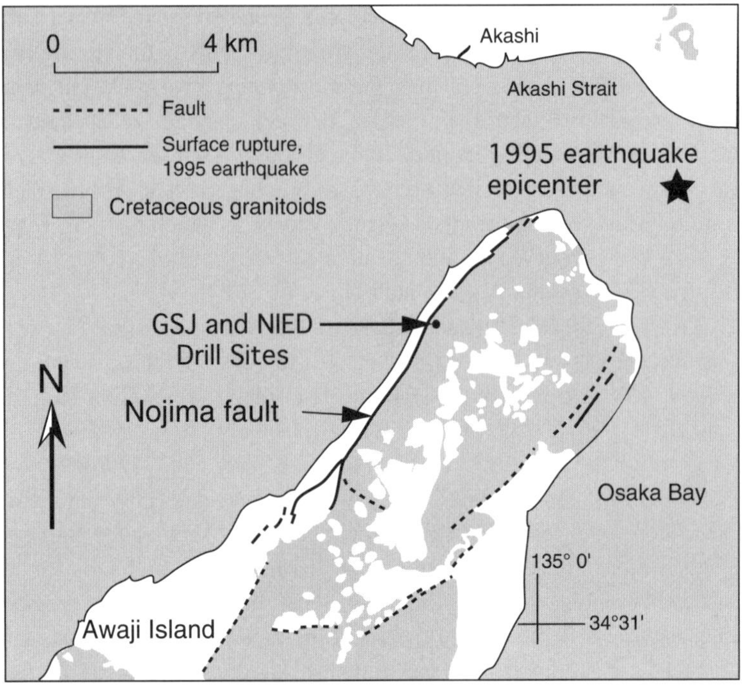

Figure 1
Map detailing 1995 Kobe earthquake epicenter (star) and surface rupture reported on Awaji Island. Rupture was on the Nojima fault following the NW shore of the island. GSJ and NIED drill sites were about 7 km from the epicenter near Nojima-Hirabayashi. The rupture also extended NE from the epicenter past Kobe (not shown on map).

2. *Experimental Technique*

Protolith rock for the Nojima fault in the vicinity of the boreholes is a biotite-hornblende granodiorite (Ohtani *et al.*, 2000; Tanaka *et al.*, 2001). A set of 22 samples was selected from the NIED and GSJ boreholes. Due to the limited number, sample selection was concentrated about the obvious shear zones in an attempt to obtain representative profiles across the fault zone structures. Depth intervals where samples were taken are indicated in Figure 2. Significant variability of rock type exists in the shear zones and a more complete evaluation of fault zone properties would require a much denser and more systematic sampling of the drill core than was possible in this study. Preferred sample dimensions were nominally 25.4 mm-diameter by 50.8 mm-length cylinders. Much of the recovered drill core (especially in the damage zones) had little or no cohesion and contained hard grain fragments. As a result, many samples were impossible to prepare to these preferred dimensions. If samples could not be prepared as cylinders an alternative shape in the form of rectangular prisms with 18.0 mm-square-cross section was attempted (Fig. 3). Most of the samples tested were of this type. These

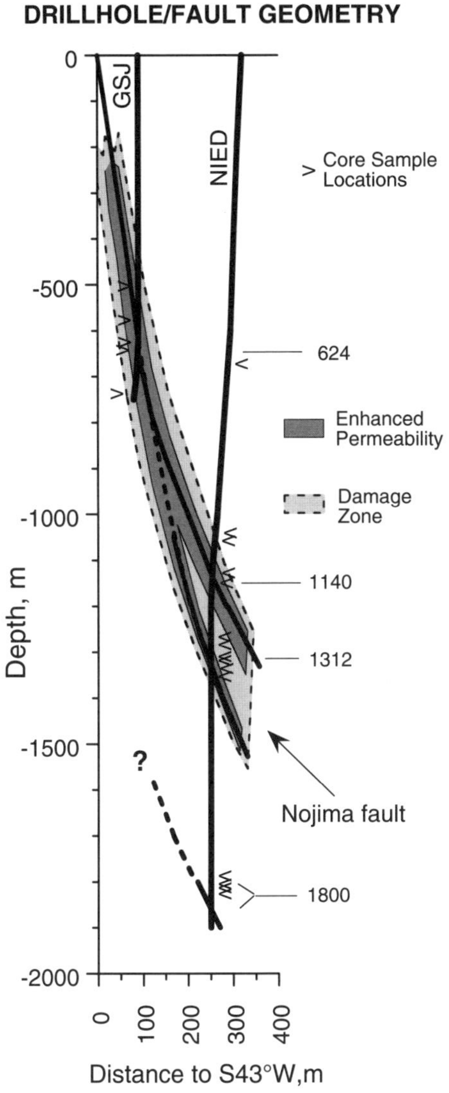

Figure 2

Schematic diagram of the Nojima fault (modified from TANAKA *et al.*, 2007) in the vicinity of the GSJ and NIED drillholes, showing the depths at which fault zones were intersected. The trend of the fault is N35E and, at the drill sites dips 84° to SE. A single fault trace was crossed by the GSJ drillhole while two activated strands were crossed by the NIED drillhole. A deeper fault zone was intersected at about 1800 m, but analysis of core samples indicates that this is a relic fault that was not reactivated by the 1995 Kobe earthquake. Sample depths tested in this study are indicated by '>'. A broad damage zone, shown in light gray, (~70 m wide) is inferred from microcracks and other petrographic evidence reported by TANAKA *et al.* (2007). The narrower damage envelopes (dark gray) indicate zones of enhanced permeability observed in this study (see Discussion).

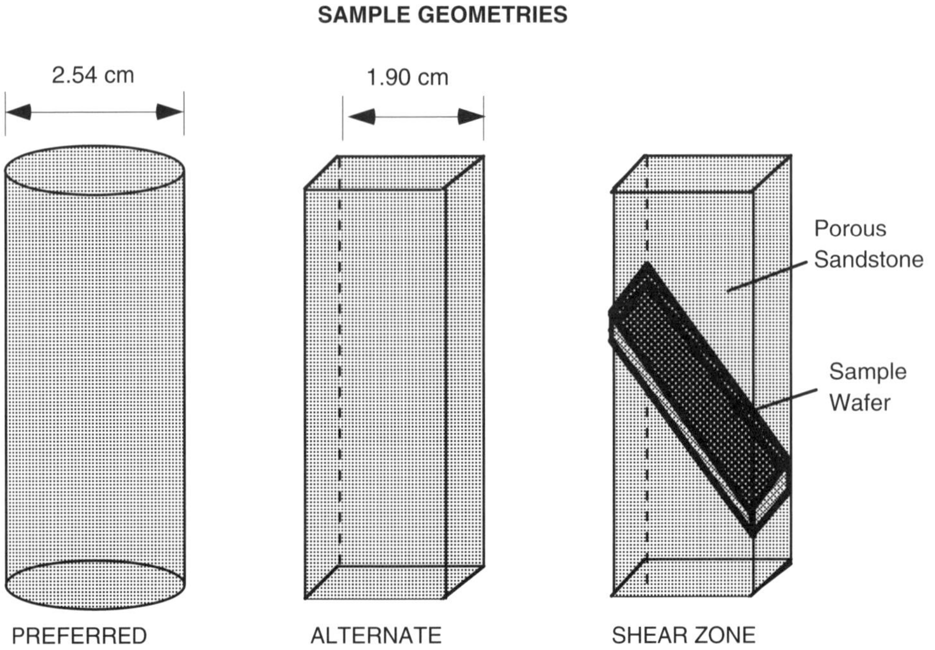

Figure 3
Diagram showing the three sample shapes used in this study. Whenever possible, right-cylinders were prepared. Most of the damage zone samples had little cohesion and could not be cored. In this case, rectangular prisms were cut from the drill core material. Only limited amounts of fault core material were available. We were able to cut wafers of this material (thicknesses of 1 to 2 cm) that were sheared between porous sandstone driving blocks with saw cut surfaces inclined by 30°.

intact samples were prepared so that their long axis was subparallel to the drill hole axis and therefore subvertical. The clay-rich shear zone core samples were available in limited quantities and sliced to provide wafers with a thickness in the 1 to 2 cm range. These were placed between porous sandstone driving blocks cut at an angle of 30° to the sample column axis.

Samples were jacketed and loaded in a standard triaxial deformation apparatus. Each sample was evacuated and then saturated with distilled, deionized water. Permeability was measured using a constant flow rate method at effective confining pressure of 10, 30, and 50 MPa. Test samples were oriented parallel to the original borehole cores so that the reported permeability and strength data are essentially for flow and maximum compressive stress in a subvertical orientation. No attempt was made to determine anisotropy in either permeability or strength. Following the hydrostatic permeability measurements, each sample was loaded incrementally at constant effective confining pressure of 50 MPa until failure occurred. Following each strain increment along the loading path, deformation was halted and permeability was determined as indicated in Figure 4. After peak stress, deformation continued to 5 mm axial shortening at a rate of 0.48 μm/sec with periodic pauses to repeat permeability measurements.

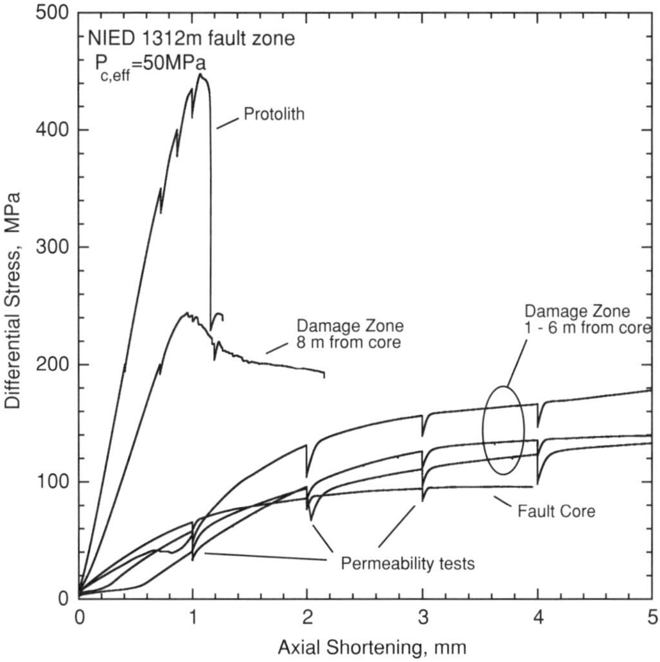

Figure 4

Differential stress plotted as a function of axial shortening for representative samples from NIED 1312 m fault zone. Periodic stress drops indicate times when deformation was stopped to measure permeability. The strongest sample is undamaged granodiorite protolith sampled outside the fault zone. The weakest sample is fault core material. Intermediate strength samples are taken from the damage zone within 8-m horizontal distance from the fault core. All deformation tests were run under drained conditions (saturated with deionized water) at constant confining pressure of 50 MPa and axial shortening rate of 0.48 μm/sec.

Accuracy of the permeability measurements is difficult to determine since some samples had less than ideal surface finishes and some permeabilities were near either the high or low limits of our measuring capabilities; sample permeabilities ranged from less than 1×10^{-9} to 3×10^{-3} Darcy (1 Darcy $= 10^{-12}$ m^2). The lower limit for our measuring system was 0.3 nDa. Absolute permeability uncertainties reported here are less than $\pm$ 25% while relative uncertainties in a single experiment are 5 to 10%.

3. *Results*

3.1. *Rock Strength*

Care was taken during sample preparation to preserve the intrinsic grain structure and cohesion to the extent possible. Obvious changes did occur in the samples due to drying and decompression of the original drill cores. However, these changes were minimized as much as possible with the expectation that the *in situ* rock properties would be preserved to large extent. Thin section observation (Moore *et al.*, 2008) revealed that while damage zone samples contained large microcrack densities, the relative position of grains remained, for the most part, intact. Therefore, any additional damage sustained during core retrieval did not result in disaggregation and loss of material. In fact, percentage of core recovery was excellent. One way to evaluate the extent of microcrack damage is to look for trends in the peak strength data for the various samples. Samples were tested at 50 MPa effective confining pressure, which is equivalent to approximately 3 km burial depth (assuming hydrostatic pore pressure gradient and average rock density of 2.7 g/cm^3), to provide an easy means of comparing sample to sample strength. A suite of stress-displacement plots for the 1312 m NIED fault trace is plotted in Figure 4. Periodic drops in strength indicate the times when deformation was stopped to make permeability measurements. In all cases, resumption of deformation resulted in strength recovery back to the original loading curve. Also included for comparison is the strongest sample, taken from a depth of 659.3 meters, well above the fault and representing strength of the undamaged protolith. The weakest sample is clay-rich fault core material taken from 1312.0 meters. Four other lines plotted in Figure 4 show strengths of samples taken from the damage zone. The sample taken 8 meters (horizontal distance) from the fault core has sustained significant damage (strength is about half of the protolith strength) and passes through a peak in strength before dropping to residual frictional strength. The initial loading slope of this sample is less than the loading slope for the protolith material, indicating that damage has reduced the elastic modulus. Other damage zone samples taken closer to the fault core undergo strain hardening without passing through a peak in strength. These samples have sustained enough damage (either *in situ* or during sample retrieval) to eliminate most of their cohesive strength. In this case, final strength is frictional sliding strength and is correlated to the relative amount of clay and other weak alteration minerals within each sample.

Peak strength data for the four shear zone crossings in the GSJ and NIED boreholes are presented in Figure 5 as the ratio of shear to normal stress (μ) resolved on a plane inclined 30° to the sample axis. The actual inclination of the experimentally induced fracture plane was difficult to determine in many of the strength tests, and 30° provides a satisfactory approximation for comparison of different samples. Calculated this way, normalized strength for the intact granodiorite protolith is $\mu_{\text{intact}} = 1.18$. (Note that this value is greater than what is commonly referred to as coefficient of internal friction, defined as the tangent to the Mohr-Coulomb failure envelope.) Peak differential stress, σ_Δ, which is the quantity measured directly in the experiments, can be recovered from the data in Figure 5 by the formula $\sigma_\Delta = 200\mu/(1.732\text{-}\mu)$ (expressed in MPa).

The horizontal axis in Figure 5 is the horizontal distance between the center of the fault zone and the *in situ* location of each core sample. Given the fault/borehole geometry shown in Figure 2, this horizontal distance is 5 to 10 percent greater than the perpendicular distance to the fault surface. No distinct fault center could be identified for the 1800 m NIED fault zone thus the horizontal axis in Figure 5d could be shifted by a

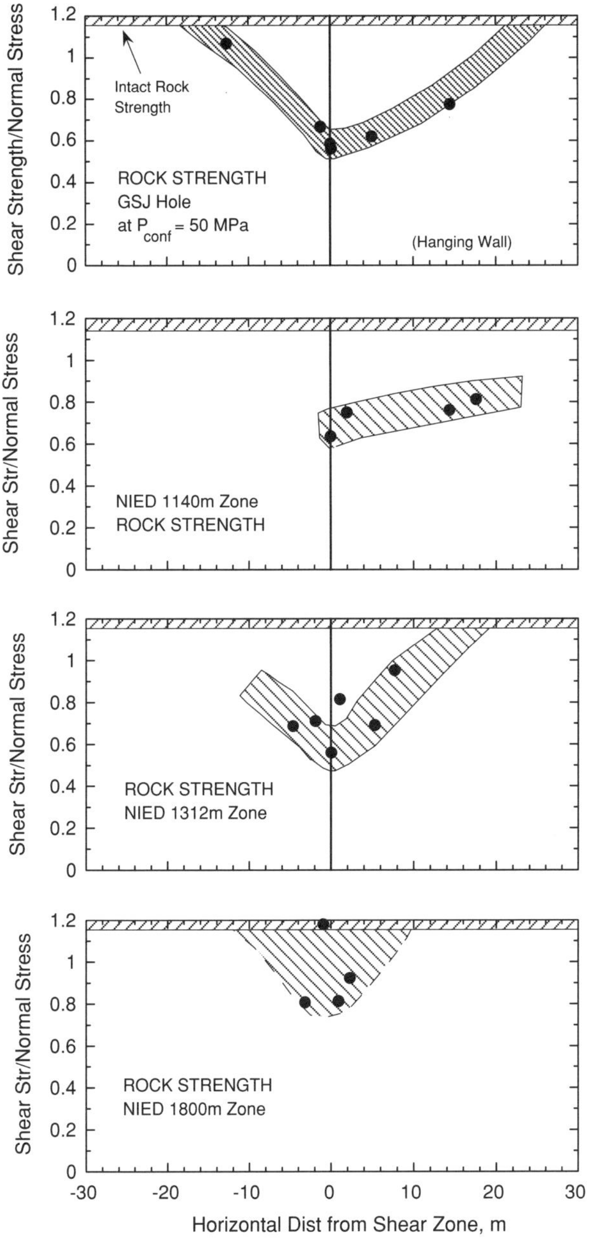

Figure 5

Profiles of peak shear strength divided by effective normal stress for the four fault zone crossings in the GSJ and NIED boreholes. Stresses are calculated for surfaces inclined 30° to maximum compressive stress direction (see text). Strength for the three active fault crossings is lowest in the fault zone core material and generally increases with distance from the core. One of the samples from the relic fault zone at 1800 m has strength similar to the protolith. The other thee deep samples have modest cohesive strength (peak strength is 5 to 20 percent higher than frictional sliding strength).

few meters. Notice that the three fault crossings with well-defined fault zone cores (Figs. 5a, 5b and 5c) all show a minimum in strength at the fault zone axis. This is consistent with the model described above in which nearly all displacement concentrates in the narrow fault core (a zone that is only a few centimeters wide in all three examples; Tanaka *et al.*, 2007). The general trend in these three examples is for strength to gradually increase with distance away from the fault zone axis. This gradual strength increase corresponds to a decrease in microcrack damage (*i.e.*, Moore and Lockner, 1995; Moore *et al.*, 2008). Damage zone half-width as suggested by the strength data in Figures 5a–5c is 15 to 25 m. While the fault core material was the weakest in all three examples, the minimum frictional sliding strength ($\mu \sim 0.55$) is significantly stronger than for smectite or other weak phyllosilicates (μ of 0.1 to 0.3) (Moore and Lockner, 2004; 2007). Tembe *et al.* (2008) have recently reported on frictional strength of quartz/smectite/illite mixtures where they found good correlation between strength of synthetic mixtures and strength of natural fault gouges. Frictional strength of 0.55 to 0.60 would imply, based on the Tembe *et al.* (2008) analysis, a dry weight clay fraction of 20 to 35 percent for the Nojima fault core material. Thus, while weak clay minerals are present in the Nojima core samples, they apparently do not dominate the fault core composition.

The 1800 m NIED shear zone (Fig. 5d) is different from the other three examples. No distinct shear zone core was observed in the borehole samples. While the three shallower fault core zones showed strength significantly less than the protolith strength (μ in the range 0.55 to 0.62 as compared to protolith strength of 1.18), rock strength from the 1800 m zone is significantly greater ($\mu \sim 0.8$). Apparently this deep zone was not activated in the Kobe earthquake and has, over time, been restrengthening by vein filling and mineral alteration (Moore *et al.*, 2008).

3.2. Rock Matrix Permeability

All permeability measurements are plotted as a function of axial stress in Figure 6. The first three measurements of each test at 10, 30, and 50 MPa were conducted without deviatoric stress and represent the permeability loss due to an increase in effective confining pressure. Permeability typically dropped by two to three orders of magnitude as microcracks closed in response to this hydrostatic loading. The remainder of each curve at axial stress above 50 MPa in Figure 6 represents the deformation tests at 50 MPa effective confining pressure. Initially, all samples showed loss of permeability with increasing deviatoric stress. By approximately 50% peak stress, the more competent samples began to exhibit increasing permeability with increasing deviatoric stress. This reversal from permeability decrease to permeability increase corresponds to the onset of dilatancy and the opening of microcracks within the sample (see, for example, Zoback and Byerlee, 1975 and similar trends in electrical conductivity, Lockner and Byerlee, 1986). The clay-rich fault core samples did not show dilatant behavior. Instead, they showed a steady loss of permeability with continued shearing and strain hardening. This behavior is similar to shear-enhanced compaction in porous sandstone as discussed by

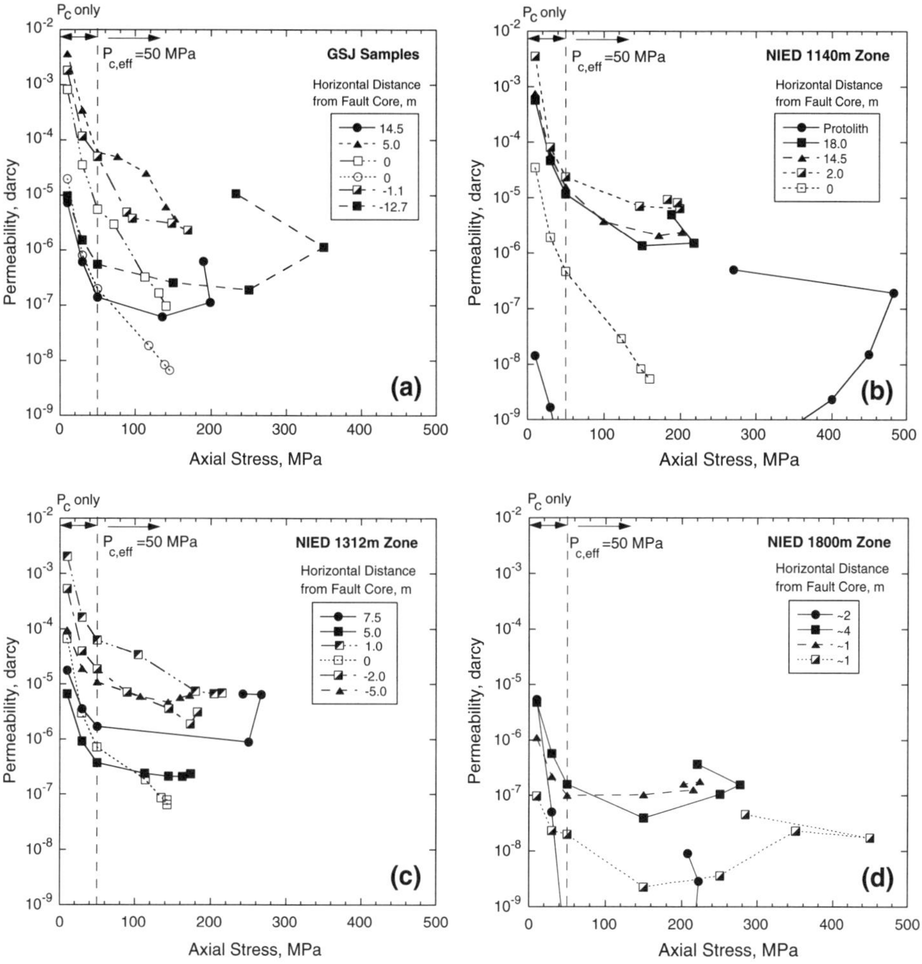

Figure 6

Permeability plotted versus axial load for GSJ and NIED samples. Pressurization to 50 MPa is hydrostatic loading without deviatoric stress. Permeability decreased by 2 to 3 orders of magnitude in response to application of 50 MPa confining pressure. Data for axial stress above 50 MPa show permeability change in response to sample deformation at 50 MPa effective confining pressure. All samples show an initial decrease in permeability with application of deviatoric stress. More competent samples sustained opening of microcracks at higher stresses and a corresponding increase in permeability. Some samples failed by forming through-going fractures. In these cases, plots show a drop in strength for the final permeability measurement that measured flow across the newly formed fault surface. Weaker samples (especially fault core samples) sustained continued grain crushing and compaction so that permeability steadily decreased with deformation.

WONG and BAUD (1999). The protolith granodiorite shown as the low permeability curve, Figure 6b, is a dense crystalline rock with low matrix permeability. In this case, application of deviatoric stress reduced the permeability below the measurement limit of our test system. By 400 MPa axial load, however, new dilatant cracks had increased

Figure 7 ▶

Profiles of matrix permeability measured at 50 MPa effective confining pressure. The three upper fault crossings show a low permeability fault core (fine-grained material containing some clay fraction) surrounded by high permeability damage zones (interlocked grains with numerous open microfractures). The deep shear zone is partially sealed and was apparently not activated by the Kobe earthquake.

permeability to a measurable level, which continued to increase to peak axial load of 483 MPa. Another trend observable in Figure 6 is a systematic decrease in peak strength of samples from the intact protolith, through the damage zone samples to the fault core samples. The low strength of the fault core is consistent with the concentration of shear deformation in this zone.

We have already noted that the permeability of all samples shows a strong sensitivity to increases in effective confining pressure. All of the examples plotted in Figure 6 show a decrease of two to three orders of magnitude for a 50 MPa pressure increase. This pressure sensitivity is indicative of rocks in which flow is through low aspect ratio cracks (long, thin microcracks) that can open or close flow paths in response to small pressure changes (Morrow *et al.*, 1994). This type of microcrack is more susceptible to changes in permeability resulting from vein filling than the more equant pores that might be found, for example, in a porous sandstone.

Because of the strong pressure sensitivity of permeability in these samples, we expect that *in situ* permeability in the Nojima fault zone will decrease rapidly with depth. For example, the GSJ samples, with a fault crossing depth of 624 m would have an effective overburden pressure of approximately 11 MPa while the 1312 m NIED crossing overburden pressure would be about 22 MPa. The maxima in the damage zone permeabilities adjusted for these effective pressures would then be 3.3 mDa and 0.45 mDa, respectively.

To provide a more direct comparison of the different fault zone permeability measurements, we have plotted matrix permeability profiles in Figure 7 using experimental values from 50 MPa effective confining pressure, equivalent to 3 km burial depth. *In situ* permeabilities for the GSJ hole (at 11 MPa) would be about 50 times larger than the values shown in Figure 7. As with the strength data, the three shallow crossings show permeability profiles similar to the idealized fault zone model described above. The fault zone axis, due to its clay content and fine gouge grain size, has relatively low permeability in the microdarcy range. Flanking the fault zone axis, the highly fractured damage zone has permeabilities in the 50 to 100 microdarcy range. Distantly from the fault zone axis, grain damage and permeability decrease until permeability returns to the protolith permeability value. The NIED borehole deep shear zone has permeability that is greater than the protolith permeability but significantly less than the shallower shear zone values. Thus both strength and permeability data suggest that the deep fault zone was not activated in the Kobe earthquake and is undergoing sealing and restrengthening.

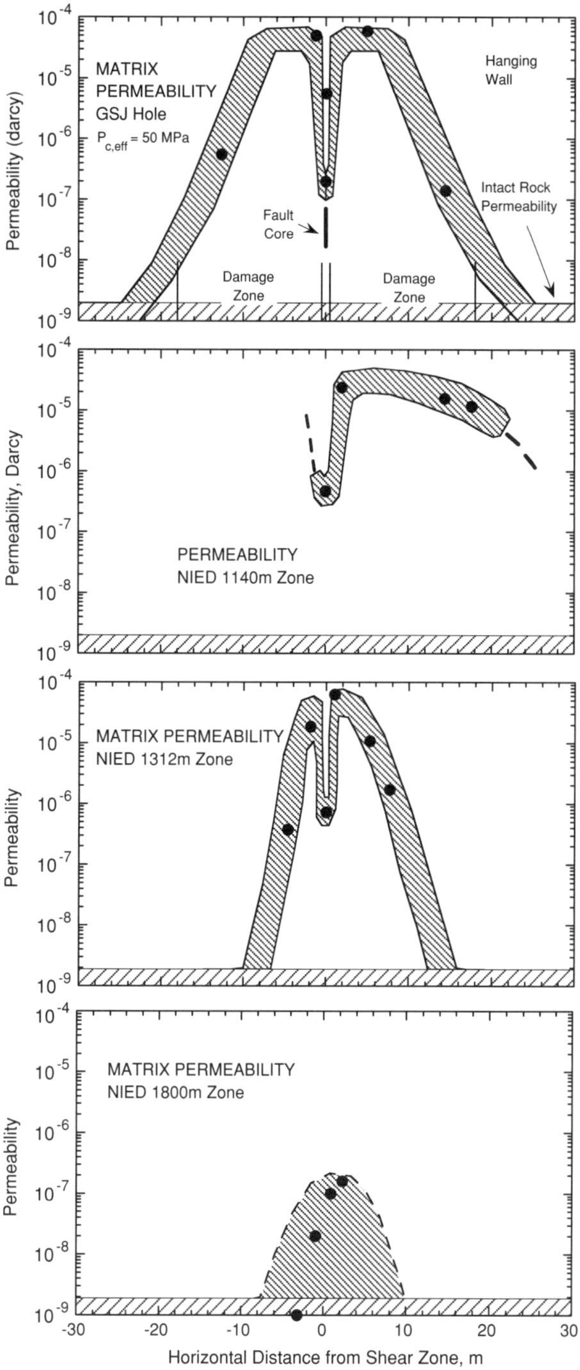
MATRIX
PERMEABILITY
GSJ Hole
$P_{c,eff}$ = 50 MPa
Hanging
Wall
Fault
Core
Intact Rock
Permeability
Damage
Zone
Damage
Zone
Permeability (darcy)
PERMEABILITY
NIED 1140m Zone
Permeability, Darcy
MATRIX PERMEABILITY
NIED 1312m Zone
Permeability
MATRIX PERMEABILITY
NIED 1800m Zone
Permeability
Horizontal Distance from Shear Zone, m

4. Discussion

Our measurements of core sample strength and permeability are in good agreement with the idealized fault zone model described, for example, by CAINE *et al.* (1996); CHESTER and LOGAN (1986) and EVANS *et al.* (1997). We observe a thin, low-strength, low-permeability fault zone core flanked by zones of high permeability rock that have undergone relatively limited total shear. These observations imply that to depths of as much as 3 to 5 km, the postseismic fault zone will act as a high permeability fluid conduit for fluid flow in the plane of the fault. Because the fault core has low permeability it is likely to act as a barrier to fluid flow across the fault. However, this barrier is notably thin and may have a complex structure. For example, the Nojima fault zone appears to have branched into two shear zones between the GSJ and NIED borehole crossings. To the degree that this thin shear zone core is spatially discontinuous and anastomosed, its ability to act as a fluid barrier will be diminished. It is difficult to determine this three-dimensional structure at depth from a limited number of (typically one or two) borehole crossings.

TANAKA *et al.* (2007) concluded that the principal slip surface for the Kobe earthquake was crossed at 1140 m in the NIED borehole. Since we find reduced strength and enhanced permeability for both the 1140 and 1312 zones, it is likely that both of these fault segments were activated in the Kobe earthquake. In this case, the 1140 and 1312 zones could represent bounding faults of a fault zone that is about 130 m wide. Unfortunately, we did not sample the intervening depth interval in this reconnaissance study to determine transport or mechanical properties. Figure 2 has been modified from the original version in TANAKA *et al.* (2007) to highlight the possible damage zone surrounding the principal slip surfaces. Our measurements do suggest a more concentrated high-permeability zone associated with each slip surface embedded in a broader damage zone as suggested by the petrographic analysis of TANAKA *et al.* (2001, 2007). The damage zone half-width, as indicated by the permeability profiles ranges from 10 to 25 m. Based on trapped seismic wave analysis, LI *et al.* (1998) found a half width of approximately 15 m for the Nojima fault. Thus, direct measurements of both the permeability and strength of the fault zone indicate a damage zone half width that is consistent with the deeper seismic observations. Indeed, a similar geometry has been observed for the San Andreas Fault near Parkfield, California. The SAFOD drillhole encountered a broad damage zone containing two actively creeping strands (one of which appears to be more active) less than 100 m apart (HICKMAN *et al.*, 2005).

The permeability data shown in Figure 7 are all measured at 50 MPa effective confining pressure to allow for a more direct sample-to-sample comparison. All samples showed a strong sensitivity of permeability to confining pressure. Pressure sensitivity can be approximated by an exponential relation of the form

$$k = k_0 \exp\left(-P_{c,eff}/P^*\right), \tag{1}$$

where k_0 is permeability at zero pressure and P^* is a constant (see, for example, RICE, 1992). Permeability data plotted in Figure 6 at 10, 30 and 50 MPa effective pressure were

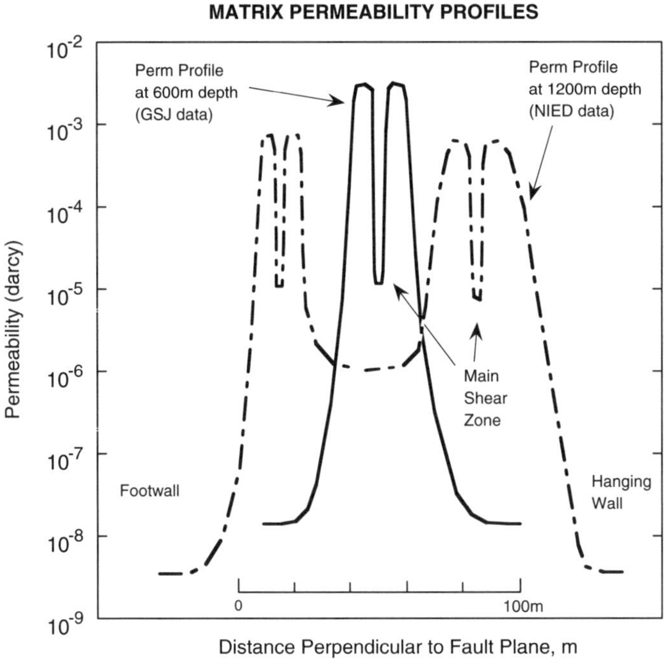

Figure 8

Interpretive profiles of permeability for Nojima fault at 600 and 1200 m depth. Permeabilities from data shown in Figure 6 have been adjusted for overburden stress using eq. (1). Additional information regarding damage zone width comes from Tanaka *et al.* (2007). Data from 1140 shear zone (Fig. 7b) and 1312 shear zone (Fig. 7c) have been combined to form the 1200 m profile. Distance is measured perpendicular to the fault plane (22° dip).

fit to (1) for each sample. All P^* values were similar and can be summarized by $P^* = 10.5 \pm 2.4$ MPa. Figure 8 shows two interpretive profiles of matrix permeability of the Nojima fault at Nojima-Hirabayashi. These profiles show permeability at depths of 600 and 1200 m for distance perpendicular to the fault plane. Effective confining pressure data from Figure 6 have been used with eq. (1) to adjust permeability for overburden pressure (assuming 2.7 g/cc average density and a hydrostatic pore pressure gradient). Additional evidence of the extent of the fault zone has come from Tanaka *et al.* (2007). The minimum in permeability at 1200 m depth between the two shear zones is not well constrained. Matrix permeability of the protolith is very low. For tight crystalline rock, bulk mesoscale permeability measured in the field is typically 1 to 2 orders of magnitude higher than matrix permeability due to dominance of fluid flow in fracture networks (Brace, 1980).

Surface offset for the Kobe earthquake in this region was restricted to a single fault trace. This simple geometry extends to at least 600 m depth. Yet by 1200 m, the fault zone appears to have doubled in total width and developed a secondary splay fault. Still, active slip surfaces are less than 1 meter in width, while the zones of intense damage, inferred from enhanced permeability, are 20 to 30 m wide and probably remain localized within the broader damage zone. Enhanced post-earthquake permeability in the damage zone is approximately 2 orders of magnitude greater that permeability of the fault core. Due to the tight crystalline protolith that hosts the Nojima fault in this region, fault core permeability is about 3 orders of magnitude greater than protolith matrix permeability and probably 1 to 2 orders of magnitude greater than protolith fracture-dominated bulk permeability.

The observations reported here provide a unique opportunity to understand fault zone properties at depth following a damaging earthquake. While we do not know the mechanical and hydraulic properties of the Nojima fault before the Kobe earthquake, it is likely that the violent rupture of the fault led to a sudden increase in permeability. This is consistent with observations following the 1989 Loma Prieta earthquake that enhanced fluid flow occurred in the epicentral region (Rojstaczer and Wolf, 1992). One important question not addressed by this study is how rapidly the enhanced fault zone permeability structure will be reduced by sealing and crack healing processes. The observation of increased strength and decreased permeability in the deep NIED fault zone indicates that these processes can significantly influence fault zone mechanics over the lifetime of an active fault. If this sealing and restrengthening process can occur over a single earthquake cycle as suggested by Sibson (1990; 1992), it could importantly influence repeat time, coseismic stress drop and rupture nucleation process. These thermally activated healing processes are expected to occur more quickly deeper in the fault zone where temperatures are higher. Laboratory observations (Moore *et al.*, 1994; 2001) indicate rapid and temperature-sensitive loss of matrix permeability in crystalline rocks under hydrothermal conditions. Observations of repeated sealing/opening cycles in thin sections taken from borehole samples (Moore *et al.*, 2008; Tanaka *et al.*, 2007) provide direct evidence that this process is operative in the Nojima fault. Observations of postseismic velocity recovery over periods of months to years (Li *et al.*, 2006; Peng and Ben-Zion, 2006) imply similar rapid changes in damage zone permeability and resistivity.

Development of off-fault damage during earthquakes is a topic of active research (Dor *et al.*, 2006). Recent studies suggest that asymmetry in off-fault damage should result from dynamic rupture propagation, either due to a contrast in material properties across the fault or the asymmetry in the dynamic stress field that propagates along the fault with the rupture front. In this study, off fault damage extending 15 to 25 meters from the fault core is clearly observed. While our core sampling is relatively sparse, we find little evidence for asymmetry in the damage zone structure as reflected by matrix permeability measurements. Since the GSJ and NIED boreholes are to the SW of the epicenter for a right-lateral fault, the tensile lobe of the dynamic stress field should be on the NW or footwall side of the fault (left-hand side of plots in Fig. 7). Figures 7a and 7c show no indication of enhanced damage on the footwall (left) side of the fault core. It is possible that the Nojima-Hirabayashi site is too near the epicenter of the earthquake for a fully developed dynamic stress wave to have occurred. Or, the significant reverse faulting stress component may inhibit development of excessive damage in the footwall.

5. *Conclusions*

Laboratory measurements of matrix permeability and strength of core samples recovered from scientific drillholes at Nojima-Hirabayashi provide direct evidence for the structure of the fault damage zone and fault core associated with the Kobe earthquake. Numerous factors contribute to making the Nojima fault an ideal location for this type of analysis. The host rock is a competent, low-permeability, crystalline granodiorite that provides a clear contrast in strength and permeability to the fault zone properties. The fault zone architecture is relatively simple and easy to interpret. Also, surface rupture provides valuable constraints on magnitude and orientation of subsurface slip. At depths to 600 meters, the Nojima fault has a classic structure consisting of a narrow core of clay-rich fine-grained gouge surrounded by a zone of highly damaged rock. By 1200 m, the fault zone has broadened and may be bounded by the main shear zone (crossed at 1140 m) and a second branch (1312 m). Enhanced permeability is restricted to narrow zones around the active slip surfaces. Triaxial deformation tests show that the fault core material that captured most of the coseismic slip has the lowest strength. Damage zone samples have high microcrack densities and low cohesive strength and tend to be weaker close to the fault core. Matrix permeability provides an excellent method for differentiating the internal structure of the fault zone. The fine grain size of the fault core results in a reduced permeability of this material relative to the damage zone rock. While the fault core does contain clay, it has a relatively high shear strength ($\mu \sim 0.55$) indicating that the fault strength is still dominated by the unaltered quartz and feldspar constituents of the host rock as well as higher-strength carbonate precipitates. This study demonstrates the importance of obtaining fresh material from depth in active fault zones to unravel the structure and fault mechanics of these systems.

Acknowledgements

We thank J. Caine and C. Morrow and an anonymous reviewer for their helpful comments as well as useful discussions with the editors.

References

Ben-Zion, Y., Peng, Z., Okaya, D., Seeber, L., Armbruster, L.G., Ozer, N., Michael, A.J., Baris, S., and Aktar, M. (2003), *A shallow fault zone structure illuminated by trapped waves in the Karadere-Duzce branch of the North Anatolian Fault, western Turkey*, Geophys. J. Int., *152*, doi:10.1046/j.1365-246X.2003.01870.x, 699-717.

Brace, W. F. (1980), *Permeability of crystalline and argillaceous rocks*, Int. J. Rock Mech. Min. Sci. Geomech. Abstr. *17,* 241–251.

Caine, J.S., Evans, J.P., and Forster, C.B. (1996), *Fault zone architecture and permeability structure*, Geology *24*, 1025–1028.

Caine, J.S. and Forster, C.B., *Fault zone architecture and fluid flow: Insights from field data and numerical modeling*. In *Faults and Subsurface Flow in the Shallow Crust: AGU Geophysical Monograph, Vol. 113* (ed. Haneberg, W.C., Mozley, P.S., Moore, J.C. and Goodwin, L.B.) (Amer. Geophys. Union, Washington, D.C 1999), pp. 101-127.

Chester, F.M., Evans, J.P., and Biegel, R.L. (1993), *Internal structure and weakening mechanisms of the San Andreas fault*, J. Geophys. Res. *98*, 771–786.

Chester, F.M. and Logan, J.M. (1986), *Implications for mechanical properties of brittle faults from observations of the Punchbowl fault zone, California*, Pure Appl. Geophys. *124*, 79–106.

Dor, O., Ben-Zion, Y., Rockwell, T.K., and Brune, J. (2006), *Pulverized rocks in the Mojave section of the San Andreas fault zone*, Earth Planet. Sci. Lett. *245*, 642–654.

Evans, J.P., Forster, C.B., and Goddard, J.V. (1997), *Permeability of fault-related rocks, and implications for hydraulic structure of fault zones*, J. Structural Geol. *19*, 1393–1404.

Faulkner, D.R., Lewis, A.C., and Rutter, E.H. (2003), *On the internal structure and mechanics of large strike-slip fault zones: field observations of the Carboneras fault in southeastern Spain*, Tectonophysics *367*, 235–251.

Hickman, S., Zoback, M., Ellsworth, W., Boness, N., Malin, P., Roecker, S. and Thurber, C. (2007), *Structure and properties of the San Andreas Fault in central California: Recent results from the SAFOD experiment*, Scientific Drilling, Special Issue 1, 29–31.

Hickman, S.H., Zoback, M.D., and Ellsworth, W.E. (2004), *Introduction to special section: Preparing for the San Andreas Fault Observatory at Depth*, Geophys. Res. Lett. *31*, L12S01.

Hickman, S.H., Zoback, M.D., and Ellsworth, W.E. (2005), *Structure and composition of the San Andreas fault zone at Parkfield: Initial results from SAFOD Phases 1 and 2*, EOS Trans. AGU, *86*(52), Fall Meeting Suppl., Abstract T23E-05.

Li, Y.-G., Aki, K., Vidale, J.E., and Alvarez, M.G. (1998), *A delineation of the Nojima fault ruptured in the M7.2 Kobe, Japan, earthquake of 1995 using fault zone trapped waves*, J. Geophys. Res. *103*, 7247–7263.

Li, Y.-G., Chen, P., Cochran, E.S., Vidale, J.E., and Burdette, T. (2006), *Seismic evidence for rock damage and healing on the San Andreas Fault associated with the 2004 M 6.0 Parkfield earthquake*, Bull. Seismol. Soc. Amer. *96, doi:*10.1785/0120050803, S349-S363.

Li, Y.-G. and Leary, P. (1990), *Fault zone seismic trapped waves*, Bull. Seismol. Soc. Am. *80*, 1245–1271.

Lin, A., Imiya, H., Uda, S., Iinuma, K., Misawa, T., Yoshida, T., Abematsu, Y., Wada, T., and Kawai, K. (1995), *Investigation of the Nojima earthquake fault occurred on Awaji Island in the southern Hyogo prefecture earthquake (in Japanese with English abstract)*, J. Geogr. *104*, 113–126.

Lin, A. and Uda, K. (1996), *Morphological characteristics of the earthquake surface ruptures on Awaji Island, associated with the 1995 southern Hyogo Prefecture earthquake*, Isl. Arc *5*, 1–15.

Lockner, D.A. and Byerlee, J.D. (1986), *Changes in complex resistivity during creep in granite*, Pure Appl. Geophys. *124*, 659–676.

Ma, K.-F., Tanaka, H., Song, S.-R., Wang, C.-Y., Hung, J.-H., Tsai, Y.-B., Mori, J., Song, Y.-F., Yeh, E.-C., Soh, W., Sone, H., Kuo, L.-W., and Wu, H.-Y. (2006), *Slip zone and energetics of a large earthquake from the Taiwan Chelungpu fault Drilling Project*, Nature *444*, 473–476.

Mizoguchi, K., Hirose, T., Shimamoto, T., and Fukuyama, E. (2008), *Internal structure and permeability of the Nojima fault, southwest Japan*, J. Struct. Geol. *30*, 513–524.

Moore, D.E. and Lockner, D.A. (1995), *The role of microcracking in shear-fracture propagation in granite*, J. Struct. Geol. *17*, 95–114.

Moore, D.E. and Lockner, D.A. (2004), *Crystallographic controls on the frictional behavior of dry and water-saturated sheet-structure minerals*, J. Geophys. Res. *109*, doi:10.1029/2003JB002582.

Moore, D.E. and Lockner, D.A., *Friction of the smectite clay montmorillonite: A review and interpretation of data*. In *The Seismogenic Zone of Subduction Thrust Faults, MARGINS Theoretical and Experimental Science Series* (eds. Dixon, T.H. and Moore, J.C.) (Columbia University Press, New York 2007), pp. 317–345.

Moore, D.E., Lockner, D.A., and Byerlee, J.D. (1994), *Reduction of permeability in granite at elevated temperatures*, Science *265*, 1558–1561.

MOORE, D.E., LOCKNER, D.A., ITO, H., IKEDA, R., TANAKA, H., and OMURA, K. (2008), *Geometry of the Nojima fault at Nojima-Hirabayashi, Japan - II. Microtructures and their implications for permeability and strength*, Pure Appl. Geophys., in press.

MORROW, C., LOCKNER, D.A., HICKMAN, S., RUSANOV, M., and RÖCKEL, T. (1994), *Effects of lithology and depth on the permeability of core samples from the Kola and KTB drill holes*, J. Geophys. Res. *99*, 7263-7274.

MORROW, C.A., MOORE, D.E., and LOCKNER, D.A. (2001), *Permeability reduction in granite under hydrothermal conditions*, J. Geophys. Res. *106*, 30,551–30,560.

NAKATA, T., YOMOGIDA, K., ODAKA, J., SAKAMOTO, T., ASAHI, K., and CHIDA, N. (1995), *Surface fault ruptures associated with the 1995 Hyogoken-Nabu earthquake (in Japanese with English abstract)*, J. Geogr. *104*, 127–142.

OHTANI, T., FUJIMOTO, K., ITO, H., TANAKA, H., TOMIDA, N., and HIGUCHI, T. (2000), *Fault rocks and past to recent fluid characteristics from the borehole survey of the Nojima fault ruptured in the 1995 Kobe earthquake, southwest Japan*, J. Geophys. Res *105*, 16,161–16,171.

PENG, Z. and BEN-ZION, Y. (2006), *Temporal changes of shallow seismic velocity around the Karadere-Duzce branch of the north Anatolian fault and strong ground motion*, Pure Appl. Geophys. *163*, doi:10.1007/s00024-005-0034-6, 567-600.

RICE, J.R., *Fault stress states, pore pressure distributions, and the weakness of the San Andreas fault.* In *Fault Mechanics and Transport Properties of Rocks* (eds. Evans, B. and Wong, T.-f.) (Academic Press, London 1992) pp. 475–503.

ROJSTACZER, S. and WOLF, S. (1992), *Permeability changes associated with large earthquakes: An example from Loma Prieta, California*, Geology, 211–214.

SIBSON, R. H. (1990), *Conditions for fault-valve behavior.* In *Deformation Mechanisms, Rheology and Tectonics* (eds. Knipe, R.J. and Rutter, E.H.) (Geol. Soc. Special Publ. 1990) pp. 15–28.

SIBSON, R.H. (1992), *Implications of fault-valve behavior for rupture nucleation and recurrence*, Tectonophys. *211*, 283–293.

TANAKA, H., FUJIMOTO, K., OHTANI, T., and ITO, H. (2001), *Structural and chemical characterization of shear zones in the freshly activated Nojima fault, Awaji Island, southwest Japan*, J. Geophys. Res. *106*, 8789–8810.

TANAKA, H., OMURA, K., MATSUDA, T., IKEDA, R., KOBAYASHI, K., MURAKAMI, M., and SHIMADA, K. (2007), *Architectural evolution of the Nojima fault and identification of the activated slip layer by Kobe earthquake*, J. Geophys. Res. *1123,* B07304, doi:10.1029/2005JB003977.

TEMBE, S., LOCKNER, D.A., and WONG, T.-f. (2008), *Effect of clay content and mineralogy on frictional sliding behavior of simulated gouges: Binary and ternary mixtures of quartz, illite and montmorillonite*, J. Geophys. Res., *submitted.*

WIBBERLEY, C.A.J. and SHIMAMOTO, T. (2003), *Internal structure and permeability of major strike-slip fault zones: The median tectonic line in Mie Prefecture, Southwest Japan*, J. Struct. Geol. *25*, 59–78.

WONG, T-f. and BAUD, P. (1999), *Mechanical compaction of porous sandstone*, Oil & Gas Sci. Tech. –Rev. IFP *54,* 715–727.

ZOBACK, M.D. and BYERLEE, J.D. (1975), *The effect of microcrack dilatancy on the permeability of Westerly granite*, J. Geophys. Res. *80*, 752–755.

(Received September 29, 2008, revised March 3, 2009)
Published Online First: June 27, 2009

To access this journal online:
www.birkhauser.ch/pageoph

Pure appl. geophys. 166 (2009) 1669–1691
0033–4553/09/101669–23
DOI 10.1007/s00024-009-0513-2

© Birkhäuser Verlag, Basel, 2009

Pure and Applied Geophysics

Geometry of the Nojima Fault at Nojima-Hirabayashi, Japan – II. Microstructures and their Implications for Permeability and Strength

Diane E. Moore,[1] David A. Lockner,[1] Hisao Ito,[2,3] Riuji Ikeda,[4,5] Hidemi Tanaka,[6,7] and Kentaro Omura[8]

Abstract—Samples of damage-zone granodiorite and fault core from two drillholes into the active, strike-slip Nojima fault zone display microstructures and alteration features that explain their measured present-day strengths and permeabilities and provide insight on the evolution of these properties in the fault zone. The least deformed damage-zone rocks contain two sets of nearly perpendicular (60–90° angles), roughly vertical fractures that are concentrated in quartz-rich areas, with one set typically dominating over the other. With increasing intensity of deformation, which corresponds generally to increasing proximity to the core, zones of heavily fragmented rock, termed microbreccia zones, develop between prominent fractures of both sets. Granodiorite adjoining intersecting microbreccia zones in the active fault strands has been repeatedly fractured and locally brecciated, accompanied by the generation of millimeter-scale voids that are partly filled with secondary minerals. Minor shear bands overprint some of the heavily deformed areas, and small-scale shear zones form from the pairing of closely spaced shear bands. Strength and permeability measurements were made on core collected from the fault within a year after a major (Kobe) earthquake. Measured strengths of the samples decrease regularly with increasing fracturing and fragmentation, such that the gouge of the fault core and completely brecciated samples from the damage zone are the weakest. Permeability increases with increasing disruption, generally reaching a peak in heavily fractured but still more or less cohesive rock at the scale of the laboratory samples. Complete loss of cohesion, as in the gouge or the interiors of large microbreccia zones, is accompanied by a reduction of permeability by 1-2 orders of magnitude below the peak values. The core samples show abundant evidence of hydrothermal alteration and mineral precipitation. Permeability is thus expected to decrease and strength to increase somewhat in active fault strands between earthquakes, as mineral deposits progressively seal fractures and fill pore spaces.

Key words: Fault zone, microstructures, secondary mineralization, permeability, strength.

[1] U. S. Geological Survey, Menlo Park, CA , USA. E-mail: dmoore@usgs.gov
[2] Geological Survey of Japan, Tsukuba, Japan.
[3] Center for Deep Earth Exploration, Japan Agency for Marine-Earth Science and Technology, Yokohama, Japan.
[4] National Research Institute for Earth Science and Disaster Prevention, Tsukuba, Japan.
[5] Department of Earth and Planetary Sciences, Hokkaido University, Hokkaido, Japan.
[6] Ehime University, Matsuyama, Japan.
[7] Department of Earth and Planetrary Sciences, University of Tokyo, Tokyo, Japan.
[8] National Research Institute for Earth Science and Disaster Prevention, Tsukuba, Japan.

1. Introduction

Most of our information regarding the internal structure of fault zones comes from the study of ancient, exhumed faults whose microscopic-scale textures may have suffered post-faulting thermal, chemical, and deformational overprints. Thus, it was especially significant that, within a year following the January, 1995 Kobe (Hyogo-ken Nanbu) earthquake in Japan, the Geological Survey of Japan (GSJ) and the National Research Institute for Earth Science and Disaster Prevention (NIED) began drilling programs into the Nojima fault zone, which ruptured during the earthquake. One of the principal goals of both groups was the recovery of core to allow petrographic, chemical, and physical examination of an active fault zone at depth immediately after an earthquake.

The results of permeability and strength tests conducted on core samples from both drillholes are reported by Lockner *et al.* (this volume, hereinafter cited as paper I). An essential part of any laboratory study is to place the measured physical properties of the samples in the context of their microstructures and mineral assemblages. In this paper, we provide an overview of the brittle deformation and secondary mineralization features of the core samples, illustrating how these processes compete to influence strength and permeability.

2. Tectonic Setting of the Nojima Fault

The approximately 8-km long Nojima fault parallels the northwest coastline of Awaji Island, located in Osaka Bay about 4 km southwest of the city of Kobe, Japan (Fig. 1 of paper I). The Nojima fault is a right-lateral strike-slip fault with a reverse component (Mizuno *et al.*, 1990) that strikes N35-40°E and dips about 83° to the southeast (e.g., Tanaka *et al.*, 2001). The fault cuts the Cretaceous (66-88 Ma; Takahashi, 1992) Nojima granodiorite (Fig. 1 of paper I) at depth. Recent activity on the Nojima fault may have begun about 1.2 Ma, associated with uplift of Awaji Island (Murata *et al.*, 2001), however, the present-day Nojima fault may have reactivated a much older, exhumed fault. Murakami and Tagami (2004) reported a cooling age of 56 Ma for pseudotachylyte from surface exposures of the fault (Otsuki *et al.*, 2003), based on fission track thermochronologic analysis of zircons. They concluded that the pseudotachylyte was generated at depths of about 8–12 km (see also Tagami and Murakami, 2007; Tanaka *et al.*, 2007). Deformation textures in older, foliated cataclasites estimated to have formed between 5 and 10 km depth indicate right-lateral offset with some vertical component (Lin *et al.*, 2001), the same as observed on the fault today.

Surface offset associated with the Hyogo-ken Nanbu (Kobe) earthquake of January 17, 1995 was concentrated on the Nojima fault and some adjoining faults at its southwest termination (Fig. 1 of paper I) (Awata *et al.*, 1996; Lin and Uda, 1996). Maximum net slip of 2.15 m was measured near the GSJ and NIED drillholes; this net slip resolves into 1.8 m right-lateral offset and 1.3 m reverse motion with east side up (Lin and Uda, 1996).

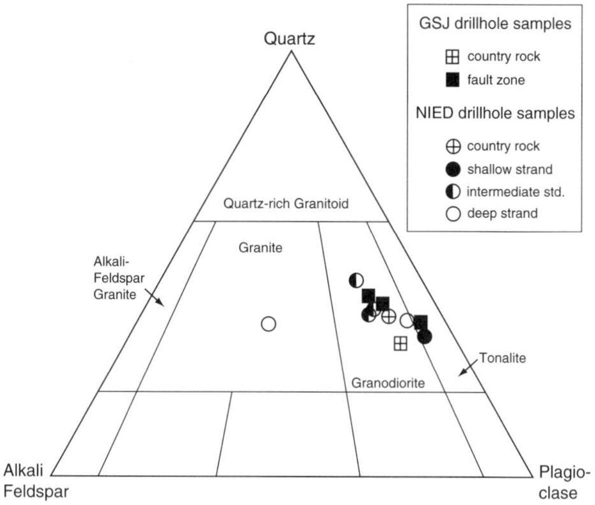

Figure 1
Classification of 11 core samples with well-preserved igneous textures, using the plutonic-rock classification scheme of STRECKEISEN (1974).

An earthquake recurrence interval of about 2000 years has been estimated from trenching studies (AWATA and SUZUKI, 1996). Based on measurements of borehole breakouts (IKEDA *et al.*, 2001) and deformation rate analysis of core samples (YAMAMOTO *et al.*, 2000), the maximum horizontal compression direction in the two drillholes is presently NW-SE (N49°W ± 15°, IKEDA *et al.*, 2001), nearly perpendicular to the strike of the fault. YAMAMOTO *et al.* (2000) noted that this orientation is consistent with the contraction measured over the past 100 years in the Osaka Bay region by the Geographical Survey Institute of Japan. Throughout much of the Quaternary Period, however, the southwestern part of Japan was characterized by E-W compression (e.g., SENO, 1999).

3. *Scope of Study*

The borehole drilled by the GSJ 75 m to the SE of the surface trace crossed a gouge-bearing fault at about 624 m measured depth, whereas the one drilled by the NIED 302 m to the SE of the fault crossed three possible fault strands at depths of about 1140 m, 1312 m, and 1800 m (Fig. 2 of paper I). Core samples from all four fault strands were obtained for laboratory measurements of strength and permeability, as detailed in paper I. A polished thin section was prepared from each sample for petrographic examination. All but two of the thin sections are oriented perpendicular to the axis of the core. Because the maximum deviation of both boreholes from vertical was about 6-7° (IKEDA, 2001; TANAKA *et al.*, 2001), these thin sections are oriented nearly horizontally. The two exceptions are

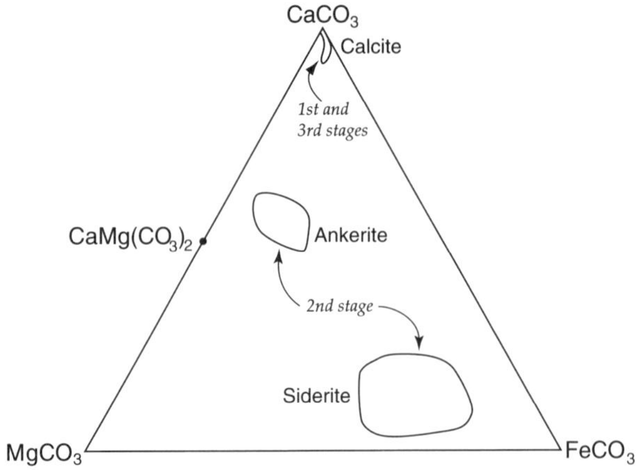

Figure 2
Compositions of carbonate minerals in Nojima fault core samples, obtained by electron microprobe techniques. Trace amounts of stage 2 dolomite, containing only about 5% substitution of Fe for Mg, were also found.

gouge samples NIED 81-22A and GSJ 99-4 which, owing to their small size, were cut parallel to the cylinder axis. The thin-section set, listed in Table 1, includes an example of the granitic country rock from shallow depths in each drillhole, although the GSJ country-rock sample was not tested by LOCKNER *et al.* The other samples are classified as damage zone or fault core, following the usage of CHESTER and LOGAN (1986; see also CHESTER *et al.*, 1993, and CAINE *et al.*, 1996). A fault core was not identified in the deepest fault strand intersected by the NIED drillhole.

This petrographic study is largely based on transmitted- and reflected-light microscopy. Additional observations were made with a scanning electron microscope (SEM) equipped with a high-resolution backscattered electron detector and an energy dispersive X-ray spectrometer (EDX) for qualitative determinations of element abundances. Compositions of carbonate minerals were obtained using electron microprobe techniques, with carbonate-mineral standards. Three sets of point counts were made with the aid of a mechanical stage, with a 1 mm × 1 mm grid. Point-count totals for the variably sized samples range from 530 to 1001. Because the core was not oriented, the relative orientations of microstructures in a given thin section were measured with reference to a designated 0° azimuth parallel to the long dimension of the thin section.

4. *Petrographic Observations*

4.1. Igneous and Secondary Mineral Assemblages

Igneous mineral modes were obtained from point counts of 11 samples with reasonably well preserved igneous textures. In the less deformed samples, it usually is

Table 1

Samples examined in this study

Sample #	Depth (m)	≈Horiz. Dist. from Gouge Core* (m)	Description	% Secondary Minerals	% Disrupted (see text)
GSJ Drillhole					
258.6	258.6		country rock; relict amphiboles	17	20
74-17	503.4	14.5	damage zone; 4 mm-wide microbreccia zone	44	40
91-6	582.0	5.0	damage zone; 10 mm-wide microbreccia zone	37	72
98-31	623.6	0	core; coarse-grained, foliated gouge (Type I**)	N A	100
99-4	624.4	0	core; fine-grained gouge (Type II**)	N A	100
101-2	634.1	1.1	damage zone; two $\leq$ 6 mm-wide shears	38 [36-41]	64
118-21	730.4	12.7	damage zone; dilational fractures, relict amph.	19	20
NIED Drillhole					
12-1A	659.3		country rock; relict amphiboles	5	2
Shallow strand:					
32-20	1063.7	18.0	damage zone; wide veins, minor brecciation	46	45
34-27	1077.9	14.5	damage zone; microbreccia zones	61 [55-64]	56
43-37-1	1131.8	2.0	damage zone; minor shear	36	70
45-23	1140.1	0	core; fine-grained gouge (Type II)	N A	100
Intermediate strand:					
74-19	1278.6	7.5	damage zone; dilational fractures	18	12
76-17	1289.0	5.0	damage zone; 20 mm-wide microbreccia zone	N A	77
80-29	1307.5	1.0	damage zone; dilational fractures	26	18
81-22A	1312.0	0	core; coarse-grained, foliated gouge (Type I)	N A	100
82-40	1320.3	2.0	damage zone; incipient shears	22 [18-35]	60
85-15B	1332.2	5.0	damage zone; large microbreccia zone	N A	100
Deep strand:					
179-33	1798.2	?	damage zone: microbreccia zones	52 [50-54]	56
181-34	1807.9	?	damage zone: microbreccia zones	38	64
183-14C	1815.9	?	damage zone: microbreccia zones	15	35
184-23	1821.9	?	damage zone: microbreccia zones	69	42

*from Lockner *et al.* (2000) **classification of Ohtani et *al.* (2000). N A = not attempted.

possible to "see" through the overprint of secondary mineralization to identify the primary igneous assemblage, because the low-temperature replacement reactions are typically pseudomorphic. Most of the samples are biotite-hornblende granodiorites, and a few are tonalites (Fig. 1; see also Ikeda, 2001 and Tanaka *et al.*, 2007). One K-feldspar

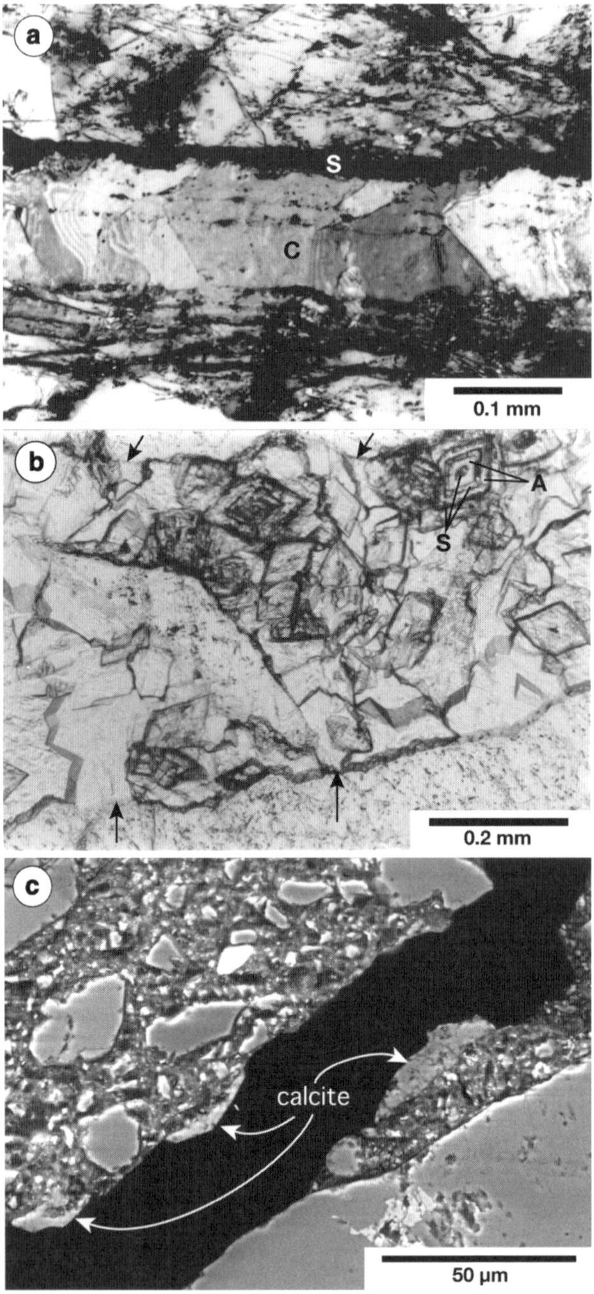

rich sample from the deep NIED fault strand has a granitic composition; a second deep NIED sample not included in Figure 1 probably also is a granite. Leaving out the granite outlier, the average igneous mode is 45% plagioclase, 33% quartz, 9% K-feldspar, 9%

◀ Figure 3
(a) First-stage calcite (C) vein with crack-seal textures indicating multiple stages of alternating fracturing and secondary-mineral crystallization. The outermost, dark band on top is second-stage, brownish-yellow siderite (S), and siderite-filled cracks also cross-cut this vein (crossed polarizers; NIED 76-17). (b) Close-up of a microbreccia zone (boundary fractures marked by arrows) from the deep NIED fault strand, whose original void spaces are filled with zoned rhombs of alternating ankerite (A) and siderite (S) (plane-polarized light; NIED 183-14C). (c) Secondary-electron SEM image of a dilational fracture in the completely brecciated sample from the intermediate NIED fault strand (NIED 85-15B). The fracture surface is lined with small crystals of stage 3 calcite.

biotite, and 4% hornblende. Subhedral, tabular plagioclase phenocrysts are as much as 6 mm long; anhedral K-feldspar crystals and clusters of anhedral quartz crystals commonly reach 5 mm in diameter. The largest amphiboles are about 1.8 mm long and 1.1 mm wide, and the largest biotites are 2.5 mm long and 2 mm wide.

The relative abundances of primary igneous and low-temperature secondary minerals in the country rock and all but two of the damage-zone samples were estimated from a second set of point counts (Table 1), in which the percentage of secondary minerals comprises those that replace igneous minerals plus those that fill fractures and voids. No attempt was made to obtain counts of the mineral assemblages of the gouge or of two heavily brecciated samples (Table 1), and fragments too small to be identified using a petrographic microscope were not included in the counts that were made. Four samples (GSJ 101-2 and NIED 34-27, 82-40, and 179-33B) contain a sufficiently large number of uncounted small grains to potentially affect the reported secondary-mineral percentages. Error estimates for these four samples were made by calculating the effect on the percentages if all uncounted points were either igneous or secondary minerals (Table 1).

Secondary-mineral contents range from about 5% in a country rock sample to nearly 70% in one of the samples from the deep NIED fault strand (Table 1). Hornblende is the most readily altered mineral in the granodiorite. Relict amphiboles were found in only three samples, the two country-rock samples and the footwall sample GSJ 118-21, which shows relatively few effects of faulting (Table 1). The hornblende is pseudomorphically replaced by fine-grained aggregates of siderite and a montmorillonitic smectite clay. Using bulk X-ray diffraction techniques, Ohtani *et al.* (2000) identified smectite clays in much of the core from the GSJ drillhole. Biotite is better preserved than hornblende, but it commonly contains lozenges of laumontite or siderite along the cleavage planes. Some biotite is altered to pale-colored kaolinite, which was also identified by Matsuda *et al.* (2004) in the NIED drillhole from XRD analysis. Other biotite is partly replaced by the smectite clay. Plagioclase is the more reactive feldspar, being replaced principally by laumontite but also by carbonate minerals. Quartz and K-feldspar are well preserved; low-temperature secondary minerals in these crystals typically fill fractures rather than replace the minerals. Flame perthite and myrmekite-like exsolution textures in K-feldspar may be faulting-induced deformation features (e.g., Moore, 1987, 1990).

Carbonates are widespread fracture-filling and alteration minerals in the Nojima fault zone, and the type of carbonate mineral that crystallized has varied with time (Figs. 2–3) (Moore *et al.*, 2000). The fracture-filling sequence comprises: 1) An early stage of calcite coeval with laumontite (Figs. 2, 3a); 2) an extensive intermediate stage of siderite + ankerite (iron-rich dolomite; Figs. 2, 3b); and 3) late-stage calcite deposition in still-open fractures (Figs. 2, 3c) associated with trace amounts of barite. The transition between stages 1 and 2 is marked by co-crystallization of calcite and siderite, without laumontite. Alteration of mafic minerals may have provided most of the Fe and Mg for siderite and ankerite crystallization. This sequence was established from cross-cutting relationships. The laumontite ± calcite veins are commonly cut by veins of siderite and/or ankerite, but do not themselves cross-cut any other veins. Stage 3 calcite-bearing fractures cross-cut all other structures. Three of the six samples in which the late-stage calcite was identified are fault gouge, one from each of the three gouge-bearing fault strands (GSJ 99-4 and NIED 45-23 and 81-22A). Present-day calcite deposits formed on a borehole seismometer that was deployed near the base of the GSJ drillhole for a year beginning in March, 2003 (Fujimoto *et al.*, 2007). Each of the first two stages comprises several episodes of crystallization, as indicated by features such as the crack-seal calcite vein in Figure 3a and the alternating concentric layers of siderite and ankerite in Figure 3b. Only a single episode of late calcite growth was seen in each of its occurrences.

4.2. Brittle Microstructures

The degree of disruption of the damage-zone samples ranges from minor fracturing to complete fragmentation. Minor shear bands and wider shear zones overprint some of the heavily fractured and brecciated areas. Microstructures are described in the order of increasing deformation intensity below.

Microfractures: In mildly deformed granitic rock crack densities are typically greater in quartz than in feldspars (e.g., Morrow *et al.*, 1994; Moore and Lockner, 1995), and this is also true for the Nojima granodiorite. Although the cracks can have any orientation, particularly in the vicinity of biotite, typically a set of elongate sub-parallel fractures dominates in quartz-rich areas, and a somewhat less prominent set of fractures is oriented at a large angle (60–90°) to the first (Fig. 4a). These two crack sets have the same orientations as older, healed cracks in quartz that are preserved as fluid-inclusion trails. Some of the microfractures associated with the Quaternary faulting have refractured portions of the healed cracks. On average, microfracture orientations in the feldspars are similar to those in the quartz, although the scatter is greater due to the influence of the two good cleavages in both feldspars.

Many individual microcracks are intragranular, transgranular, or grain-boundary cracks $\leq$5 mm in length (Fig. 4a). Longer, commonly mineral-lined intergranular fractures form as a result of the linkage of shorter cracks. Fractures in the less deformed damage-zone samples are generally $\leq$0.04 mm wide, and the majority are hairline cracks (e.g., Fig. 4a). Some relatively narrow fractures are lined with multiple rows of euhedral crystals (Fig. 4b). Most of the mineral-filled fractures or veins in the heavily deformed samples are also narrow, although a few are significantly wider, up to 3.6 mm for

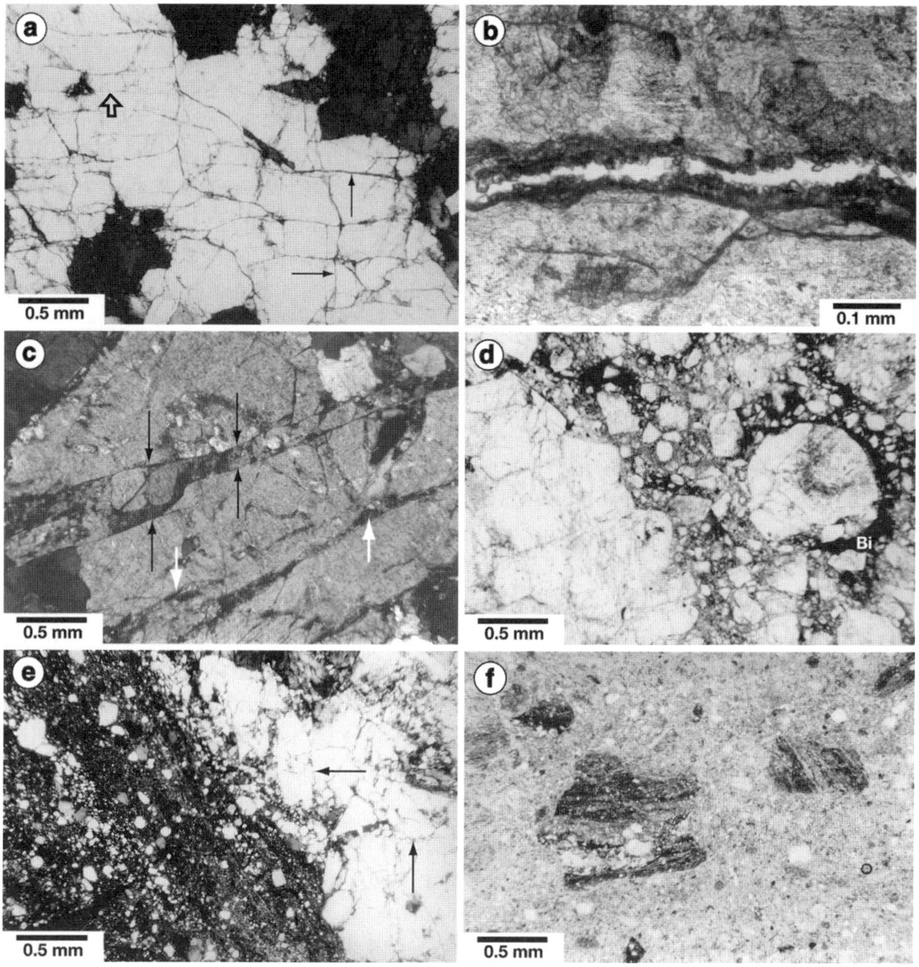

Figure 4
Photomicrographs illustrating the range of deformation textures in the samples. (a) Fractures in quartz form two, nearly perpendicular sets (examples indicated by long arrows), with the set trending left-to-right being the more prominent one. The quartz also contains older, healed fractures (the short, wide arrow points to one) preserved as fluid inclusion trails, with orientations similar to those of the more recent fractures (crossed polarizers; GSJ 118-21). (b) Open fracture lined with multiple rows of euhedral siderite crystals (plane-polarized light; GSJ 74-17). (c) Microbreccia zone in a plagioclase phenocryst. The pairs of black arrows indicate the change in width of the microbreccia zone where the boundary steps over to another microfracture. The plagioclase surrounding the microbreccia zone is undeformed except for a few mineral-filled fractures (marked by white arrows) (crossed polarizers; NIED 181-34). (d) Thoroughly disaggregated granodiorite within a 20-mm wide microbreccia zone. Biotite (Bi) wraps around variably sized, angular fragments of quartz and feldspar (plane-polarized light; NIED 76-17). (e) Boundary between a 6-mm wide minor shear band (left) and fractured damage-zone rock (right). The two principal fracture sets in the damage-zone rock (marked by arrows) are at moderate angles to the slip surface (see Fig. 8). The granodiorite within the shear is finely ground, and many of the larger grains have rounded rather than angular shapes. Alternating finer and coarser-grained bands define a foliation parallel to the shear planes (crossed polarizers; GSJ 101-2). (f) Gouge from the shallow NIED fault strand contains clasts of dark, foliated gouge, suggesting multiple episodes of shear and granulation (plane polarized light; NIED 45-23).

laumontite-filled veins and 1.5 mm for carbonate veins. The wider veins record multiple episodes of fracturing and sealing, as illustrated in Figure 3a. The orientation of the most prominent veins in a thin section usually corresponds to that of the dominant set of fractures in quartz, with some variability resulting from the way shorter cracks became linked to form them. Rough estimates of the amount of extension in the three most heavily veined thin sections were made by summing the vein widths along traverses parallel to the principal extension direction. The traverses yielded local extensional strains of 9% (NIED 43-37-1), 12.5% (NIED 32-20) and 15% (NIED 184-23).

Microbreccia Zones: The dominant features of the more highly deformed damage zone samples are microbreccia zones in which the rock between closely spaced fractures has been fragmented (Fig. 4c). Microbreccia zones can form from either set of microcracks, although the larger zones are usually parallel to the dominant set. Microbreccia zones as wide as 20 mm were identified (Table 1), and one sample (NIED 85-15B) is entirely microbreccia. Different stages of microbreccia zone development were identified in different thin sections. For example, the width of the microbreccia zone in Figure 4c doubles where it steps over to the next closest fracture, hence the wider microbreccia zones represent brecciation across an array of closely spaced fractures. Microbreccia zones extend their lengths by means of linkages to nearby zones in the same way that individual long fractures are formed. This is well illustrated by a composite microbreccia zone from the deep NIED strand that is up to 5 mm in width and extends the length of the thin section (Fig. 5a). Fracture spacing varies across the width of this zone, and the slivers of granodiorite between them are fragmented to varying degrees. Subsidiary microbreccia zones can form between pairs of cross fractures in the larger zones, as seen in the center of Figure 5a (white arrows).

The fragmented granodiorite within the wider microbreccia zones resembles a coarse-grained gouge (Fig. 4d), with fragments ranging from $\approx$8 mm in diameter down to clay size. Grain-size reduction in the microbreccia zones appears to be principally the product of intense dilational fracturing, such that the original distribution of igneous minerals in the brecciated regions can often be discerned. Many of the pore spaces in the microbreccia zones have been filled with secondary minerals (Fig. 6a). Fragments of quartz and feldspar have euhedral overgrowths, and neoblasts of these crystals fill pore spaces. Many plagioclase fragments have been partly albitized, and newly crystallized plagioclase is albite. Zeolites, carbonate minerals (e.g., Fig. 3b), and clay minerals are also common pore-filling secondary minerals.

Granodiorite adjacent to some of the larger microbreccia zones is strongly fractured and locally brecciated, particularly near intersecting microbreccia zones. For example, the sample containing the 20-mm wide microbreccia zone also has a second, nearly perpendicular zone on one side that is $\geq$1.5 mm wide. Biotites in the granodiorite bounded by these two zones have been heavily deformed and altered to siderite and clays. Large quartz and feldspar grains between the deformed biotites have been pulled apart along the two main fracturing directions, and the fractures are filled with siderite with crack-seal textures. Large void spaces partly filled with siderite are also present. On the

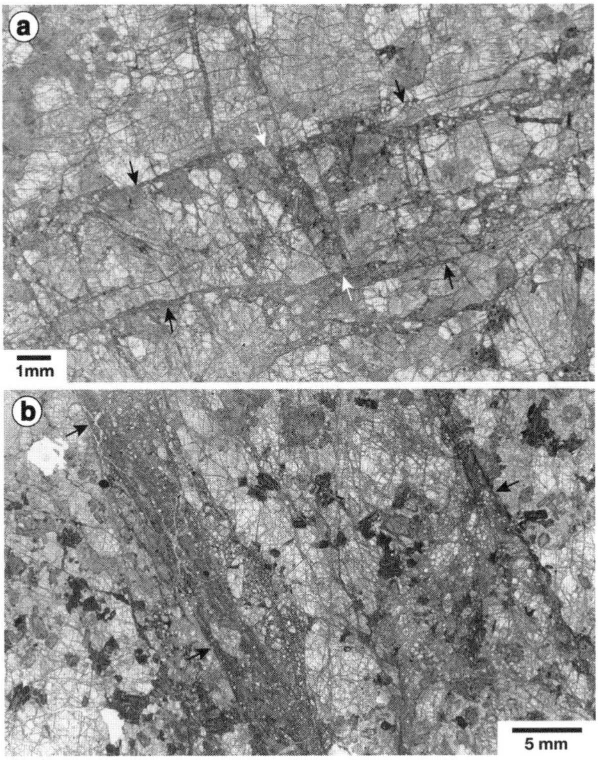

Figure 5
Views of larger structures from scans of thin sections. (a) Larger-scale microbreccia zone (NIED 183-14C); (b) Heavily damaged area between two minor shears (GSJ 101-2). Black arrows denote the borders of the zones; white arrows in (a) mark a band of concentrated fracturing between two cross fractures in the microbreccia zone.

other side of the large microbreccia zone, where the perpendicular microbreccia zone is not present, the granodiorite is less heavily fractured and biotite shows little replacement by carbonate minerals and clays.

The heavily deformed granodiorite in the GSJ fault strand and the shallow and intermediate NIED fault strands typically contain open fractures and open voids in brecciated areas. In contrast, the deep NIED fault strand contains few open fractures. Fractures in samples from that strand have been sealed with laumontite or carbonate minerals, and nearly all the void space in the microbreccia zones is filled (Fig. 3b).

Minor Shear Bands: Some of the samples located within 2-m horizontal distance of a gouge core contain minor shear bands (Table 1). The grains in the shear bands are much more strongly comminuted and typically more rounded than those in the microbreccia zones. In sample GSJ 101-2 (Fig. 4e) bands of alternating coarser and finer grained material are oriented roughly parallel to the strike of a given slip surface. The transitions between the well-developed shear bands and the adjoining rock in this sample are sharp,

Figure 6 ►
Backscattered-electron SEM images. (a) Void space in a microbreccia zone is largely filled with euhedral, secondary K-feldspar (probably adularia, labelled K), albite (Ab), and interstitial montmorillonite (M). Some K-feldspar crystals show two stages of growth; the boundary between stages in one crystal is marked by arrows. Anhedral feldspars around the edges of the photo are partly altered breccia fragments (GSJ 91-6). (b) Clast of pseudotachylyte from the shallow NIED fault strand, characterized by round pores and partly resorbed quartz crystals (Q) in a light-gray, cryptocrystalline matrix. Possible euhedral crystallites now contain calcite (C) and montmorillonite (M) (NIED 45-23). (c) Closer view of the fine-grained gouge in (b), with a clast of gouge (G) in the center. Many quartz (Q) grains are present in this view, along with K-feldspar (K), ankerite (A), and albite (Ab) (NIED 45-23).

and the shear-band orientations roughly bisect the trends of the perpendicular sets of fractures and microbreccias. The 4-mm wide shear band shown in Figure 4e contains clasts of foliated rock, suggesting multiple episodes of slip. Offset across the wider shear bands could not be determined; however, a narrow shear in sample NIED 43-37-1 offsets separate ankerite and laumontite veins 9 mm right-laterally. This shear band varies in texture from a 0.3-mm wide band that is very fine-grained but not well foliated to a wider but more diffuse zone. No evidence for more than one episode of slip was found in this shear band. Two other relatively fine-grained bands that cut across microbreccia zones may be incipient shear bands.

The 4-mm wide shear band and a second, less well defined one bound a roughly 10-mm wide sliver of granodiorite that is more heavily broken up than the rock outside the area between the two shear bands (Fig. 5b). The overall structure is interpreted as a minor shear zone.

Fault Gouge: The approximately 30-cm thick gouge core identified in the GSJ drillhole is divided into three subzones that were active at different times (Fujimoto *et al.*, 2000; Ohtani *et al.*, 2000): the Type I and Type II subzones are the locations of former principal slip surfaces, whereas the Type III gouge is inferred to have taken up the offset during the 1995 earthquake. The four gouge samples that we obtained from the GSJ and NIED drillholes correspond to the older, Type I and Type II subzones:

- Type I — GSJ 98-31 and NIED 81-22a (intermediate strand, Fig. 2 of paper I).
- Type II — GSJ 99-4 and NIED 45-23 (shallow strand).

The Type I GSJ sample is a coarse-grained, foliated gouge that is cross-cut by a network of fractures filled with either the early laumontite + calcite or the subsequent siderite + ankerite mineral assemblages. These textures suggest that this gouge sublayer has long been abandoned as a site of localized shear and now behaves as part of the damage zone (see also Boullier *et al.*, 2004). The gouge sample from the intermediate NIED fault strand is a coarse-grained, foliated gouge that has some textural similarities to the Type I GSJ gouge sample and that also is cut by laumontite-lined fractures.

The Type II GSJ gouge and the gouge sample from the shallow NIED fault strand (Fig. 4f) are very similar, fine-grained, massive gouges that contain fragments of earlier-formed, foliated gouge (Fig. 4f), indicating multiple episodes of localized shear. The foliation within these clasts of gouge is similar to that seen in the minor shear bands

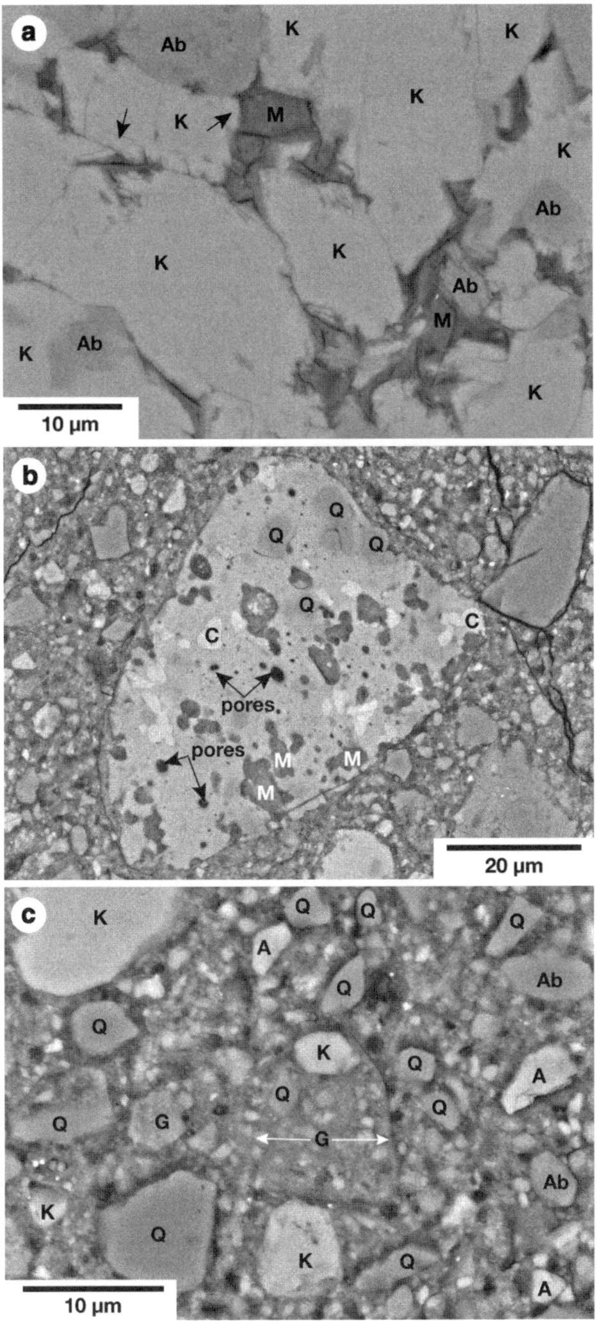
a
Ab
K
K
K
M
K
K
Ab
K
K
Ab
M
Ab
K
K
10 µm
b
Q
Q
Q
Q
C
C
pores
pores
M
M
M
20 µm
c
K
Q
Q
Q
A
Q
Ab
Q
K
Q
A
Q
Q
Q
G
G
Ab
K
Q
K
Q
A
10 µm

Figure 7 ▶

Comparison of the degree of deformation of a given sample with its: (a) secondary-mineral content (degree of alteration), (b) peak shear strength, and (c) permeability (k); reported laboratory data were measured at 50 MPa effective pressure. In (a), error bars were added to the four samples for which small to significant numbers of unidentifiable grains were not counted (Table 1). Quartz and K-feldspar are resistant to alteration in these samples, and an approximate upper limit to the proportion of secondary minerals in a given sample plotted in (a) would be determined by its original quartz + K-feldspar content, adjusted for the amount of dilation and in-filling by secondary minerals. The secondary-mineral content of the gouge samples may approach the upper limit. Samples containing only dilational fractures were $\leq$ 45% deformed, and samples containing microbreccia zones were $\geq$ 35% deformed.

(Fig. 4e). Fujimoto *et al.* (2000) and Ohtani *et al.* (2000) described clasts of the $\approx$56 Ma (Murakami and Tagami, 2004) pseudotachylyte in the Type II GSJ gouge, and we found a similar clast in the gouge layer of the shallow NIED fault strand (Fig. 6b; compare with Fig. 3f of Fujimoto *et al.*, 2000, p. 108). Tanaka *et al.* (2007) also reported clasts of pseudotachylyte in an ultracataclasite subzone that is cut by younger shears in the core of the shallow NIED strand. Many grains of quartz, K-feldspar, unaltered and albitized plagioclase, laumontite, calcite, ankerite, and siderite were identified in the fine-grained gouge samples with the SEM (e.g., Fig. 6c). The abundance of siderite and ankerite occurring as clasts in the gouge suggests that these two fine-grained gouge samples were sheared considerably more recently than was the type I gouge. Both gouge samples, however, do contain late-stage calcite $\pm$ barite-lined fractures.

4.3. Deformation Intensity

Deformation intensity was estimated from the degree of disruption of each sample, determined by a point count of "intact" versus "disrupted" granite, in which the "disrupted" counts comprise open and filled microfractures, microbreccia zones, and minor shear bands (Table 1). This method provides a simple way to assess and compare the cumulative damage in different granodiorite samples, and the reported percentages represent the total damage of various types incurred during the faulting history. In gouge, the deformation processes resulted in the complete disaggregation of the rock, and all such samples were assigned a value of 100% disrupted. The large microbreccia zone that constitutes all of sample NIED 85-15B (Table 1) is also considered 100% disrupted. All of the samples in Table 1 that are listed as < 100% disrupted contain at least some areas of relatively intact igneous rock. These counts represent the deformation intensity at the scale of the thin sections and the tested laboratory samples.

The percentage of secondary minerals in a sample (Table 1) is plotted against its degree of disruption in Figure 7a. Error bars are included for the four samples with small to significant numbers of unidentifiable, uncounted grains. In general, the abundance of secondary minerals increases with increasing deformation in country-rock and damage-zone samples. Because quartz and K-feldspar are resistant to alteration in these samples, there will be some upper limit to the secondary-mineral content in a sample that is a function of its igneous mode (Fig. 1). The amount of secondary minerals will vary with

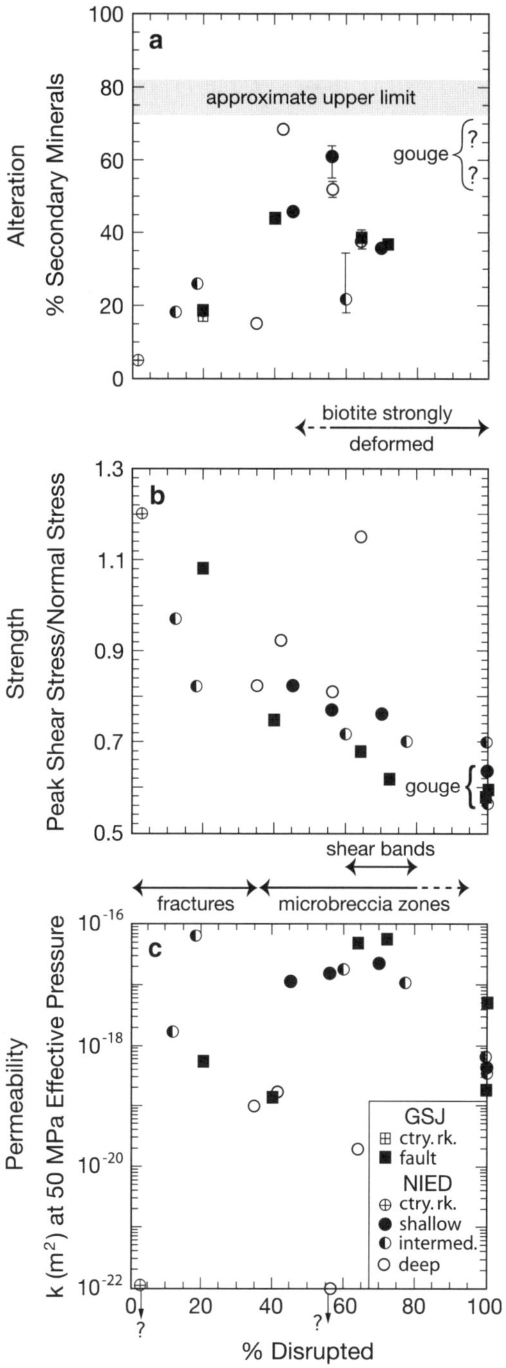
a
approximate upper limit
gouge
?
?
Alteration
% Secondary Minerals
100
80
60
40
20
0
biotite strongly
deformed
b
gouge
Strength
Peak Shear Stress/Normal Stress
1.3
1.1
0.9
0.7
0.5
shear bands
fractures
microbreccia zones
c
Permeability
k (m²) at 50 MPa Effective Pressure
10⁻¹⁶
10⁻¹⁸
10⁻²⁰
10⁻²²
GSJ
ctry. rk.
fault
NIED
ctry. rk.
shallow
intermed.
deep
0
20
40
60
80
100
?
?
% Disrupted

the amount of dilation of the sample and the degree to which the opened cracks and voids have been filled with secondary minerals. The degree of disruption of one sample from the deep NIED fault strand (NIED 184-23, Table 1) could not be determined because of the extensive alteration overprint, but its estimated 69% secondary-mineral content is probably near the limiting value. Most of the plagioclase and biotite and all of the hornblende in that sample have been altered, and the numerous wide laumontite veins indicate substantial extensional strain. The quartz is heavily fractured, and essentially all the fractures are filled, predominantly with minerals other than quartz. In contrast, the minor K-feldspar in the sample is only sparsely fractured and little altered. Gouge samples may also have secondary-mineral contents near the upper limit (Fig. 7a).

Relationships between degree of deformation and the strength and permeability data presented in paper I are shown in Figures 7b and 7c, respectively. The strength data plotted in Figure 7b are the peak strengths at 50 MPa effective confining pressure. With one marked exception, a strong inverse relationship exists between peak strength of a rock sample and its degree of disruption. The tested intact-rock sample is the strongest with a ratio of peak shear stress to normal stress of ≈ 1.2, and the gouge-core samples are the weakest at ≈ 0.56–0.64. The exception to the trend in Figure 7b is a heavily deformed sample from the deep NIED fault strand whose peak strength nevertheless approaches that of the country rock. A correlation between permeability and degree of disruption (Fig. 7c) is weaker than that between peak strength and deformation intensity, nonetheless the relationship is more pronounced if the deep NIED samples are not considered. Compared to the country-rock sample, permeability is at least 3 orders of magnitude higher for all the GSJ and the shallow- and intermediate-strand NIED samples. With one exception, samples that are $\leq 40\%$ disrupted have lower permeability than those in the range 40–80% disrupted. Permeability of the completely disaggregated samples is 1–2 orders of magnitude lower than that of the 40–80% disrupted samples. The samples from the deep NIED fault strand have lower permeabilities than all but one of the damage-zone samples from the other strands.

5. *Discussion*

5.1. *Microstructure Orientations*

Although the core samples were not oriented, the relative orientations of the different types of microstructures are consistent from sample to sample in the two boreholes; a representative example is shown in Figure 8. Summarizing the microstructural descriptions presented in previous sections, microfractures concentrated in quartz form at the lowest level of deformation in the Nojima fault, and two nearly perpendicular sets of fractures commonly can be identified, with one set dominating over the other. The widest veins and the largest microbreccia zones are oriented parallel to the dominant fracture set, whereas minor shears cross-cut both sets (Fig. 8).

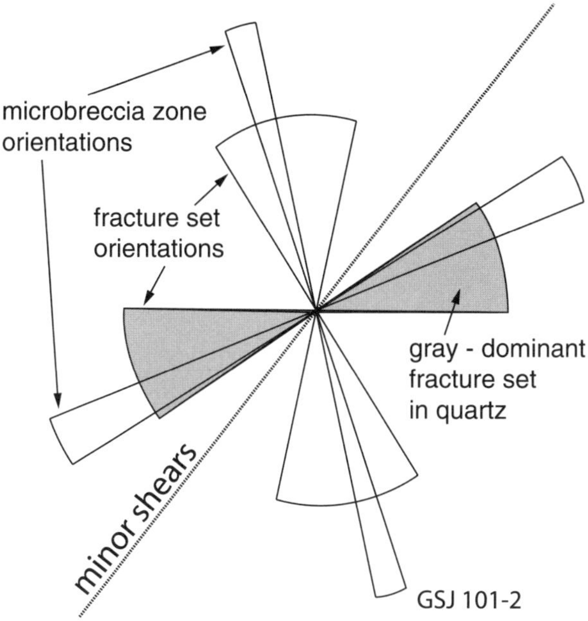

Figure 8
Relative strikes of microstructural elements in damage-zone sample GSJ 101-2. The figure was oriented such that the dominant fracture set in quartz has the roughly EW (to NE-SW) trend measured by TAKESHITA and YAGI (2000), and the minor shear is sub-parallel to the strike ($\approx$N36°E) of the Nojima fault (that is, a north arrow would point upwards in this interpretation).

CÉLÉRIER *et al.* (2000) determined that subsidiary shear planes identified in borehole televiewer images from the GSJ drillhole principally have NE–SW trending strikes, subparallel to the strike of the fault. TAKESHITA and YAGI (2000, 2001) measured the orientations of healed microcracks and cracks sealed with carbonate minerals in quartz in oriented core samples from the GSJ drillhole and from a second drillhole that crosses the Nojima fault near its southern end. The sealed fractures generally have the same orientations as the healed cracks, and they form three groups: one set of sub-horizontal fractures and two sets of vertical cracks oriented N-S (to NW-SE) and E-W (to NE-SW) (TAKESHITA and YAGI, 2000, 2001). Among the vertical microcracks, E-W orientations generally predominate for the sealed cracks and N-S orientations for the older, healed cracks. Mutually perpendicular, healed crack sets in quartz in plutonic rocks have been attributed to contraction of the quartz during cooling of the plutons (e.g., VOLLBRECHT *et al.*, 1991), and the prominence of the N-S vertical set was interpreted by TAKESHITA and YAGI (2000, 2001) to reflect N-S compression during solidification of the Nojima granodiorite. The dominance of E-W orientations among the carbonate-sealed vertical cracks is consistent with the E-W compression that dominated during the Quaternary (SENO, 1999).

The diagram in Figure 8 was rotated to a position in which the dominant set of fault-related microfractures in quartz would have the roughly E-W (to NE-SW) orientation determined by TAKESHITA and YAGI (2001). This positioning also causes the minor shear to be oriented subparallel to the strike of the Nojima fault, consistent with the results of CÉLÉRIER *et al.*, (2000). The arrangement of microstructures in Figure 8 looks the same if the figure is rotated 180°C; thus two possible matches exist for each thin section (see also MOORE *et al.*, 2001).

5.2. Microstructural Evolution

WONG (1982) described how long, closely spaced dilational fractures in quartz isolate narrow columns that buckle and fragment accompanying failure of granitic rock in laboratory loading experiments. KING and SAMMIS (1992) incorporated these and other observations into a model of shear-zone development and the formation of fault gouge. In their model, clustered dilational fractures that initiated from flaws in the rock expand their lengths, until the aspect ratio of the rock slivers between the fractures reaches a critical value beyond which the columns bend and then break along transverse, tensile fractures. This process, repeated at smaller and larger scales at clusters of dilational fractures, eventually produces localized zones of fragmented rock that KING and SAMMIS (1992) call 'superdefects'. Closely-spaced superdefects may interact and link together over time to form a shear band.

The KING and SAMMIS (1992) model provides a framework for interpreting the microstructural features of the core samples from the Nojima fault zone. In particular, the microbreccia zones correspond to their superdefects, and the sequence of development of the microbreccia zones, described previously, is essentially the same. Initial microbreccia zone lengths are on the order of a few millimeters, corresponding to the grain sizes of igneous feldspars and quartz (Fig. 4c). Some of them have scaled upwards by stepping across to adjoining fractures to increase width and by linking up with nearby zones to increase length (Fig. 5a). Downward scaling eventually leads to complete fragmentation of the rock within the zones (Fig. 4d). Initiation of the microbreccia zones occurs in rock that has been significantly affected by dilational fracturing. Formation of a minor shear band requires even greater amounts of disruption, which for this sample set is restricted to $\leq$ 2 m horizontal distance from the fault cores (Table 1). Once the shear bands form, deformation may be localized within them, so that the surrounding rock is not completely disrupted.

5.3. Physical Properties Relative to Microstructural Development

The interpretation of peak strength in terms of deformation intensity is straightforward, with strength decreasing continuously with increasing cumulative damage (Fig. 7b). The greatest strength decreases accompany the initial increases of fracture density, and the heavily deformed damage-zone rocks (>60% disrupted) are not significantly stronger than the completely disaggregated microbreccia and gouge samples.

Shear strength reaches a minimum in the gouge core and the one completely brecciated damage-zone sample, but the coefficient of friction does not go below $\approx$0.55. Such friction values are consistent with the abundance of hard grains such as quartz and feldspars in the gouge (Fig. 6c). Laumontite and calcite also are strong minerals at room temperature (MORROW and BYERLEE, 1991; MORROW *et al.*, 2000); other zeolites and carbonate minerals are likely to have similar strengths. These strong minerals are mixed with varying amounts of weaker clay minerals, although smectite clays will contribute more to the weakening than kaolinite (MOORE and LOCKNER, 2004).

Some of the implications of the shear-forming process that KING and SAMMIS (1992) outline are applicable to the permeability trends of these samples (Fig. 7c). Tensile cracking will dilate the rock, and the Nojima samples characterized by dilational fracturing alone have permeabilities at least 1-2 orders of magnitude greater than the country-rock permeability. However, the most permeable samples are principally those that contain microbreccia zones. Buckling and rotation of narrow rock columns between adjoining fractures will initially generate large fragments that rotate and create additional void space in the microbreccia zones. Continued fracturing at smaller scales in the developing microbreccia zones will gradually reduce the grain size of the breccia, leading to greater compaction and consequently a reduction of porosity and permeability in the zones. The smaller microbreccia zones with relatively large fragments have substantial voids that serve as permeability conduits. The larger microbreccia zones and especially the minor shear bands are characterized by finer-grained, more densely packed fragments. For the samples containing these structures, enhanced fracturing in the vicinity of intersecting microbreccia zones or between minor shear bands is the probable cause of their high net permeability values. A connected fracture network appears to be critical to maintaining elevated permeability, such that the completely brecciated sample has permeability comparable to that of the gouge samples. The gouge sample whose permeability is about an order of magnitude higher than that of the others (Fig. 7c) is the type I GSJ gouge sample (GSJ 98-31), which has been extensively fractured (see also BOULLIER *et al.*, 2004). The permeability measurements for that sample may thus be a reflection more of microfracture permeability than of matrix permeability.

As the secondary mineralization textures of these samples indicate, permeability of the granodiorite in the damage zone changes continually, as dilational fracturing processes accompanying earthquakes create pathways for fluid flow that subsequently become clogged with newly precipitated minerals. Accordingly, the samples with the highest permeabilities also contain $\approx$30–60% secondary minerals (Fig. 7). Samples from the GSJ drillhole and the shallow and intermediate fault strands from the NIED drillhole characteristically contain fresh-looking dilational fractures and brecciated areas. In contrast, there is little evidence of recent fracturing in the samples from the deep NIED fault strand.

The deep NIED fault strand is characterized by permeability values that are closer to the country rock than to the other three fault strands, and the strength of one of the three samples approaches that of the country rock (Figs. 7b-c). The textures in the examined samples from the deep NIED fault strand support the conclusion of LOCKNER *et al.* (paper I) that this strand is an older, abandoned fault trace. Deformation largely ceased during the early laumontite-forming stage, but fluids continued to migrate through these rocks. As a result, nearly all the pore spaces in the microbreccia zones have been filled and most fractures are sealed. Unless active fracturing continues to create new void space, permeability will gradually decrease. In these samples from the deep NIED strand, permeability is returning to the original values of the granodiorite country rock more readily than strength is restored (Fig. 7). This implies that a sealed fracture or a cemented microbreccia zone is not particularly strong, as also illustrated by the crack-seal veins that have repeatedly fractured along the vein walls. Refracturing of some of the older microfractures in quartz suggests that even healed cracks are not as strong as the original quartz grains.

6. Concluding Remarks

Core samples obtained from deep drillholes that cross the active Nojima fault of Japan provide important new information about fault-zone processes in crystalline rock. This petrographic study has presented some general aspects of the microstructural evolution of the damage zone and the effects that the microstructures have on physical properties such as permeability and strength. Offset is concentrated within the core of a strand, although minor sheared zones are present in damage-zone rocks immediately adjacent to the fault cores. Deformation in the damage zone is largely the result of brittle fracturing processes that, at their peak, produce wide microbreccia zones and associated intensely fractured rocks. Fractures and pore spaces are partly filled with secondary minerals deposited from migrating solutions that have also partly altered the igneous mafic minerals and plagioclase. This scenario is consistent with the model of fault zone architecture and the systematic variations in fault-zone properties and processes over the seismic cycle described by CHESTER *et al.* (1993) from detailed observations of the San Gabriel and Punchbowl faults. The measured strength and permeability data for the core samples thus represent just one moment in a continually evolving system. Abandonment of a fault strand is accompanied by a gradual return of permeability and strength towards the country-rock values, although permeability recovers more readily than strength, and strength may never attain the original value.

The nearly continuous core from the GSJ and NIED drillholes is an extremely valuable resource for understanding the deformational and mineralogical evolution of a fault zone. The fault has not experienced considerable offset and the fault-zone geometry is reasonably simple. The deep NIED fault strand in particular is a critical source of information on the early stages of microstructural development of the fault zone that can be used to interpret the more heavily deformed rocks associated with the active fault strands.

References

Awata, Y., and Suzuki, Y. (1996), *Paleoseismology and activity study of the Nojima fault system, which generated the Hyogo-ken Nanbu earthquake of January 17, 1995*, Geol. Surv. Japan Open-File Report *250*, 4–8 (in Japanese).

Awata, Y., Mizuno, K., Sugiyama, Y., Imura, R., Shimokawa, K., Okumura, K., and Tsukuda, E. (1996), *Surface fault ruptures on the northwest coast of Awaji Island associated with the Hyogoken Nanbu earthquake of 1995*, J. Seismol. Soc. Japan *49*, 113–124 (in Japanese with English abstract).

Boullier, A.-M., Fujimoto, K., Ohtani, T., Roman-Ross, G., Lewin, E., Ito, H., Pezard, P., and Idlefonse, B. (2004), *Textural evidence for recent co-seismic circulation of fluids in the Nojima fault zone*, Awaji island, Japan, Tectonophysics *378*, 165–181.

Caine, J.S., Evans, J.P., and Forster, C.B. (1996), *Fault zone architecture and permeability structure*, Geology *24*, 1025–1028.

Célérier, B.P., Pezard, P.A., Ito, H., and Kiguchi, T. (2000), *Borehole wall geometry across the Nojima fault: BHTV acoustic scans analysis from the GSJ Hirabayashi hole, Japan.* In *The International Workshop on the Nojima Fault Core and Borehole Data Analysis* (eds. Ito, H., Fujimoto, K., Tanaka, H., and Lockner, D.), U. S. Geol. Surv. Open-File Report 00-129, pp. 233–238.

Chester, F.M., and Logan, J.M. (1986), *Implications for mechanical properties of brittle faults from observations of the Punchbowl fault zone, California.* In *Internal Structure of Fault Zones* (ed. Wang, C.-Y.), Pure Appl. Geophys. *124*, 79–106.

Chester, F.M., Evans, J.P., and Biegel, R.L. (1993), *Internal structure and weakening mechanisms of the San Andreas fault*, J. Geophys. Res. *98*, 771–786.

Fujimoto, K., Tanaka, H., Tomida, N., Ohtani, T., and Ito, H. (2000), *Characterization of fault gouge from GSJ Hirabayashi core samples and implications for the activity of the Nojima fault.* In *The International Workshop on the Nojima Fault Core and Borehole Data Analysis* (eds. Ito, H., Fujimoto, K., Tanaka, H., and Lockner, D.), U. S. Geol. Surv. Open-File Report 00-129, pp. 103–109.

Fujimoto, K., Ueda, A., Ohtani, T., Takahashi, M., Ito, H., Tanaka, H., and Boullier, A.-M. (2007), *Borehole water and hydrologic model around the Nojima fault, SW Japan*, Tectonophysics *443*, 174–182.

Ikeda, R. (2001), *Outline of the fault zone drilling project by NIED in the vicinity of the 1995 Hyogo-ken Nanbu earthquake, Japan*, Island Arc *10*, 199–205.

Ikeda, R., Iio, Y., and Omura, K. (2001), *In situ stress measurements in NIED boreholes in and around the fault zone near the 1995 Hyogo-ken Nanbu earthquake, Japan*, Island Arc *10*, 252–260.

King, G.C.P., and Sammis, C.G. (1992), *The mechanics of finite brittle strain,* Pure Appl. Geophys. *138*, 611–640.

Lin, A., and Uda, S. (1996), *Morphological characteristics of the earthquake surface rupture which occurred on Awaji Island, associated with the 1995 southern Hyogo Prefecture earthquake*, Island Arc *5*, 1–5.

Lin, A., Shimamoto, T., Maruyama, T., Sigetomi, M., Miyata, T., Takemura, K., Tanaka, H., Uda, S., and Murata, A. (2001), *Comparative study of cataclastic rocks from a drill core and outcrops of the Nojima Fault zone on Awaji Isand, Japan*, Island Arc *10*, 368–380.

Lockner, D., Naka, H., Tanaka, H., Ito, H., Ikeda, R., and Omura, K. (2009), *Geometry of the Nojima fault at Nojima-Hirabayashi, Japan – I. A simple damage structure inferred from borehole core permeability*, this volume.

Matsuda, T., Omura, K., Ikeda, R., Arai, T., Kobayashi, K., Shimada, K., Tanaka, H., Tomita, T., and Hirano, S. (2004), *Fracture-zone conditions on a recently active fault: Insights from mineralogical and geochemical analyses of the Hirabayashi NIED drill core on the Nojima fault, southwest Japan, which ruptured in the 1995 Kobe earthquake*, Tectonophysics *378*, 143–163.

Mizuno, K., Hattori, H., Sangawa, A., and Takahashi, Y. (1990), *Geology of the Akashi district*, quadrangle series, Geol. Surv. Japan, scale 1:50,000, 90 pp. (in Japanese with English abstract).

MOORE, D.E. (1987), *Syndeformational metamorphic myrmekite in granodiorite of the Sierra Nevada, California*, Geol. Soc. Am. Abstr. with Progr. *19*(7), 776.

MOORE, D.E. (1990), *Flame perthite associated with faulting in granodiorite, Mt. Abbot quadrangle, California*, EOS, Trans. Am. Geophys. Union *71*, 1596.

MOORE, D.E., and LOCKNER, D.A. (1995), *The role of microcracking in shear-fracture propagation in granite*, J. Struct. Geol. *17*, 95–114.

MOORE, D.E., and LOCKNER, D.A. (2004), *Crystallographic controls on the frictional behavior of dry and water-saturated sheet structure minerals*, J. Geophys. Res. *109*, B03401, doi: 10.1029/2003JB002582.

MOORE, D.E., HICKMAN, S., LOCKNER, D.A., and DOBSON, P.F. (2001), *Hydrothermal minerals and microstructures in the Silangkitang geothermal field along the Great Sumatran fault zone, Sumatra*, Geol. Soc. Am. Bull. *113*, 1179–1192.

MOORE, D.E., LOCKNER, D.A., ITO, H., and IKEDA, R. (2000), *Carbonate mineralization sequence and the earthquake history of the Nojima fault zone, Japan*, EOS, Trans. Am. Geophys. Union *81*, F1099–F1100.

MORROW, C.A., and BYERLEE, J.D. (1991), *A note on the frictional strength of laumontite from Cajon Pass, California*, Geophys. Res. Lett. *18*, 211–214.

MORROW, C., LOCKNER, D., HICKMAN, S., RUSANOV, M., and RÖCKEL, T. (1994), *Effects of lithology and depth on the permeability of core samples from the Kola and KTB drill holes*, J. Geophys. Res. *99*, 7263–7274.

MORROW, C.A., MOORE, D.E., and LOCKNER, D.A. (2000), *The effect of mineral bond strength and adsorbed water on fault gouge frictional strength,* Geophys. Res. Lett. *27*, 815–818.

MURAKAMI, M., and TAGAMI, T. (2004), Dating pseudotachylyte of the Nojima fault using the zircon fission-track method, Geophys. Res. Lett. *31*, L12604, doi:10.1029/2004GL020211.

MURATA, A., TAKEMURA, K., MIYATA, T., and LIN, A. (2001), *Quaternary vertical offset and average slip rate of the Nojima Fault on Awaji Island, Japan*, Island Arc *10*, 360–367.

OHTANI, T., FUJIMOTO, K., ITO, H., TANAKA, H., TOMIDA, N., and HIGUCHI, T. (2000), *Fault rocks and past to recent fluid characteristics from the borehole survey of the Nojima fault ruptured in the 1995 Kobe earthquake, southwest Japan*, J. Geophys. Res. *105*, 16,161–16,171.

OTSUKI, K., MONZAWA, N., and NAGASE, T. (2003), *Fluidization and melting of fault gouge during seismic slip: Identification in the Nojima fault zone and implications for focal earthquake mechanisms*, J. Geophys. Res. *108*, B4, 2192, doi:10.1029/2001JB001711.

SENO, T. (1999), *Syntheses of the regional stress field of the Japanese islands*, Island Arc *8*, 66–79.

STRECKEISEN, A. (1974), *Classification and nomenclature of plutonic rocks*, Geol. Rundschau *63*, 773–785.

TAGAMI, T., and MURAKAMI, M. (2007), *Probing fault zone heterogeneity on the Nojima fault: Constraints from zircon fission-track analysis of borehole samples*, Tectonophysics *443*, 139–152.

TAKAHASHI, Y. (1992), *K-Ar ages of the granitic rocks in Awaji Island — with an emphasis on timing of mylonitization*, Gankou *87*, 291–299 (in Japanese with English abstract).

TAKESHITA, T., and YAGI, K. (2000), *Dynamic analysis based on 3-D orientation distribution of microcracks in quartz from the Cretaceous granodiorite core samples drilled along the Nojima fault, southwest Japan*. In *The International Workshop on the Nojima Fault Core and Borehole Data Analysis* (eds. Ito, H., Fujimoto, K., Tanaka, H., and Lockner, D.), U. S. Geol. Surv. Open-File Report 00-129, pp. 133–140.

TAKESHITA, T., and YAGI, K. (2001), *Paleostress orientation from 3-D orientation distribution of microcracks in quartz from the Cretaceous granodiorite core samples drilled through the Nojima fault, southwest Japan*, Island Arc *10*, 495–505.

TANAKA, H., FUJIMOTO, K., OHTANI, T., and ITO, H. (2001), *Structural and chemical characterization of shear zones in the freshly activated Nojima fault, Awaji Island, southwest Japan*, J. Geophys. Res. *106*, 8789–8810.

TANAKA, H., OMURA, K., MATSUDA, T., IKEDA, R., KOBAYASHI, K., MURAKAMI, M., and SHIMADA, K. (2007), *Architectural evolution of the Nojima fault and identification of the activated slip layer by Kobe earthquake*, J. Geophys. Res. *112*, B07304, doi:10.1029/2005JB003977.

VOLLBRECHT, A., RUST, S., and WEBER, K. (1991), *Development of microcracks in granites during cooling and uplift: Examples from the Variscan basement in NE Bavaria, Germany*, J. Struct. Geol. *13*, 787–799.

WONG, T.-F. (1982), *Micromechanics of faulting in Westerly granite*, Int. J. Rock Mech. Mining Sci. Geomech. Abstr. *19*, 49–64.

YAMAMOTO, K., SATO, N., and YABE, Y. (2000), *Stress state around the Nojima fault estimated from core measurements*. In *The International Workshop on the Nojima Fault Core and Borehole Data Analysis* (eds. Ito, H., Fujimoto, K., Tanaka, H., and Lockner, D.), U. S. Geol. Surv. Open-File Report 00-129, pp. 239–246.

(Received September 29, 2008, revised March 6, 2009)
Published Online First: June 30, 2009

To access this journal online:
www.birkhauser.ch/pageoph

Pure appl. geophys. 166 (2009) 1693–1724
0033–4553/09/101693–32
DOI 10.1007/s00024-009-0518-x

© Birkhäuser Verlag, Basel, 2009

Pure and Applied Geophysics

The Energetics of Cataclasis Based on Breakage Mechanics

GIANG D. NGUYEN and ITAI EINAV

Abstract—We develop a constitutive model for rocks that are constituted from brittle particles, based on the theory of breakage mechanics. The model connects between the energetics and the micromechanics that drive the process of confined comminution. Given this ability, our model not only describes the entire stress-strain response of the material, but also connects this response to predicting the evolution of the grain size distribution. The latter fact enables us to quantify how the permeability reduces within cataclasite zones, in relation to aspects of grain crushing. Finally, our paper focuses on setting a framework for quantifying how the energy budget of earthquakes is expensed in relation to dissipation events in cataclasis. We specifically distinguish between the dissipation directly from the creation of new surface area, which causes further breakage dissipation from the redistribution of locked-in stored energy from surrounding particles, dissipations from friction and from the configurational reorganisation of particles.

Key words: Cataclasis, fault gouge, breakage, permeability, energy balance, fracture energy.

1. Introduction

The stability of fault gouges holds the key to understanding earthquake dynamics (e.g., MARONE *et al.*, 1990; SCHOLZ, 1990; BEN-ZION and SAMMIS, 2003), and the associated release of energy (e.g., SLEEP and BLANPIED, 1992; RECHES and DEWERS, 2005). Fault gouges evolve in high pressure or even ultrahigh-pressure environments of cataclasis (e.g., MORROW *et al.*, 1981; LUND and AUSTRHEIM, 2003), commonly because of the severe motion of the lithospheric plate boundaries. In fault gouges, cataclastic rocks undergo lifecycles of healing and granulation (MARONE *et al.*, 1995; OLSEN *et al.*, 1998; RENARD *et al.*, 2000; TENTHOREY *et al.*, 2003; HEILBRONNER and KEULEN, 2006; BEN-ZION, 2008). During the granulation stage the parent rocks initially disintegrate into the product of granular minerals, which may then be furthered crushed during the tectonic slip. The grain size distribution (*gsd*) is therefore an evolving property during the slip, which starts from a healed agglomerated rock mass and becomes attracted towards an ultimate grading that is often found (or conveniently assumed) to be fractal, complying both self-similarity and power-law scaling (SAMMIS *et al.*, 1986; STEACY and SAMMIS, 1991; An and SAMMIS, 1994). Experiments show that this continuous shift of the *gsd* towards

School of Civil Engineering J05, The University of Sydney, Sydney, NSW 2006, Australia.
E-mail: i.einav@civil.usyd.edu.au

smaller-sized particles governs the inelastic behaviour of the material. Dilation shear bands are observed in geomaterials at low confining pressure, however in high-pressure cataclastic shear, the material contracts because it is easier for the particles to crush than dilate; instead shear-enhanced compaction bands tend to form (e.g., MANDL *et al.*, 1977; SCHULTZ and SIDDHARTHAN, 2005; AYDIN *et al.*, 2006). This contraction effect is important to quantify as it may result in the build-up of internal pore pressures, reduction of effective pressures and possibly the trigger of fault instability (BEELER *et al.*, 1996; ZHANG and TULLIS, 1998; GUO and MORGAN, 2007).

Cataclasis is a severe tectonic process that includes the distortion, sliding, fracture and bending of sublayers. These mechanisms dissipate energy, which controls the amount of energy release. A prevailing assumption was made that in cataclasis grain-crushing is a negligible contributor, compared to the earthquake energy budget in general and frictional shear dissipation in specific (e.g. OLGAARD and BRACE, 1983; SCHOLZ, 1990; CHESTER *et al.*, 2005). We examine the energy balance issue from a mechanistic direction. In particular, we uniquely distinguish between two seemingly similar, but different entities: the dissipation from grain crushing, or as we call it 'breakage dissipation', and from the creation of new surface area, or 'surface area dissipation'. Breakage dissipation embodies the surface area dissipation, but also the dissipation to the surrounding from the redistribution of locked-in strain energy; this latter factor, in fact, appears more significant in value, though triggered by the same grain crushing event. Using a mechanical analog, a simple demonstration of this point is given in the Appendix.

A more general account of these issues of cataclasis is explored here however primarily using the theory of 'breakage mechanics' (EINAV, 2007a,b). While the breakage mechanics theory bears striking links to fracture mechanics (EINAV, 2007c), it is directly linked to the statistics of the *gsd*. The introduction of the concept of Breakage, *B*, allows the development of a new hierarchy of constitutive models, which for the first time can consistently account for the effect of the *gsd* as a predictable constitutive variable. An alternative understanding of energy release is therefore motivated herein by advancing this hierarchical development for the study of cataclasis problems, through the establishment of a realistic and useful model version. The model connects between the energetics and the micromechanics that drive the process of confined comminution.

Models based on damage mechanics have been used extensively (e.g., LYAKHOVSKY *et al.*, 1997; RICARD and BERCOVICI, 2003; HAMIEL *et al.*, 2004) to study the mechanical behaviour of rocks and the formation of shear bands at appropriate scales (e.g. Fig. 1). In these models, the use of a scalar damage variable may characterize the underlying evolution of the micro-structures, if the representative volume can be considered large enough to allow viewing the distribution of the internal flaws as homogeneous (HAMIEL *et al.*, 2004). On the other hand the use of Critical State Soil Mechanics (CSSM) has also been adopted (e.g. SHAH, 1997; CUSS *et al.*, 2003; SHELDON *et al.*, 2006) to analyse the mechanical and permeability evolution of faulted rocks. Our study focuses on the material within the cataclasite layer (Fig. 1), where the emergence of granular-like behaviour is expected. Therefore, rather than stretching the limits of CSSM, by

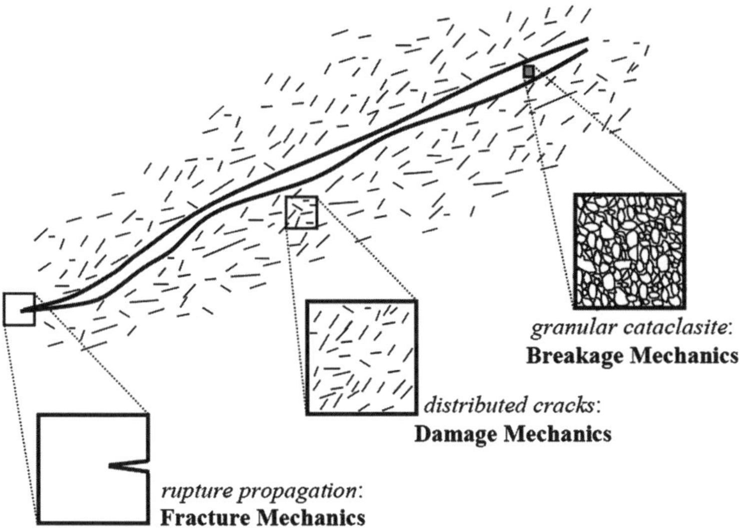

Figure 1
Fault gouge, cataclasis and the scope of breakage mechanics.

introducing *ad hoc* equations and parameters, we attack the problem from a new and rather more physical direction, illustrating how breakage mechanics allows prediction of the permeability reduction in a natural way by tracking the *gsd*. The constitutive behaviour and the particular fabric formation in the deformation bands of geomaterials depend on the loading paths (SCHULTZ and SIDDHARTHAN, 2005; AYDIN *et al.*, 2006). For that purpose our predictions will cover a wide range of loading conditions, without adjusting a few physically meaningful parameters used in breakage mechanics (EINAV, 2007b,c). Our mathematical advances of the model concentrate on developing Hertzian-like (hyper)elastic stiffness matrix and the related stored free energy potential. We then introduce this into the model of the breakage mechanics formulation, and quantify the energetics in cataclasis.

2. A Model Based on Breakage Mechanics

2.1. Fundamental Constitutive Equations

The evolution of *gsd* governs the mechanical properties of crushable geomaterials in shear-enhanced cataclastic bands (e.g., WONG *et al.*, 1997; ZHU and WONG, 1997; OLSSON, 1999; SCHULTZ and SIDDHARTHAN, 2005). An ideal constitutive model for cataclastic materials should be able to predict the evolution of the *gsd*, which will guide the predictions of the mechanical behaviour. This is an essential feature that, at least until recently, has not been addressed appropriately.

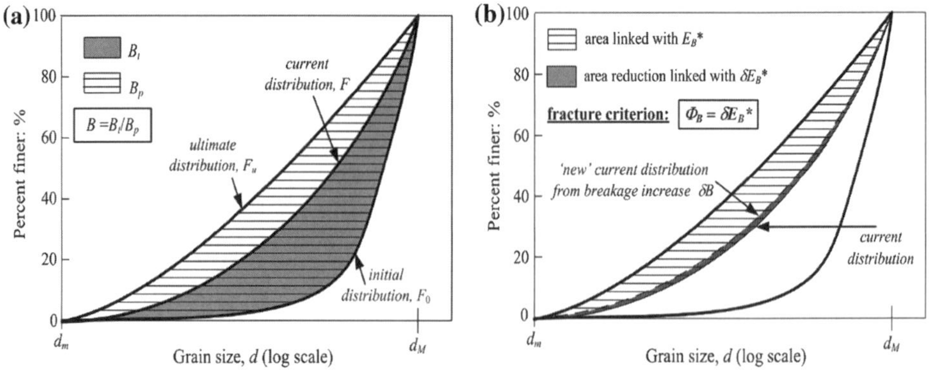

Figure 2
The breakage measurement and evolution law (EINAV, 2007a,c). The left figure (a) portrays the measurable definition of breakage. The right diagram (b) presents the breakage propagation criterion for granular materials. Φ_B is the breakage dissipation, denoting the energy consumption from incremental increase of breakage. δE_B^* is the incremental reduction in the residual breakage energy.

The formulation of a thermodynamically consistent constitutive model that is able to predict the evolving *gsd* in granular materials has been proposed by EINAV (2007a,b). The theory allows tracking the relative distance of the current (evolving) *gsd* ($p(d)$; d is the grain diameter) from the initial ($p_0(d)$) and ultimate ($p_u(d)$) *gsd*'s via the use of a macroscopic internal variable B called 'breakage' (see Fig. 2a):

$$p(d) = (1 - B)p_0(d) + Bp_u(d). \tag{1}$$

It was therefore termed "continuum breakage mechanics", in analogy with the term "continuum damage mechanics" in solid mechanics. In turn, predicting the breakage allows estimating the current *gsd*, given assumed initial and ultimate *gsd*'s. Breakage is therefore a measurable state quantity that could be directly deduced by measuring the *gsd*.

In thermodynamic context, the form of the Helmholtz energy potential for breakage mechanics models is obtained from a statistical homogenization process, taking into account the effects of the *gsd* on the macroscopic behaviour of the material model (EINAV, 2007a):

$$\Psi = (1 - \vartheta B)\psi_r(\boldsymbol{\varepsilon}), \tag{2}$$

where $\psi_r(\boldsymbol{\varepsilon})$ is expressed in terms of the total strain tensor $\boldsymbol{\varepsilon}$ and represents the elastic strain energy in a reference grain size

$$d_r = \sqrt{\int_{d_m}^{d_M} d^2 p_0(d)\, \mathrm{d}d} = \langle d^2 \rangle_0^{1/2}. \tag{3}$$

It can be seen from Eq. (2) that the reference elastic strain energy $\psi_r(\boldsymbol{\varepsilon})$ represents the virgin elastic strain energy before crushing commences. The special form of the reference particle size d_r was obtained via the execution of statistical homogenization, averaging the elastic strain energy stored in the various fraction sizes. The proximity index property ϑ in (2) is a result of the same statistical homogenization procedure:

$$\vartheta = 1 - \left\langle d^2 \right\rangle_u \Big/ \left\langle d^2 \right\rangle_0 \tag{4}$$

and can be seen as a "distance" measure between the initial and ultimate *gsd*'s, with $\left\langle d^2 \right\rangle_u$ and $\left\langle d^2 \right\rangle_0$ being the second order moments of the ultimate and initial grain size distribution functions (by mass).

From (2), the energy conjugated to the breakage variable is defined as:

$$E_B = -\frac{\partial \Psi}{\partial B} = -\vartheta \psi_r(\boldsymbol{\varepsilon}). \tag{5}$$

The crushing of particles creates new surface areas which leads to energy release via the dissipation (termed Φ_B) of the breakage process. It was proposed that the breakage dissipation is driven by the loss of strain energy in the particles, as schematically portrayed in Figure 2b and mathematically expressed by:

$$\Phi_B = \delta E_B^*, \tag{6}$$

where

$$E_B^* = (1 - B)E_B \tag{7}$$

is the residual breakage energy, that has the physical meaning of the *available* energy in the system for the crushing process (EINAV, 2007a). Eq. (6) is an important feature in exploring the relationship between the input energy (e.g., from an earthquake or a mechanical process) and the energy released due to crushing of particles.

The yield criterion, which signifies the onset and then governs the evolution of breakage can be readily worked out from (6) and (7) (EINAV, 2007c):

$$y_B = (1 - B)^2 E_B - E_c = 0, \tag{8a}$$

where E_c, a constant arising from the derivation process and termed the critical breakage energy, is an energy constant (with stress units) of the granular material. Later in section 2.3 (see eqs. (26) and (34)) we can see that E_c in the formula for critical pressure has a role similar to Griffith's fracture energy constant G_c (GRIFFITH, 1921) in the critical stress formula in fracture mechanics. However, their physical interpretations are different. The introduction of the yield criterion required the specification of the analysis to rate-independent processes, suggesting that during dissipation $E_B = \partial \Phi_B / \partial B$, and $\Phi_B = (\partial \Phi_B / \partial B)\delta B$.

In addition, a link between the breakage measure and the changes of the specific surface areas was derived (EINAV, 2007c), suggesting a deeper look at the model from the viewpoint of fracture mechanics:

$$B = \frac{S - S_0}{S_u - S_0}, \tag{8b}$$

where the initial and ultimate specific surface areas (the averaged surface area of particles divided by their volumes; unit of 1/length, e.g. 1/mm) are defined respectively as $S_0 = 6\langle d^{-1}\rangle_0$ and $S_u = 6\langle d^{-1}\rangle_u$. This relationship between B and S, as two dependent internal variables of a thermo-mechanical process, suggests that the breakage dissipation Φ_B can be viewed in terms of the current specific surface area S and its associated energy $E_S = \partial\Psi/\partial S = E_B(S_u - S_0)$:

$$\Phi_B = E_B \delta B = E_S \delta S, \tag{9}$$

where E_B has the unit of energy (e.g., N/mm^2), and E_S unit of energy times length (e.g., N/mm^2*mm). The last relation resembles the increment of non-negative entropy production, as defined by RICE (1978). The property E_S is parallel to specific surface energy γ as in Griffith's analysis. However, we note that breakage mechanics accounts for the redistribution of the elastic stored energy in the particles, as we will further discuss in the following.

2.2. *Model Formulation*

Based on the above fundamentals and earlier developments in EINAV (2007b,c), our current formulation will account for the coupling between the release of surface energy and other dissipative mechanisms, namely the dissipations due to frictional shear and the reorganisation of the granular ensemble following crushing. A triaxial exposition is used in the formulation, for which standard notations in soil mechanics are employed as follows: Mean effective stress p (positive in compression), shear stress q, total volumetric strain ε_v and elastic volumetric strain ε_v^e (positive in contraction), total shear strain ε_s and elastic shear strain ε_s^e. The Helmholtz free energy potential Ψ and dissipation potential Φ, from which all other constitutive equations are derived, take the following general forms:

$$\Psi = (1 - \vartheta B)\left[\psi_v\left(\varepsilon_v^e\right) + \psi_s\left(\varepsilon_v^e, \varepsilon_s^e\right)\right], \tag{10}$$

$$\Phi = \sqrt{\Phi_B^2 + \Phi_p^{v2} + \Phi_p^{s2}}. \tag{11}$$

Use of Ψ in its general form (10) allows the incorporation of both pressure-independent and pressure-dependent features into the elastic behaviour of the material. Functions ψ_v and ψ_s in (10) govern the elastic volumetric and shear behaviours of the model, and represent the 'unbroken' stored energy in a reference particle size d_r; this paper places particular attention on refining the particular structures of ψ_v and ψ_s.

In this regard breakage can be experimentally measured with reference to Figure 2a. The dissipation potential Φ comprises three parts corresponding to breakage dissipation Φ_B, plastic volumetric dissipation Φ_p^v and plastic shear dissipation Φ_p^s :

$$\Phi_B = \frac{\sqrt{2E_B E_c}}{(1-B)}\delta B, \tag{12}$$

$$\Phi_p^v = \frac{p}{(1-B)}\sqrt{\frac{2E_c}{E_B}}\delta\varepsilon_v^p, \tag{13}$$

$$\Phi_p^s = Mp\left|\delta\varepsilon_s^p\right|, \tag{14}$$

where Φ_B was derived from the energy balanced criterion of Figure 2b, Φ_p^v was obtained by applying the understanding that in isotropic compression particle reorganisation is only passive to breakage, and Φ_p^s was derived by consulting with Coulomb's friction law. In the above equations $M = q_u/p_u$ is the ratio between the ultimate shear stress q_u and ultimate volumetric stress p_u at failure (i.e., related to the ultimate mobilised friction angle); finally, ε_v^p and ε_s^p are the volumetric and shear plastic strains. Note that the above model does not involve the coupling angle ω, as explored by EINAV (2007b); instead, this angle is conveniently taken as 45°. It will be shown later that this simplification results in the same relationship between the incremental plastic volumetric and shear strains as in the modified Cam clay model (ROSCOE and SCHOFIELD, 1963).

Via Eqns. (11)–(14), we identify three components of dissipation: frictional dissipation Φ_p^s, dissipation from the configurational reorganisation of particles Φ_p^v, and from breakage Φ_B. Here, the 'reorganisation dissipation' Φ_p^v is simply represented by the volumetric plastic dissipation, which relates to the increasing capacity of the particles to organise within fixed volumes, since the sub-elements are growing in numbers, i.e., representing dissipation in the sense of increasing configurational entropy via changing granular compactivity (EDWARDS and GRINEV, 2001). We note that while the breakage dissipation Φ_B can be calculated alternatively in relation to the increasing specific surface area (EINAV, 2007c), its value does not correspond entirely to 'surface area dissipation'. Most of the breakage dissipation arises from the redistribution of locked-in stored energy. An analog that portrays this point is given in the Appendix.

The constitutive equations can be defined as the derivatives of the potentials as follows:

$$p = \frac{\partial\Psi}{\partial\varepsilon_v^e} = (1-\vartheta B)\left(\frac{\partial\psi_v}{\partial\varepsilon_v^e} + \frac{\partial\psi_s}{\partial\varepsilon_v^e}\right), \tag{15}$$

$$q = \frac{\partial\Psi}{\partial\varepsilon_s^e} = (1-\vartheta B)\frac{\partial\psi_s}{\partial\varepsilon_s^e} \tag{16}$$

$$E_B = -\frac{\partial\Psi}{\partial B} = \vartheta\left[\psi_v\left(\varepsilon_v^e\right) + \psi_s\left(\varepsilon_v^e, \varepsilon_s^e\right)\right]. \tag{17}$$

The yield function y^* in dissipative stress space is obtained as a result of the degenerate Legendre transformation of the dissipation potential (11):

$$y^* = \left(\frac{E_B}{\partial\Phi_B/\partial\delta B}\right)^2 + \left(\frac{p}{\partial\Phi_p^v/\partial\delta\varepsilon_v^p}\right)^2 + \left(\frac{q}{\partial\Phi_p^s/\partial\delta\varepsilon_s^p}\right)^2 - 1 \leq 0 \tag{18}$$

with the following flow rules ($\delta\lambda$ is a common non-negative multiplier):

$$\delta B = \delta\lambda\frac{\partial y^*}{\partial E_B} = 2\delta\lambda\frac{E_B}{(\partial\Phi_B/\partial\delta B)^2} = 2\delta\lambda\frac{(1-B)^2}{2E_c}, \tag{19}$$

$$\delta\varepsilon_v^p = \delta\lambda\frac{\partial y^*}{\partial p} = 2\delta\lambda\frac{p}{\left(\partial\Phi_p^v/\partial\delta\varepsilon_v^p\right)^2} = 2\delta\lambda\frac{E_B(1-B)^2}{2pE_c}, \tag{20}$$

$$\delta\varepsilon_s^p = \delta\lambda\frac{\partial y^*}{\partial q} = 2\delta\lambda\frac{q}{\left(\partial\Phi_p^s/\partial\delta\varepsilon_s^p\right)^2} = 2\delta\lambda\frac{q}{(Mp)^2}. \tag{21}$$

Using (12)–(14), y^* can be replaced by the mixed breakage/yield function:

$$y_{mix} = \frac{E_B(1-B)^2}{E_c} + \left(\frac{q}{Mp}\right)^2 - 1 \leq 0. \tag{22}$$

It should be noted that in principle y_{mix} can be further rewritten, fully in terms of triaxial stresses p and q, and the breakage B, provided the explicit expressions of ε_v^e and ε_s^e in terms of p and q are obtained from (15) and (16). In this case, we will redesignate the yield function again by replacing the notation y_{mix} with y. Non-associated flow rules for breakage and plastic shear strain are a direct consequence of the derivation procedure and no assumption relative to the existence of a plasticity/breakage potential is required.

Using (20), (21) and the yield condition (22), the ratio between the incremental plastic volumetric and shear strains is obtained as:

$$\frac{\delta\varepsilon_v^p}{\delta\varepsilon_s^p} = \frac{M^2 - \eta^2}{2\eta} \quad \text{with} \quad \eta = \frac{q}{p}.$$

Perhaps attractive, in structure the above ratio coincides with that of the modified Cam clay model (Roscoe and Schofield, 1963), but the current model is obviously different. On the other hand, this result motivates us to take ω as 45°, as discussed earlier.

2.3. *Hierarchy in Model Development*

It is interesting to note that in breakage mechanics the elastic properties directly affect post-yield behaviour. This is due to the presence of the energy-like quantity E_B in the mixed breakage/yield function (22). E_B is a function of the elastic stored energy, and governs the rate of dissipation during the crushing process. Consequently, the hardening of the material during comminution is directly driven by the competition between E_B and

$(1-B)^2$: as B grows towards unity, E_B must increase to balance the equality in eq. (22). Since E_B depends on the elastic strains via eq. (17), changing the elastic law will directly change the model response in the inelastic regime. Subsequently, improving the model behaviour in the elastic regime may automatically result in positive improvements in the inelastic regime. We believe that this aspect is a genuine step ahead, which allows us to be more economical in introducing model parameters, all by applying first principles. The best example is our capability to eliminate parameters such as the so-called normally-consolidated compression index λ in CSSM, which bears no direct physical meaning but is purely a fitting parameter. This economy in model parameters is discussed in length by EINAV (2007d), highlighting that the modelling of crushable granular materials using CSSM should require at least four additional *adhoc* parameters (slope λ of the isotropic hardening compression line, specific volume parameter N, and two additional parameters to span the range of λ and N) to achieve the same level of detail.

In the following we extend the hierarchy of Einav's breakage model by refining the structure of the reference stored energy functions ψ_v and ψ_s. We start by illustrating the effects of using linear elasticity, but then replace this by developing a nonlinear hyperelastic Hertzian-like model of ψ_v and ψ_s. Both models require only 4 physical parameters, however applying the more realistic elastic law clearly improves the predictions.

2.3.1 Use of linear isotropic elasticity. Linear elasticity is the simplest elastic law possible, which requires two constant parameters, e.g., the bulk and shear moduli K and G. For this purpose, ψ_v and ψ_s are defined as follows:

$$\psi_v = \frac{1}{2}K\varepsilon_v^{e^2}, \tag{23}$$

$$\psi_s = \frac{3}{2}G\varepsilon_s^{e^2}. \tag{24}$$

In this case, the breakage/yield surface in true triaxial stress space is written as:

$$y = \frac{\vartheta}{2E_c}\left(\frac{1-B}{1-\vartheta B}\right)^2\left(\frac{p^2}{K}+\frac{q^2}{3G}\right)+\left(\frac{q}{Mp}\right)^2-1 \leq 0. \tag{25}$$

Solving Eq. (25) for $q = 0$ and $B = 0$, the crushing pressure p_c in isotropic compression is given by:

$$p_c = \sqrt{\frac{2KE_c}{\vartheta}}. \tag{26}$$

As discussed by EINAV (2000c), this relation bears a striking relation to Griffith's criterion in fracture mechanics. Figure 3a shows the breakage/yield surface and its evolution in triaxial stress space. This simple model was motivated by EINAV (2007d) as a "student model" to illustrate certain features of breakage mechanics models. In the next section

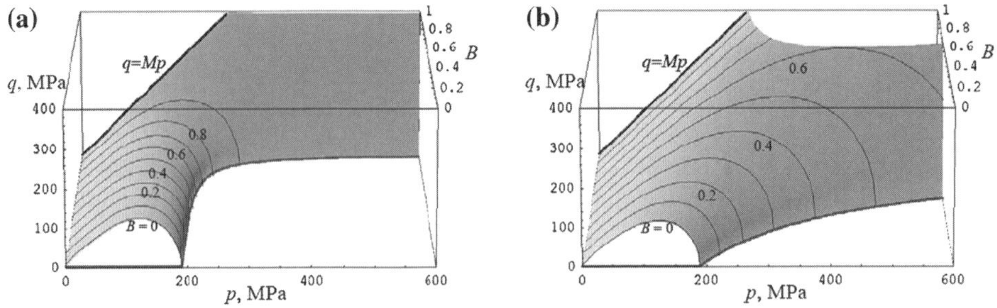

Figure 3
Yield/breakage surface in *p-q-B* space, using: (a) linear elasticity (Einav, 2007d), and (b) pressure-dependent elasticity. While the $B = 0$ projection planes are quite similar, the projection planes along progressing B values are of different sizes. (To generate these plots, parameters were taken in relation to the Adamswiller sandstone, as listed in Table 1).

we further advance the model, slightly increasing the complexity, but not the number of parameters.

2.3.2 Use of pressure dependent elasticity. Enhancement to the model response can be made by developing appropriate nonlinear (hyper)elastic terms ψ_v and ψ_s, which account for pressure-dependent bulk and shear moduli. From eqs. (15) and (16), the incremental stress-strain behaviour of the model in the elastic regime ($\delta B = 0$) is given as $\{\delta p, \delta q\}^T = (1 - \vartheta B)\mathbf{D}\{\delta\varepsilon_v^e, \delta\varepsilon_s^e\}^T$, where $\mathbf{D}$ is the elastic tangent stiffness matrix in triaxial stress space, defined by:

$$\mathbf{D} = \begin{bmatrix} \left(\dfrac{\partial^2\psi_v}{\partial\varepsilon_v^{e^2}} + \dfrac{\partial^2\psi_s}{\partial\varepsilon_v^{e^2}}\right) & \dfrac{\partial^2\psi_s}{\partial\varepsilon_v^e\partial\varepsilon_s^e} \\ \dfrac{\partial^2\psi_s}{\partial\varepsilon_s^e\partial\varepsilon_v^e} & \dfrac{\partial^2\psi_s}{\partial\varepsilon_s^{e^2}} \end{bmatrix}. \tag{27}$$

Therefore the general form of ψ_v and ψ_s introduces both the presence of cross-coupling and additional bulk terms in the elastic stiffness tensor. These terms are absent in the linear elastic model, but present in non-linear elastic models, and correspond to stress-induced anisotropy and elastic dilation, which is experimentally observed in granular materials (e.g., LoPresti and O'Neill, 1991; Hoque and Tatsuoka, 1998). Similar understanding has been motivated in previous studies of the (hyper)elasticity of granular materials (Zytynski *et al.*, 1978; Houlsby, 1985; Einav and Puzrin, 2004; Houlsby *et al.*, 2005).

In the development of the pressure-dependent model, we first assume the function ψ_v to describe the behaviour of the model in isotropic compression, which is then used to guide the determination of function ψ_s, thus revealing the appropriate stress-induced cross anisotropy terms. The following form of ψ_v is adopted from Einav (2007b):

$$\psi_v = p_r \frac{\left[\xi\left(\varepsilon_v^e\right)/p_r\right]^{2-m}}{\bar{K}(2-m)}, \tag{28}$$

where $\xi\left(\varepsilon_v^e\right) = p_r \sqrt[1-m]{\bar{K}(1-m)\left(\varepsilon_v^e - \varepsilon_{vr}\right)+1}$ and p_r is defined as a reference pressure conveniently taken as 1 kPa; $m = 0.5$ (for Hertzian contact), $\bar{K}$ and $\bar{G}$ are nondimensional material constants, which replace the dimensional K and G, and ε_{vr} the initial volumetric strain dependent on the maximum $e_{\max}$ and initial e_0 void ratios of the sample (EINAV, 2007d): $\varepsilon_{vr} = -\log((1+e_0)/(1+e_{\max}))$. Since experimental data for $e_{\max}$ are not available in our comparisons, for simplicity, we take the reference strain as $\varepsilon_{vr} = 0$.

The reference shear stored energy ψ_s can be conveniently assumed to decompose into two multiplicative parts, with the shear modulus being pressure-dependent:

$$\psi_s = \frac{3}{2} p_r \bar{G} \left[\frac{\xi\left(\varepsilon_v^e\right)}{p_r}\right]^m \varepsilon_s^{e^2}. \tag{29}$$

The stress-dependent elastic stiffness matrix (27) in this case ($m = 0.5$) becomes:

$$\mathbf{D} = \begin{bmatrix} p_r\bar{K}A & \frac{q\bar{K}}{2A} \\ \frac{q\bar{K}}{2A} & 3p_r\bar{G}A \end{bmatrix}, \tag{30}$$

where $A = \sqrt{\xi/p_r} = \bar{K}\varepsilon_v^e/2+1$ is purely stress-dependent, obtained by replacing the volumetric elastic strain by solving the system of eqs. (15)–(16) in terms of triaxial stresses p and q:

$$A = \sqrt{\frac{p^*}{2} + \sqrt{\left(\frac{p^*}{2}\right)^2 - \frac{1}{12}\frac{\bar{K}}{\bar{G}} q^{*2}}}, \tag{31}$$

where $p^* = p/p_r$ and $q^* = q/p_r$. Along the isotropic compression axis ($q = 0$), the apparent shear and bulk modulii are given as:

$$K = \bar{K}\sqrt{p^*} \quad \text{and} \quad G = \bar{G}\sqrt{p^*}. \tag{32}$$

Using (28), (29) and (31), the breakage/yield surface can be rewritten in terms of triaxial stresses as:

$$y = \vartheta\left[\frac{2p_rA^3}{3\bar{K}} + \frac{q^2}{6p_r\bar{G}A(1-\vartheta B)^2}\right]\frac{(1-B)^2}{E_c} + \left(\frac{q}{Mp}\right)^2 - 1 \leq 0. \tag{33}$$

In this case, the crushing pressure p_c in isotropic compression becomes:

$$p_c = p_r\left(\frac{3\bar{K}E_c}{2p_r\vartheta}\right)^{2/3} \tag{34}$$

corresponding to a relation given by EINAV (2007c), though here completed through eq. (33). The effects of change in elastic behaviour on the shape of yield/breakage surface can be seen in Figure 3. Along with the difference observed in the evolution of the breakage/yield surface (Fig. 3b) with increasing breakage, compared to the case of linear isotropic elasticity (Fig. 3a), the response of the model is greatly improved, as we will demonstrate in the next sections.

The use of pressure-dependent (hyper)elasticity requires the elimination of instability issues, even in elastic regime. This is due to the presence of cross-coupling terms in the elastic stiffness matrix (HOULSBY, 1985; BORJA *et al.*, 1997; EINAV and PUZRIN, 2004; HOULSBY *et al.*, 2005). Using (22), the stability condition det $(\mathbf{D}) = 0$ results in:

$$\eta = \frac{q}{p} = \sqrt{\frac{3\bar{G}}{2m\bar{K}}}. \tag{35}$$

It can be shown that the above expression coincides with the expression for the maximum attainable stress ratio $\eta_{\max}$, which represents the tangent of all stress curves in elastic undrained loading. Since the complete breakage mechanics model response is bounded by the friction law $q = Mp$ (see eq. (14) and Fig. 3), it seems reasonable to restrict the choice of the parameters $\bar{G}$ and $\bar{K}$ via:

$$\frac{\bar{G}}{\bar{K}} > \frac{2mM^2}{3} = \frac{M^2}{3}(\text{for } m = 0.5). \tag{36}$$

Combining the two alternative forms of ψ_v and ψ_s with the general formulation structure completes the representation of two models.

2.4. *Model Parameters*

Besides the two parameters that determine the elastic behaviour (G and K with unit of stress; or $\bar{G}$ and $\bar{K}$ without a unit), which can be determined based on the stress-strain behaviour of the material in elastic regime, the model depends on only two additional parameters (E_c and M), both of which possess physical meaning. Additionally, the model behaviour also depends on the criticality proximity index property (i.e., not a parameter) $\vartheta = 1 - \langle d^2 \rangle_u / \langle d^2 \rangle_0$ (EINAV, 2007a). The initial *gsd* can most often be assumed comfortably, if not by using one of the various sieving techniques, by using one of the image analysis techniques. The ultimate *gsd* can often be assumed fractal, with a fractal dimension 2.5–2.8 (SAMMIS *et al.*, 1986), although in pathological cases it may be different (EINAV, 2007a).

The critical breakage energy E_c, which represents the elastic strain energy per unit volume stored in the material before crushing commences and governs the activation of the grain breaking process in the model, may be determined via a single isotropic compression test (eqs. (26) and (34)). In particular, the limiting breakage/yield pressure p_c in isotropic compression is directly related to E_c, through the expression of the yield condition (22) in the absence of the triaxial shear stress q. It is noted that E_c only decides the onset but not the evolution of the dissipative process, which is in fact driven by the breakage energy E_B. The total dissipation can be determined by integrating the explicitly defined dissipation potential (eq. (11); see also section 5.1).

The last parameter is the slope M in (22), which is in fact the ratio between the triaxial shear stress q and the pressure p at ultimate critical state conditions. This parameter relates to the mobilised friction angle of the material.

In brief, only four parameters: G (or $\bar{G}$), K or $(\bar{K})$, E_c, and M are needed to complete the definition of the model and its behaviour in both elastic and inelastic regimes.

3. Permeability Reduction

The permeability in cataclasite zones tends to reduce, both from volumetric changes and the increasing surface area of the particles, which can thereafter act as a permeability barrier for the adjacent flow regime (CAINE *et al.*, 1996; EVANS *et al.*, 1997; TENTHOREY *et al.*, 2003). This factor is extremely important since the changes to the flow regimes may affect the surrounding pore fluid pressures, hence the local strengths, and overall stability of the fault during earthquake cycles (SEGALL and RICE, 1995; ZHANG and TULLIS, 1998). Experiments on the material behaviour in shear bands have indicated the importance of the *gsd* in governing the evolution of the permeability (WONG *et al.*, 1997; ZHU and WONG, 1997; OKADA *et al.*, 2004), i.e., from the increasing surface area. Although constitutive models based on CSSM can be adapted, calibrated and then used for the numerical analysis of crushable geomaterials (SHAH, 1997; SHELDON *et al.*, 2006), their application for predicting the associated permeability changes neglects the effect of the evolving *gsd* (e.g., in SHELDON *et al.*, 2006).

Here this problem is circumvented by adopting the merits of the breakage mechanics formulation. We follow an approach which we recently developed with Prof. Ioannis Vardoulakis, from the National Technical University of Athens, to describe the permeability reduction as related to the so-called sand production problem in the petroleum industry (EINAV *et al.*, 2008). Accordingly, we first express the reduction in the cataclasis problem by applying the modified Kozeny-Carman's formula (MATYKA *et al.*, 2008):

$$k = c_0 \frac{\phi^3}{T^2 S^2}, \tag{37}$$

where c_0 is a constant; $T = T(\phi)$ is the tortuosity as a function of the porosity ϕ (Matyka *et al.*, 2008); S is the specific surface area of the grains. For spherical grains, $S = 6/D_H$, and (37) can be expressed in terms of the harmonic mean grain size D_H:

$$k = \frac{c_0 \phi^3}{36T^2} D_H^2. \tag{38}$$

The harmonic mean grain size D_H is defined as:

$$\frac{1}{D_H} = \int_{D_m}^{D_M} p(D) \frac{\mathrm{d}D}{D} = (1 - B) \frac{1}{D_{H0}} + B \frac{1}{D_{Hu}} \tag{39}$$

in which D_H and D_m are the maximum and minimum grain sizes, respectively. Due to the grain crushing, the *gsd* $p(D)$ is also evolving, resulting in an evolving harmonic mean grain size D_H. Using relationship (1) of breakage mechanics, we can express D_H in terms of the breakage variable B, via the second equality in (39), while D_{H0} and D_{Hu} are the initial and ultimate harmonic mean grain sizes, obtained by replacing $p(d)$ with $p_0(d)$ and $p_u(d)$, respectively, in eq. (39). The permeability reduction can now be defined as:

$$R_k = \frac{k}{k_0} = \frac{(\phi/\phi_0)^3}{(T/T_0)^2} \left(\frac{D_H}{D_{H0}} \right)^2. \tag{40}$$

As can be seen, R_k composes two different contributions from the changes in the porosity ϕ and the grain sizes.

For an ultimate *gsd*, we follow Einav (2007a), who describes an ultimate cumulative fractal grading by mass that accounts for the minimum possible grain size D_m (Kendall, 1978; Sammis and Ben-Zion, 2008):

$$F(D) = \frac{D^{3-\alpha} - D_m^{3-\alpha}}{D_M^{3-\alpha} - D_m^{3-\alpha}} \tag{41}$$

from which the ultimate *gsd*, as the probability density function, can be deduced:

$$p(D) = F'(D) = \frac{(3-\alpha) D^{2-\alpha}}{D_m^{3-\alpha} - D_m^{3-\alpha}}. \tag{42}$$

As discussed by Einav *et al.* (2008), the incorporation of the minimum possible grain size D_m in the distribution proves critical for evaluating the harmonic mean grain size D_H. Without including this factor (i.e., by taking $D_m = 0$), the harmonic mean size would always be zero (i.e., getting $D_H = 0$), implicitly suggesting unrealistically that rocks with fractal *gsd* should be impermeable.

4. Numerical Simulation of Shear Tests

The behaviour of porous sandstones under drained loading (Wong *et al.*, 1997) is numerically simulated, with the use of either linear elasticity or pressure-dependent elasticity in the constitutive model. The experimental tests were conducted on various kinds of sandstones at different confining pressures. The breakage model described in section 2 is used here in the numerical simulation. For simplicity, we neglect the cohesive strength of the material model. Taking into account cohesion would require the incorporation of another internal variable to govern decohesion (or damage) due to material degradation. The coupling between damage and breakage within a thermo-mechanical framework will obviously broaden the capability of the model. It requires further work on the coupling effects which would inevitably increase the complexity of the model while having little effect on predicting the material behaviour under high confining pressures. This incorporation of cohesive strength therefore falls beyond the scope of this study.

In the calibration of the model parameter, the linear elasticity constants (for model 1) are determined based on the experimental data and formulae described in WONG *et al.* (1997), at lateral stress $\sigma_r = 150$ MPa for the Adamswiller sandstone and $\sigma_r = 250$ MPa for the Berea sandstone. In the case of pressure-dependent elasticity, the elastic parameters were determined to provide the best fit predictions in the elastic regime, but taking into account the stability condition (36).

Since the initial and ultimate *gsd*'s were not available in the paper by WONG *et al.* (1997), the index property ϑ for the Adamswiller sandstone is calculated as follows. From eq. (3), for an assumed initial poorly-graded material we have $\vartheta \approx 1 - \langle d^2 \rangle_u / d_M^2$. Assuming that the ultimate *gsd* is fractal, i.e., that $\langle d^2 \rangle_u = d_M^2(3 - \alpha)/(5 - \alpha)$ (see EINAV, 2007a), $\vartheta \approx 2/(5 - \alpha)$. Using the fractal dimension $\alpha = 2.7$, which falls well inside the proposed range of 2.5–2.8 (SAMMIS *et al.*, 1986), we have $\vartheta \approx 0.9$. For the Berea sandstone, ϑ can be determined based on an assumed fractal dimension $\alpha = 2.7$ for the ultimate *gsd* and an initial *gsd* from another source (SAMMIS *et al.*, 1986), and with the ultimate *gsd* by mass accounting for the minimum grain size D_m of 1 μm (see eqs. (3) and (4) in Section 2.1). Due to the irregular shape of the experimental initial *gsd* of Berea sandstone, which leads to difficulties in calculating ϑ in breakage mechanics, an assumed initial *gsd* is used, on condition that it produces the same harmonic mean grain size D_{H0} (see eq. (39)) as that of the experimental one.

In both model cases, the energy threshold E_c is determined based on the corresponding formula for the critical breakage-pressure (see EINAV, 2007c, and herein Eqs. (26) and (34)), using the same crushing pressure p_c. Since the underlying elastic behaviour is different for models 1 and 2, the values of E_c are therefore different in both cases. Table 1 summarizes the parameters for the models used. Note that besides higher stiffness (compared to that of sand) the energy-like parameter E_c for sandstones, which has the meaning of energy stored in the material before crushing, is about one to two orders of magnitude higher than that for sand (see EINAV, 2007b).

A quick assessment of the parameters K or $\bar{K}$ and the corresponding E_c can be done by plotting the normal compression curve against the experimental counterpart (Fig. 4). We can see that the inelastic behaviour is greatly improved just by correcting the underlying elasticity of the model. It is stressed that this correction is done without having to introduce additional parameters as required in CSSM models for fitting

Table 1

Model parameters for sandstones

Model	Adamswiller sandstone ($\vartheta = 0.9$; $p_c = 190$ MPa)	Berea sandstone ($\vartheta = 0.6$;$p_c = 380$ MPa)
Using linear elasticity (model 1)	$G = 4710$ MPa; $K = 4608$ MPa; $M = 1.5$; $E_c = 3.53$ MPa	$G = 4637$ MPa; $K = 19109$ MPa; $M = 1.5$; $E_c = 2.27$ MPa
Using pressure-dependent elasticity (model 2)	$\bar{G} = 16500$; $\bar{K} = 22000$; $M = 1.5$; $E_c = 2.26$ MPa	$\bar{G} = 14250$; $\bar{K} = 19000$; $M = 1.5$; $E_c = 4.93$ MPa

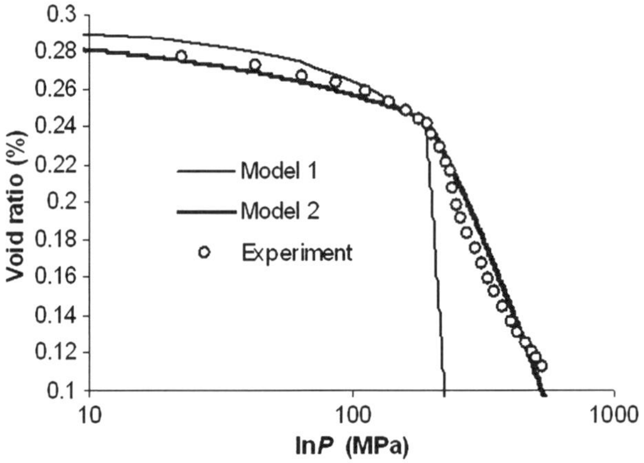

Figure 4
Normal compression curves (Adamswiller sandstone; experimental data from Wong *et al.*, 1997).

experimentally-based normal compression curves (e.g., as in Sheldon *et al.*, 2006). Model 2 of the pressure-dependent elasticity improves the numerical predictions in both pre- and post-breakage/yielding regimes.

Figure 5 shows the model and experimental yield/breakage surfaces for the Adamswiller and Berea sandstones in triaxial stress space. We remind that the evolution of the surfaces with growing B is given along a 3-D space in Figure 3 for the Adamswiller sandstone. Both brittle failure (denoted as “b” in Fig. 5) and shear-enhanced compaction (denoted as “c”) regimes of the experimental initial yield/breakage surfaces are shown. Good agreements between numerical and experimental data for both kinds of sandstones can be observed in shear-enhanced compaction regime (open symbols, Fig. 5), where the model is aimed for. It is also noted here that the chosen pressure-dependent elastic parameters for the Berea sandstone lead to better prediction of the experimental surface, compared to that of the linear elastic model. In the regime of brittle failure (solid symbols, Fig. 5) the theoretical breakage surface is restricted by the critical state line $q = Mp$ and hence does not yield a perfect match with the experimental values. Continuing along the hierarchical structure of the breakage mechanics, model

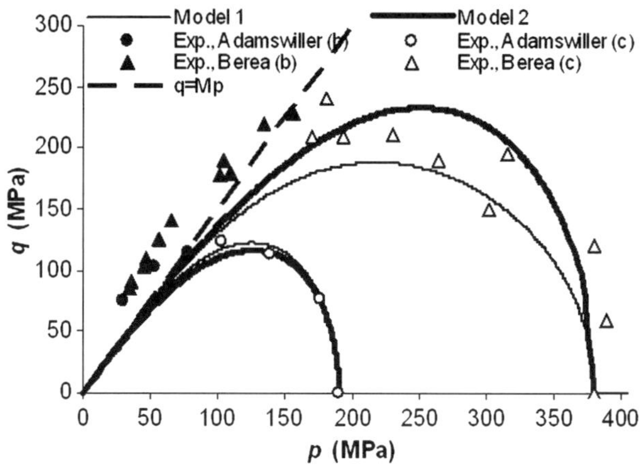

Figure 5
Yield/breakage surfaces for two kinds of sandstones (experimental data from WONG *et al.*, 1997).

derivation can enable to derive a more sophisticated model that focuses further on the behaviour above the critical state line, prior to reaching failure, but falls beyond our current aim of forming the simplest possible exposition.

The behaviour of the sandstones under different confining stress, in drained triaxial tests, is illustrated in Figure 6. The qualitative trends in the response of the material are well predicted by both models.

The experimental evolution of the porosity changes under different loading conditions (WONG *et al.*, 1997) is depicted in Figure 7. It can be seen that the model can capture rather well the trends in porosity changes. In addition the degree of matching of the models with the experimental results is improved considerably if the elastic law is taken as pressure-dependence (model 2).

For brevity, in the following we will continue with the predictions using model 2 only, as that seems to be superior. The predictions of the permeability evolution are based on the formula in Section 3 which includes changes in both porosity and grain sizes. The porosity change is illustrated in Figure 8a, agreeing very nicely with the experimental data for the Berea sandstone, considering the small number of parameters. Note, however, that the predictions of the permeability reduction rely on the dependency of the formula on the porosity. Good agreement in the first stage (before any grain crushing), where the reduction in the permeability is only due to the change in the porosity, is obtained by assuming $T \propto \phi^{-p}$ (with $p = 5$) in the modified Kozeny-Carman's formula (MATYKA *et al.*, 2008), while the original formula $(T \propto 1 - \phi)$ fails. A similar expression for describing the permeability reduction in fault gouges was made by TENTHOREY *et al.* (2003). In post breakage/yielding, the permeability reduction curves abruptly change their slope due to the grain size reduction (i.e., the increasing specific surface area). This rapid change in the permeability is clearly portrayed via the experiments in Figure 8b,

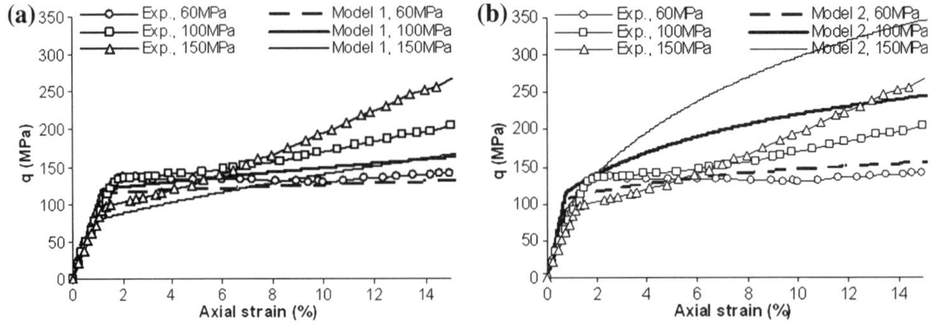

Figure 6
Stress-strain behaviour at different levels of effective lateral stress. (a): Model 1 (linear elasticity); (b): Model 2 (pressure-dependent elasticity). (Adamswiller sandstone; experimental data from WONG *et al.*, 1997).

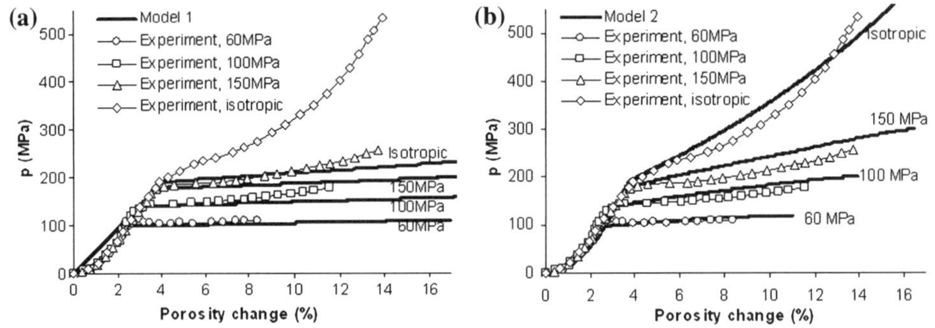

Figure 7
Porosity changes in different loading conditions. (a): Model 1); (b): Model 2. (Adamswiller sandstone; experimental data from WONG *et al.*, 1997).

against the evolving effective mean stress. The effect of grain size reduction on the permeability reduction is novel in continuum mechanics, as the theory of breakage mechanics provides the first access of continuum mechanics to modelling confined comminution.

5. *A look at energetics*

We now turn our attention to a study of the energy balance in fault gouges. We focus on the dissipation from the particulate gouge material, rather than aspects of fracture and damage mechanics, such as the scaling of fracture energy (e.g., MCGARR *et al.*, 2004; XIA, 2006). There is a growing appreciation that the ratio between the dissipated surface energy and dissipated frictional energy from the gouge is a significant factor for understanding earthquake dynamics (e.g., OLGAARD and BRACE, 1983; CHESTER *et al.*,

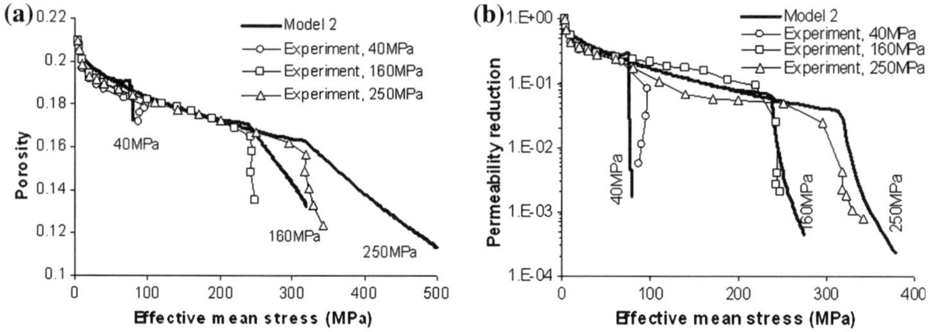

Figure 8
Porosity change and permeability reduction in drained shearing at different effective confining pressures (Berea sandstone; experimental data from Zhu and Wong, 1997). (a): Porosity change; (b): Effective mean stress – permeability reduction.

2005; Reches and Dewers, 2005; Wilson *et al.*, 2005; Pittarello *et al.*, 2008; Shi *et al.*, 2008). To examine this, we provide a rather different view based on breakage mechanics that is supported by a mechanical analog, shown in the Appendix. The breakage models in this study will be used to characterize fault gouge materials (Fig. 1), which involves the crushing of rock grains and consequently the rearrangement of fragments, which may induce severe compaction. They are not aimed at modelling the full scale fault zone, which includes both the damage zone and cataclasite layer, as seen in Figure 1. We therefore wish to classify our following results, on aspects of energy partition, to be more relevant to cataclasite layers rather than the whole fault zone. It is however noted that most of the earthquake energy consumed for creating new surface areas (more than 90%, as in Chester *et al.*, 2005) is due to the crushing of grains within the cataclasite layer.

Figure 9 illustrates the various compressive dissipative mechanisms (i.e., not representing frictional dissipation, that is pivotal in shear) in terms of a single crushing grain and its first ring of neighbours. The breakage dissipation is not only due to freeing surface energy, directly from the split of the particles, i.e., namely the 'surface area dissipation', but also from the redistribution of locked-in stored (strain) energy in the surrounding particles; the latter appears much larger in magnitude. Therefore, experimental deductions using measurements of energy changes through BET surface area (Brunauer *et al.*, 1934) may not represent the entire dissipation from grain crushing, as they miss the contribution from energy redistribution. We wish to clarify, however, that this does not mean that surface area creation is not important. On the contrary, without liberating the surface area (and freeing surface energy), the other component can not be activated, i.e., dissipation from energy redistribution. Therefore, a more significant dissipation ratio would be between the breakage (i.e., not only the surface area) dissipation and the frictional dissipation. It is important to note that while the surface area dissipation does not principally depend on the stress or strain paths, the breakage dissipation does, since it also involves the redistribution of the strain energy (which is path-dependent).

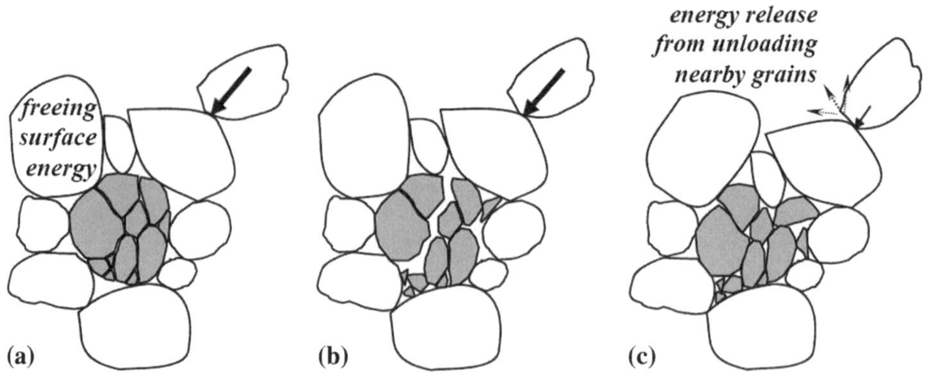

Figure 9
The dissipative events following the split of a particle. (a) new surface area is suddenly liberated, freeing surface energy; (b) the particle is replaced by fragments; (c) reorganisation of the fragments and their neighbours, leading to volumetric plastic dissipation (changes to configurational granular entropy) and dissipation from release of stored energy (Appendix) in nearby particles.

Therefore, for completion, it is important to study the dissipation ratio in relation to the load path. Finally, we remind that the overall dissipation in Eq. (11) includes not only the breakage dissipation and the frictional shear dissipation, but also the dissipation from the reorganisation of the surrounding particles, effectively represented by the plastic volumetric dissipation of Eq. (13).

5.1. Energy Balance Equation

The triaxial model described in Section 2 will be used as a basis for the following investigation. For isothermal processes (i.e., ignoring thermal temperature changes at this stage), the energy balance equation in incremental form is given as (Ziegler, 1983):

$$\tilde{W} = \delta\Psi + \tilde{\Phi}, \tag{43}$$

where $\tilde{W}$ is the rate of mechanical work input, and Ψ and $\tilde{\Phi}$ are the stored energy and dissipation potentials defined in Section 2. The tilde symbol $\sim$ is used here to highlight that the stored energy potential Ψ has a proper differential and is path-independent, while the other energy quantities are path-dependent and therefore only their rates can be defined, as $\tilde{W}$ and $\tilde{\Phi}$. Integration of this incremental equation over time results in the following energy balance equation in total form:

$$\Delta W = \Delta\Psi + \Delta\Phi \tag{44}$$

although we should highlight again the path-dependency of the first and third terms (i.e., not proper potentials). The total dissipated energy from a gouge element during the course of an earthquake is denoted here as $\Delta\Phi$, and the earthquake energy budget as the cumulative work ΔW.

While the overall dissipation involves three coupled mechanisms, as captured by Eq. (11), it may still be represented by three additive components (Einav, 2007b), corresponding to the breakage, particles reorganisation and frictional shear:

$$\Delta\Phi(t) = \int_0^t E_B\dot{B}dt + \int_0^t p\dot{\varepsilon}_v^p dt + \int_0^t q\dot{\varepsilon}_s^p dt = \Delta\Phi_{surface} + \Delta\Phi_{redist} + \Delta\Phi_{frict}. \tag{45}$$

However it should be noted that the strong coupling (Einav, 2007b) enforces the connection of these three additive components through the intrinsic dependence of E_B, p, and q on each other and also on δB, $\delta\varepsilon_v^p$, and $\delta\varepsilon_s^p$, besides their interaction through the constitutive equations. Based on the development in section 2, we can work out the ratio between the increments of energy dissipations:

$$\frac{E_B \delta B}{q\delta\varepsilon_s^p} = \frac{1}{2}\left(\frac{M^2}{\eta^2} - 1\right). \tag{46}$$

Although this is not the ratio between the total dissipations, it hints at the effect of loading path on the dissipation, through the stress ratio η between triaxial shear stress q and pressure p.

Furthermore, note that because the coupling angle in the current model is taken by $\omega = 45^\circ$, the energy losses from the breakage and the reorganization, through plastic volumetric strain, are equal for the sake of simplicity

$$p\delta\varepsilon_v^p = E_B \delta B. \tag{47}$$

In the current study, we focus on the dissipation ratio between the cumulative breakage and frictional shear terms, which is given as:

$$\text{dissipation ratio} = \frac{\int_0^t E_B\dot{B}\,\delta t}{\int_0^t q\dot{\varepsilon}_s^p\,\delta t} = \frac{\int_0^{B_t} E_B \delta\, B}{\int_0^{\varepsilon_{st}^p} q\,\delta\varepsilon_s^p}. \tag{48}$$

The above indicates the time-dependence of the dissipation ratio during the course of an earthquake.

5.2. *Numerical Approach to the Earthquake Energy Balance*

Numerical simulations are used here to explore the dissipation ratio. We consider four factors:

- initial stress state: through the ratio p_0/q_0, represented by the angle φ_0 (Fig. 10);
- the index property ϑ: measuring how far the initial *gsd* is from the ultimate *gsd*;
- trends of loading paths: through the loading angle φ (Fig. 10); and
- shear strain levels ε_s: the signature of a slip event.

The relative movement of the fault planes changes the triaxial stresses, initially represented here via the loading angle φ, with $\tan\varphi = \delta q/\delta p$. This presents only the

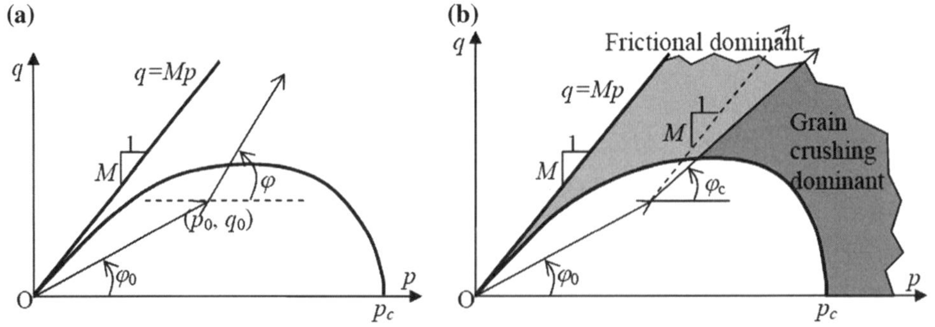

Figure 10
Schematic representation of loading trend during the course of an earthquake. (a): loading trend; (b): effect of loading trends on the dissipation.

trend of the loading during the course of an earthquake (Fig. 10). We note that the stress path depends on the far-field boundary conditions and the position of the granular element within the fault gouge, similar to those points that exist out of the gouge (Templeton and Rice, 2008). Both models 1 and 2 (see Section 2) are used here, with the following elasticity parameters: Young modulus $E = 70000$ MPa, Poisson's ratio $v = 0.15$ (Model 1); and $\bar{K} = \bar{G} = 255150$ (Model 2). Other parameters governing the inelastic response of the model are the same: crushing pressure $p_c = 300$ MPa, $M = 1.5$, $\vartheta = 0.5 \div 0.99$. The higher values of ϑ indicates a substantial "distance" between the initial and ultimate *gsd*'s, as experimentally observed in Reches and Dewers (2005). Other parameters in the parametric study are: initial pressure $p_0 = 30$ MPa, $\varphi = 0 \div \pi$, $\varphi_0 \approx 27°$ and $50°$. The time at which all dissipated energies are measured is set for a cumulative shear strain of $\varepsilon_s^{\max} = 20$. In all numerical simulations, we get $B \approx 1$ at the final strain value of $\varepsilon_s^{\max} = 20$.

The breakage energy is higher than the frictional dissipation when φ is below a critical value φ_c determined as follows. For loading angle φ small enough, e.g., $\varphi < \operatorname{atan}(M)$, the pressure p is very high at $\varepsilon_s = \varepsilon_s^{\max}$. In such cases the effect of the initial value $\eta_0 = q_0/p_0$ disappears. Therefore:

$$\tan\varphi = \frac{\delta q}{\delta p} \approx \frac{q|_{\text{at } \varepsilon_s^{\max}}}{p|_{\text{at } \varepsilon_s^{\max}}} = \eta|_{\text{at } \varepsilon_s^{\max}}. \tag{49}$$

From (46) and (48), the dissipation ratio has an upper bound (for $\varphi < \operatorname{atan}(M)$):

$$\text{dissipation ratio} < \frac{1}{2}\left(\frac{M^2}{\left(\eta|_{\text{at } \varepsilon_s^{\max}}\right)^2} - 1\right) \approx \frac{1}{2}\left(\frac{M^2}{\tan^2\varphi} - 1\right). \tag{50}$$

This upper bound is independent of the index property ϑ, initial stress state represented by the angle φ_0, and even the elasticity model used. The numerical results in

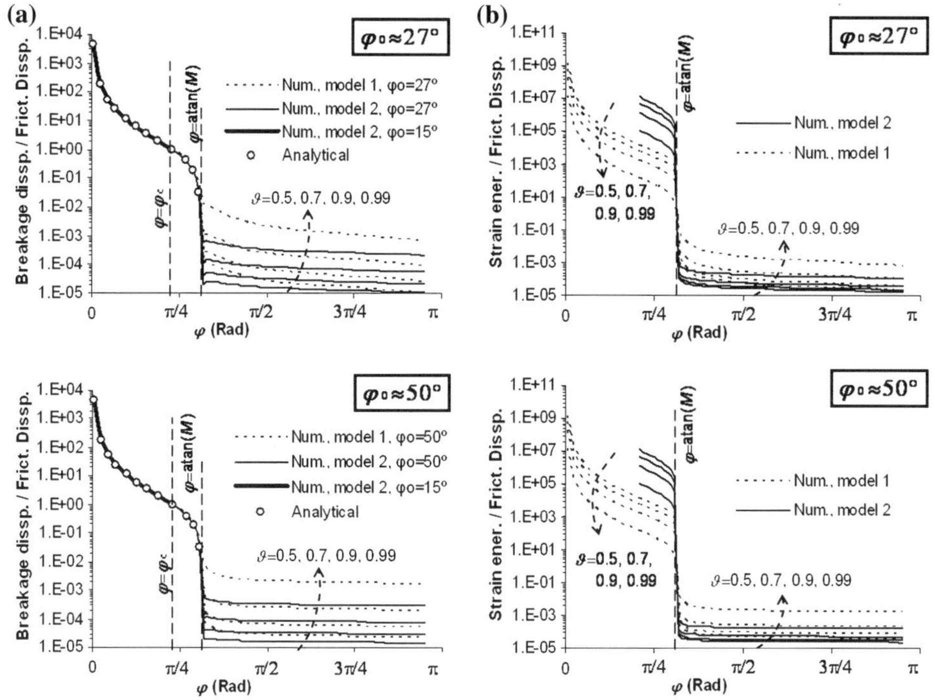

Figure 11
Effects of loading angle φ, initial stress state φ_0 and index property ϑ (using the constraint $\delta q/\delta p = \tan\,\varphi$). (a): on dissipation ratio; (b): on stored energy against friction dissipation.

Figure 11 confirm the above analytical approximation, which almost coincide with the dissipation ratio for $\varphi < \text{atan}(M)$.

From Eq. (50), for equal breakage and frictional shear dissipations we have $\eta = M/\sqrt{3}$, which results in an estimate for the critical angle φ_c separating frictional dominant and grain crushing dominant regimes (Fig. 10): $\varphi_c \approx \text{atan}\left(M/\sqrt{3}\right)$. In such cases, the critical state line can never be reached and the ratio between the breakage and frictional dissipation keeps increasing during the course of an earthquake. In addition, the elastic strain energy stored in the fault gouge material is also high (Fig. 11b), due to the confinement conditions.

The model behaviour under the constraint $\delta q/\delta p = \tan\,\varphi$ can result in different stress paths: either $\delta p > 0$ and $\delta q > 0$ (negative stress path) or $\delta p > 0$ and $\delta q > 0$ (positive stress path), depending on the initial stress state φ_0. For the pressure-dependent model, coupling terms in the constitutive equations (Eq. 30) affect the model responses under the constraint $\delta q/\delta p = \tan\,\varphi$. Therefore shearing too much (through high initial angle φ_0), and then reducing the shearing rate (through low loading angle φ) results in the model following negative stress path to conform with the enforced constraint. Since the effect of initial angle φ_0 on the energy balance under high pressure (e.g., for $\varphi < \text{atan}(M)$) is

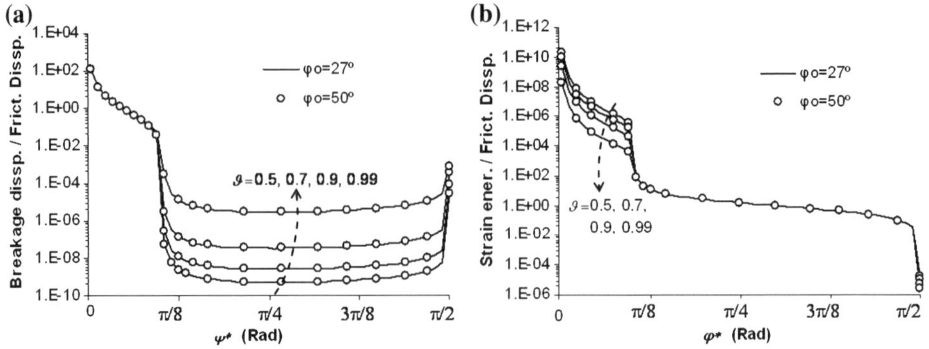

Figure 12
Effects of loading angle φ^* (in strain space), initial stress state φ_0 and index property ϑ (using the constraint $\delta\varepsilon_s/\delta\varepsilon_v = \tan\varphi^*$ in strain space). (a): on dissipation ratio; (b): on stored energy against friction dissipation.

negligible (Fig. 11a, and Eq. (50)) $\varphi_0 = 15°$ was used to complete the energy balance curve for the pressure-dependent model (Fig. 11a). To complement the results obtained in stress space, loading paths with constraint $\delta\varepsilon_s/\delta\varepsilon_v = \tan\varphi^*$ in strain space were also used, resulting in only negligible difference for two different initial states with $\varphi_0 \approx 27°$ and $\varphi_0 \approx 50°$ (Fig. 12).

The initial states of the material seem to have only marginal effects on the way the earthquake energy is dissipated, as seen in Figures 11 and 12. The dissipation ratio and the ratio between the stored energy against the friction dissipation, for the varying initial states ($\varphi_0 = 27°$ and $\varphi_0 = 50°$), do not show significant difference. Conversely, the index property ϑ has a major effect on the dissipation ratios only when $\varphi > \text{atan}(M)$. Below that, it is mainly governed by the loading path (Fig. 11). Similar conclusions can be withdrawn for the ratio between the stored strain energy Ψ and the frictional dissipation, except the fact that the index property ϑ has effects on this ratio in the entire range of loading angles φ (Fig. 11b).

The effects of shear-strain levels on the dissipative processes can be seen in Figure 13. For load paths with low loading angles φ (in grain crushing dominant regime; Fig. 10) the effects of shear-strain level on the dissipation ratio are negligible. In such cases, regardless of the amount of shear or to what extent the *gsd* progresses, the breakage energy always dominates the earthquake energy budget. For φ higher than the critical angle φ_c, the dissipation ratio decreases with increasing shear-strain level, indicating that frictional shear dissipation is the dominant dissipation mechanism.

Based on the above numerical investigation, we conclude that the loading trend, which governs the stress field (e.g., shear-dominant or compression-dominant), has the most important effect on the consumption of the energy budget. If the material is subjected to compression-dominant stress fields, we find that the crushing of rock grains (including both the creation of new surface area and release of locked-in strain energy) consumes most of the energy budget. It is noted that these results are subjected to the

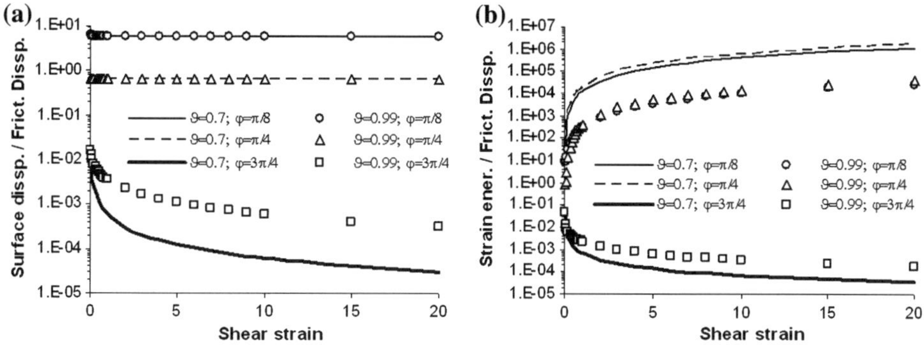

Figure 13
Effects of shear strain levels. (a): On dissipation ratio; (b): on stored energy against friction dissipation.

condition that the *gsd* at the end of the faulting process has reached its ultimate state, characterized in the analysis by breakage variable B reaching one. This reflects a state at which the cataclastic flow occurs without further grain crushing. Therefore, conclusions based on experimental analysis of cataclasite material, that the surface area dissipation can account for 50% or more of the earthquake energy, should be carefully examined (ROCKWELL *et al.*, 2008).

We note that the surface area dissipation is only the smaller component of the breakage dissipation (see the Appendix); the other bigger component, indirectly related to the production of new surface, is the energy release from the redistribution of locked-in strain energy in the surrounding grains. In shear-dominant faults, we find that the sum of these two crushing dissipation terms (giving the breakage dissipation) is negligible compared to the frictional dissipation, which dominates the earthquake energy budget. This in agreement with the common observations related to fault zone environments (e.g., in OLGAARD and BRACE, 1983; CHESTER *et al.*, 2005; and ROCKWELL *et al.*, 2008).

6. Conclusions

We studied several aspects of faulting using a breakage mechanics constitutive model, with only four physically identifiable parameters. The intrinsic capability of the model to predict the aspect of evolving *gsd* opens up opportunities in characterising well both the mechanical responses and permeability reduction of rocks during cataclasis events. Our study highlighted the importance of understanding the various elements of energy dissipation mechanisms for studying fault gouge energetics. In particular, the total energy dissipation from crushing of a confined granular ensemble, modelled here using the breakage energy dissipation, was found to contain both the surface area dissipation and the redistribution of locked-in strain energy. This understanding raises a question regarding the relevance of the BET surface area (Brunauer *et al.*, 1938) for quantifying the energy released during an confined pulverization, since the redistribution of locked-in strain energy is more prominent compared to the surface area dissipation. Based on this finding, the issue of energy balance in fault zones during an earthquake was tackled differently from previous research. The consumption of earthquake energy budget was found to depend mostly on a loading path governed by the relative movement of fault planes, resulting in different scenarios of the energy balance.

Acknowledgements

The authors would like to express their gratitude to Professor Teng-fong Wong (State University of New York) for the permission to use his experimental data on sandstones. The supply of experimental data on Cambria sand from Professor Poul Lade (The Catholic University of America), although not used in the paper, is also acknowledged. This research was supported under the Australian Research Council's *Discovery Projects* funding scheme (project number DP0774006).

Appendix: Mechanical Analog of Breakage Dissipation

A simple mechanical analog is set up here to illustrate the conceptual difference between the surface area dissipation, i.e., the part which relates linearly to changes in BET surface area (Brunauer *et al.*, 1938), and the breakage dissipation in breakage mechanics. Two sets of particle blocks are stacked alongside in a one-dimensional configuration, separated by a single particle (blackened in Fig. A1). For simplicity we assume that each of the separated sets contains a similar number of particles, each having a stiffness K_p. The particle that separates between these packs, i.e., the blackened particle, is assumed to be fully rigid but connected via two flexible bonds to the ground. The bond stiffness is assumed to be $K_b = aK_p$. Given N equally long particles in each of the blocks, with each particle having a length l, a single particle block has the stiffness $K = K_p/N$ and length $L = Nl$. The result is that the overall combined stiffness of the system is initially given by:

$$\bar{K} = \frac{(1 + 2aN)}{2aN(1 + aN)} K_b. \tag{A1}$$

This suggests that as the number of particles (N) in a stack is assumed sufficiently large, the overall stiffness becomes much smaller compared to the bond stiffness K_b (for a finite coefficient a).

The characteristic number N of the particle blocks in a stack can be seen as the number of particles in natural granular agglomerates that transmit a single force chain, with a mean value $N = 5 \div 6$ (Muthuswamy and Tordesillas, 2006). The characteristic

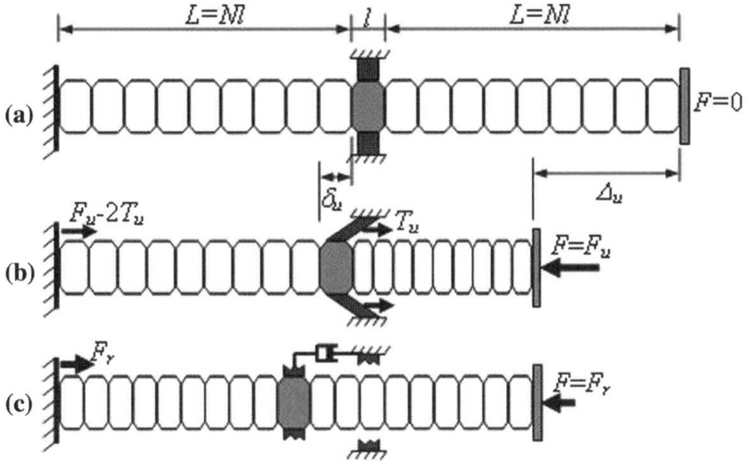

Figure A1
Mechanical analog: (a) initial configuration; (b) ultimate state; (c) bond breaking → release of locked-in strain energy and stress redistribution. The damper is shown only in (c), as it is irrelevant in (a) and (b) during quasi-static loading.

length of a particle block in the analog may be taken at the order of a harmonic grain diameter D_H in the natural agglomerate, since the analog sets grain blocks in a series. Finally, the coefficient a is of the order of one, to allow seeing the analog bonds as representative of a crushable particle.

We start to load the analog, under sufficiently slow displacement controlled conditions, to eliminate dynamic damping effects before the bonds open up. The whole system of particle blocks and bonds compress, as shown in Figure A1-b. When the bonds reach their ultimate shear force capacity T_u, and their ultimate elastic slip δ_u (see inset in Fig. A2), the system reaches an ultimate state represented by $F = F_u$ and $\Delta = \Delta_u$ (see Fig. A2), where $\Delta \ll L$ for small strain deformations. Right before the opening of the bonds at the ultimate state, the total elastic strain energy in the system, denoted here as Ψ_u, is:

$$\Psi_u = \frac{K}{2}(\Delta_u - \delta_u)^2 + K_b\delta_u^2 + \frac{K}{2}\delta_u^2. \tag{A2}$$

The bond breaking is abrupt (Fig. A2), and therefore the elastically stored energy in the bonds ($K_b\delta_u^2 =$ double the shaded areas in Fig. A2) is totally released for the creation of surface area:

$$\Delta\Phi_{\text{surface}} = K_b\delta_u^2. \tag{A3}$$

On the other hand, at this point the overall displacement freezes at $\Delta = \Delta_u$, which represents the confining condition of the granular ensemble. However, the force F fluctuates around an equilibrium force F_r, with a velocity that can no longer be

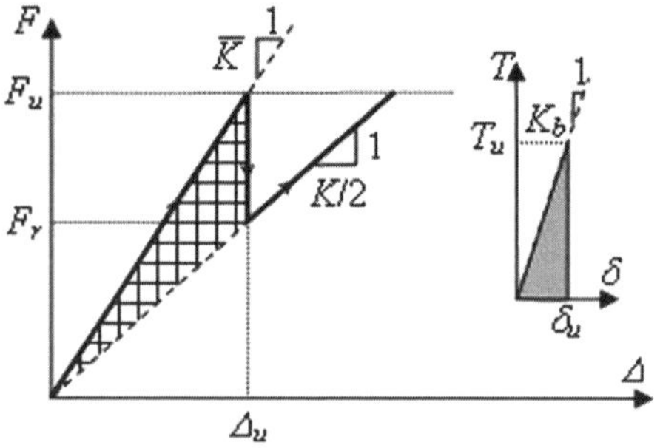

Figure A2
Load-displacement curves of the system response (hatched area designates the overall breakage dissipation, including both surface area dissipation and dissipation from redistribution of locked-in strain energy). Inset picture shows the response of an individual breakable bond (in this case the shaded area designates only the surface area dissipation of a single bond).

considered negligible. Therefore, the notional damper, shown in Figure A1-c, is activated for the first time, providing an additional reaction against the motion of the rigid particle, until this gets into static equilibrium and the force fluctuations are eliminated, i.e., when $F = F_r$. In analogy with natural agglomerates, the damper reflects heat production.

After the oscillations stop (which could notionally be considered quite rapidly, given a sufficiently large damping coefficient[1]), the system could again be seen in a quasi-static condition and the loading is still under an overall displacement $\Delta = \Delta_u$. In other words, the system travels along the minimum of the global potential energy. As a consequence, during the entire dynamic equalisation towards the static equilibrium condition, the change in the mechanical work input could be considered zero $\Delta W = 0$, and the elastically stored energy in the system at static equilibrium condition can be worked out:

$$\Psi_r = \frac{K}{4}\Delta_u^2. \tag{A4}$$

Recalling now the energy balance equation in thermodynamics (Eq. (44)):

$$\Delta\Phi = \Delta W - \Delta\Psi, \tag{A5}$$

the energy dissipation is:

$$\Delta\Phi = -\Delta\Psi = \Psi_u - \Psi_r = K_b\delta_u^2 + \frac{K_b}{K}K_b\delta_u^2 = \Delta\Phi_{\text{surface}}\left(1 + \frac{K_b}{K}\right). \tag{A6}$$

In other words, we find that in this analog the total dissipation is the sum of two dissipation terms:

[1] as found useful for simulating quasi-static loading of granular assemblies using the discrete element method (Cundall and Strack, 1979).

$$\Delta\Phi = \Delta\Phi_{\text{surface}} + \Delta\Phi_{\text{redist}}, \tag{A7}$$

where we designate the dissipation from the redistribution of initially locked-in strain energy:

$$\Delta\Phi_{\text{redist}} = aN\Delta\Phi_{\text{surface}} \tag{A8}$$

recalling that $K_b = aK_p$ and $K = K_p/N$.

For a growing number of particles N in the block, $\Delta\Phi_{\text{redist}} \gg \Delta\Phi_{\text{surface}}$, since a is in the order of one. As can be seen via eq. (A7), the total energy dissipation (hatched area in Fig. A2) contains both the dissipation from the creation of new surface area (shaded area in inset, Fig. A2), and the dissipation from the redistribution of the locked-in strain energy.

Note that in this example the total dissipation $\Delta\Phi$ depends only on the difference between two (static) equilibrium states of the system before and after the bond breaking process. Independently from the details that define the dynamics during the intermediate process, i.e., the choice of the damping coefficient, the total mechanical dissipation $\Delta\Phi$ remains unchanged and can be readily worked out from the difference between minimum potential energies in static states before and after the breaking of bonds. Therefore the notional damper illustrates that the release of "locked-in strain energy" goes to the production of heat.

The above analysis shows that under confining conditions there is a fundamental difference between the breakage energy dissipation and the surface area dissipation. It is the entire breakage energy dissipation which was shown to be primarily related to the redistribution of locked-in strain energy, and not only the surface area dissipation, that represents the grain crushing. Therefore, estimates of dissipation energetics in cataclasis must account for the path dependency of the material behaviour, in addition to measurements of BET surface area. However, we note that in shear-dominant faults, our analysis in the main body of the paper shows that the sum of these two crushing dissipation terms (giving the breakage dissipation) is negligible compared to the frictional dissipation, which dominates the earthquake energy budget. This is in agreement with the common observations related to fault zone environments (e.g., in OLGAARD and BRACE, 1983; CHESTER *et al.*, 2005; and ROCKWELL *et al.*, 2008).

REFERENCES

AN, L.-J. and SAMMIS, C.G. (1994), *Particle size distribution of cataclastic fault materials from Southern California: A 3-D study*, Pure Appl. Geophys. *143*(1), 203–227.

AYDIN, A., BORJA, R.I., and EICHHUBL, P. (2006), *Geological and mathematical framework for failure modes in granular rock*, J. Struct. Geol. *28*(1), 83–98.

BEELER, N.M., TULLIS, T.E., BLANPIED, M.L., and WEEKS, J.D. (1996), *Frictional behavior of large displacement experimental faults*, J. Geophys. Res. *101*(B4), 8697–8715.

BEN-ZION, Y. (2008), *Collective behavior of earthquakes and faults: Continuum-discrete transitions, progressive evolutionary changes and different dynamic regimes*, Rev. Geophys. *46*, RG4006, doi:10.1029/2008RG000260.

Ben-Zion, Y. and Sammis, C.G. (2003), *Characterization of fault zones*, Pure Appl. Geophys. *160*(3), 677–715.

Borja, R.I., Tamagnini, C., and Amorosi, A. (1997), *Coupling plasticity and energy-conserving elasticity models for clays*, J. Geotech. Geoenviron. Engin. *123*(10), 948–957.

Brunauer, S., Emmett, P.H., and Teller, E. (1938), *Adsorption of gases in multimolecular layers*, J. Am. Chem. Soc. *60*(2), 309–319.

Caine, J.S., Evans, J.P. and Forster, C.B. (1996), *Fault zone architecture and permeability structure*, Geology *24*(11), 1025–1028.

Chester, J.S., Chester, F.M., and Kronenberg, A.K. (2005), *Fracture surface energy of the Punchbowl fault, San Andreas system*, Nature *437*(7055), 133–136.

Cundall, P. and Strack, O. (1979), *A discrete numerical model for granular assemblies*, Geotechnique *29*, 47–65.

Cuss, R.J., Rutter, E.H. and Holloway, R.F. (2003), *The application of critical state soil mechanics to the mechanical behaviour of porous sandstones*, Internat. J. Rock Mech. and Mining Sci. *40*, 847–862.

Edwards, S.F. and Grinev, D.V. (2001), *Transmission of stress in granular materials as a problem of statistical mechanics*, Physica A *302*, 162–186

Einav, I. (2007a), *Breakage mechanics–Part I: Theory*, J. Mechan. Phys. Sol. *55*(6), 1274–1297.

Einav, I. (2007b), *Breakage mechanics–Part II: Modelling granular materials*, J. Mech. Phys. Sol. *55*(6), 1298–1320.

Einav, I. (2007c), *Fracture propagation in brittle granular matter*, Proc. Roy. Soc. A: Math., Phys. Eng. Sci. *463*(2087), 3021–3035.

Einav, I. (2007d), *Soil mechanics: breaking ground*, Philosoph. Transac. Roy. Soc. A: Math., Phys. Engin. Sci. *365*(1861), 2985–3002.

Einav, I., Vardoulakis, I., and Chan, A.H.C. (2008), *Confined comminution and permeability reduction in expanding perforations*, to be submitted.

Einav, I. and Puzrin, A.M. (2004), *Pressure-dependent elasticity and energy conservation in elastoplastic models for soils*, J. Geotechn. Geoenviron. Engin. *130*(1), 81–92.

Evans, J.P., Forster, C.B., and Goddard, J.V. (1997), *Permeability of fault-related rocks, and implications for hydraulic structure of fault zones*, J. Struct. Geol. *19*(11), 1393–1404.

Griffith, A.A. (1921), *The phenomena of rupture and flow in solids*, Philosoph. Transact. Roy. Soc. London. Series A, Containing Papers of a Math. or Phys. Character (1896–1934) 221(-1), 163–198.

Guo, Y. and Morgan, J.K. (2007), *Fault gouge evolution and its dependence on normal stress and rock strength—Results of discrete element simulations: Gouge zone properties*, J. Geophys. Res. *112*(B10403), 1–17.

Hamiel, Y., Lyakhovsky, V., and Agnon, A. (2004), *Coupled evolution of damage and porosity in poroelastic media: theory and applications to deformation of porous rocks*, Geophys. J. Internatl. *156*, 701–713

Heilbronner, R. and Keulen, N. (2006), *Grain size and grain shape analysis of fault rocks*, Tectonophysics *427*(1–4), 199–216.

Hoque, E. and Tatsuoka, F. (1998), *Anisotropy in elastic deformation of granular materials*, Soils Found. *38*(1), 163–179.

Houlsby, G.T. (1985), *The use of a variable shear modulus in elastic-plastic models for clays*, Comp. Geotechn. *1*(1), 3–13.

Houlsby, G.T., Amorosi, A., and Rojas, E. (2005), *Elastic moduli of soils dependent on pressure: A hyperelastic formulation*, Geotechnique *55*(5), 383–392.

Kendall, K. (1978), *The impossibility of comminuting small particles by compression*, Nature *272*, 710–711.

LoPresti, D.C.F. and O'Neill, D.A. (1991), *Laboratory investigation of small strain modulus anisotropy in sands*, Proc. ISOCCTI, Clarkson University, Potsdam, 1991, pp. 213–224.

Lund, M.G. and Austrheim, H. (2003), *High-pressure metamorphism and deep-crustal seismicity: Evidence from contemporaneous formation of pseudotachylytes and eclogite facies coronas*, Tectonophysics *372*(1–2), 59–83.

Lyakhovsky, V., Ben-Zion, Y., and Agnon, A. (1997), *Distributed damage, faulting, and friction*, J. Geophys. Res. *102*(B12), 27,635–27,649.

Mandl, G., de Jong, L.N.J., and Maltha, A. (1977), *Shear zones in granular material*, Rock Mech. Rock Engin. *9*(2), 95–144.

MARONE, C., RALEIGH, C.B., and SCHOLZ, C.H. (1990), *Frictional behavior and constitutive modeling of simulated fault gouge*, J. Geophys. Res. *95*(B5), 7007–7025.

MARONE, C., VIDALE, J.E., and ELLSWORTH, W.L. (1995), *Fault healing inferred from time-dependent variations in source properties of repeating earthquakes*, Geophys. Res. Lett. *22*(22), 3095–3098.

MATYKA, M., KHALILI, A., and KOZA, Z. (2008), *Tortuosity-porosity relation in porous media flow*, Phys. Rev. E *78*, 026306.

MCGARR, A., FLETCHER, J.B., and BEELER, N.M. (2004), *Attempting to bridge the gap between laboratory and seismic estimates of fracture energy*, Geophys. Res. Lett. *31*, L14606.

MORROW, C., SHI, L.Q., and BYERLEE, J. (1981), *Permeability and strength of San Andreas fault gouge under high pressure*, Geophys. Res. Lett. *8*(4), 325–328.

MUTHUSWAMY, M. and TORDESILLAS, A. (2006), *How do interparticle contact friction, packing density and degree of polydispersity affect force propagation in particulate assemblies?*, J. Stat. Mech. - Theory and Experiment, P09003.

OKADA, Y., SASSA, K., and FUKUOKA, H. (2004), *Excess pore pressure and grain crushing of sands by means of undrained and naturally drained ring-shear tests*, Engin. Geol. *75*(3–4), 325–343.

OLGAARD, D.L. and BRACE, W.F. (1983), *The microstructure of gouge from a mining-induced seismic shear zone*, Internatl. J. Rock Mechan. and Mining Sci. Geomechan. Abstracts 20(1), 11–19.

OLSEN, M.P., SCHOLZ, C.H., and LÉGER, A. (1998), *Healing and sealing of a simulated fault gouge under hydrothermal conditions: Implications for fault healing*, J. Geophys. Res. *103*(B4), 7421–7430.

OLSSON, W.A. (1999), *Theoretical and experimental investigation of compaction bands in porous rock*, J. Geophys. Res. *104*(B4), 7219–7228.

PITTARELLO, L., DI TORO, G., BIZZARRI, A., PENNACCHIONI, G., HADIZADEH, J. and COCCO, M. (2008), *Energy partitioning during seismic slip in pseudotachylyte-bearing faults (Gole Larghe Fault, Adamello, Italy)*, Earth Planet. Sci. Lett. *269*(1–2), 131–139.

RECHES, Z. and DEWERS, T.A. (2005), *Gouge formation by dynamic pulverization during earthquake rupture*, Earth Planet. Sci. Lett. *235*(1-2), 361–374.

RENARD, F., GRATIER, J.-P., and JAMTVEIT, B. (2000), *Kinetics of crack-sealing, intergranular pressure solution, and compaction around active faults*, J. Struct. Geol. *22*(10), 1395–1407.

RICARD, Y. and BERCOVICI, D. (2003), *Two-phase damage theory and crustal rock failure: The theoretical 'void' limit, and the prediction of experimental data*, Geophys. J. Internatil. *155*, 1057–1064.

RICE, J.R. (1978), *Thermodynamics of the quasi-static growth of Griffith cracks*, J. Mechan. Phys. Sol. *26*, 61–78.

ROCKWELL, T., SISK, M., GIRTY, G., DOR, O., WECHSLER, N., and BEN-ZION, Y. (2009), *Chemical and physical characteristics of pulverized Tejon lookout Granite Adjacent to the San Andreas and Garlock faults: implications for earthquake physics*, Pure Appl. Geophys., in press.

ROSCOE, K.H. and SCHOFIELD, A.N. (1963), *Mechanical behaviour of an idealised 'wet' clay*, European Conf. Soil Mechan. Foundation Engin. Essen, Germany: Deutsche Gesellschaft für Erd- und Grundbau, E.v., Wiesbaden, 46–54.

SAMMIS, C.G. and BEN-ZION, Y. (2008), *Mechanics of grain-size reduction in fault zones*, J. Geophy. Res. *113*, B02306.

SAMMIS, C.G., OSBORNE, R.H., ANDERSON, J.L., BANERDT, M., and WHITE, P. (1986), *Self-similar cataclasis in the formation of fault gouge*, Pure Appl. Geophys. *124*(1), 53–78.

SCHOLZ, C. *The Mechanics of Earthquakes and Faulting* (Cambridge University Press 1990).

SCHULTZ, R.A. and SIDDHARTHAN, R. (2005), *A general framework for the occurrence and faulting of deformation bands in porous granular rocks*, Tectonophysics *411*(1–4), 1–18.

SEGALL, P. and RICE, J.R. (1995), *Dilatancy, compaction, and slip instability of a fluid-infiltrated fault*, J. Geophys. Res. *100*(B11), 22,155–122,171.

SHAH, K.R. (1997), *An elasto-plastic constitutive model for brittle-ductile transition in porous rocks*, Internatl. J. Rock Mechan. Mining Sci. *34*(3–4), 283.e281–283.e213.

SHELDON, H.A., BARNICOAT, A.C., and ORD, A., (2006), *Numerical modelling of faulting and fluid flow in porous rocks: An approach based on critical state soil mechanics*, J. Struct. Geol. *28*(8), 1468–1482.

SHI, Z., BEN-ZION, Y. and NEEDLEMAN, A. (2008), *Properties of dynamic rupture and energy partition in a solid with a frictional interface*, J. Mechan. Phys. Sol. *56*, 5–24.

SLEEP, N.H. and BLANPIED, M.L. (1992), *Creep, compaction and the weak rheology of major faults*, Nature *359*(6397), 687–692.

STEACY, S.J. and SAMMIS, C.G. (1991), *An automaton for fractal patterns of fragmentation*, Nature *353*(6341), 250–252.

TEMPLETON, E.L. and RICE, J.R., (2008), *Off-fault plasticity and earthquake rupture dynamics, 1. Dry materials or neglect of fluid pressure changes*, J. Geophys. Res. B (Solid Earth), *113*, B09306, doi:10.1029/2007JB005529.

TENTHOREY, E., COX, S.F., and TODD, H.F. (2003), *Evolution of strength recovery and permeability during fluid-rock reaction in experimental fault zones*, Earth Planet. Sci. Lett. *206*(1–2), 161–172.

WILSON, B., DEWERS, T., RECHES, Z.E. and BRUNE, J. (2005), *Particle size and energetics of gouge from earthquake rupture zones*, Nature *434*(7034), 749–752.

WONG, T., DAVID, C., and ZHU, W. (1997), *The transition from brittle faulting to cataclastic flow in porous sandstones: Mechanical deformation*, J. Geophys. Res. *102*(B2), 3009–3025.

XIA, K. (2006), *Scaling of fracture energies: The rationalization of different laboratory measurements*, Geophys. Res. Lett. *33*, L01305.

ZHANG, S. and TULLIS, T.E. (1998), *The effect of fault slip on permeability and permeability anisotropy in quartz gouge*, Tectonophysics *295*(1–2), 41–52.

ZHU, W. and WONG, T. (1997), *The transition from brittle faulting to cataclastic flow: Permeability evolution*, J. Geophys. Res. *102*(B2), 3027–3041.

ZIEGLER, H. *An introduction to thermomechanics*, Second Ed. (North Holland, Amsterdam 1983).

ZYTYNSKI, M., RANDOLPH, M.F., NOVA, R. and WROTH, C.P. (1978), *On modelling the unloading-reloading behaviour of soils*, Internatl. J. Numer. Analyt. Methods in Geomechan. *2*(1), 87–93.

(Received August 1, 2008, accepted January 10, 2009)
Published Online First: June 30, 2009

To access this journal online:
www.birkhauser.ch/pageoph

Pure appl. geophys. 166 (2009) 1725–1746
0033–4553/09/101725–22
DOI 10.1007/s00024-009-0514-1
© Birkhäuser Verlag, Basel, 2009
Pure and Applied Geophysics

Chemical and Physical Characteristics of Pulverized Tejon Lookout Granite Adjacent to the San Andreas and Garlock Faults: Implications for Earthquake Physics

Thomas Rockwell,[1] Matthew Sisk,[1] Gary Girty,[1] Ory Dor,[2,3] Neta Wechsler,[2] and Yehuda Ben-Zion[2]

Abstract—We present new detailed analyses of samples of pulverized Tejon Lookout granite collected from sections adjacent to the San Andreas and Garlock faults in southern California. The Tejon Lookout granite is pulverized in all exposures within about 100 m from both faults. Chemical analyses indicate no or little weathering in the collected samples, although XRD analysis shows the presence of smectite, illite, and minor kaolinite in the clay-size fraction. Weathering products may dominate in the less than 1 micron fraction. The average grain size in all samples of pulverized Tejon Lookout granite ranges between 26 and 208 microns (silt to fine sand), with the particle size distribution in part a function of proximity to the primary slip zone. The San Andreas fault samples that we studied are generally finer grained than those collected from adjacent to the Garlock fault. The particle size distribution for each studied sample from both faults follows a pseudo-power law with a continuously changing exponent, which suggests that pulverization is not simply a consequence of direct shear. The average particle size that we determined for our samples is considerably coarser than reported in previous investigations, which we attribute to possible measurement errors in the prior work. Our data and observations suggest that dynamic fracturing in the wall rock of the San Andreas and Garlock faults only accounts for about 1% or less of the earthquake energy budget.

Key words: Pulverized rocks, particle size distribution, rock weathering, fault-zone structure, earthquake physics.

1. Introduction

The structural and material properties of damaged fault zone rocks can reveal important information pertaining to various aspects of earthquake mechanics, including the amount of fracture surface energy contained in the rocks, the type (shear vs. tensile) and amplitude of the likely generating stress field, and the possible relationship to the statistically-preferred propagation direction of earthquake ruptures. Geological mapping indicates that fault zones contain a hierarchical damage structure, with a narrow core of

[1] Department of Geological Sciences, San Diego State University, San Diego, CA 92182, U.S.A. E-mail: trockwell@geology.sdsu.edu
[2] Department of Earth Sciences, University of Southern California, Los Angeles, CA 90089, U.S.A.
[3] Department of Geological Sciences, Brown University, Providence, RI 02912, U.S.A.

slip localization that accommodates most of the motion across the fault, and damaged rocks with various crack densities in the adjacent regions (e.g., BEN-ZION and SAMMIS, 2003; CHESTER *et al.*, 2005; ROCKWELL and BEN-ZION, 2007).

Recent studies have shown that large continental strike-slip fault structures have belts of pulverized rocks that are about ~100–300 m wide and exist primarily on one side of the fault. This was documented along the Mojave section of the San Andreas fault (WILSON *et al.*, 2005; DOR *et al.*, 2006a), portions of the Garlock (SISK *et al.*, 2005) and San Jacinto faults (STILLINGS, 2007) in California, and in a section of the Arima-Takatsuki fault in Japan (MITCHELL *et al.*, 2008). Rock pulverization is assumed to be the result of a dynamic reduction of normal stress during earthquake ruptures (e.g., BRUNE *et al.*, 1993; BEN-ZION, 2001), and its asymmetric distribution along a fault may indicate a preferred direction of rupture propagation (BEN-ZION and SHI, 2005; DOR *et al.*, 2006b, 2008).

Studies of the Particle Size Distribution (PSD) of fault zone rocks have tended to fit the results to power laws. SAMMIS *et al.* (1987) and later studies concluded that the PSD of damaged rocks is associated with fractal statistics, having exponent values of non-cumulative distribution in planar sections of 1.6. CHESTER *et al.* (2005) analyzed the PSD of grains in the slip localization zone of the Punchbowl fault in California and obtained an exponent of non-cumulative statistics in planar sections of 2.0. HEILBRONNER and KEULEN (2006) and KEULEN *et al.* (2007) found that the PSD of naturally and experimentally deformed fault rocks requires two slopes, with the break in slope marking the transition between shearing and fracturing. SAMMIS and BEN-ZION (2008) provided a review of various mechanisms that can lead to grain size reduction.

It is generally accepted (e.g., MCGARR, 1999; ABERCROMBIE *et al.*, 2006; SHI *et al.*, 2008) that over 90% of the change of the elastic strain energy during earthquake ruptures is converted to heat associated with dissipative motions. Such motions occur primarily in the slip localization region but may include sliding between grain boundaries and crack surfaces in the entire damage structure. The portions of energy involving the radiation and fracturing processes are on the order of a few percent each. However, WILSON *et al.* (2005) argued, based on analysis of PSD from an outcrop of pulverized Tejon Lookout granite in Tejon Pass, adjacent to the slip zone of the San Andreas fault, that the energy dissipated in fracturing surfaces may be 50% or more of the earthquake energy budget.

The analysis of WILSON *et al.* (2005) of pulverized Tejon Lookout granite led them to conclude that: (1) Pulverization is apparently the result of a mechanical process since no significant amount of weathering product was found in the analyzed samples; and (2) the PSD of these rocks is time dependent when measured in a laser particle analyzer. WILSON *et al.* (2005) attributed this latter result to an apparent slow disaggregation of particle aggregates in the circulating fluids. The final PSD obtained by WILSON *et al.* (2005) was peaked and centered in the sub-micron scale. This result was the basis for their inference that the fracture energy that is contained in fault zone rocks is considerably larger than prior estimates.

In the present work we provide a detailed characterization of the chemistry and PSD of pulverized Tejon Lookout granite along the San Andreas fault, and from Tejon Ranch along the Garlock fault. The former locality is the same studied by WILSON *et al.* (2005). The results of our work indicate that weathering products account for some of the smaller (<4 microns) fraction, but that overall the rocks are not significantly weathered. Our PSD analyses, which include measurements using sieves, pipette, and laser analyzer methods, show that the distributions of the examined samples have exponents that change with increasing particle size (rather than being associated with a single power-law exponent), and that the mean grain sizes of the pulverized rocks are significantly larger than were reported by WILSON *et al.* (2005). The temporal evolution of PSD observed by WILSON *et al.* (2005) appears to result from sedimentation of the heavier particles during the long experiment rather than disaggregation. Our results are in agreement with many previous studies, that only minor dissipation of energy into surface area creation occurs during seismic ruptures.

2. *Methods and Results*

2.1. *Sampling and Laboratory Techniques*

Along the Garlock fault in Bear Trap Canyon, we collected nonoriented samples from trenches that were excavated by Earth Consultants International (ECI) as part of a geotechnical study for Tejon Ranch (Fig. 1). In granite along the south side of the fault, samples were collected along three fault normal traverses beginning at the principal slip surface and then moving outward to the ends of the trenches. These three separate transects are referred to herein as T-1, T-2, and T-3 (Table 1). North of the fault, non pulverized foliated metamorphic rocks (Pelona Schist) were exposed. Though non pulverized, the schists are extensively fractured at approximately the 5–10 cm scale. Such rocks do not afford the same type of analysis reported herein as they are still lithified and relatively undamaged.

Along the San Andreas fault, we sampled the same locality studied by WILSON *et al.* (2005) at Tejon Pass: A road cut along the old highway above the I-5 freeway (Fig. 1). Our initial study of the properties of pulverized Tejon Lookout granite in Bear Trap Canyon along the Garlock fault began as an attempt to compare our results with the previously published results from Tejon Pass. However, after preliminary particle distributions were determined with a laser particle analyzer, the results were significantly different from the previous work. Hence, we began an investigation into different methods used to determine the PSD of pulverized rocks. We recollected four non oriented samples of Tejon Lookout granite from the Tejon Pass locality for direct comparison with the Bear Trap Canyon samples and the study of WILSON *et al.* (2005). We split the field samples from both localities into ~40 g splits using a mechanical splitter, and used different splits for the various analyses conducted in this work.

For the study of particle size distribution, we utilized a Horiba LA 930 Laser Particle Size Analyzer and a pipette rack (classical particle size distribution). We used both

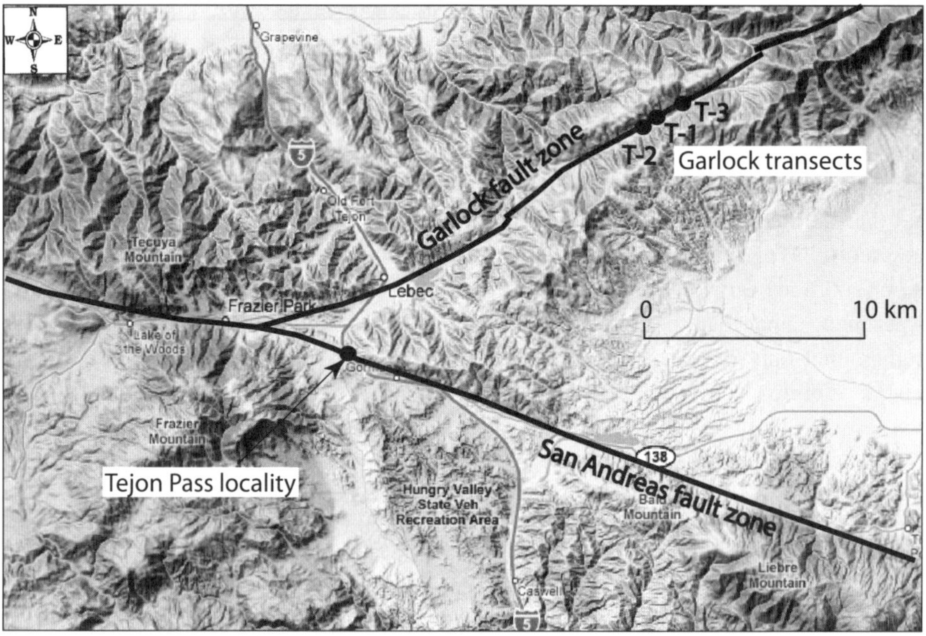

Figure 1

Sampling localities used in this study. The site at Tejon Pass is identical to that sampled by Wilson *et al.* (2005). The three transects along the Garlock fault were sampled in trenches excavated by Earth Consultants International as part of a seismic hazard study for Tejon Ranch.

Table 1

Sample transects for both study areas. T-1, T-2 and T-3 are from the Garlock fault in Bear Trap Canyon, and SAF indicates the transect from the Wilson locality in Tejon Pass

	Distance from the PSS			
GF				
T-1	**GF T1a**	**GF T1b**	**GF T1c**	
	2.7 m	4.6 m	61.0 m	
T-2	**GF T2a**	**GF T2b**	**GF T2c**	
	2.4 m	5.5 m	7.6 m	
T-3	**GF T3a**	**GF T3b**	**GF T3c**	**GF T3d**
	0.3 m	2.4 m	7.6 m	45.7 m
SAF	**SAFa**	**SAFb**	**SAFc**	**SAFd**
	0.1 m	1.0 m	5.0 m	20.0 m

methods to crosscheck the laser analyzer against traditional particle size analysis using settling time (Folk, 1974).

Elemental concentrations were analyzed with a Phillips MagiX Pro instrument to investigate the degree of weathering. For this aspect of our study, we determined major elemental concentrations of samples extracted from both field sites using fused discs. Trace elements were analyzed using powdered pressed pellets. Loss on Ignition (LOI) (representing the sum of all volatile components in each sample) was also determined for each sample.

The mineralogy of the clay-sized sample fraction (<4 μm) was analyzed using a Phillips Xpert Plus X-ray multipurpose diffractometer with Kα radiation of 1.5405 A, and 45 kV and 40 mA settings. We used a pipette to extract clay-sized particles, and this mixture was then used to produce standard clay smears on glass slides. The clay smears were first analyzed in their unheated and untreated states, and then were run glycolated after they were heated to 350°C, and finally after being heated to 550°C.

2.2. *Rock Chemistry*

XRF Elemental Analysis. The results of XRF analysis are shown in the ternary diagrams in Figure 2. Note that all samples in A-CN-K space plot relatively tightly about the join between plagioclase and K-feldspar (Nesbitt and Young, 1982). This result implies a lack of severe weathering in all samples collected at both the Tejon Pass and Tejon Ranch field areas. Such an interpretation is supported by CIA values (A/[A + C+N + K]) ranging from 46 to 54 (Table 2), well within the CIA values published for unweathered granite (Nesbitt and Young, 1982; Nesbitt and Markovics, 1997). This result suggests that particle size distributions are largely not influenced by chemical weathering. Nevertheless, some samples from both studied locations plot slightly above the "feldspar join," a relationship that may suggest incipient or minor weathering.

In A-CNK-FM space, our samples plot about the join between plagioclase and biotite (Fig. 2). However, some samples plot slightly above this join; a result that is consistent with minor weathering.

XRD Mineralogy. The diffractograms from all <4 μm samples clearly demonstrate the presence of secondary clay minerals (Fig. 3). For example, the samples from along the Garlock fault in Bear Trap Canyon (Fig. 3a) display peaks indicative of the presence of smectite, illite and kaolinite in various quantities, with the later (kaolinite) being the least prominent, as expected for this weathering environment. Samples taken from the Tejon Pass locality along the San Andreas fault revealed a similar suite of clay minerals, along with the presence of clinochlore (Fig 3b). The presence of these weathering products clearly demonstrates that some alteration is present in all samples, which is not surprising, as all samples were recovered from shallow exposures. These results are also consistent with the XRF analyses which indicate the possible presence of minor weathering.

2.3. *Particle Size Distribution*

To establish the complete and most reproducible methods for determination of particle size distribution of pulverized granitic materials, we ran a series of baseline tests

Table 2

Major element contents, LOI and CIA of bulk samples from the Tejon Look granite

Samples	GF T1c	GF T1b	GF T1a	GF T2a	GF T2b	GF T2c	GF T3a	GF T3b	GF T3c	GF T3d	SAFa	SAFb	SAFc	SAFd
SiO_2	67.54	68.30	73.75	72.00	72.93	71.90	73.71	73.77	68.38	68.16	68.73	71.27	62.93	59.89
Al_2O_3	15.55	15.18	11.66	13.86	14.01	11.24	12.79	12.35	15.13	15.49	14.32	13.32	16.48	15.87
Fe_2O_3	3.25	2.79	1.37	1.73	1.47	0.55	0.51	0.51	2.84	3.02	2.43	1.96	4.42	4.54
CaO	2.90	2.26	1.72	1.33	1.27	4.33	2.00	1.49	2.61	2.86	2.10	2.03	4.46	6.92
MgO	0.71	0.58	0.27	0.30	0.22	0.26	0.28	0.20	0.64	0.72	0.98	0.60	1.70	1.72
K_2O	2.84	3.38	4.63	3.59	3.83	2.99	3.33	5.23	3.35	3.00	4.07	3.92	2.42	2.00
Na_2O	3.66	3.61	2.67	3.65	3.65	2.96	3.49	3.08	3.43	3.69	3.57	3.12	3.91	3.68
MnO	0.05	0.04	0.02	0.03	0.03	0.01	0.02	0.02	0.04	0.05	0.05	0.03	0.08	0.07
TiO_2	0.60	0.52	0.09	0.23	0.20	0.03	0.04	0.04	0.53	0.57	0.36	0.29	0.86	0.91
P_2O_5	0.12	0.13	0.01	0.06	0.05	0.01	0.01	0.01	0.14	0.15	0.09	0.06	0.26	0.27
LOI	2.19	1.56	1.24	4.31	1.76	1.66	1.32	1.12	1.25	1.74	2.15	1.96	1.27	2.81
CIA	52.49	53.12	48.27	53.25	53.17	46.42	49.66	47.93	52.49	52.23	50.75	50.73	51.55	52.65

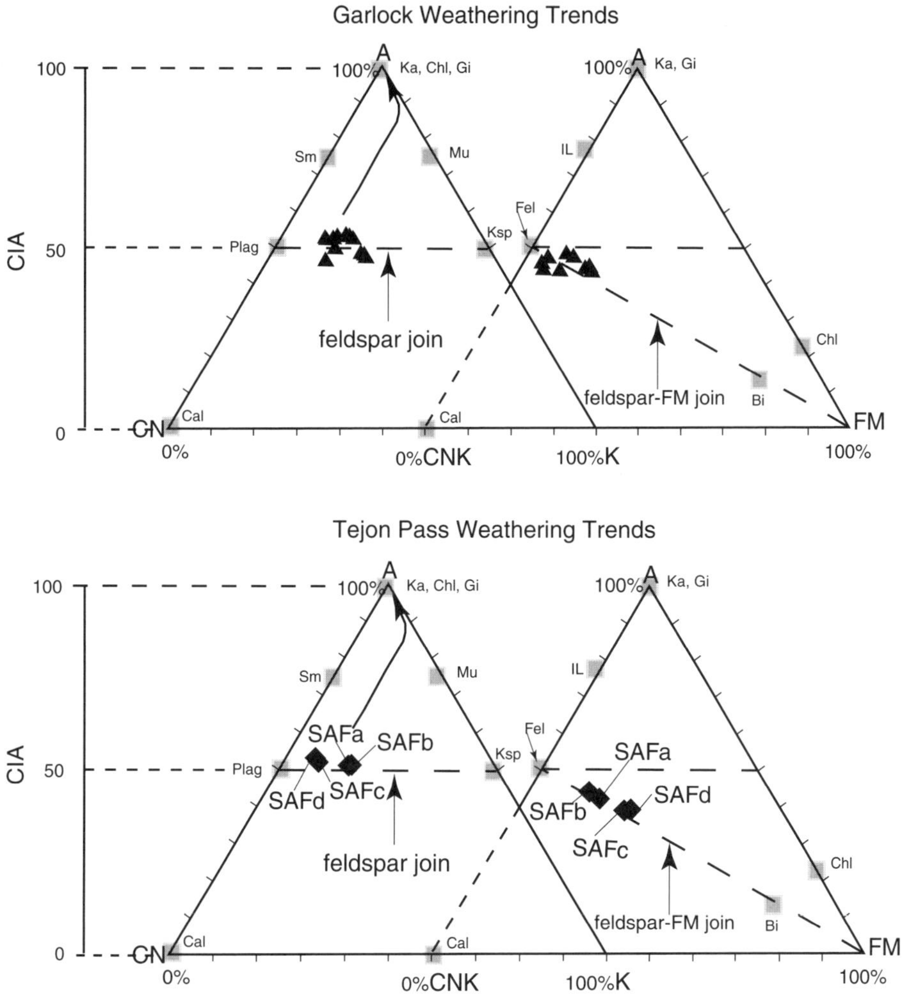

Figure 2

Tejon Lookout Granite bulk compositions plotted in A-CN-K and A-CNK-FM space. The arrow depicts the predicted weathering trend for the Tejon Lookout Granite. Corresponding CIA values are denoted on the left. Ka = kaolinite; Chl = chlorite; Gi = gibbsite; Sm = smectites; Mu = muscovite; Plag = plagioclase; Ksp = K-feldspar; Fel = feldspar; Il = Illite; Bi = biotite; Cal = calcite. The triangles in the upper diagram represent 10 samples from the Garlock fault on Tejon Ranch. The triangles in the lower diagram represent the four samples from the San Andreas fault at Tejon Pass.

on some of the sample splits to assess the affect of clay flocculation, clumping of pulverized grains, length of shaking time with and without a dispersant, and to compare the results obtained from the laser analyzer with those obtained from standard sieve and pipette methods.

Shaking with sodium hexametaphosphate is the standard method for dispersing clay in soil studies (Rockwell, 2000; Sperazza *et al.*, 2004), and typical shaking durations are

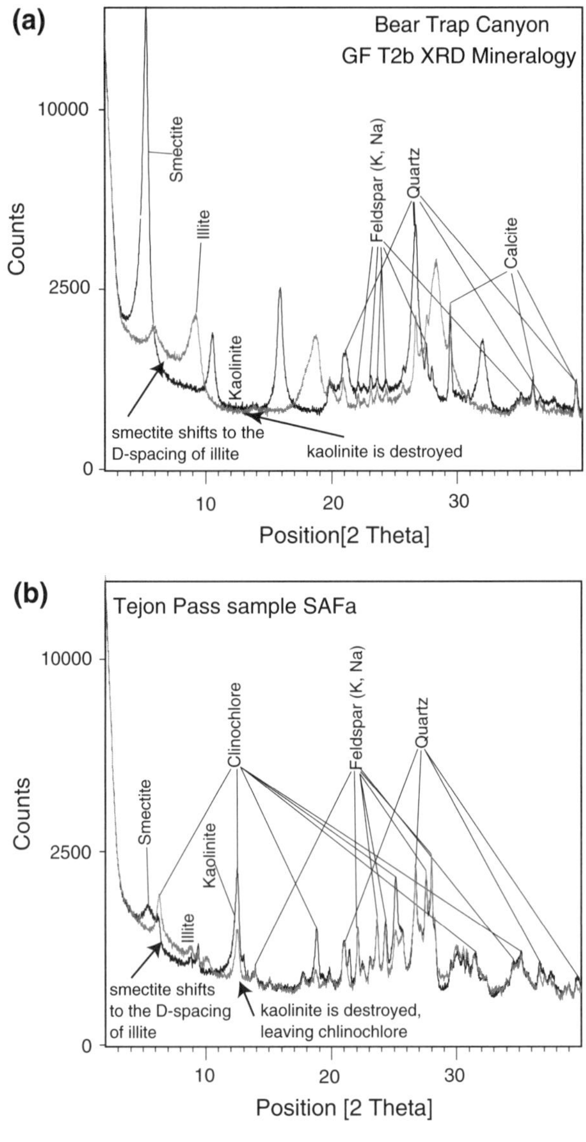

Figure 3

a) XRD trace from sample GF T2b, which is representative of the mineralogy of the pulverized Tejon Lookout Granite in Bear Trap Canyon along the Garlock fault. Note the presence of smectitie, illite and minor kaolinite, in addition to quartz and feldspar. Calcite was also present, although no secondary pedogenic carbonate was observed in the field. b) XRD trace from sample SAFa from Tejon Pass locality along the San Andreas fault. Note the presence of smectite, illite and kaolinite, in addition to quartz, feldspar and chlinochlore.

about 8–24 hours to disperse pedogenic clay. For this analysis, two splits of each Tejon Pass sample were initially shaken in a horizontal box shaker for 24 and 96 hours in deionized water without any dispersant, and analyzed using the standard pipette method

(Folk, 1974). We repeated this experiment with an additional two splits of each of the four samples mixed with an 0.05% solution of sodium hexa metaphosphate as a dispersant and shaken for 24 and 96 hours. The samples shaken for 96 hours show slightly finer particle distributions than those shaken for 24 hours, although the decrease in grain size was not significant. In contrast, the samples run without any dispersant displayed significantly coarser distributions indicating that a substantial proportion of the sample remained aggregated, even after 96 hours of vigorous shaking. We conclude from this analysis that the use of a dispersant may be critical in disaggregating near-surface samples of pulverized granite when even small amounts of weathering product are present, as with the samples we analyzed of the Tejon Lookout Granite.

To compare results from the Horiba laser particle size analyzer and data generated using the standard pipette method of Folk (1974), we used identical splits [splitting errors are approximately ±1%, based on comparative analysis exceeding a thousand standard runs (Rockwell, 2000)], and processed both sets with sodium hexametaphosphate dispersant. The samples were wet-sieved through a 63 μm sieve to separate the sand from the silt and clay. This is required for the standard pipette technique, as well as the Horiba, as laser particle size analyzers are only appropriate for determination of particle size distribution in fine sand and finer materials. Consequently, all laser particle size and pipette data discussed below are for the silt and clay fractions only.

The coarse material (> 63 μm) was dried, sieved at standard phi fractions (−1 through 4), and weighed to determine the percent of each sand fraction. These data were then combined with the settling tube data to determine the full particle range. Sieving also allowed for the visual inspection of each of the coarse fractions to assess for grain damage and to verify that the coarser grains were, in fact, not agglomerates of finely-shattered material. In the dry-sieving process, a small amount of material passed through the 63 μm sieve, which indicates incomplete rinsing of the sample during the initial wet-sieving process. This "pan fraction" was reanalyzed and found to be entirely in the 45–63 micron size range, so the weight of the pan fraction was binned with the coarse silt in the final particle size assessment.

For the fine fraction analyses, one split was analyzed by the standard pipette technique (Folk, 1974), and one was analyzed with the Horiba instrument. For the laser analyzer, we started the analyses using the manufacturer's default circulation speed (preset at 2: flow rate = 3 l/min), and a refractive index = 1.163 based on the composition of the material being measured (primarily quartz and feldspar) and the solution in which the particles are suspended (deionized water with a 0.05% concentration of sodium hexa-metaphosphate for clay dispersion).

For a group of selected samples from Bear Trap Canyon (GF T1c, GF T1a, GF T2b, GF T3a, and GF T3d) we prepared each sample with dispersant, sieved at 63 μm, hand-shook the samples and immediately drew an aliquot for analysis with the Horiba. The laser analyzer produces results that are binned at much finer fractions than the phi intervals we used during the pipette analysis. Consequently, the two data sets cannot be directly compared without common binning. Once this is done, an interesting artifact of

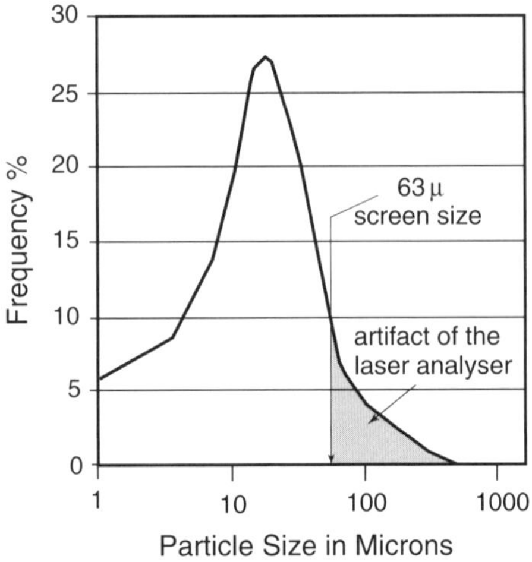

Figure 4
Horiba laser particle size analyzer data for a representative sample of pulverized granite along the Garlock fault. Note that this sample was wet-seived at 63 μ, so all of the apparent material above 63 μ in size (the "tail" to the right) is a result of the laser analyzer placing a Gaussian distribution on the particle size estimate. As the sample was seived at 63 μ, there can be no material of this size present in the analyzed sample.

the laser analyzer is the "tail" of coarse fraction that is distributed out from 63 μm into the few hundred micron range (Fig. 4). As all of the samples were initially passed through a 63 μm sieve, there can be no real data representing particles greater in size than 63 μm. Discussions with the manufacturer revealed that the Horiba LA 930 analyzer software utilizes an algorithm that smoothes the results and displays them as a Gaussian distribution (T. Shimamoto, pers. comm., 2006). In the case of our samples, it redistributed data and apparently added a coarse fraction that is not present in our sample, and we must assume that it has had a similar affect on the finest fraction.

We also tested the time dependant method for PSD of Wilson (2004) and Wilson *et al.* (2005), using the Horiba LA 930 laser analyzer (they employed a Beckman Coulter LS 230 Particle Size Analyzer) with measurements taken at 0, 24, 48, 72 and 96 hour intervals. A similar marked decrease in grain size was observed beginning from the initial measurement (0 hours) and at each subsequent measurement (Fig. 5). However, following the measurement taken at the 96-hour interval, the circulation speed was increased stepwise from speed 2 (flow rate = 3 l/min) to 3 (flow rate = 3.5 l/min) and finally to 4 (flow rate = 4 l/min), and at each increase in circulation speed there was a noticeable increase in the grain sizes of the PSD of the sample, with the entire initial PSD recovered at the final measurement (Fig. 5). These observations demonstrate that the coarse fraction of the sample settled out of suspension due to slow circulation over long periods of time, and was recovered when the circulation rate was increased.

We note that the final measurement has a 12% increase in mean grain size over the first measurement. We attribute this result to either rapid sedimentation of the coarsest

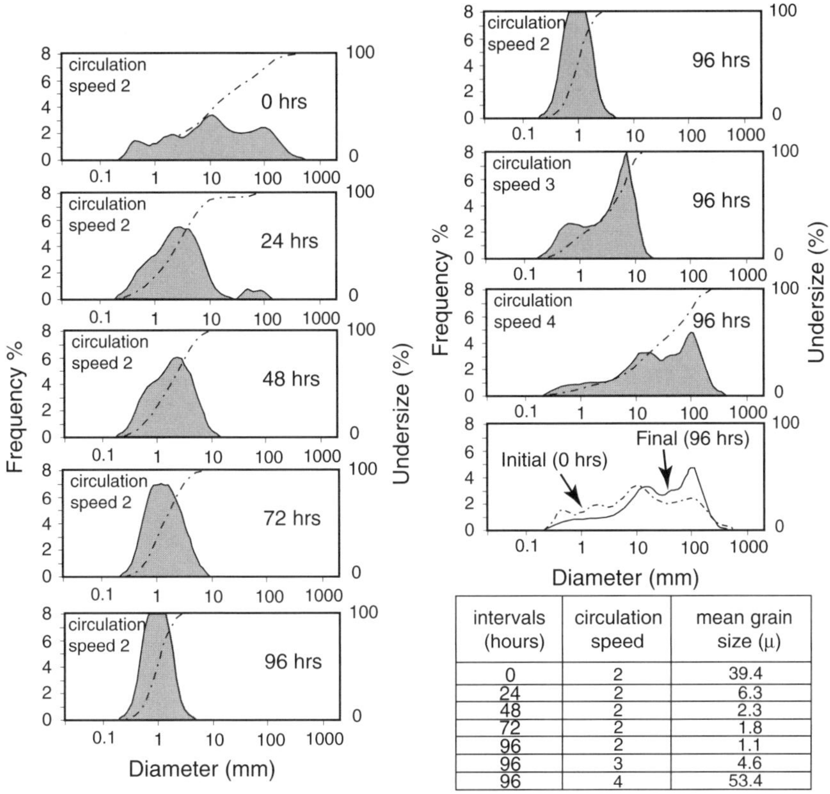

intervals (hours)	circulation speed	mean grain size (μ)
0	2	39.4
24	2	6.3
48	2	2.3
72	2	1.8
96	2	1.1
96	3	4.6
96	4	53.4

Figure 5
Analysis of spin time and spin speed on a pulverized granite sample (SAFb) from Tejon Pass using the Horiba LA 930 Laser Particle Size Analyzer. Measurements were taken at 24 hour intervals, with the analyzer set at the default circulation speed (2). Apparent grain size reduction occurs as a function of time in the machine, whereas the apparent reduction is recovered as the circulation speed is increased. The last graph depicts the initial vs final measurements, and the mean grain size is indicated in the table.

particles prior to the initial measurement (0 hours) or possible agglomeration during the sedimentation process at low circulation speeds. Allowing the sample to run at a circulation speed of 4 for an extended period of time may have allowed for re-entrainment of coarse particles, or dispersement of agglomerated particles. We, in affect, reproduced the previous observation of very fine particle size of WILSON *et al.* (2005), but believe it to be an artifact of the method employed. Note that the tail produced by the algorithm has also been added to these sample data, as all were sieved at 63 μm prior to analysis, so although the measurements taken immediately after the sample is introduced into the machine appear to be closest to the true representation of the sample PSD, they still have the software artifact of imposing a Gaussian distribution to the "estimated" PSD.

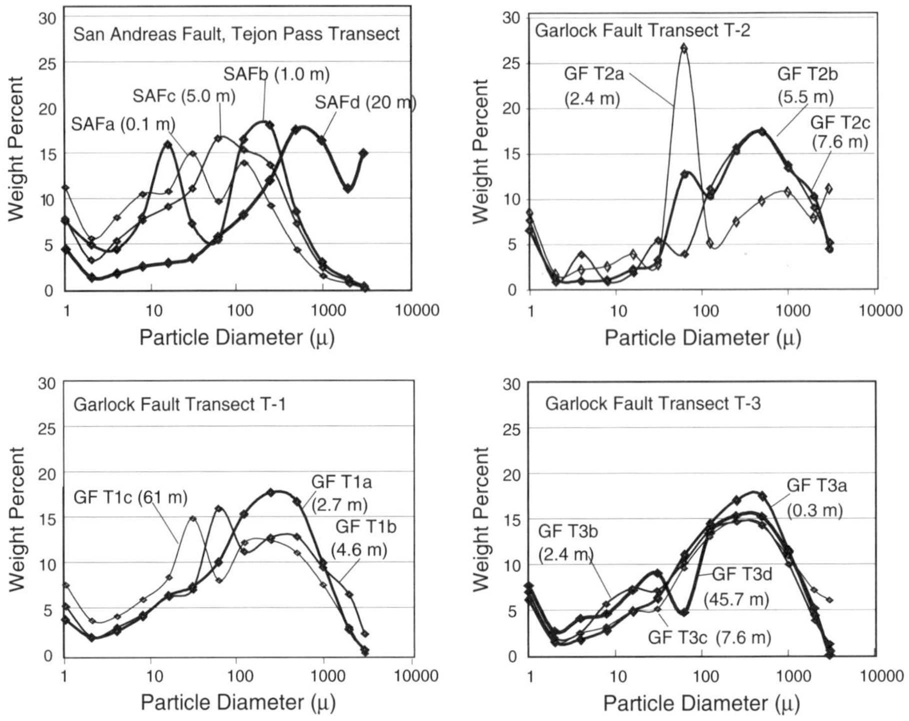

Figure 6
PSD for Tejon Pass and Garlock pulverized Tejon Lookout Granite samples by pipette and seive method. All samples were shaken in sodium hexametaphosphate for 96 hours prior to wet seiving. Note that the peak wt. % of sample mass falls in the 20–200 micron range, with the finest distributions from samples immediately adjacent to the San Andreas fault slip surface.

We conclude that laser particle size results, although quick and easy, may not provide an accurate representation of the particle distribution in pulverized rocks, although we are continuing to work on this problem. Consequently, we ran all samples using the standard pipette method of FOLK (1974), which is generally believed to be accurate to about 1%, and we took aliquot samples to determine the 31–63 μm, 16–31 μm, 8–16 μm, 4–8 μm, 2–4 μm, 1–2 μm and <1 μm fractions (phi 4 through 10, respectively).

The particle size distributions from the combined sieve and pipette runs are displayed in Figure 6, and the mean particle sizes are displayed as a function of proximity to the primary fault core in Figure 7. For all samples that we analyzed, the mean particle size lies between 26 μm and 208 μm, in the silt to fine sand range. Of note is the apparent coarsening of average grain size outward from the fault core at the Tejon Pass locality, consistent with the idea that comminution is more intense close to the active rupture surface. However, the differences out to a few meters are minor when compared to 20 m from the principal slip zone.

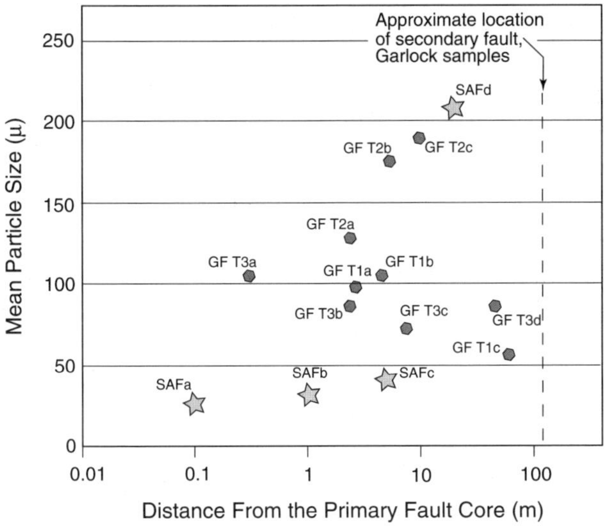

Figure 7

Plot of mean particle size versus distance from the primary fault core for the San Andreas and Garlock samples. Note that the San Andreas samples become finer, on average, as the fault core is approached, whereas the Garlock samples do not show as clear a trend. This may be due, in part, to the presence of a second significant fault strand about a hundred meters south of the main fault, and some of the samples were recovered from near that second fault.

In contrast, samples recovered from along the Garlock fault in Bear Trap Canyon do not show as clear a trend outward from the principal fault core. In transect 1, all three samples have a peak in the few hundred micron range, whereas the samples taken farther from the principal slip surface have finer particle distributions. In transect 2, the samples display a more normal trend, with the finest pulverized sample closest (2.4 m) to the principal slip surface. Transect 3 shows all samples damaged to about the same degree, but again with the samples taken away from the principal slip surface displaying slightly finer particle distributions. A possible explanation is the presence of a secondary break out thrust at Bear Trap Canyon that is a few tens of meters or less south of the three sampling transects. The presence of this structure means that the Garlock samples were collected from an actively deforming wedge between two fault strands. Although we see no evidence of shear dislocations in the thin sections, it is likely that the presence of a nearby secondary strand has affected the local pulverization signal. It is also possible that the thrust was a primary slip zone, which may explain the apparent reversal in particle size trend. In any case, more examples of fault perpendicular transects are clearly warranted to test the variability of damage relative to proximity to the active slip zone.

Microscope Comparison of Particle Size. To further confirm that coarse fragments make up the dominant particle size (by weight) in these samples of pulverized granite, we cut

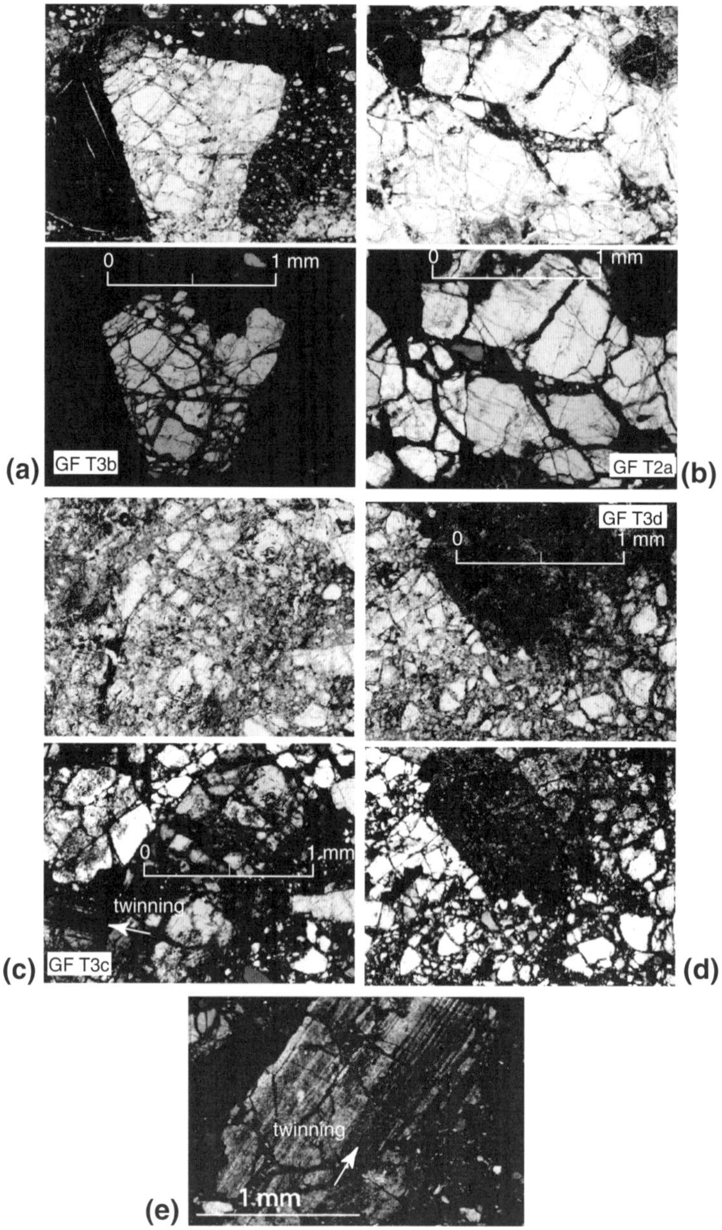

Figure 8

Photomicrographs of representative thin section from the Tejon Lookout Granite along the Garlock fault. All sections showed the extensive *in situ* shattering of the crystalline components of the granite. a and b demonstrate the abundance of residual coarse fragments of quartz and feldspar that are broken to the $\sim 100\mu$ scale. c and d show more extensively damaged portions from samples GF T3c and GF T3d, and c also shows twinned feldspar with no evidence of shear, as does the lower photomicrograph (e), which exhibits fine twinning.

thin sections of each sample. We also use the sections to describe additional evidence of weathering, and to describe the presence or lack of shear.

Figures 8a through 8e are photomicrographs (most in plain and polarized light) of representative thin sections from along the Garlock fault. From these photomicrographs, a number of general observations can be made. First, there are numerous large fragments preserved throughout all samples (Figs. 8a, b). As large particles will dominate when measured by weight, this observation is consistent with the particle size results. Although the presence of micro-fractures may be obscured even in polarized light, we see no evidence that the large fragments are shattered to the degree suggested by WILSON *et al.* (2005). In fact, when we ran the sieve analyses and separated out the coarse sand material, we found that sand-sized fragments, although smaller than their original 1–4 mm grain size, were hard and essentially undamaged and did not yield under finger pressure. We take both observations as support for the measured particle size distributions discussed earlier in this report.

All thin sections revealed that most grains are broken (comminuted), but with the original grain boundaries still recognizable (Figs. 8a, b, c, d). In most such cases, the entire crystal goes extinct at the same time when the stage is rotated under polarized light, indicating little or no rotation of grain fragments after shattering. In some cases, very small fragments may show a different extinction, indicating rotation of the smaller pieces in between the large, unrotated fragments. Also of interest is the apparent expansion or dilation in many of these grains, which suggests that dilation may be an important mechanism in the comminution process, although we are not certain that this is not an artifact of the thin-section making process.

Another potentially important observation is that twin planes in plagioclase show little or no evidence for offset (Figs. 8c, e). Twin planes can be resolved in some cases to a few microns in width and their lack of offset indicates a lack of *in situ* shear at this scale. We searched for twinned plagioclase grains in many orientations and we observed minor shear in only a few of them, including the thin section shown in Figure 8e. Thus, most of the comminution does not appear to be the result of direct shearing of the grains, at least at the scale observable in standard thin sections.

Finally, there was secondary clay observable in many thin sections, along with alteration rims on some feldspar grains. These observations are consistent with the presence of some pedogenic products, as indicated by our XRD work.

3. *Discussion*

We present new results on the degree of weathering and extent of comminution in pulverized Tejon Lookout granite adjacent to the San Andreas and Garlock faults. Although weathering is minor, there is clear evidence of weathered products dominated by smectite and illite in the <4 μm size fraction, which add to the cumulative weight of the finest material that we measured. We found evidence of minor weathering along all three of the Garlock transects as well as at the Tejon Pass locality. This is not unexpected as all sites are within a few meters of the modern ground surface and potentially are within the weathering zone. These results differ from the previous conclusion of WILSON (2004) on the relative absence of weathered material. To what degree the weathering has affected other aspects of the analysis, such as particle size distribution, is not clear. However, as the degree of weathering appears to be minor, we believe that its overall affect on the particle size distribution is also small.

Based on the prior results of WILSON *et al.* (2005), the particle sizes that we measured are markedly different than we expected. In fact, part of the reason we recollected the Tejon Pass locality was because the initial Garlock fault results, as well as preliminary results from samples of pulverized granitic rock collected from along the San Jacinto fault (STILLINGS, 2007), displayed much coarser overall grain size distributions. This led to our methodological experiments on the laser particle size analyzer, to which we attribute most of the differences.

Our first-order conclusion is that the Tejon Lookout granite is pulverized along both the San Andreas and Garlock faults, but that the average grain size ranges between about 26 and 208 microns (silt to fine sand), with the size distribution partly a function of distance from the principal fault core, at least for the Tejon Pass sample suite. This average grain size is much coarser than the previously reported data on pulverized rocks, a result that we attribute to sedimentation of the relatively coarse particles in the measurements of WILSON *et al.* (2005) with the laser particle size analyzer. Although we cannot be certain of the methodology employed by WILSON *et al.* (2005), as we did not use the same laser analyzer and the specific details of the previous analysis were not clearly laid out, we were able to reproduce their result of decreasing average particle size with increasing circulation time, and then we were able to recover essentially instantaneously the initial particle size distribution by increasing the circulation speed. We confirmed our particle distributions through analysis of thin sections, and we confirmed the presence of very coarse material (0.5–5 mm in some samples) by testing the coherence and strength of the coarse fraction after it had been separated by dry sieving. In all cases, the remaining coarse sand and pebble-sized particles were found to be fresh and unbroken and could not be further broken without use of a hammer.

We attempted to correct for the "Gaussian artifact" of the laser particle size analyzer by taking the coarse tail fraction (> 63 μm) and rebinning it with the next lower fraction (coarse silt, 31–63 μm). This resulted in a final distribution that is quite similar to the settling tube results, but not the same. More work is warranted in this area.

The implications for the amount of fracture energy and type of stress field required to have produced the observed damage are significant, as the inferred energy depends strongly on the grain size distribution, and the existence of very small grains may imply tensile stress (e.g., RECHES and DEWERS, 2005; SAMMIS and BEN-ZION, 2008). Rather than $\geq$ 50% of the earthquake energy budget being consumed by dynamic fracturing and pulverization as suggested by WILSON *et al.*, (2005), we find that the average value is closer to 1%. This value is compatible with estimates of fracture energy needed to comminute the fault core rock (CHESTER *et al.*, 2005), as well as estimates based on seismological data (e.g., KANAMORI *et al.* 1998; MCGARR, 1999), rock mechanics experiments (e.g., LOCKNER and OKUBO, 1983), and measurements of energy components in computer simulations (e.g., SHI *et al.*, 2008).

The actual mechanism of fracturing in the damage zone needed to produce pulverized fault zone rock is still not clear, but our data shed light on the process. The study of CHESTER *et al.* (2005) of thin sections in the ultra-cataclasite core of the Punchbowl fault indicates that the particles in the fault core follow a power-law distribution from the micron to the nanoscale. SAMMIS and KING (2007) suggest, based on model considerations, that a fractal distribution is mechanically favored for situations in which the primary mechanism for grain-size reduction is by shear and grinding assumed to characterize fault cores. However, several recent studies on natural and experimentally deformed rocks show (e.g., HEILBRONNER and KEULEN, 2006; KEULEN *et al.*, 2007) that the grain-size distribution follows a strict power law only over a narrow range, and has a pronounced slope break that may be related to the grinding limit of quartz. HEILBRONNER and KEULEN (2006) and KEULEN *et al.* (2007) also found differences in the slope of the particle distribution between quartz and feldspar, a result that supports the idea of a mineralogical control on grain size distribution. These authors suggest that the slope break common in their experimental results represents a change in process from grain cracking produced by rupturing to comminution by wear and attrition.

For the pulverized Tejon Lookout granite, it is not clear that shear is a significant contributor to grain size reduction, as there is little evidence of shear in any of the thin sections: Plagioclase twin planes show little to no offset in any orientation and large quartz grains, although shattered, have similar extinction in cross-polars. This is consistent with previous suggestions that pulverized rocks are generated by dynamic reduction of normal stress during successive earthquake ruptures, possibly aided by strong dynamic dilation and transient fault opening at the tip of ruptures propagating along bimaterial fault interfaces in a preferred direction (e.g., BRUNE, 2001; BEN-ZION and HUANG, 2002). We note, however, that pulverized sandstones in the damage zone of the San Andreas fault near Little Rock, California, show evidence of preferred fracture orientations along with radial and conical Hertzian cracking (DOR *et al.*, 2009). These observations in porous sandstones suggest that pulverization may have a contribution from transient compressive stressing associated with earthquake ruptures on the San Andreas fault.

To compare our mechanically-derived particle size data to those from studies using thin sections, we converted the particle data into the standard format of log particle size density versus log particle class (SAMMIS and KING, 2007). Figure 9 shows the Tejon Pass particle size data, with a best-fit line for each sample that follows a power-law distribution and a best-fit curve that has a continuously varying slope with increasing particle size. Sample SAFd recovered from about 20 m from the fault core can be fitted

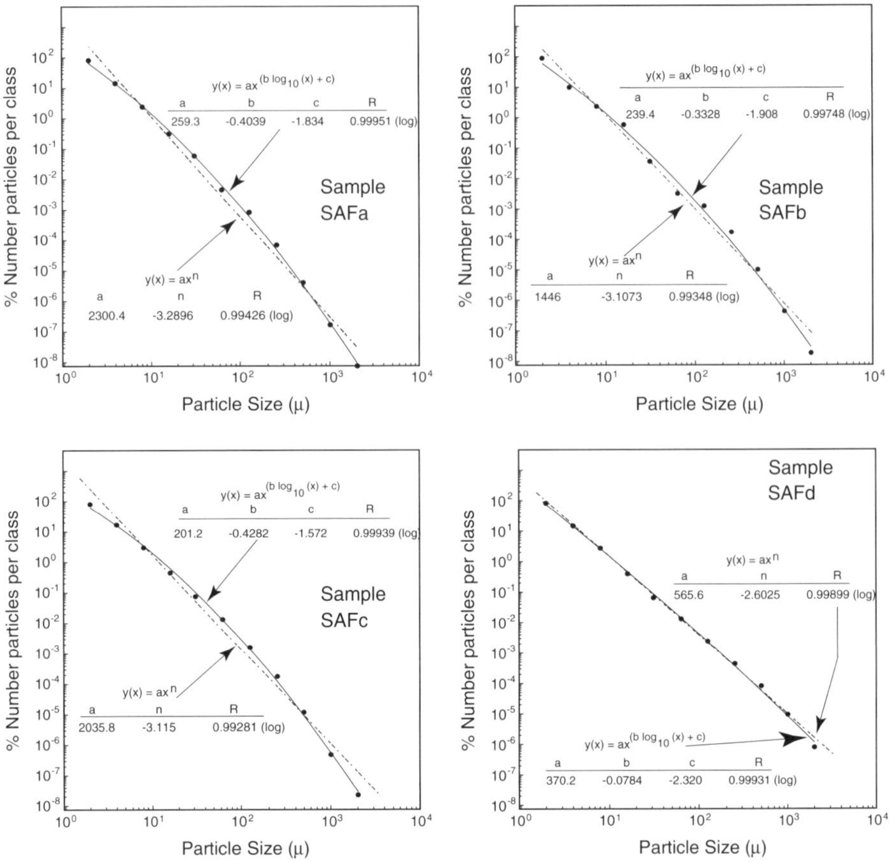

Figure 9
Particle size plots of the four samples collected from along the San Andreas fault at Tejon Pass (the Wilson locality), plotted as particle diameter class against the number of particles in that class. These data are derived directly from the particle size data determined by pippette analyses. The best-fit curves for each sample are plotted, with one that follows a power-law distribution and one that has a continuously varying slope with increasing particle size. Note that the data from the more pulverized samples from within 10 m of the fault core are best fit by the continuously changing slope, although the sample from 20 m (SAFd) can be well fit with either regression.

well using a power-law distribution with a single exponent, however, the other three samples recovered from closer distances to the fault core are better fit by a continuously varying slope.

The four Tejon Pass samples exhibit a clear trend towards finer particle sizes as the primary fault core is approached. This relationship is seen clearly in Figure 10a, where we plot together the data for the different SAF samples, as well as in the particle size plots of Figures 6 and 7. In contrast, the Garlock samples (Figs. 10b–10d) do not show as clear a trend. We do not use the material that was coarser then 2 mm as it is all grouped as > 2 mm rather than by phi scale dimensions. The finest fraction generally falls along

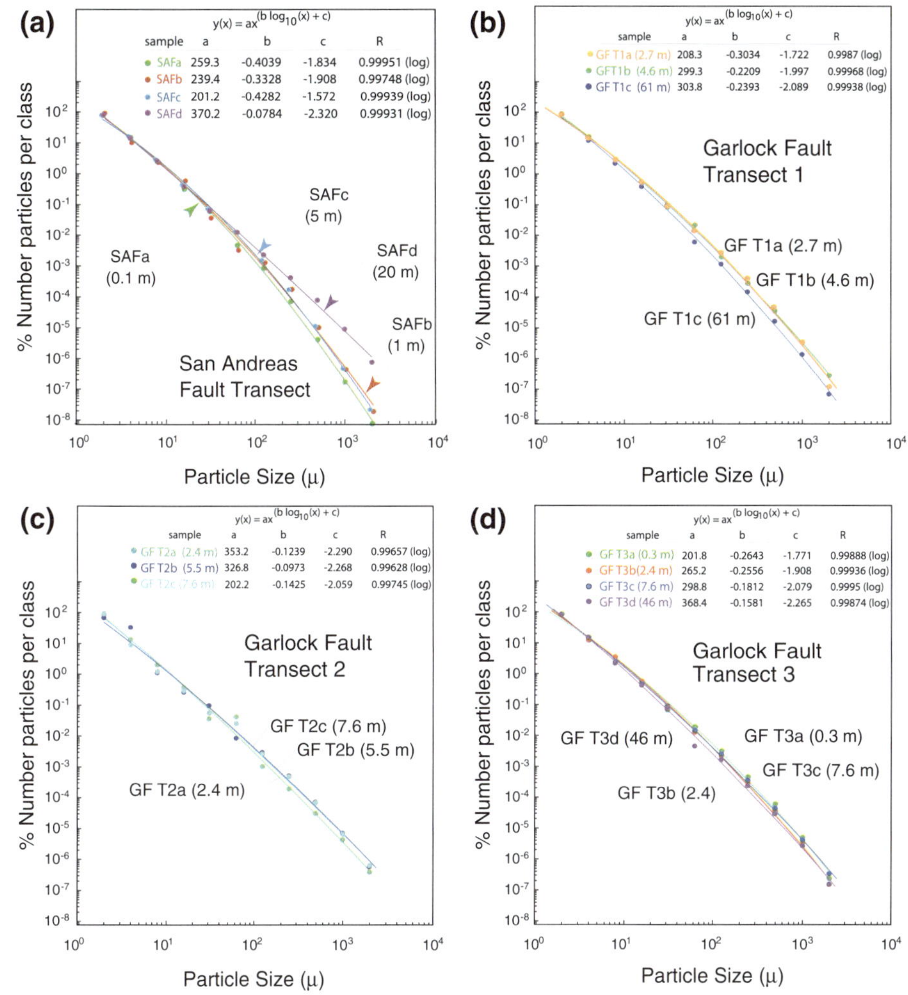

Figure 10

Particle size plots for samples from San Andreas fault (a) and the three Garlock transects (b-d), plotted as particle diameter class against the number of particles in that class. For these samples, we plot only the best fit lines with continuously varying slope.

trend but was not used in the slope determinations because it is also grouped as $<1\mu$ and not by phi scale, and because it contains weathering products (pedogenic clay) and may be affected by secondary processes. Nevertheless, these data show clear trends that can be compared with other published data.

The general first-order observation is that the pulverized granitic rocks that we analyzed do not display a true power-law particle size distribution, but rather follow a distribution with varying exponent. It is interesting to note that KEULEN *et al.* (2007)

report the main slope break (r_K) around 1–2 μm, whereas the pulverized granites from this study show a progressive slope reduction towards the finer material. The primary difference, of course, is that they looked at sheared rock of the fault core while we are looking at pulverized rocks within the broader damage zone. Also, in our case we analyzed the pulverized rock primarily over the dimensions of 1–1000 μm, which is appropriate for the methods we employed. In contrast, thin section studies can quantify the finer material that is typically less than a micron in size. In any case, if the SAMMIS and KING (2007) model is applicable, the observed distributions in our study suggest that direct shear is not likely the primary cause of the shattering. Our observations that most grains appear to have deformed at least partially by dilation are consistent with some shattering during dynamic rupture under locally tensile stress (e.g., BRUNE *et al.*, 1993; BEN-ZION, 2001).

4. Conclusions

Pulverization along large continental strike-slip faults may be quite common, as there are now well-documented observations of pulverized granitic rock along the San Andreas, Garlock, and San Jacinto faults in southern California (WILSON *et al.*, 2005; DOR *et al.*, 2006a; STILLINGS, 2007, this study), and the Arima-Takatsuki fault in Japan (MITCHELL *et al.*, 2008). These faults have tens of kilometers of lateral offset, with the extent of pulverization possibly related to the magnitude of fault offset, the size of earthquake that can be generated, and the possible existence of a bimaterial interface in the fault structure. Of the damage zones studied in California, the Tejon Pass site along the San Andreas fault exhibits the most intense damage, and this fault arguably is capable of producing the largest earthquakes among the strike-slip faults in southern California. It is also the fault with the most cumulative slip and the most pronounced lithology contrast between the bounding crustal blocks. If varying degrees of displacement, seismic potential, and lithologic contrast are factors controlling the degree and extent of pulverization, then more work on faults with these characteristics is warranted.

Pulverization appears to be spatially related to the active fault core however, direct evidence for significant shear is lacking. At Tejon Pass, the comminution is most intense close to the fault core, but the particle size distribution follows a pseudo power law with a continuously changing exponent and suggests a shattering mechanism other than direct shear. In all thin sections open cracks were observed in many grains, suggesting some degree of dilatancy, although we cannot be sure that some of these open cracks formed during the thin-section process. Nevertheless, such an observation, along with the lack of evidence for significant shear, suggests that at least part of the shattering and comminution of grains may result from a rapid decrease in normal stress and perhaps transient fault opening during passage of the dynamic rupture front, as suggested by BRUNE (2001) and BEN-ZION and SHI (2005).

Acknowledgements

We thank Tejon Ranch and Earth Consultants International for access to trench exposures along the Garlock fault. The study benefited from highly informative discussions with Toshi Shimamoto and Oliver Spieler regarding measurements of particle-size distributions with the laser analyzer. We thank Charlie Sammis and Tom Mitchell for comments that improved the presentation of this paper. This research was supported by the Southern California Earthquake Center. SCEC is funded by NSF Cooperative Agreement EAR-0106924 and USGS Cooperative Agreement 02HQAG0008. The SCEC contribution number for this paper is 1258.

References

Abercrombie, R., McGarr, A., Kanamori, H., and Di Toro, G. (eds.) *Earthquakes: Radiated Energy and the Physics of Faulting*, Geophysical Monograph Series, Vol. *170*, (2006) 327 pages, hardbound.

Ben-Zion, Y. (2001), *Dynamic rupture in recent models of earthquake faults*, J. Mechan. Phys. Solids *49*, 2209–2244.

Ben-Zion Y. and Huang, Y. (2002), *Dynamic rupture on an interface between a compliant fault zone layer and a stiffer surrounding solid*, J. Geophys. Res. *107* (B2): Art. No. 2042, doi 10.1029/2001JB000254.

Ben-Zion, Y. and. Sammis, C. G (2003), *Characterization of fault zones*, Pure Appl. Geophys. *160*, 677–715.

Ben-Zion, Y. and Shi, Z. (2005), *Dynamic rupture on a material interface with spontaneous generation of plastic strain in the bulk*, Earth Planet. Sci. Lett. *236*, 486-496, doi: 10.1016/j.epsl.2005.03.025.

Brune, J.N. (2001), *Fault normal dynamic loading and unloading: an explanation for "non-gouge" rock powder and lack of fault-parallel shear bands along the San Andreas fault*, EOS Trans. Am. Geophys. Union, 82.

Brune, J. N., Brown, S. and Johnson P.A. (1993), *Rupture mechanism and interface separation in foam rubber models of earthquakes; a possible solution to the heat flow paradox and the paradox of large overthrusts*, Tectonophys. *218*, 1–3.

Chester, J. S., Chester, F. M., and Kronenberg, A. K. (2005), *Fracture surface energy of the Punchbowl fault, San Andreas system*, Nature, *437*, 133–136.

Dor, O., Ben-Zion, Y, Rockwell, T. K. and Brune, J. (2006a), *Pulverized Rocks in the Mojave section of the San Andreas Fault Zone*, Earth Planet. Sci. Lett. *245*, 642–654.

Dor, O., Rockwell, T. K., and Ben-Zion, Y. (2006b), *Geological observations of damage asymmetry in the structure of the San Jacinto, San Andreas and Punchbowl Faults in southern California: A possible indicator for preferred rupture propagation direction*, Pure Appl. Geophys. *163* (2–3), 301–349.

Dor, O., Yildirim, C., Rockwell, T. K., Ben-Zion, Y., Emre, O., Sisk, M., Duman, T. Y. (2008), *Geologic and geomorphologic asymmetry across the rupture zones of the 1943 and 1944 earthquakes on the North Anatolian Fault: Possible signals for preferred earthquake propagation direction*, Geophys. J. Int. 173, 483–504, doi: 10.1111/j.1365-246X.2008.03709.x.

Dor, O., Chester J. Ben-Zion Y., Rockwell T., Brune J. (2009), *Characterization of damage in sandstones along the Mojave section of the San Andreas Fault: Implications for the shallow extent of damage generation*, Pure Appl. Geophys. *166*, (10).

Folk, R. L., *Petrology of Sedimentary Rocks*, (Hemphill Publishing Company1974).

Heilbronner, R., and N. Keulen, (2006), *Grain size and grain shape analysis of fault rocks, Tectonophys. 427* (1–4), 199–216.

Kanamori, H., Anderson, D.L., and Heaton, T.H. (1998), Frictional melting during the rupture of the 1994 Bolivian Earthquake. Science *279*, 839–842.

Keulen, N., Heilbronner, R., Stünits, H., Boullier, A. M., and Ito, H. (2007), *Grain size distributions of fault rocks: A comparison between experimentally and naturally deformed granitoids*, J. Struct. Geol. *29*, 1282–1300.

LOCKNER, D. A., and OKUBO, P. G. (1983), *Measurements of frictional heating in granite*, J. Geophys. Res. *88*, B5, 4313–4320.

MCGARR, A. (1999), *On relating apparent stress to the stress causing earthquake fault slip*, J. Geophys. Res. *104*, 3003–3011.

MITCHELL, T.M., SHIMAMOTO, T. ANDO, J. BEN-ZION, Y. (2008), *The seismic velocity and permeability properties of pulverized rocks*, Geophys. Res. Abstracts *10*, EGU2008-A-00000, EGU General Assembly.

NESBITT, H. W., and YOUNG, G. M. (1982), *Early Proterozoic climates and plate motions inferred from major element chemistry of lutites*, Nature *299*, 715–717.

NESBITT, H. W., and MARKOVICS, G. (1997), *Weathering of Granodioritic crust, long-term storage of elements in weathering profiles, and petrogenesis of siliciclastic sediments*, Geochimica et Cosmochimica Acta *61* (8), 1653–1670.

RECHES, Z., and DEWERS, T. A. (2005), *Gouge formation by dynamic pulverization during earthquake rupture*, Earth Planet. Sci. Lett. *235*, 361–374.

ROCKWELL, T.K. (2000), *Use of soil geomorphology in fault studies: In Quaternary Geochronology: Methods and Applications.* J.S. Noller, J.M. Sowers, and W.R. Lettis, eds. AGU Reference Shelf 4 Am. Geophys. Union, Washington D.C., 2000, pp 273–292.

ROCKWELL, T. K. and BEN-ZION Y. (2007), *High localization of primary slip zones in large earthquakes from paleoseismic trenches: Observations and implications for earthquake physics*, J. Geophys. Res. *112*, B10304, doi:10.1029/2006JB004764.

SAMMIS, C. G. and BEN-ZION, Y. (2008), *The mechanics of grain-size reduction in fault zones*, J. Geophys. Res. *113*, B02306, doi:10.1029/2006JB004892.

SAMMIS, C. G., KING, G. C. P., and BIEGEL, R. (1987), *The kinematics of gouge deformation*, Pure. Appl. Geophys. *125*, 777–812.

SAMMIS, C. G., and KING, G. C. P. (2007), *Mechanical origin of power law scaling in fault zone rock*, Geophys. Res. Lett. *34*, L04312.

SHI, Z., BEN-ZION, Y., and NEEDLEMAN, A. (2008), *Properties of dynamic rupture and energy partition in a two-dimensional elastic solid with a frictional interface*, J. Mech. Phys. Solids *56*, 5–24, doi:10.1016/j.jmps.2007.04.006.

SISK, M., ROCKWELL, T., GIRTY, G., DOR, O., and BEN-ZION, Y. (2005), *Potentially pulverized granites along the Garlock fault: An analysis into their physical and chemical properties, EOS Trans.* Amer. Geophys. Union *86* (52), Fall Meet. Suppl., Abstract S41B-0994,.

SPERAZZA, M., MOORE, J.N., and HENDRIX, M.S. (2004), *High-resolution particle size analysis of naturally occurring very fine-grained sediment through laser diffractometry.* J. Sedimentary Res. *74*, (5), 736–743.

STILLINGS, M. (2007), *Structural, textural, and geochemical analyses of fault damage zones: Clark strand, San Jacinto fault zone, southern California*, M.S. Thesis, San Diego State University, San Diego, California.

WILSON, B. T. (2004), *Meso- and micro-structural Analysis of the San Andreas Fault at Tejon Pass, California*, USA, M.S. Thesis, University of Oklahoma, Norman, Oklahoma.

WILSON, B., DEWERS, T., RECHES, Z., and BRUNE, J. (2005), *Particle size and energetics of gouge from earthquake rupture zones*, Nature *434* (7034), 749–752.

(Received September 3, 2008, accepted February 12, 2009)
Published Online First: June 30, 2009

To access this journal online:
www.birkhauser.ch/pageoph

Pure appl. geophys. 166 (2009) 1747–1773
0033–4553/09/101747–27
DOI 10.1007/s00024-009-0516-z
© Birkhäuser Verlag, Basel, 2009
Pure and Applied Geophysics

Characterization of Damage in Sandstones along the Mojave Section of the San Andreas Fault: Implications for the Shallow Extent of Damage Generation

ORY DOR,[1,5] JUDITH S. CHESTER,[2] YEHUDA BEN-ZION,[1] JAMES N. BRUNE,[3] and THOMAS K. ROCKWELL[4]

Abstract—Following theoretical calculations that suggest shallow generation of rock damage during an earthquake rupture, we measure the degree of fracture damage in young sedimentary rocks from the Juniper Hills Formation (JHF) that were displaced 21 km along the Mojave section of San Andreas Fault (SAF) and were not exhumed significantly during their displacement. In exposures adjacent to the fault, the JHF typically displays original sedimentary fabrics and little evidence of bulk shear strain at the mesoscopic scale. The formation is, however, pervasively fractured at the microscopic scale over a zone that is about a 100 m wide on the southwest side of the SAF near Little Rock. The abundance of open fractures, the poor consolidation, and the shallow inferred burial depth imply that the damage was generated close to the surface of the Earth. The spatial correlation of this damage with a seismically active trace of the SAF suggests that it was generated by SAF slip events that by assumption were of a seismic nature throughout the displacement history of the JHF. Thus the JHF provides a very shallow upper bound for the generation of brittle damage in a seismic fault zone. The fracture fabric is characterized by preferred orientations of fractures that split grains between contact points and is consistent with overall deformation under directed compression. However, the available results cannot be used to distinguish between proposed off-fault damage mechanisms. Fracture orientations are compatible with a maximum compressive stress oriented at a high angle to the fault at about 10 m, and at a lower, more variable angle farther away from the fault. The fracture distribution and fabric are consistent with observations made of the microscale damage characteristics of the Hungry Valley Formation in the northwestern section of the SAF in the Mojave, and with previous observations of exhumed, ancestral strands of the SAF.

Key words: Fault-zone structure, rock damage, San Andreas fault, earthquake rupture mechanism, mode I fractures, Sedimentary rocks.

[1] Department of Earth Sciences, University of Southern California, Los Angeles, CA 90089-0740, U.S.A. E-mail: dor@brown.edu

[2] Department of Geology and Geophysics, Texas A&M University, College Station, TX 77843-3115, U.S.A.

[3] Nevada Seismological Laboratory, University of Nevada, Reno, NV 89557, U.S.A.

[4] Department of Geological Sciences, San Diego State University, San Diego, CA 92182-1020, U.S.A.

[5] Now at Department of Geological Sciences, Brown University, Providence, RI 02912, U.S.A

1. Introduction

1.1. The Origin and Depth of Pulverized Fault Zone Rocks

The key characteristics of pulverized fault zone rocks found along the San Andreas Fault (SAF) and elsewhere are (WILSON *et al.*, 2005; DOR *et al.*, 2006a; ROCKWELL *et al.*, 2009) reduced grain size by fracture with minimal distortion of the primary rock structure on many scales. WILSON *et al.* (2005) and ROCKWELL *et al.* (2009) have shown that the pulverized Tejon Lookout Granite at Tejon Pass lacks significant weathering products, suggesting that the pulverization occurred primarily by a mechanical process. DOR *et al.* (2006a) mapped the distribution of pulverized crystalline rocks along the Mojave section of the SAF and found that they occupy a tabular zone parallel to the fault that is 100–300 m wide. Based on evidence for the exhumation depth of the SAF in the Mojave area, they inferred that the observed pulverization occurred in the top 1–3 km, in contrast to the conclusion reached earlier by WILSON *et al.* (2005).

Our investigation of potential geological signals constraining the depth and mechanism of rock pulverization discussed in this paper is driven by theoretical expectations that we review here briefly. Lab-constrained damage rheology (LYAKHOVSKY *et al.*, 1997; HAMIEL *et al.*, 2004; LYAKHOVSKY and BEN-ZION, 2008), analytical work (RICE *et al.*, 2005), and numerical simulations of spontaneous damage generation during dynamic rupture on a bimaterial interface (BEN-ZION and SHI, 2005) indicate that pervasive fracturing is likely confined to the top few km of the crust. In the case of ruptures along faults that separate different elastic materials at depth, the shallow-generated damage is expected to have an asymmetric pattern and be concentrated on the side of the fault with higher seismic velocity at depth (BEN-ZION and SHI, 2005). These predictions are compatible with properties of the pulverized rock layer observed by DOR *et al.* (2006a), and with the symmetry properties of smaller-scale damage products mapped by DOR *et al.* (2006b) along several fault sections of the San Andreas system in southern California (including the Mojave section of the SAF).

Brittle fractures in the damage zone adjacent to large displacement seismogenic faults can result from numerous loading conditions at dynamic and quasi-static deformation rates. In particular, the orientation of microfractures (given that the fracture pattern was not extensively overprinted by various other processes) may reflect the orientation of the near-field stresses developed during the passage of an earthquake rupture (e.g., DI TORO *et al.*, 2005; RICE *et al.*, 2005). BRUNE *et al.* (1993) and Brune (2001) suggested that the crushed rocks adjacent to the SAF resulted from fault-normal loading and unloading due to a dynamic reduction of normal stress associated with the passage of successive rupture tips. Ruptures propagating on a bimaterial interface may produce local fault-normal openings associated with transient absolute tension (BEN-ZION, 2001; BEN-ZION and HUANG, 2002), which may also lead to variability in the orientation of stresses, and fractures, throughout the damage zone. Directed stress concentrations may develop along a rough fault surface during cyclic frictional sliding, depending on the far field stress direction and the friction on the fault (CHESTER and CHESTER, 2000). Damage developed due to the geometrical effect of the fault roughness will fall off linearly as a function of the log-distance from the fault (e.g., SCHOLZ *et al.*, 1993). A summary of several models and their predicted microfracture orientations is given by WILSON *et al.* (2003). Damage generated during the propagation of successive earthquakes will overshadow the initial damage produced by the growth of the fault (CHESTER *et al.*, 2004b).

1.2. Research Rational

To further assess the shallow depth extent and origin of pulverization in the damage structure of the SAF, and test model predictions, we examine the presence and type of damage in sedimentary rocks, mainly from the Plio-Pleistocene Juniper Hills Formation (JHF). These sedimentary rocks were never deeply buried while being displaced along the SAF (Section 1.3). A spatial correlation between the level of damage in those rocks, if found, and the trace of the SAF, should indicate that the damage is associated with SAF slip events. Paleoseismic studies show evidence for large seismic events on the SAF in the Mojave during the last 6000 years (WELDON *et al.*, 2004). We assume that this was the case during the entire displacement history of the JHF (total of 19–21 km of strike-slip, BARROWS *et al.*, 1985). The presence of SAF-related damage in these sandstones can therefore provide an estimate for the upper boundary of the depth range of damage generation during seismic faulting events. The properties of the damage can help identify the possible damage mechanisms operating during seismic rupture.

1.3. Characteristics and Exhumation of the Juniper Hills Formation

The JHF was derived, in part, from the Punchbowl Formation and contains several distinctive clasts not found in older sedimentary rocks (BARROWS, 1980; 1985). It consists of elongated tectonostratigraphic units that parallel the SAF and its subsidiary faults. Distinctive JHF units containing the unique clasts are offset 13 to 16 km by the Northern Nadeau fault, and 19 to 21 km by the SAF in this and the surrounding regions. Samples were collected from the TQjh member (Juniper Hills formation, undifferentiated), a poorly to moderately indurated fluvial deposit of coarse arkosic sandstone, lesser conglomerate and thin-bedded shale that commonly exhibits distinct bedding and often is poorly sorted (BARROWS *et al.*, 1985). Clasts are subangular to well-rounded and varicolored. The JHF is characterized by considerable variation in bedding, sorting and induration properties between exposures and even between adjacent layers. Compositional variations also have been noted (BARROWS, 1980, 1985). The poorly indurated layers are sometimes powdery even far from a fault. Therefore, and in contrast to crystalline rocks studied previously (e.g., DOR *et al.*, 2006a), the texture in a hand sample of the JHF or its appearance in the field does not provide a systematic indication of microscopic fracture damage.

Exhumation of the SAF in the study area was inferred by DOR *et al.* (2006a) to be minimal, and certainly below the maximum uplift of 2–4 km inferred for the Punchbowl fault (CHESTER *et al.*, 1993). SPOTILA *et al.* (2007) show that exhumation of the SAF in the vicinity of the San Gabriel Mountains ranges 0.03 to 0.5 mm/y with an exceptional rate of up to 1 mm/y in a few places based on theormochronological data. Similar rates are proposed for blocks bounding the SAF in the relevant time frame (BLYTHE *et al.*, 2002). These rates imply tens of meters up to about 1 km of uplift of the SAF in the last 2 Ma. While there is no direct evidence for the maximum burial depth of the JHF, the poor consolidation state is consistent with shallow burial. WEBER (1999; see also KENNY, 2000) suggests that uplift of the southeast portion of the San Gabriel mountains, initiating 0.75 to 0.5 Ma, raised beds of the middle (?) Pleistocene Shoemaker Gravel Formation (SGF) in the Big Pine area 1100 m with respect to equivalent units near Big Rock Creek (inset in Fig. 1). The easternmost zone of our sampling area is 12 km farther to the northwest of Big Rock Creek, and 25 km northwest of Big Pine. Although the JHF is older than the SGF, its distance from the locus of uplift (farther away from the correlative lower body of the SGF), and the inference that uplift had intensified toward the end of the Pleistocene also support considerably lower exhumation values for the JHF.

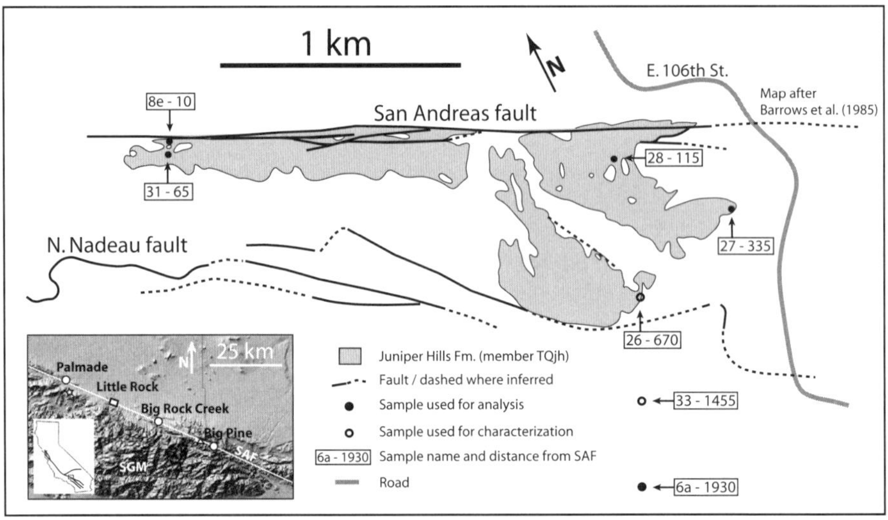

Figure 1

The inner-inset (lower left corner) shows a map of California with main faults (thick black lines). Rectangle marks the section of the SAF shown in the larger inset. Larger inset: Digital elevation model of the southwestern Mojave with the SAF marked as a white line. Location of geological map is shown as a rectangle near the town of Little Rock. The star indicates the location of sample 6a. SGM: San Gabriel Mountains. The geological map (after BARROWS *et al.*, 1985) shows the distribution of some of the JHF bodies in the area with the location of our sampling sites. Samples 6a and 33 are marked not in their geographic location. Note that the JHF here is bounded by the SAF and the N. Nadeau fault.

Our two sampling points farthest from each other are near E. 106th St. (e.g., point 28–115 in Figure 1) and near the intersection of Pearblossom Hwy and Sierra Hwy (point 6a, shown as a star in the inset in Fig. 1), with the first being 16 km east-southeast and 190 m higher than the second one. An additional 22 km to the southeast, the same JHF member is exposed near Jackson Lake (shown as a triangle in the inset of Fig. 1), 790 m higher than the E. 106th St. sampling point. The elevation gradient changes from 12 m/km to 34 m/km between these two sections, and the trace of the SAF responds to it by an abrupt change in its elevation gain 12 km to the southeast of 106th St., immediately east of Big Rock Creek. These topographical features are compatible with a substantial drop in uplift rates in the vicinity of Big Rock Creek and to the northwest, also consistent with minimal exhumation in the study area.

2. *Methodology*

2.1. *Research Approach*

The microscale damage we observe in the JHF is characterized by an inhomogeneous fragmentation of grains without apparent distortion of their shape, while the matrix includes detrital grain fragments that are not considered damage products. In addition, there are variations in the original modal and grain size distribution between the sampling locations (Section 3.1). Consequently, damage analysis of the JHF rocks cannot take advantage of 'whole rock' particle size distribution analyses methods (e.g., image segmentation, particle laser analyzer, etc.) as a means of comparison between the samples, and therefore targets the individual grains by comparing their current damage content to their original, predamage state. Since most of the microfractures observed are open, it is impossible to separate the grains from the matrix without breaking them apart, and so the damage analysis is best done optically on photomicrographs using the image analysis methods discussed below.

2.2. *Sampling*

Damage estimates are based on 13 JHF samples collected at seven stations along a fault-normal traverse southwest of the SAF (Fig. 1). We compare the results to observations from three samples of the Hungry Valley Formation collected at distances of 125, 1055 and 3380 m from the SAF in the northwest portion of the Mojave Desert and in Ridge Basin (Barrows *et al.*, 1985; Crowell, 1982).

To ensure as much uniformity as possible between our samples, specimens were collected from layers having similar grain sizes and colors, and as far as possible from mesoscale faults, veins, fine layering, large clasts and other obvious structural and sedimentary elements in the TQjh member of the JHF. Samples were impregnated in the lab with a low viscosity epoxy (Epo-Tek 301©) under a vacuum and then sectioned and polished for optical analysis.

2.3. *Image Analysis*

The fracture damage analysis was performed on a representative population of grains in each thin section under cross-polarized, plane-polarized and reflected-light conditions. We first mapped the original grain boundaries of 170–320 grains per section on plane-polarized light images. We confined our analysis to a grain size range of ~150 – 2000 microns (with some exceptional larger grains), avoiding large rock fragments, grains showing significant alteration and grains with many subgrains. The minimum size was set to avoid host-rock matrix particles and because grains smaller than 150-micron show very little fracturing. The maximum size is the actual maximum original grain size in the samples, excluding large (up to 3–4 mm) rock fragments. We digitized the overall outlines of the mapped qualified grains to estimate the original (prefracturing) grain size distribution for this range using Image J (http://rsbweb.nih.gov/ij/). A random selection of grains for the analysis from the entire mapped grain population would result in a non-representative subset of mainly small grains. Such a subset would underestimate the degree of damage. To address this, we divided the grain population into four size bins and chose 10–12 random grains from each bin (Fig. 2a) to obtain a subset of the initial population that consists of 40–48 grains per thin section.

Each selected grain was photographed at 100X magnification under reflected light (Fig. 2b). We used this photograph to generate a grayscale image of the fractured grain, masked from the background (Fig. 2c), and a second image showing a digitized and filled trace of the original grain to represent the intact grain in its predamaged form (Fig. 2d). The image of the masked grain was transformed into a bitmap (binary) using a 50% threshold value and then inverted so that the grain and its fragments are black, and the

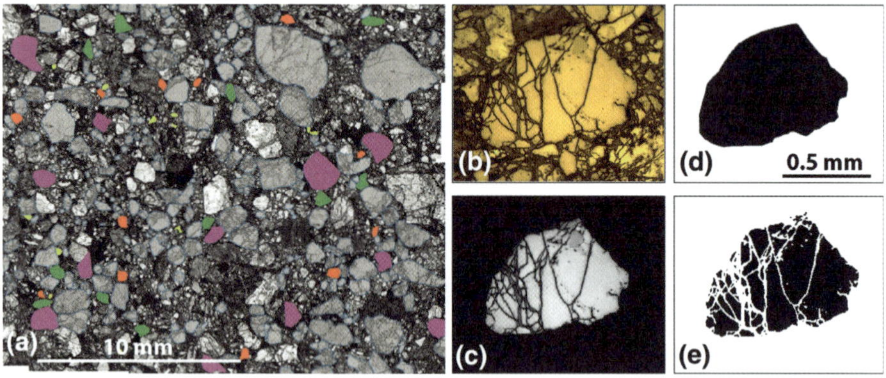

Figure 2
Intermediate and final products of the analysis process: a. Forty grains in four size bins (color coded) are chosen from a population of mapped grains (light transparent gray). b. An image of a single quartz grain taken under reflected light, optical microscope. c. Grayscale version of the same grain, masked from its environment. d. A map of the grain's outline, filled, representing the predamaged state of the grain. e. The product of (c): binary (bitmap) image of the fragments that belong to the same grain. Their total perimeter length is divided in the perimeter length of the "intact" grain in (d) to give the FIPL (see text).

background is white (Fig. 2e). Epoxy-filled gaps that passed the threshold process were masked manually in Adobe Photoshop.

The total perimeter length of each grain was determined using Image J. The "intact" grain image (Fig. 2d) gives the length of the original grain perimeter before it was fractured, and the "damaged" grain image (Fig. 2e) gives the cumulative perimeter length of all the fragments that belong to the grain. We define a Factor of Increase in Perimeter Length (FIPL) as the ratio between the "damaged" and the "intact" perimeter lengths, and the FIPL for the entire sample is calculated as the area-weighted average FIPL of all grains that were analyzed in that sample. The perimeter length for the original grains is not corrected for dilation of the grain caused by fracture opening, consequently our FIPL measurements slightly underestimate the actual values.

For modal distribution analysis we used backscatter images of the thin sections acquired by a Cameca SX100 Electron-probe with a current of 15KeV. We used Image SXM (http://www.liv.ac.uk/wsdb/ImageSXM/) to generate bitmap images of each of the phases in the sample by density slicing of the grayscale images at the corresponding gray levels, and calculated the proportions between the phases from their relative area on the images. "Noise" (very small particles) was subtracted from the bitmap of each phase and was added to the total area of the matrix bitmap.

The trends of microfractures were determined for selected samples on a flat microscope stage. In these samples, we measured the orientation of fractures in all the grains chosen for the perimeter length analysis. We classified the measured fractures into the fracture types described in Section 3.2.

3. *Observations*

3.1. *Juniper Hills Formation Host Rock*

A representative exposure of the JHF at 670 m from the fault shows a light brown to yellow arkosic sandstone with layering defined by grain size variations. The formation is tilted shallowly at this location and the surrounding regions, and is cut by calcite veins and scattered mesoscale faults. Overall, the host rock is friable and most layers can be disaggregated fairly easily down to the grain scale with a rock hammer (sometimes yielding a powder), reflecting poor cementation. The formation typically is grain supported, containing angular to subrounded quartz and feldspar grains that are either intact or cut by open fractures and by fluid inclusion plains.

The granulometric and mineral composition properties of the five samples used for the FIPL and fracture orientation measurements (Sections 3.2–3.5) are shown in Table 1. The grain size data are based on the population of grains that were mapped for the image analysis purpose (i.e., data include the original, "intact" size of qualified grains) and is not an absolute measure for particle-size distribution in the samples. The smaller grains tend to be more angular, while larger, more rounded grains dominate the thin section area

Table 1

Grain Size and Composition of Samples Used for FIPL Measurements

Sample	8E-b	31	28	27	6a
Distance from SAF	10	65	115	335	1930
Granulometric properties					
No. grains analyzed	189	247	210	205	310
Diameter[I] range (μm)	160-2296	151-1680	198-1444	187-1957	151-2526
Mean grain diameter (μm)	816	488	530	566	366
Standard deviation (μm)	423	241	224	289	229
Skewness of distribution	1.29	1.76	1.33	1.77	4.15[II]
Phases area percent on thin sections (normalized to 100% without rock fragments, epoxy)[III]					
Quartz	36	38	31	46	No data
Plagioclase	9	14	16	5	No data
K-Feldspar	8	11	11	12	No data
Others phases[IV]	2	1	1	1	No data
Matrix[V]	45	36	41	36	No data
Total	100	100	100	100	No data
Epoxy gaps	2	2	8[VI]	2	No data
Rock fragments	18	9	3	6	No data

Comments:

[I] The diameter of a circle with an area identical to that of the particle

[II] The distribution of mapped grains in sample 6a is more linear on a log-log scale than in the other samples, with a slope of ~ 1.7

[III] Due to the large size of individual rock fragments with respect to the analyzed image frame, their percentage area may not represent their actual proportion in the sample. Their percentage area (before normalization) is given as a rough estimation for their area proportion

[IV] Mainly biotite and iron oxides

[V] The original proportion of matrix in the samples is probably larger since it likely was a major component of the material replaced by epoxy

[VI] The area used for this analysis included a macroscopic wide fracture filled with epoxy, hence this anomalous proportion

of each sample. The matrix is partially opaque to transmitted light, but has an average visible grain size of 5–20 microns including, in addition to quartz and feldspar fragments, grains of micas and oxides. The JHF was deposited in a local fluvial system in the vicinity of the source rocks. As such, some granulometric and compositional variations are expected between its various exposures, and this is manifested in the data in Table 1.

3.2. Microscale Damage Characteristics

The dominant microscale damage observed is the *in situ* shattering of grains without significant distortion of the original grain shape. Fragments fit together in a hierarchical fashion, and grains, even when extensively shattered, have uniform extinction in cross-polars (Fig. 3a and 3b). These features suggest that minimal or no shear and rotation are involved in the microscale deformation process of the grains. Similar damage was observed by ROCKWELL *et al.* (2009) in samples from the pulverized Tejon Lookout Granite in Tejon Pass on the SAF and in Tejon Ranch on the Garlock fault. The

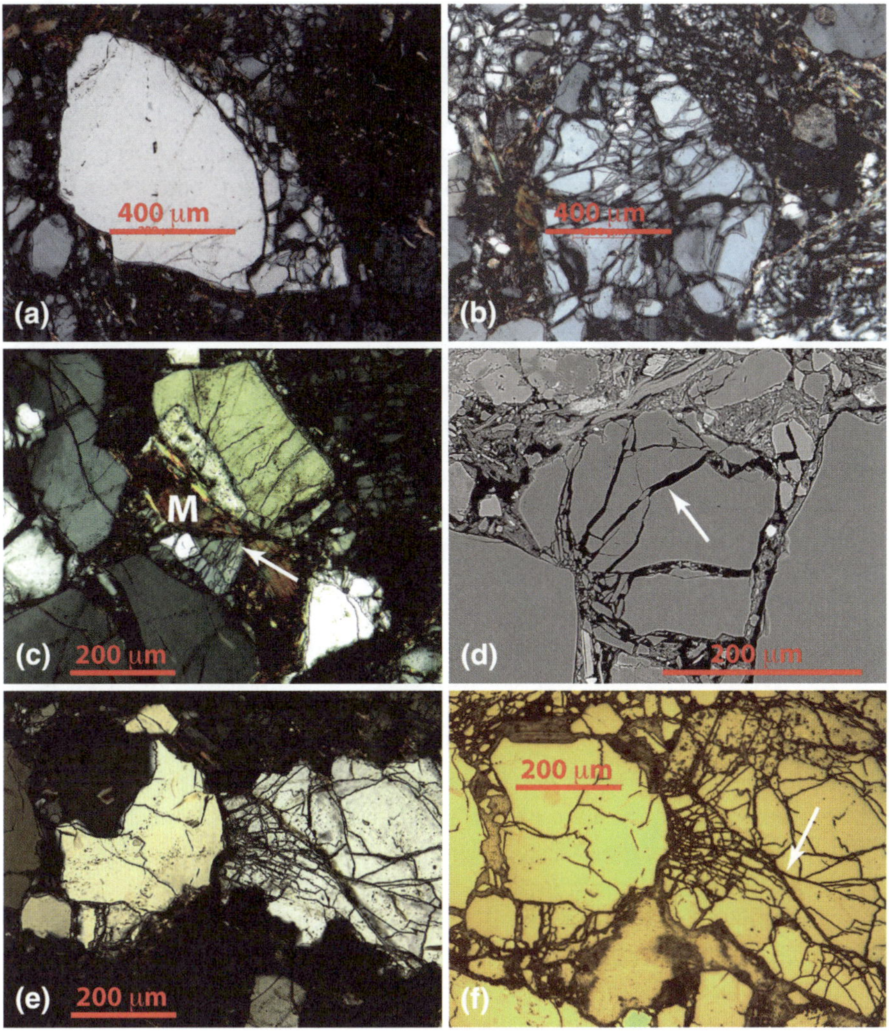

Figure 3
Microscale damage features: a. Partially fractured grain. b. Heavily fractured grain. Note the three columns of fragments: may be caused by impingement, like in (e, f). The grains in (a) and (b) are identified due to simultaneous extinction of the fragments, suggesting minimal or no shear and rotation of the grains. c. Cross-polarized photo of bended and sheared mica grain (marked "M"), in this case clearly due to an impingement between the two grains bounding the mica grain: Note the fractures radiating from the contact point between the grains (white arrow). d. Grain-scale fractures emanating from a contact point (white arrow) between two quartz grains, BSE image. Note intense small scale fragmentation near contact point. e, f. Cross-polarized (e) and reflected light (f) photos of fractures radiating in a Hertzian pattern from a contact zone between two grains. White arrow marks one of those fractures.

distribution of fracturing is heterogeneous in all samples, with some grains fragmented down to the micron scale whereas others are intact. This type of damage is somewhat similar to fault zone cataclasites classified by Sibson (1986) as crushed and implosion breccias. In many samples there is clear evidence of intragranular and intergranular cracks that have relatively sharp traces displaying opening-mode, relatively narrow apertures. The amount of fracture opening requires some dilation of the grains. These fractures often emanate from grain-grain contact points and have parallel or radiating geometries that sometimes link opposing points of contacts (Figs. 3 c-f, e.g., arrows in Fig. 3d, 3f); they often are associated with small conical cracks at contact sites (arrow in Fig. 3c). These features are consistent with Hertzian-type stressing in unconsolidated porous aggregates (e.g., CHESTER *et al.*, 2004; GALLAGHER *et al.* ,1974). Some mica crystals ("M" in Fig. 3c) are kinked and squeezed due to a displacement of another more rigid grain.

Several types of microfractures are observed in the JHF. These are distinguished according to their relations to grain size and to their hierarchical internal arrangement within and with respect to the grain (Fig. 4):

Type I: Transgranular fractures, extending beyond grain boundaries into the matrix and neighboring grains (not common).

Type II: Intragranular fractures that cut an entire grain. They are frequently parallel to each other, dissecting the grain or part of it to elongated columnar fragments.

Type III: Intragranular fractures that cut at least half of the grain's width in the direction of the fracture, and terminate within a grain. Some of them taper into the grain while others connect fractures to other fractures or to grain boundaries.

Type IV: Intragranular fractures that are considerably shorter than the grain's average axis length. Those are usually connecting fractures that terminate against type II or III fractures. They often cut elongated columnar fragments into rectangular or otherwise angular fragments, creating webs in crisscrossing relationships (e.g., category II fractures of LAUBACH, 1997).

Type V: Subgrain boundary and grain boundary fractures.

Type VI: Fluid inclusion planes (always intragranular); these are shown as linear traces of bubbles, marking the location of healed (or sealed) fractures (e.g., TUTTLE, 1949).

The vast majority of all fractures observed optically are joints (mode I fractures), associated with dilation normal to the fracture wall. Most fractures appear to be open (not healed or sealed), except fractures of Type VI. For our fracture orientation analysis (Section 3.5) we measured the orientation of fracture Types II-IV because these are less likely inherited from the source rock. The average orientation of Type I fractures is recorded, even though the number captured in thin sections is not statistically significant. We also measured the orientation of Type VI fractures. Type V fractures are curved in most cases, making their strike determination problematic.

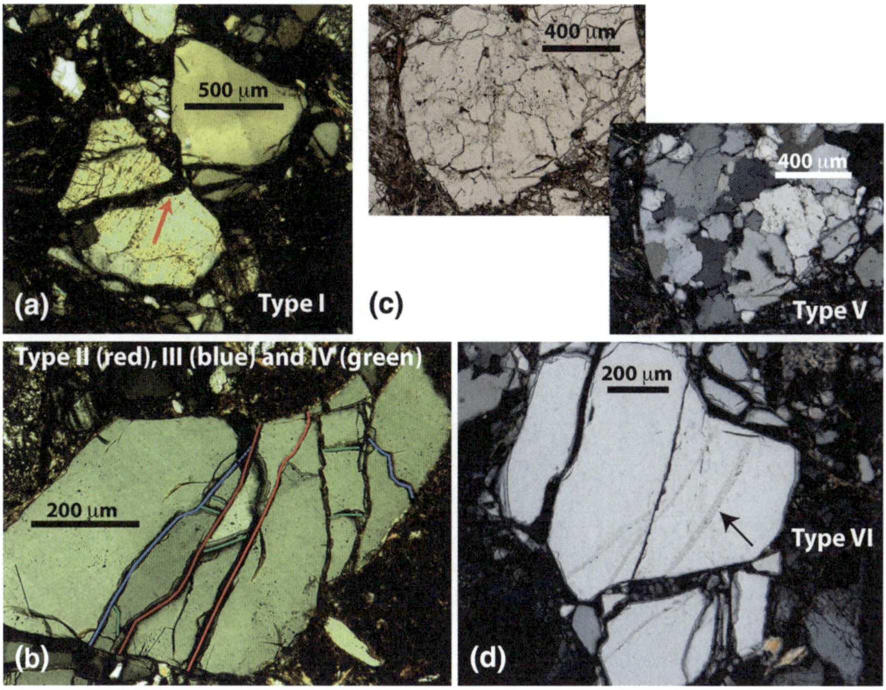

Figure 4
Fracture types (see text for complete definition): a. Type I — transgranular fractures. Not common in our samples and therefore were not considered for the fracture analysis (such possible fracture is marked with a red arrow. These, however, could also be two aligned intragranular fractures). b. Type II (red) — fractures that cut the entire grain; Type III (blue, doted were projected) — cut at least half of the grain's length in the direction of the fracture, taper into the grain or terminate against other fracture; IV (green) — short, connecting fractures. c. Type V — sub grain and grain boundary fractures. d. Type VI — fluid inclusion plains (marked by an arrow), typically cross-cut by other (open) types of fractures. Type V fractures were not included in our analysis because they are hard to measure and to interpret.

3.3. Statistical Significance of the FIPL Method

The statistical significance of the FIPL method was tested on a population of 104 grains out of the 189 grains that were mapped from sample 8E-b. This sample was taken 10 m from the fault and displays a wide range of grain sizes and a heterogeneous fracture distribution. The area-weighted average FIPL for the 104 grains is 8.14 and the area-weighted standard deviation is 5.34 (Table 2). A histogram for the FIPL values from this population is presented in Figure 5. FIPL values are plotted against original grain size in Figure 6. These data show that the original grain size limits the FIPL value, although there is no strong correlation between them ($R^2 = 0.37$).

We assume that the damage content of the 104-grain population represents the damage state of the rock because it consists of more than half of the 'qualified grains' in the sample. To determine if a smaller subset of the population is sufficient to define the

Table 2

Area-weighted Average FIPL and Standard Deviation for the 104-grain Population and the ten 40-grain subsets from sample 8E-b

40 grains Set	1	2	3	4	5	6	7	8	9	10	Average	*all 104 grains*
FIPL	7.40	8.06	7.09	6.10	8.72	8.40	7.99	7.15	8.92	7.29	7.71	8.14
A. weighted STDV	4.45	5.84	4.35	3.48	6.01	5.25	5.03	4.37	5.02	4.18	4.8	*5.34*
t-test *P* value	0.8	0.92	0.86	0.86	0.82	0.95	0.72	0.28	0.42	0.54		

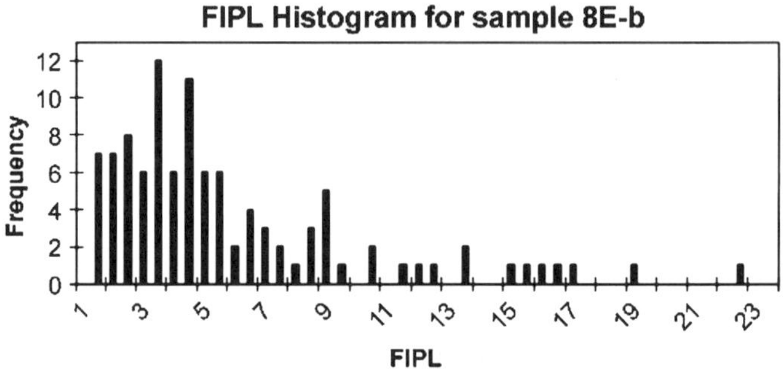

Figure 5
Frequency-distribution of FIPL values for 104 grains in sample 8E-b.

damage we randomly picked 10 subsets, each composed of 40 grains. We then calculated the area-weighted average FIPL and area-weighted standard deviation for each subset (Table 2). We performed a t-test between the FIPL result of each of the 40-grain subsets and the FIPL of the 104-grain population in order to verify that the chosen subsets are indeed representative of the larger population. The results (Table 2) show that the subsets are not statistically different than the parent population. The *P* values of all are larger than 0.05, and all subsets pass the Shapiro-Wilk and Shapiro-Francia W normality tests with over 95% probability (Appendix 1). The statistical tests suggest that our sampling method captures the damage properties of the entire grain population.

3.4. *FIPL as a Function of Distance from the SAF*

A qualitative examination of the thin sections before the measurements suggested that fracturing and fragmentation of grains are very intense in the samples taken 10 m from the fault, high in the sample taken 65 m from the fault, and significantly less farther away from the fault. We observe similar patterns in other sedimentary formations along the Mojave section of the SAF.

The FIPL measurements for five samples taken at 10 m, 65 m, 115 m, 335 m, and 1930 m from the fault confirm this trend (Table 3). The FIPL values decrease with

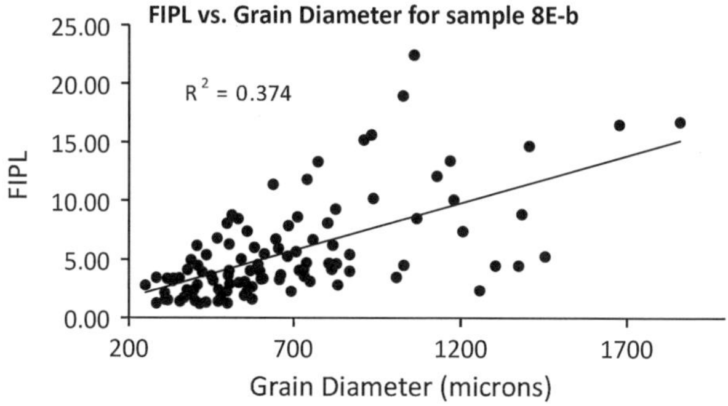

Figure 6
FIPL as a function of original grain diameter for 104 grains in sample 8E-b. Although FIPL values have no strong correlation with the grain sizes (note small R^2 value), grain size seems to control the upper limit of the FIPL values.

distance from the fault in all samples except the one at 335 m. A t-test between the FIPL values of each sample and the next sample closer to the fault shows that the samples taken at 10, 65 and 115 m are statistically different from each other, while the samples taken at 115, 335 and 1930 m are statistically indistinguishable from each other in their FIPL values (Table 3). All sample results passed the Shapiro-Wilk and Shapiro-Francia W normality tests with over 95% probability (Appendix 1). These results suggest that the damage gradient is significant from 0 to ~100 m, after which FIPL values probably approach a regional background damage state consistent with previous studies (e.g., Wilson *et al.*, 2003).

To test the effect of the mineralogy of the measured grain population on the results, we obtained FIPL values for each sample using only quartz grains from the subset of the grain population used for the analysis presented in Table 3. The existence of elevated densities of fractures in the damage zone may enhance weathering processes, particularly along the cleavage planes of feldspar grains. If this process is significant in extending existing fractures or creating new fractures, FIPL values obtained using quartz grains

Table 3

FIPL values for JHF samples

Sample	8E-b	31	28	27	6a
distance from fault (m)	**10**	**65**	**115**	**335**	**1930**
FIPL	**8.14**	**5.21**	**2.42**	**3.81**	**2.38**
FIPL (not weighted)[I]	5.61	3.85	2.16	2.43	2.28
Section orientation	28/240	70/028	80/256	46/145	83/175
t-test *P* value[II]		0.00	0.00	0.4	0.38

Comments:
[I] Calculated as a simple average between FIPL values of the individual grains
[II] For each sample with respect to the next sample closer to the fault

only ($FIPL_{quartz}$) should be substantially lower with respect to FIPL values obtained from all mineral phases ($FIPL_{all\text{-}phases}$). The results, presented in Table 4, reveal no significant difference between $FIPL_{all\text{-}phases}$ and $FIPL_{quartz}$. This is particularly true in sample 8E-b where more than 1/3 of the grains in the analyzed subset are feldspars. Quartz grains can be fragmented due to weathering alone (e.g., FRAZIER and GRAHAM, 2000). In such cases, products of pedogenesis, such as clay minerals, will be present between fragments; for the most part, we do not see these products in the open fractures cutting quartz in our samples (e.g., Fig. 3d). The similarity between the $FIPL_{all\text{-}phases}$ and the $FIPL_{quartz}$ values (Table 4), and the lack of significant weathering products in open fractures in quartz grains suggest that the damage we observe is mostly due to mechanical breakage.

We note that the proportions between the mineral phases in the analyzed subsets (Table 4) are different than the apparent absolute proportions of those phases in the samples (Table 1), usually with higher proportion of quartz in the subset (except for sample 8E-b). One reason for this is that during the initial mapping of grains on the thin sections we disqualified grains that are especially weathered, mostly feldspars, so that the proportion of quartz in the mapped grain population is larger in most cases than its proportion in the sample.

The mean grain diameter in sample 8E-b is larger than the mean grain diameter in the other samples (Table 1). A weak positive dependence of FIPL values in grain size is shown in Figure 6, although the relationships presented there alone cannot explain the significantly higher FIPL value in sample 8E-b with respect to the other samples.

3.5. *Microfracture Trends*

The orientation of fracture types II and III was measured on three mutually perpendicular sections made from sample 8E (8E-a, 8E-b and 8E-c), and on a single section from samples 31, 28, 27 (the four samples correspond to distances of 10, 65, 115 and 335 m from the fault). The orientation of fracture types IV and VI was measured on sections 8E-b and 31. All measurement results are presented in Figure 7. Section 8E-c is vertical with a strike of N60E, and the data for this section in Figure 7 correspond to the inclination of the fractures on the plane of the section, so that "north" in the projection is vertical. Besides fracture type VI (fluid inclusion planes) and fracture type II in sections 8E-a and 28, fractures in all other sections show preferred orientations with various degrees of

Table 4

The Partition of Analyzed Grains Between Quartz, Plagioclase and K- feldspar, and its Effect on FIPL values

Sample		8E-b	31	28	27	6a
Number of grains analyzed	Quartz	31	40	45	41	No data
	Palgioglase	8	2	1	1	No data
	K-Feldspar	8	3	1	6	No data
$FIPL_{quartz}$		7.9	4.89	2.33	3.78	No data
$FIPL_{quartz}$ as a fraction of $FIPL_{all\text{-}phases}$		0.97	0.94	0.96	0.99	No data

scattering. Fractures in sections 8E-a, 31, 28 and 27 are oriented at a low angle with respect to the SAF (13°– 31°). Fractures in section 8E-b are oriented with a high angle to the fault (74°–88°) and have the lowest 95% confidence margins (least scattered).

The measurements on the thin sections of sample 8E allow deciphering the approximate three-dimensional mean orientation of fractures at 10 m from the fault. The orientations of the three thin sections from sample 8E are projected in Figure 8 as gray lines. The combined mean orientation for fracture types II and III for each thin section is marked as a black circle and its 95% confidence interval serves as an error bar and is marked as a thick black line. The best-fit great circle representing the inferred mean fracture orientation for the sample is constrained by the mean orientations and error ranges from each of the thin sections, and is marked on the projection as a black line. Its orientation is 57/132, with its strike 14° off the normal to the strike of the SAF in this area (black arrow in Figure 8).

3.6. The Hungry Valley Formation

The majority of the Hungry Valley Formation (HVF) is widespread in Ridge Basin and is truncated on the north by the SAF. Tectonic slivers of the formation appear north of the SAF west of Three Points. The HVF is highly variable in texture and composition (Barrows *et al.*, 1985). Both the age and displacement of the formation are poorly constrained and highly debated in the literature. The age may range between late Miocene (Miller and Downs, 1974) to Pleistocene (Kahle, 1979) and the displacement is estimated to be on the order of 12 to 27 km by Barrows *et al.* (1985) and about 220 km by Ramirez (1983). There are no direct constraints on the maximum burial depth, nonetheless it is probably not significant as there are indications for only moderate exhumation of the area since the late Miocene (Dor *et al.*, 2006a; Spotila *et al.*, 2007). In addition, consolidation of the formation in most places is poor.

We compare three samples of the HVF (Fig. 9a): One collected 125 m northeast of the fault west of Three Points (22a); one collected 1055 m southwest of the fault in a road cut of Hwy 138 near its intersection with the I-5 (24); and one collected in the heart of the Ridge Basin 3380 m southwest of the SAF and in the immediate vicinity of no other known faults (25a). The HVF west of Three Points is an extremely heterogeneous fine- to coarse-grained arkosic sandstone that is poorly to moderately cemented (although it can be well indurated locally), bedded to massive, and poorly to moderately sorted (Barrows *et al.*, 1985). Typically, the formation contains subrounded quartz and feldspar grains, and large varicolored, sub- to well-rounded clasts of granite, volcanic rock and some metamorphic pebbles and cobbles. We find this description appropriate for the area from which the sample at 125 m from the fault (22a) was collected.

Other locations sampled are more similar to each other and are well-bedded, well-sorted, and show less lateral lithologic variation when compared to the Three Points location. The samples at 1055 and 3380 m from the fault were collected from very poorly cemented layers of uniform material with mm-size clasts of mainly quartz and feldspar.

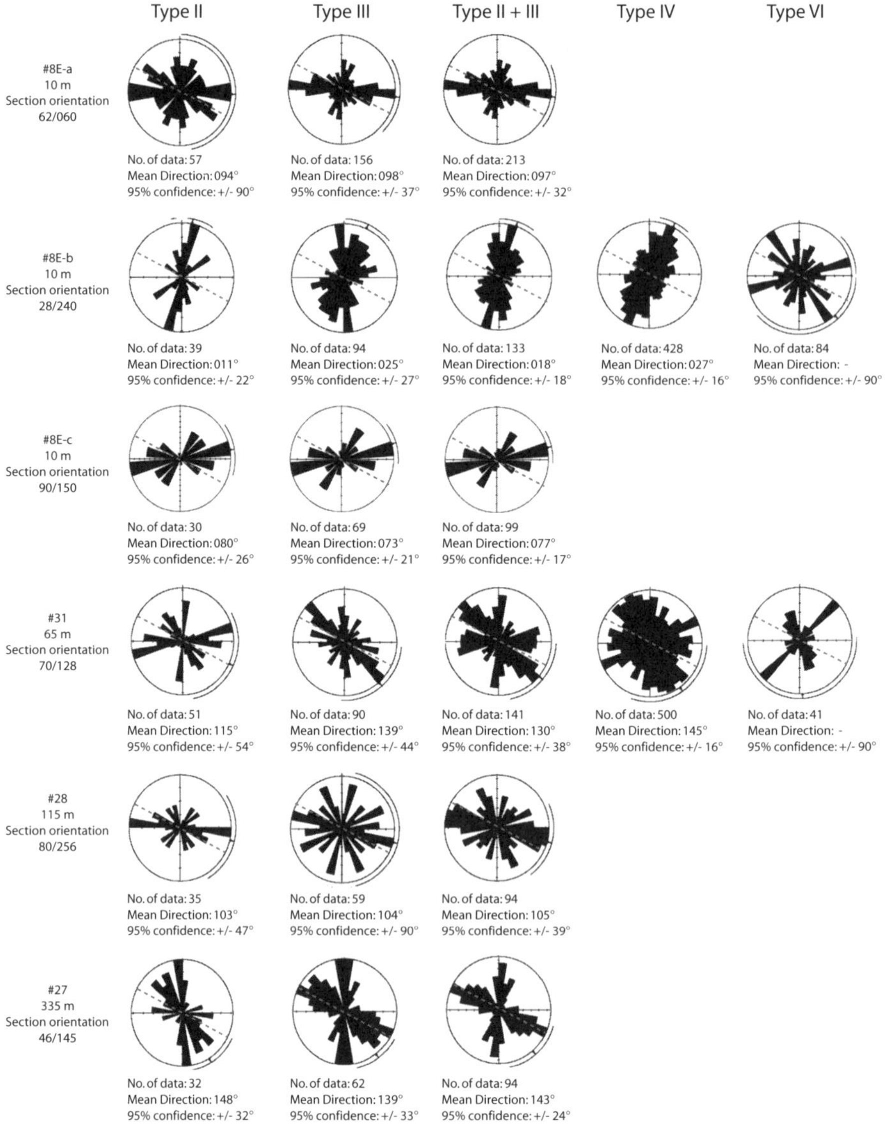

Figure 7
Rose diagrams (area weighted projections) of fracture orientation data, classified according to fracture type. Dashed lines show the strike orientation of the SAF. Section 8E-c is vertical and therefore data represent an apparent dip of the fractures (or their 2-D projection on a vertical plane), with north in the projections being "up".

Figure 9b shows cross-polarized and reflected light photomicrographs of samples 22a and 25a (125 and 3380 m from the fault, respectively). At 3380 m from the fault the grains are subrounded to subangular, with quartz and feldspar as main constituents. The rock is grain supported, containing a significant volume of calcite cement. The grains are

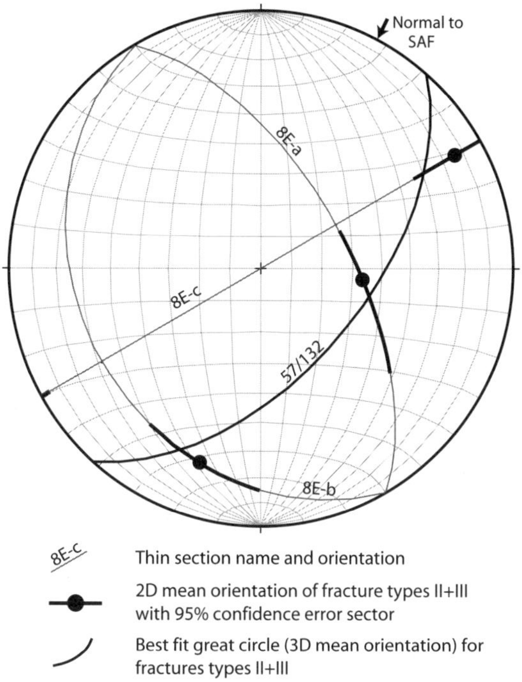

Figure 8
A stereonet (equal area) showing the orientation of the mutually perpendicular sections from sample 8E (gray lines), the combined mean preferred direction of fracture types II and III from each of these sections (black circles), the 95% confidence interval for this mean direction (thick black lines) and the best-fit combined mean orientation for fracture types II and III in the sample (black lines).

intact or only weakly fractured. Some post-cementation compaction has occurred as is evident by the presence of open fractures and twinned calcite (marked by green arrows on Fig. 9b). The host rock mineralogy at 125 m (Fig. 9b) is similar to that at 3380 m, but the intensity of fracturing is significantly greater, and the volume of calcite cement is considerably lower. Original variations in the porosity between the various parts of the formation are possible, although it is likely that at least some of these differences reflect damage-related volumetric strain. The sample from 1055 m shows characteristics intermediate to the other samples bounding it in terms of its fragmentation intensity and volume of cement.

The existence of additional fault strands parallel to the SAF up to about 1 km southwest of its currently active trace (Brune, personal comm.), folding in the Ridge Basin area or thrusting in Frazier Park could affect the intermediate and minor damage contents of samples 24 and 25a, respectively. In such a case, the SAF slip-related damage gradient suggested by our observations may be an overestimate.

In comparison with the JHF, the intensity of damage in the HVF is substantially higher. At 115 m from the fault, damage in the JHF is interpreted to reflect background levels (Table 2). In contrast, the HVF sample collected at 125 m displays fracture levels

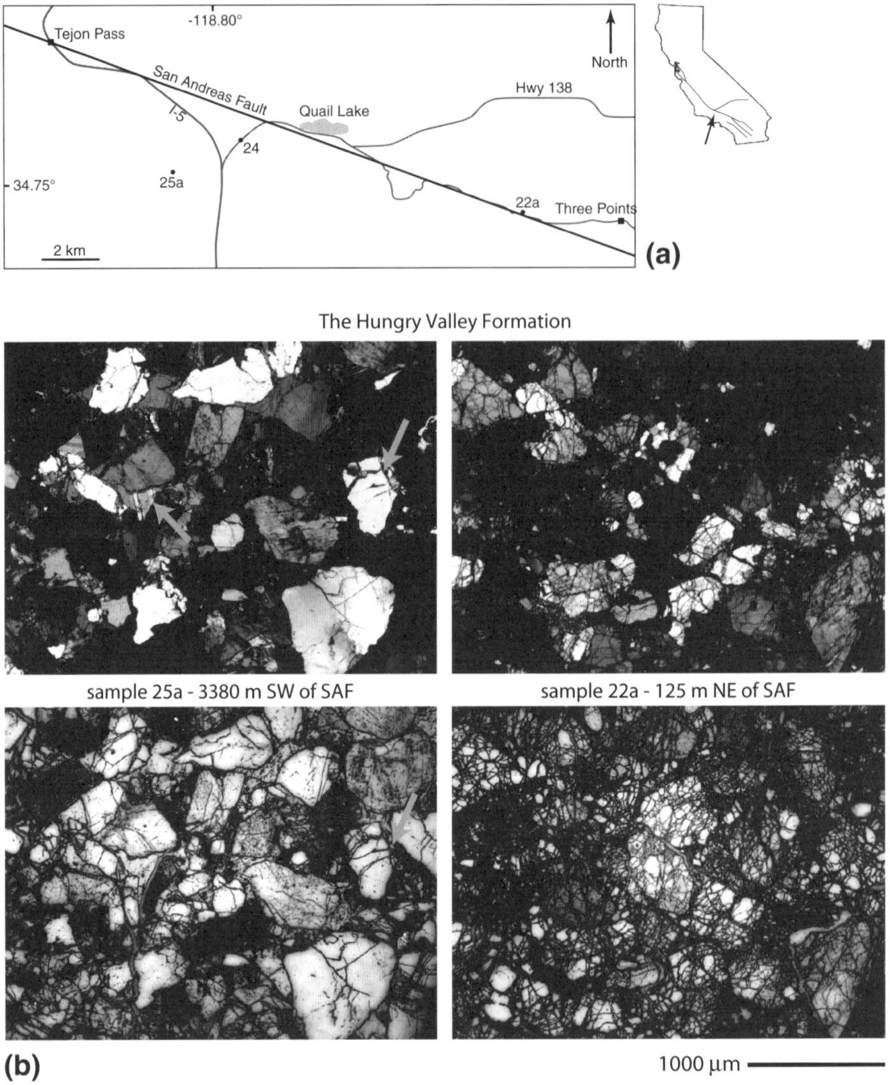

Figure 9

a. Location map of the HVF sampling points. Arrow on the small index map shows the location of the sampling area on the SAF. b. Cross-polarized (upper panels) and reflected light (lower panels) photomicrographs of Hungry Valley formation samples taken 3380 m from the fault (left) and 125 m from the fault (right). The far-field sample is only mildly damaged, containing some post-depositional compression features such as open cracks associated with Hertzian-type contacts and calcite twinning of the cement (green arrows). The near-field sample is intensely damaged although grain boundaries are not distorted.

well above the background and appears to be more damaged than the JHF sample at 10 m from the fault. The lack of a direct relationship between degree of fragmentation and shape distortion, however, is shared by both formations.

4. *Discussion*

4.1. *Precision of Measurements*

To evaluate the precision of the semi-automated image analysis technique, we compare the total perimeter length obtained for a single fractured grain (Fig. 10a) by using the technique described above (Fig. 10b), with the total perimeter length obtained by manual digitization of the grain (Fig. 10d). Figures 10c and 10e are the automatic and manually digitized grain, respectively, from which the noise (epoxy spots that passed the threshold filter in the automatic process and were also manually digitized) was removed. The FIPL obtained from the automatically digitized image (Fig. 10b) is higher than the FIPL obtained from the manually digitized image (Fig. 10d) because the manual digitization does not account for very small pixel clusters of gray values (could also be epoxy spots) that passed the automatic thresholding process (the amount of particles in Figure 10b is 1161 vs. 337 particles that were manually digitized in Figure 10d). An opposite trend occurs in the images that were cleaned from noise: The total perimeter length of all the fragments in the manually digitized image without the noise (Fig. 10e) is 7% higher than the corresponding value obtained for the autodigitized, noise-free image (Fig. 10c). Once the noise (pixel-scale spots as well as larger epoxy spots that were digitized in Fig. 10b) is removed and only actual rock fragments are analyzed for perimeter length, the image obtained by manual digitization yields a higher FIPL value because the automatic detection thresholding process tends to shrink fragments (as they are darker at the edges) that are otherwise mapped accurately by hand.

We also compare selected areas of the optically photographed grains (white frames in Fig. 10a) with SEM images of the same areas (Fig. 11). The fragments observed under reflected light do not appear to be more fragmented in the SEM images. Particles that are not observed under reflected light (with the optical limit for X100 magnification at about 5 microns) appear to be restricted mostly to fractures, floating in the epoxy. Some zones in the epoxy are densely populated with such particles while others contain few or no particles.

We mapped and calculated the total perimeter length of the small particles in Figure 11b (white lines in lower panel) and estimated the corresponding value for the entire grain. When we add the total perimeter length of these small fragments to the autodigitized, noise-free total perimeter length (Fig. 10c), the FIPL rises from 7.2 to 9.6. When we add the total perimeter length of the small fragments to the manually-digitized, noise-free total perimeter length (Fig. 10e), the FIPL rises from 7.7 to 10.1.

The origin of the SEM observed particles in the epoxy-filled fractures is uncertain. During the impregnation process, epoxy flows through the fractures and may carry and deposit grain fragments from the same grain, bring fragments in from neighboring grains, or redistribute matrix material from the host-rock. The original amount of small particles inside the fractures could have been larger *or* smaller than the current amount.

Calculating adjustments associated with the small (<5 μm) particles in the fractures, with noise removal and with corrections due to manual digitization for other grains may

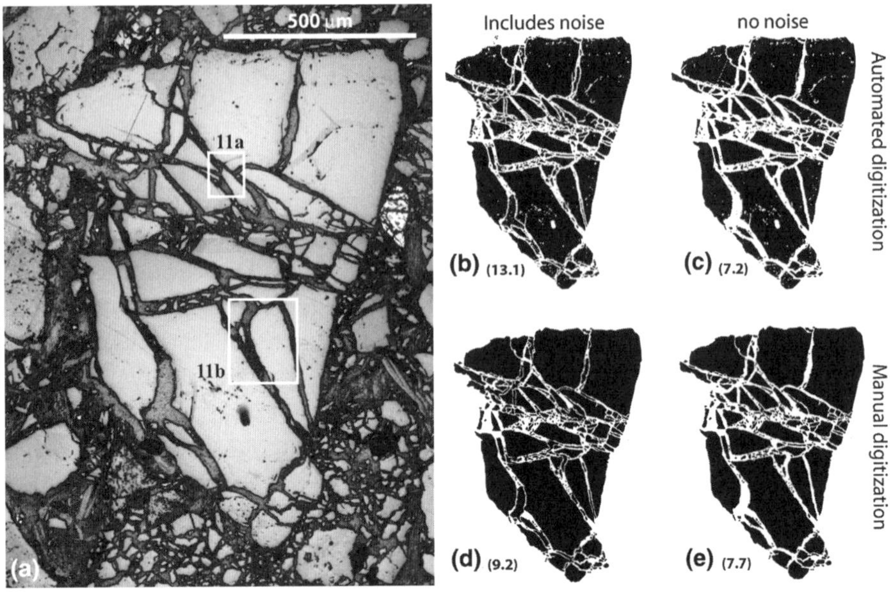

Figure 10

a. A gray scale image of a quartz grain from sample 8E-a. b, c. An automatically digitized fragment map of the grain in (a). (b) includes noise and (c) is manually cleaned from the noise. The image analysis program does not measure holes. d, e. A manually digitized fragment map of the grain in (a). (d) includes noise and (e) cleaned from the noise. FIPL values appear in parenthesis.

result in different values of increase (or decrease) of the FIPL values, but is unlikely to be different by more than a factor of two.

4.2. *Width of the Damage Zone*

For the portion of the SAF fault that displaces the JHF at the surface, the reduction in FIPL values between the samples in the first 115 m suggests that the fault at this location may be characterized by a microfractured damage zone of about 100–200 m. This scale is commonly observed in geological studies of fault zones (e.g., SCHULZ and EVANS, 2000), and is consistent with the zone of pulverized crystalline rocks along the same section of the fault (DOR *et al.*, 2006a) and the width of seismic fault zone trapping structures (e.g., LI *et al.*, 1990; PENG *et al.*, 2003). The JHF data are consistent with a linear decrease in damage intensity with the logarithm of distance from the SAF (Fig. 12) as was found for faults with comparable displacements (e.g., CHESTER *et al.*, 2004b) and smaller magnitudes of displacement (e.g., VERMILYE and SCHOLZ 1998). The damage zone in the HVF appears to be wider than about 100 m, possibly because of its association with the Big Bend in the SAF, however, our width constraints at this location are relatively poor.

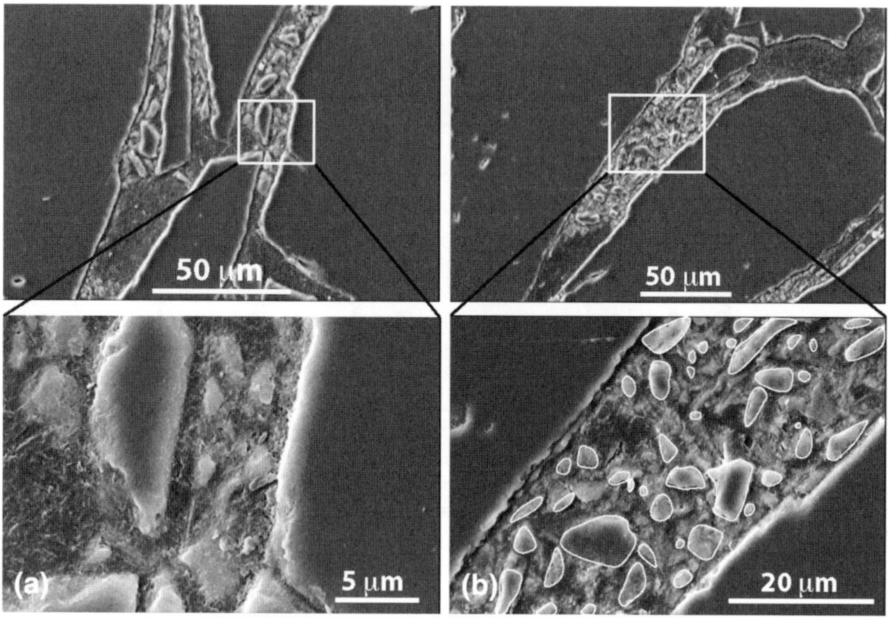

Figure 11
SEM (Backscatter electron mode) images corresponding to frames in Figure 10a. Both images are rotated clockwise with respect to their orientation in Figure 10a. The images show many fragments that were not detected by the optical microscope analysis, floating in an epoxy matrix. The small fragments are restricted to certain zones within open fractures between larger fragments. They are mapped (white lines) in (b), lower panel.

The damage zones of ancestral, exhumed strands of the SAF system show similar distribution, mode and fabric of microfractures (WILSON *et al.*, 2003; CHESTER *et al.*, 2004b), lending support to our inferences about the width of the SAF damage zone.

4.3. Depth of Damage Generation

Fluid inclusion planes (healed fractures) that form a small minority in the fracture population are cut by open fractures, have no preferred orientation, and are likely inherited from the source rock. The fact that most of the remaining fractures are open is consistent with a near surface environment in which temperatures are low and fracture sealing is not favored (e.g., SPRUNT and NUR, 1979). Although some grains can survive transportation while they are partially fractured, the extensive open-mode fractures that transect entire grains indicate that these latter fractures occurred in-situ and were not inherited from the source rock. This inference is supported by the presence of a preferred orientation in our data and by the overall poor consolidation state of the JHF.

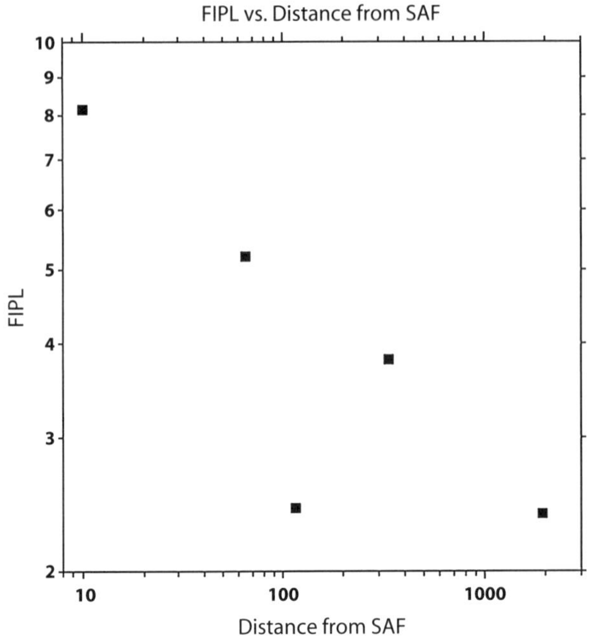

Figure 12
Area-weighted averaged FIPL values plotted against log-distance from the SAF. With the exception of the sample at 115 m from the SAF, the data seem to have a linear correlation in this space.

These observations imply that the damage zone within the JHF extends to the ground surface or very close to it. This conclusion combined with the inference about the shallow exhumation depth of the JHF suggests that the observed damage was generated close to the ground surface. Similar conclusions apply to the HVF based on the mode of fractures and assuming that it was never deeply buried as suggested by the state of the rocks (Barrows *et al.*, 1985).

The similarity between the fracture fabric and the width of the damage zone between the SAF and its exhumed ancestral faults (Section 4.2) brings up the possibility that the observed damage near the SAF may extend to a depth of a few km.

4.4. Mechanical Interpretation of Damage Fabric

Surface exposures of the sedimentary rocks we studied adjacent to the SAF, away from mesoscale shears, are pervasively and penetratively microfractured at the grain scale. These fractures are dominantly opening-mode features that display nominal shear displacements and indicate that tensile stresses existed locally (i.e., at the grain- or crack-tip-scale) during their formation. This type of damage was observed also in crystalline rocks spatially associated with the SAF (Rockwell *et al.*, 2009). Fractures associated with Hertzian-type contacts along the SAF were, however, observed only in the sandstones, and are not expected in tight crystalline rocks. Overall, the microstructures morphologies suggest that they formed under a state of macroscopic compression, i.e., under a compressive mean stress, consistent with conical, axial and radial fracture geometries emanating from grain-grain contacts (Fig. 3) and with the JHF behaving as a porous, poorly consolidated granular aggregate during deformation (CHESTER *et al.*, 2004a). However, the fact that microfracture orientation measurements suggest a preferred fabric within the fault damage zone further suggests that these fractures formed under a directed macroscopic stress that was not isotropic, i.e., not lithostatic or uniform tension.

The transform plate boundary of the SAF system has been at its current location on the north side of the San Gabriel Mountains since about 4–5 Ma (e.g., POWELL and WELDON, 1992). If the currently active trace of the SAF participated in accommodating the slip since its onset in this area, which is likely the case (POWELL and WELDON, 1992), then the JHF, being only late Pliocene-Pleistocene in age, was deposited over an existing structure. In this case the damage that we see is associated primarily with slip events on the SAF and not with the inelastic deformation associated with the propagation of the fault process zone during its formation (e.g., SCHOLZ *et al.*, 1993).

4.5. Implications of Results for Dynamic Rock Failure

The JHF was deposited in the vicinity, and displaced by strands of the SAF system and therefore its entire deformation history is likely associated with the activity of those fault strands. We assume that displacement of the JHF occurred by seismic slip (Section 1.2). If this is correct, concentrated damage within $\sim$ 100–200 m of the active trace of the SAF is most likely the product of SAF earthquake ruptures. The inference for the depth of damage generation (Section 4.3) implies that dynamic generation of rock damage along this section of the SAF occurs very close to the surface of the earth.

Fault-related microfracture fabrics may have been modified and weakened by local tilting and folding of the JHF layers. Nevertheless, orientation data at 10 m from the fault suggest that the fractures have a moderate preferred orientation consistent with extensive fracture under conditions close to a fault-normal compressive stress state in the near surface environment. CHESTER and CHESTER (2000) presented a mechanical model of stress and deformation in the vicinity of a wavy frictional surface and showed that stress is heterogeneous near the fault due to the juxtaposition of geometric irregularities. With a far-field stress oriented at high angles to a low friction master fault, fault-normal compression is generated locally during cyclic slip events and inelastic deformation is expected. This deformation is partially consistent with our data at 10 m from the fault in the JHF, and with previous studies on faults of the same system (WILSON *et al.*, 2003).

Fault opening and significant reduction of normal stress associated with the tip zone of the propagating earthquake rupture are expected to produce a change in the orientation of stresses as a function of distance from the fault. This may lead to a variability in the preferred orientation of microfractures within the damage zone that reflect changes from the background stress to near-zero transient shear stress on the fault. Our limited orientation data show changes in the preferred orientation of microfractures throughout the damage zone that is therefore compatible with dynamic fault opening produced by the rupture tip (e.g., BRUNE *et al.*, 1993; BEN-ZION and HUANG, 2002).

5. *Conclusions*

Our observations show that the JHF sandstone contains a considerable amount of damage that is spatially associated with the trace of the SAF, demonstrating that significant fracture damage can be produced in the near surface portion of a seismic fault zone. These inferences are supported by additional microscale observations on the damage content of the HVF along the northwest Mojave section of the SAF. The width of the damage zone within the JHF is similar to the 100–200 m width of the crystalline pulverized rock layer (DOR *et al.*, 2006a). In both rock units the grains are shattered *in situ*, and distortion of original fabrics is minimal in all observational scales. A distinctive feature of the porous sandstones described here, however, is the abundance of fractures associated with Hertzian-type contacts. The damage fabric and its spatial association with the fault are consistent with overall formation under directed compressional tectonic stressing and not burial-induced stressing, and cannot be used to distinguish between proposed off-fault damage mechanisms. The distribution of microfracture fabric and density in the active SAF damage zone is consistent with the microfracture fabrics and density distribution of the exhumed North-branch San Gabriel fault and the Punchbowl fault (CHESTER *et al.*, 2004b).

Acknowledgments

We thank two anonymous referees for constructive comments that improved the quality of this manuscript. This study was funded by the Southern California Earthquake Center (based on NSF Cooperative Agreement EAR-0106924 and USGS Cooperative Agreement 02HQAG0008), and by the National Science grant EAR-0510892 to JSC.

Appendix 1

Table 5

Shapiro-Wilk and Shapiro-Francia W tests for normality of the distribution of FIPL results in Table 2, 3

Shapiro-Wilk W' test for normal data				z	Prob > z
Variable	Obs	W	V		
8E-b_104	104	0.82	15.08	6.03	0
set_10	40	0.88	4.55	3.19	0.00071
set_9	40	0.87	5.03	3.40	0.00034
set_8	40	0.83	6.54	3.95	0.00004
set_7	40	0.85	5.94	3.75	0.00009
set_6	40	0.83	6.54	3.95	0.00004
set_5	40	0.76	9.47	4.73	0
set_4	40	0.87	5.09	3.42	0.00031
set_3	40	0.86	5.52	3.60	0.00016
set_2	40	0.75	9.69	4.78	0
set_1	40	0.82	7.00	4.10	0.00002
8E-a	43	0.94	2.64	2.05	0.02006
8E-c	43	0.74	10.89	5.05	0
31	47	0.88	5.58	3.65	0.00013
28	47	0.83	7.63	4.32	0.00001
27	46	0.75	10.87	5.06	0
6a	41	0.92	3.18	2.44	0.00737
Shapiro-Francia W' test for normal data				**z**	**Prob > z**
Variable	Obs	W'	V'		
8E-b_104	104	0.82	16.60	5.35	0.00001
set_10	40	0.88	5.16	3.01	0.00132
set_9	40	0.88	5.43	3.10	0.00098
set_8	40	0.83	7.36	3.63	0.00014
set_7	40	0.85	6.58	3.43	0.0003
set_6	40	0.83	7.27	3.61	0.00015
set_5	40	0.76	10.69	4.27	0.00001
set_4	40	0.87	5.47	3.11	0.00093
set_3	40	0.86	6.23	3.34	0.00042
set_2	40	0.75	10.99	4.32	0.00001
set_1	40	0.82	7.94	3.76	0.00008
8E-a	43	0.94	2.62	1.80	0.0359
8E-c	43	0.73	12.23	4.51	0.00001
31	47	0.88	6.15	3.35	0.00041
28	47	0.83	8.30	3.87	0.00005
27	46	0.75	11.94	4.49	0.00001
6a	41	0.92	3.61	2.38	0.00871

References

Barrows, A.G., Kahle, J.E., and Beeby, D.J. (1985), *Earthquake Hazard and tectonic history of the San Andreas fault zone, Los Angeles County, California*, Open File report 85-10 LA., California Department of Conservation, Division of Mines and Geology.

BARROWS, A.G. (1980), *Geologic map of the San Andreas fault zone and adjoining terrains, Juniper Hills and vicinity, Los Angeles County, California*, California Division of Mines and Geology Open-file Report 80-2 LA, map scale 1:9000.

BARROWS, A.G. (1980), *Geology of the San Andreas fault zone and adjoining terrains, Juniper Hills and vicinity, Los Angeles County, California*, California Division of Mines and Geology Open-file Report 85- LA.

BEN-ZION (2001).

BEN-ZION, Y. and HUANG, Y. (2002), *Dynamic rupture on an interface between a compliant fault zone layer and a stiffer surrounding solid*, J. Geophys. Res. *107*, 2042, doi: 10.1029/2001JB000254.

BEN-ZION, Y. and SHI, Z. (2005), *Dynamic rupture on a material interface with spontaneous generation of plastic strain in the bulk*. Earth. Planet. Sci. Lett. *236*, 486–496, doi: 10.1016/j.epsl.2005.03.025.

BLYTHE, A.E., HOUSE, M.A., and SPOTILA, J.A. (2002), *Low temperature thermochronology of the San Gabriel and San Bernardino Mountains, southern California: Constraining structural evolution*, Geol. Soc. Am. Special Paper *365*, 231–250.

BRUNE, J.N. (2001), *Fault normal dynamic loading and unloading: an explanation for "non-gouge" rock powder and lack of fault-parallel shear bands along the San Andreas fault*, EOS Trans. Am. Geophys. Union, 82, Abstract.

BRUNE, J. N., BROWN, S., and JOHNSON P.A. (1993), *Rupture mechanism and interface separation in foam rubber models of earthquakes; a possible solution to the heat flow paradox and the paradox of large overthrusts*, Tectonophysics *218*, 1–3.

CHESTER, F.M. and CHESTER, J.S. (2000), *Stress and deformation along wavy frictional faults*, J. Geophys. Res. *105*, 23,421–23,430.

CHESTER, F.M., CHESTER, J.S., KIRSCHNER, D.L., SCHULZ, S.E., and EVANS, J.P. *Structure of large-displacement strike-slip fault zones in the brittle continental crust*. In (Karner, Gary D., Taylor, Brian, Driscoll, Neal W., Kohlstedt, David L., eds.), *Rheology and Deformation in the Lithosphere at Continental Margins*, (Columbia University Press, New York, 2004b) MARGINS Theoretical and Experimental Earth Science Series 1, pp. 223–260.

CHESTER, J.S., LENZ, S.C., CHESTER, F.M., and LANG, R.A. (2004a), *Mechanisms of compaction of quartz sand at diagenetic conditions*, Earth Planet. Sci. Lett. *220*, 435–451.

CROWELL, J.C. (1982). *The Pliocene Hungry Valley Formation, Ridge Basin, southern California*. In (Crowell, J.C., and Link, M.H., eds.) *Geologic History of Ridge Basin, Southern California*, (Los Angeles, Pacific Section, Society of Economic Paleontologists and Mineralogists, 1982) pp. 143–150.

DOR, O., BEN-ZION, Y., ROCKWELL, T.K., and BRUNE, J.N. (2006a), *Pulverized Rocks in the Mojave section of the San Andreas FZ*, Earth Planet. Sci. Lett. *245*, 642–654, doi:10.1016/j.epsl.2006.03.034.

DOR, O., ROCKWELL, T.K., and BEN-ZION, Y. (2006b), *Geologic observations of damage asymmetry in the structure of the San Jacinto, San Andreas and Punchbowl faults in southern California: A possible indicator for preferred rupture propagation direction*, Pure Appl. Geophys. *163*, doi 10.1007/s00024-005-0023-9.

DI TORO, G., NIELSEN, S., and PENNACCHIONI, G. (2005). *Earthquake rupture frozen in exhumed ancient faults*. Nature *436*, 1009–1012.

FRAZIER, C.F. and GRAHAM, R.C. (2000), *Pedogenic transformation of fractured granitic bedrock, Southern California*, Soil Sci. Soc. Am. J. *64*, 2057–2069.

GALLAGHER, J.J., Friedman, M., Handin, J., and Sowers, G.M. (1974), *Experimental studies relating to microfracture in sandstone*, Tectonophysics *21*, 203–247.

HAMIEL, Y., LIU, Y., LYAKHOVSKY, V., BEN-ZION, Y., and LOCKNER, D. (2004), *A Visco-elastic damage model with applications to sTable and unsTable fracturing*, Geophys. J. Int. *159*, 1155–1165, doi: 10.1111/j.1365-246X.2004.02452.x.

KAHLE, J.E. (1979), *Geology and fault activity of the San Andreas fault zone between Quail Lake and Three Points, Los Angeles County, California*, California Division of Mines and Geology Open-File Report 79-3 LA, 42 p., 5 plates, map scale 1:12,000.

KENNY, M. (2000), *Quaternary uplift of the eastern San Gabriel Mountains: A case for crystalline basement rock folding*. 2000 AAPG Pacific Section and Western Region Society of Petroleum Engineers Meeting; abstracts AAPG Bulletin *84*, 6, 878, doi: 10.1306/A9673770-1738-11D7-8645000102C1865D.

LAUBACH, S.E. (1997), *A method to detect natural fracture strike in sandstones*, Am. Assoc. Petrol. Geologist Bull. *81*, 604–623.

LI, Y.-G., LEARY, P.C., AKI, K., and MALIN, P. (1990), SEISMIC TRAPPED MODES IN THE ORVILLE AND SAN ANDREAS FAULT ZONES, Science *249*, 763–766.

Lyakhovsky, V., Ben-Zion, Y., and Agnon, A. (1997), *Distributed damage, faulting and friction*, J. Geophys. Res. *102*, 27, 635–27, 649.

Lyakhovsky, V. and Ben-Zion, Y. (2008), *Scaling relations of earthquakes and aseismic deformation in a damage rheology model*, Geophys. J. Int. *172*, 651–662, doi: 10.1111/j.1365-246X.2007.03652.x.

Miller and Downs (1974).

Peng, Z., Ben-Zion, Y., Michael, A. J., and Zhu, L. (2003), *Quantitative analysis of seismic trapped waves in the rupture zone of the 1992 Landers, California earthquake: Evidence for a shallow trapping structure*, Geophys. J. Int. *155*, 1021–1041.

Powell, R.E., and Weldon, R.J. (1992), *Evolution of the San Andreas fault*, Annu. Rev. Earth Planet. Sci. *20*, 431–468.

Ramirez, V.R. (1983). *Hungry Valley Formation; evidence for 220 kilometers of post Miocene offset on the San Andreas fault.* In (Andersen and Rymer eds) *Tectonics and Sedimentation along Faults of the San Andreas System*, Soc. Econ. Paleontol. and Mineral., pp. 33–44.

Rice, J. R., Sammis, C. G., and Parsons, R. (2005), *Off-fault secondary failure induced by a dynamic slip pulse*, Seismol. Soc. Am. Bull. *95*, 109–134.

Rockwell *et al.* (2009), *Granulometric and mineralogical properties of pulverized rocks from Tejon Pass on the San Andreas Fault and from Tejon Ranch on the Garlock Fault, California*, Pure Appl. Geophys., in press.

Scholz, C.H., Dawers, N.H., Yu, J.Z., Anders, M.H., and Cowie, P.A. (1993), *Fault growth and fault scaling laws: Preliminary results*, J. Geophys. Res. *98*, 21951–21961.

Schultz, S.E. and Evans, J.P. (2000), *Mesoscopic structure of the Punchbowl Fault, Southern California and the geologic and geophysical structure of active strike slip faults*, J. Struct. Geol. 913–930.

Sibson, R. H. (1986), *Brecciation processes in fault zone: Inferences from earthquake rupturing*, Pure Appl. Geophys. *124*, 159–175.

Spotila, J., House, M., Niemi, N., Brady, R., Oskin, M., and Buscher, J. (2007), *Patterns of bedrock uplift along the San Andreas fault and implications for mechanisms of transpression.* In (Till, A., Roeske, S., Foster, D., and Sample, J., eds.), *Uplift and Extension along Continental Strike-slip Faults*, Geol. Soc. Am. Special Paper *434*, pp. 15–33.

Sprunt, E.S. and Nur, A. (1979), *Microcracking and healing in granites: new evidence from Cathodolumin-iescence*, Science, *205*, 4405, 495–497, doi: 10.1126/science.205.4405.

Tuttle, O.F. (1949), *Structural petrology of planes of liquid inclusions*, J. Geology, *57*, 331–356; Pure Appl. Geophys. *124*, 159–175.

Vermilye, J. M. and Scholz, C. H. (1998), *The process zone: A microstructural view of fault growth*, J. Geophys. Res. *103*, 12223–12237.

Weber, F.H., Jr. (1999), *Right-lateral displacement of Pleistocene sedimentary deposits along the San Andreas Fault, Palmdale to Cajon Pass, Southern California*, USGS, Final Technical Report *103*, 4 sheets.

Weldon, R., Scharer, K., Fumal, T., and Biasi, G. (2004), *Wrightwood and the earthquake cycle: What a long recurrence record tells us about how faults work*, GSA Today; *14*, 9, doi: 10.1130/1052-5173(2004)014.

Wilson, J.E., Chester, J.S., and Chester, F.M. (2003), *Microfracture analysis of fault growth and wear processes, Punchbowl Fault, San Andreas System, California*, J. Struct. Geol. *25*, 1855–1873.

Wilson, B., Dewers, T., Reches, Z., and Brune, J. (2005), *Particle size and energetics of gouge from earthquake rupture zones.* Nature, *434*, 749–752.

(Received August 25, 2008, accepted March 17, 2009)
Published Online First: June 30, 2009

To access this journal online:
www.birkhauser.ch/pageoph

Pure appl. geophys. 166 (2009) 1775–1798
0033–4553/09/101775–24
DOI 10.1007/s00024-009-0523-0
© Birkhäuser Verlag, Basel, 2009
Pure and Applied Geophysics

Constructing Constitutive Relationships for Seismic and Aseismic Fault Slip

N. M. Beeler[1,2,3]

Abstract—For the purpose of modeling natural fault slip, a useful result from an experimental fault mechanics study would be a physically-based constitutive relation that well characterizes all the relevant observations. This report describes an approach for constructing such equations. Where possible the construction intends to identify or, at least, attribute physical processes and contact scale physics to the observations such that the resulting relations can be extrapolated in conditions and scale between the laboratory and the Earth. The approach is developed as an alternative but is based on Ruina (1983) and is illustrated initially by constructing a couple of relations from that study. In addition, two example constitutive relationships are constructed; these describe laboratory observations not well-modeled by Ruina's equations: the unexpected shear-induced weakening of silica-rich rocks at high slip speed (Goldsby and Tullis, 2002) and fault strength in the brittle ductile transition zone (Shimamoto, 1986). The examples, provided as illustration, may also be useful for quantitative modeling.

Key words: Friction, dynamic fault slip, brittle ductile transition.

1. *Introduction*

In laboratory faulting experiments on quartzofeldspathic rock surfaces where temperatures are less than a few hundred °C, slip rates are below a few mm/s, and total slip is limited to a few 100 μm, coseismic strength losses are typically less than 10% of the ambient shear stress (Fig. 1a) (Wong, 1986). The maximum slip velocity is proportional to the fault strength loss approximately by twice the ratio of the shear-wave speed to the shear modulus (e.g., Johnson *et al.*, 1973), consistent with theory of Brune (1970) (Fig. 1b). When these observations are extrapolated to natural conditions (e.g., McGarr, 1994, 1999; McGarr and Fletcher, 2003; McGarr *et al.*, 2004) they are consistent with typical natural earthquake properties: stress drops of a few tenths to a few MPa, low seismic efficiency, slip rates up to a few m/s and rupture propagation at rates approaching the shear-wave speed. The cause of the low seismic efficiency, small stress

[1] U. S. Geological Survey, Cascades Observatory, 1300 Cardinal Court, Bldg 10, Suite 100, Vancouver, Washington 98683, U.S.A. E-mail: nbeeler@usgs.gov
U. S. Geological Survey, Menlo Park, California 94025, U.S.A.
Brown University, Providence, Rhode Island 02712, U.S.A.

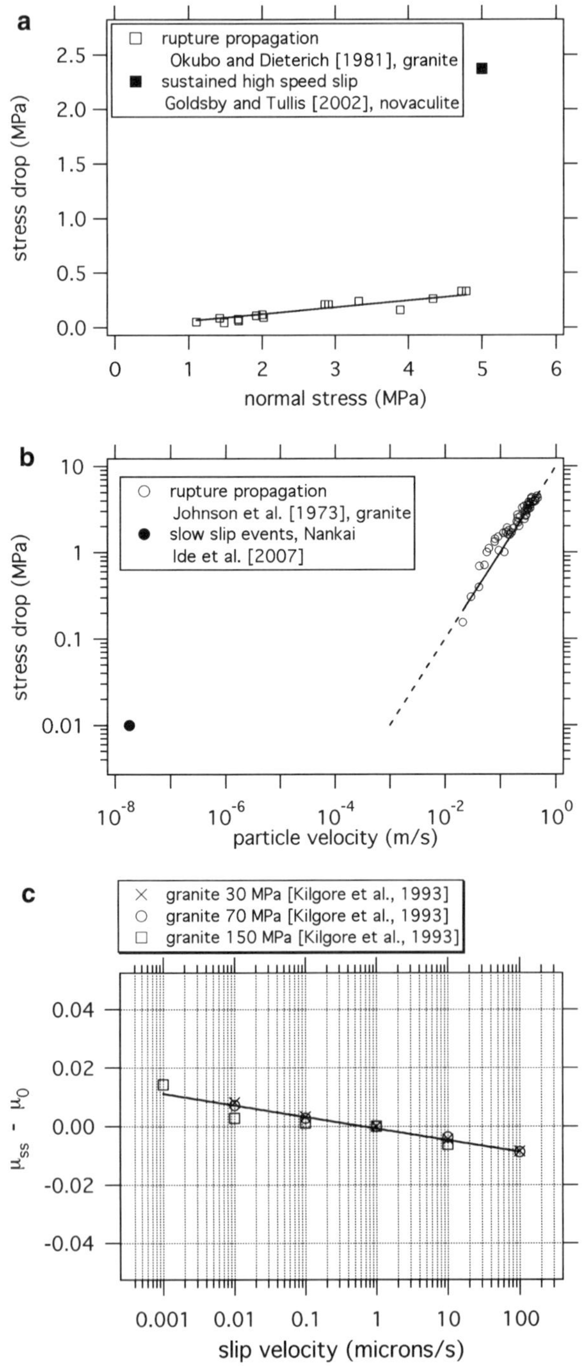
a
rupture propagation
Okubo and Dieterich [1981], granite
sustained high speed slip
Goldsby and Tullis [2002], novaculite
stress drop (MPa)
normal stress (MPa)
b
rupture propagation
Johnson et al. [1973], granite
slow slip events, Nankai
Ide et al. [2007]
stress drop (MPa)
particle velocity (m/s)
c
granite 30 MPa [Kilgore et al., 1993]
granite 70 MPa [Kilgore et al., 1993]
granite 150 MPa [Kilgore et al., 1993]
μss - μ0
slip velocity (microns/s)

◀ Figure 1

Select frictional properties from rupture propagation and quasi-static sliding experiments. a) Variation of strength loss, with normal stress during slip between granite surfaces from OKUBO and DIETERICH (1981) (open symbols). The line is a fit constrained to past through (0,0) with resulting slope 0.06 MPa/MPa. Also shown is data from sustained slip at high speed on novaculite (black) from GOLDSBY and TULLIS (2002). b) Variation of peak near-fault particle velocity with strength loss during rupture propagation between granite surfaces from JOHNSON *et al.* (1973). Line is the expression of BRUNE (1970) relating stress drop $\Delta\tau$ to particle velocity $\dot{u}, \Delta\tau = G\dot{u}/\beta$, with G = 30,000 MPa and $\beta = 3000$ m/s. Also shown are typical source properties of slow slip events in Nankai, Japan (black). Particle velocity is taken as half the ratio of average slip (0.016 m) to average event duration (4.5×10^5 s) (IDE *et al.*, 2007, Table 1) and the strength loss is 10 kPa (IDE *et al.*, 2007). c) Variation of friction with sliding velocity at room temperature for typical bare surfaces of quartzofeldspathic rock at normal stresses of 30, 70 and 150 MPa (from KILGORE *et al.*, 1993). Data are offset by $\mu_{v=1}$ friction at 1 micron/s. These are from rate-stepping tests of one order of magnitude. They are plotted at the lower velocity, thus the point at 100 microns/s represents a rate step from 100 to 1000 microns/s. The fit to the granite data assumes constant rate dependence over the entire range of sliding rates.

drops and high slip rates in laboratory faulting tests is a second order, negative dependence of fault strength on the logarithm of slip velocity. This 'rate dependence' is sustained over many orders of magnitude in velocity, typically from 0.001 μm/s up to 0.75 mm/s at low normal stress and up to at least 1 mm/s at higher stress (Fig. 1c) (OKUBO and DIETERICH, 1986; KILGORE *et al.,* 1993). The lab observations are well represented by rate and state variable constitutive equations (DIETERICH, 1978, 1979; RUINA, 1983). Though largely empirical, numerous advances in earthquake research have followed the development of these equations. Such work continues and these rate and state relations are used routinely in models of the earthquake cycle, rupture propagation, triggered seismicity, time-dependent hazard, etc.

However, there are aspects of lab and natural faulting that are not consistent with these particular rate and state equations. For example, in controlled laboratory tests at slip speeds up to 1 m/s (e.g., TSUTSUMI and SHIMAMOTO, 1997; GOLDSBY and TULLIS, 2002), stress drops can be much larger (Fig. 1a), having greater than 50% strength loss. These are thought to result from phase changes on the fault surface, not present in the low speed experiments on which the rate and state equations were based. A separate example is the periodic slow slip events (SSE) in subduction zones (DRAGERT *et al.*, 2004; ITO *et al.*, 2006). SSE have stress drops that are small relative to earthquakes, extremely low average slip speeds in the range of 0.01 to 1 μm/s (Fig. 1b), and average rupture propagation speeds of no more than a few meters per second. Since rate-weakening extends up to 1 mm/s in typical stick-slip faulting tests, the characteristics of SSEs are not consistent with the slip rate being limited by rate & state fault properties alone. It should be noted that SSE source properties can be explained by a rate-weakening friction that extends to seismic slip rates (LUI and RICE, 2005; AMPUERO and RUBIN, 2008) if factors other than the intrinsic rate dependence are chosen carefully. A slow slip event may be adequately simulated by standard rate and state equations if the entire event is contained within the nucleation phase of rupture development (AMPUERO and RUBIN,

2008). For typical SSE dimensions (>100 km) this requires choosing the effective normal stress to be on the order of a few MPa.

In addition to these two examples, other phenomena such as deep fault-associated non-volcanic tremors (OBARA, 2002) are not easily represented by existing laboratory-based constitutive equations, therefore there is some need for new models of these and other poorly understood faulting phenomena. In this report I describe an approach for constructing semi-empirical constitutive equations and apply the approach. An objective is to follow RUINA (1983) and RICE (1983) to produce relatively simple relations that are slip- and time-differentiable, and that can be deployed in whole-cycle numerical faulting models and calculations of rupture propagation. The equations are based on physical processes and contact scale physics where possible and 'state' variables will have a specific physical interpretation. In the following discussion I briefly summarize and contrast Ruina's and the approach advocated in this study, re-derive two familiar state variable constitutive equations, and then construct two example constitutive relationships. These new relations are illustrative, qualitative representations of the unexpected weakening of silica-rich rocks observed by GOLDSBY and TULLIS (2002) and faulting in the crustal transition zone where slip shows both brittle and ductile responses (e.g., SHIMAMOTO, 1986; MOORE *et al.*, 1997). These constitutive equations primarily demonstrate the approach but may also be useful for quantitative data modeling.

2. *Background*

Ruina's constitutive equations for low-temperature frictional sliding of rocks represent friction as the sum of a first-order constant μ_0, a second-order instantaneous dependence of friction on log velocity V and another second-order term containing a state variable ψ,

$$\tau = \mu\sigma_e = \sigma_e\left(\mu_o + a\ln\frac{V}{V_o} + \mathrm{b}\psi\right). \tag{0}$$

Here and throughout τ is shear stress, σ_e is effective normal stress and friction is μ is the ratio of shear to normal stress. Ruina assumed that the state variable is associated with a fault surface having a characteristic length scale d_c, and whose strength depends on current slip speed and the history of previous sliding. d_c is the sliding distance controlling the memory of prior sliding. Ruina further assumed that the change of state with time t is some function of state and slip rate V

$$\frac{d\psi}{dt} = G(\psi, V). \tag{1a}$$

The dependence of state on previous slip history is incorporated by requiring that strength depend on the displacement-weighted average w of some function f of slip speed

$$\psi = \int_{-\infty}^{\delta} w\left(\delta - \delta'\right) f\left[\frac{V}{V_0}\right] d\delta'. \tag{1b}$$

Equation (1b) can satisfy (1a) and incorporate the characteristic length by allowing w to be an exponential decay over length d_c

$$\psi = \frac{1}{d_c} \int_{-\infty}^{\delta} \exp \frac{-\left(\delta - \delta'\right)}{d_c} f\left[\frac{V}{V_0}\right] d\delta'. \tag{2a}$$

Noting that (2a) can be expressed in the form of (1a) by differentiating,

$$\frac{d\psi}{dt} = \frac{V}{d_c}\left(f\left[\frac{V}{V_0}\right] - \psi\right), \tag{2b}$$

RUINA's (1980, 1983) equations for evolution of the state variable were constructed by choosing different functions for f. For example, taking f to be logarithmic produces an expression for the 'slip' state variable,

$$\frac{d\psi}{dt} = -\frac{V}{d_c}\left(\ln\left[\frac{V}{V_0}\right] + \psi\right). \tag{3}$$

While this approach was a breakthrough, leading to advances in modeling earthquake occurrence, it is likely that not all valuable state-variable relationships for friction are of the form (1) and (2). For example, the evolution equation of PERRIN *et al.* (1995) satisfies Ruina's condition equation (1a) but not equations (2), suggesting that (2) may be too restrictive in some cases. Another region of possible improvement over (1) and (2) is that ideally the constitutive relationship should have physical mechanisms attributed to state so that the equations can be extrapolated beyond the conditions (e.g., strain rate, temperature, pressure) of the experiments on which they are based. Unfortunately, the mechanisms underlying the RUINA (1980) and DIETERICH (1972, 1978, 1979) experiments were not completely understood at the time and perhaps because of this Ruina's state variables intentionally were not attributed to physical mechanisms. In intervening years there has been some but not complete clarification of the underlying mechanisms. For instance, state in equations such as (3) can be used to represent time-dependent strengthening observed at vanishingly low slip rates in laboratory tests. This effect had long been suspected to result from time-dependent yielding of asperity contacts (DIETERICH, 1972), and direct observations of frictional surfaces have shown this to be the correct interpretation under particular circumstances (DIETERICH and KILGORE, 1994). At the same time, the same state variable also produces the apparent slip-weakening which accompanies increases in sliding rate in rate-stepping experiments. This weakening seems to correspond to a 'brittle' behavior not easily or directly related to plastic yielding of asperities. Thus, the single state variable (2) represents more than one process in the experiments these equations describe. Such attribution of state to multiple processes is necessary and Ruina's state could represent surface separation (dilation), perhaps as suggested by TOLSTOI (1967) and as indicated by limited experimental observations (MARONE *et al.*, 1990), equivalently real area of contact, or redimensioned to have a temporal interpretation as age of contact (RUINA, 1983; LINKER and DIETERICH, 1990). Therefore, in the present we are at an advantage over Ruina in that we have a better idea of what processes are involved in 'state'. As I intend to show in this study, in the context of developing new constitutive equations from lab data, there can be further advantages in explicitly identifying the processes associated with state. In any case, without sound physical interpretation and a mechanistic basis, with associated material constants and activation energies, constitutive descriptions of laboratory observations cannot be easily extrapolated to natural conditions and used in earth models.

As a variation on Ruina's, the approach in this study is to allow that changes in fault strength depend on particular selected variables, such as slip, time, slip rate and shear stress, in the particular ways shown in laboratory tests or as indicated by theory or natural observations. As an example illustration of the approach consider the original observations of rate-dependent steady-state strength and time-dependent changes in 'static' friction from low-temperature studies (DIETERICH, 1972, 1978, 1979, RUINA, 1980) as resulting from a state variable that represents inelastic contacting area within the fault zone. Though the motivating laboratory observations imply slip and time dependencies, the examples in this section and throughout follow Ruina closely and changes of state with time depend explicitly on slip rate and on state. Thus, I follow RUINA (1983) by casting *G,* the time derivative of the state variable, to depend on state and slip velocity (1a) but rather than explicitly considering the dependence of state on the history of previous sliding (1b) I represent G as a function with two or more terms that will be selected based directly on lab-observed behaviors. In this paper I use two terms

$$\frac{d\psi}{dt} = G(V, \psi) = g_1 + g_2. \tag{4a}$$

Accordingly, g_1 and g_2 may each depend on state and velocity as required by the observations of interest. Ideally the terms might represent particular physical processes that change contact area such as contact scale plasticity and shear-induced fracture dilatancy, or, for the sake of practicality, the approach can simply be to choose the terms explicitly to match particular laboratory observations. As a specific example, as noted by DIETERICH (1972), at nominally zero slip rate fault strength increases with the logarithm of time. g_1 can be selected to allow state (contact area) to have a component that increases with time, for instance $g_1 = (V_o \exp -\psi)/d_c$ integrates in time to a logarithmic increase. Conceptually we follow Dieterich and attribute this increase in contact area to plastic yielding. Using this choice (4a) is then

$$\frac{d\psi}{dt} = \frac{V_o}{d_c} \exp -\psi + g_2. \tag{4b}$$

g_2 can be determined by using another observational constraint, namely that at steady-state contact area depends on the logarithm of slip rate $\psi_{ss} = \ln(V_0/V)$ (DIETERICH, 1979; RUINA, 1983)

$$0 = \frac{V_o}{d_c}\left(\frac{V}{V_0}\right) + g_2. \tag{4c}$$

This is Ruina's slowness relation,

$$\frac{d\psi}{dt} = \frac{V_o}{d_c} \exp(-\psi) - \frac{V}{d_c}. \tag{4d}$$

Ruina's slip state variable can also be constructed using this approach. Again assume that changes in inelastic contact area with time result from dependencies on state and velocity (4a). From RUINA's (1980) experiments, a key observation is that strength changes nearly exponentially with displacement following changes in slip speed, presumed to represent changes of contact area due to varying amounts of dilatancy at different sliding rates. To represent this, take $g_2 = -\psi V/d_c$ which integrates in slip to exponential slip weakening if slip rate is increased, thus

$$\frac{d\psi}{dt} = g_1 - \frac{V\psi}{d_c}. \tag{5a}$$

The remaining term can be deduced again by using the observational constraint that at steady state $\psi_{ss} = \ln(V_0/V)$

$$0 = g_1 - \frac{V}{d_c} \ln\left[\frac{V_0}{V}\right], \tag{5b}$$

leading to Ruina's slip relation, equation (3),

$$\frac{d\psi}{dt} = \frac{V}{d_c}\left(\ln\left[\frac{V}{V_0}\right] - \psi\right). \tag{5c}$$

PERRIN *et al.*'s (1995) quadratic relationship can be developed using still different combinations of laboratory constraints, as might other potentially useful state variable equations for low-temperature friction. In the following Sections 3, 4 and 5, I apply the approach used to construct equations (4d) and (5c) to lab observations that are not well-represented by Ruina's state variable equations.

3. A Trial Constitutive Relation for Unexpected Weakening

To derive another example lab-based constitutive relation, consider friction at sustained slip speeds higher than in the experiments of DIETERICH (1978,1979) and RUINA (1980), i.e., above a few mm/s. For initially bare surfaces of silica-rich rocks (quartzite, novaculite, feldspar, and granite) at slips greater than a few meters at slip speeds from 10 mm to 0.1 m/s, there is considerable strength loss (Figs. 1a, 2a) (GOLDSBY and TULLIS, 2002; DITORO *et al.*, 2004), whereas silica-poor rocks (gabbro) and silica-absent rocks (marble) show no weakening. Measurements and calculations indicate that asperity contact temperatures are <140 °C, and average temperatures on the sliding surface are 10s of degrees less than this, therefore melting is thought not to be involved in the weakening (GOLDSBY and TULLIS, 2002). Furthermore, changes in strength occur more slowly than shear induced temperature changes on the fault surface, consequently temperature does not directly influence shear strength and appears to play no important role in the observed weakening. It has been proposed, on the basis of inference and microstructure observations, that shear lubricates the surface through production of a highly comminuted, comparatively wet, amorphous, gel-like material, i.e., "silica gel" (GOLDSBY and TULLIS, 2002).

3.1. Interpretation of Weakening

The primary observations of the weakened fault are well explained by a 'thixotropic' material, one whose strength is controlled by a near-fluid that becomes more fluid-like (weaker) when strained. In the case of a thixotropic gel, the weakening may result from dynamic reduction of the number of chemical bonds between particles. Think of this as the progressive replacement of strong bonds between solid particles with weaker bonds. The transition from high friction at low-speeds to this unexpected weakening behavior at higher slip rates can be represented as a reduction of the number of active strong bonds in the shear zone. When shearing ceases, or shearing rates are decreased, the material regains strength with time at a rate controlled by a chemical reaction that reforms the strong bonds. The net strength represents a dynamic balance between shear which decreases the number of strong bonds and time which increases their number. Therefore, we expect the steady-state strength to decrease with increasing slip rate. Given these constraints, I'll now use the approach developed in Section 2 and apply it to the experimental observations of GOLDSBY and TULLIS (2002).

3.2. Some Equations

Three assumptions are used to construct the fault strength relations:

(1) Shearing produces a dynamic reduction of strong bonds.
(2) Time increases the number of strong bonds.
(3) The steady-state number of strong bonds decreases with increasing slip speed.

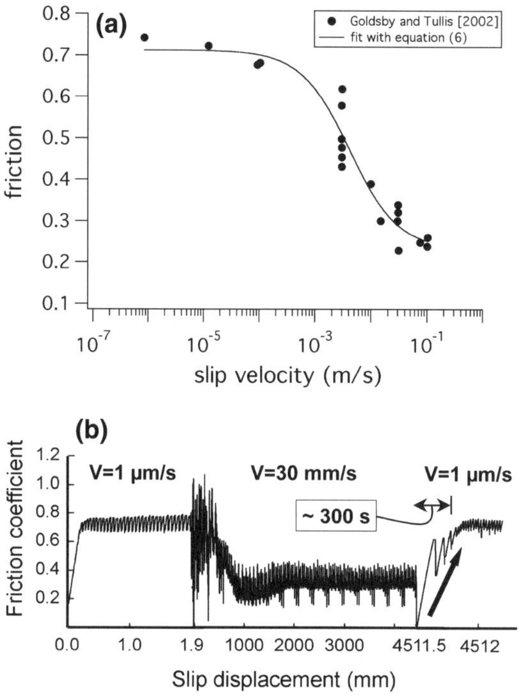

Figure 2
Unexpected and extraordinary rate weakening. a) GOLDSBY and TULLIS (2002) data from novaculite at 5 MPa normal stress showing the dependence of steady-state friction on slip rate. Shown for reference is a fit to the data using equations (6a) and (6e), resulting in $\mu_O = 0.71$, $\mu_w = 0.23$, and $V_c = 0.0041$ m/s. b) Summary of strength loss and strength recovery due to unexpected weakening of novaculite. Friction with slip at 1 μm/s, 30 mm/s, and again at 1 μm/s. Note the scale of the horizontal axis changes with velocity (DITORO *et al.*, 2004).

Call the ratio of strong bonds to the total number of bonds state, ψ. Take μ_0 as the strength of the strong bonds, and μ_{w} as the lower strength of bonds within the gel. The combined macroscopic strength is

$$\tau = \mu\sigma_e = \sigma_e[\mu_0\psi + \mu_w(1 - \psi)]. \tag{6a}$$

Assume that the rate that state changes with time depends on state itself and slip velocity (4a). The observations suggest an approximately exponential slip-weakening following an increase in sliding rate (Fig. 2b), so using a characteristic distance d_c that scales with the layer thickness w_c allow $g_2 = -\psi V/d_c$ so that

$$\frac{d\psi}{dt} = g_1 - \frac{V\psi}{d_c}. \tag{6b}$$

By assuming that in the absence of slip the interparticle bonds are formed at a rate controlled by a chemical reaction, at constant temperature and zero slip rate the number of bonds increases with time. A reasonable representation is $g_1 = (1 - \psi)/t_c$, where t_c is the reaction time constant. Integrating this choice of g_I in time yields the first-order chemical reaction rate equation that is exponential in time (RIMSTIDT and BARNES, 1980; STUMM and MORGAN, 1981), $\psi = 1 - \exp(-t/t_c)$. The state variable equation is

$$\frac{d\psi}{dt} = \frac{1}{t_c}(1 - \psi) - \frac{V\psi}{d_c}, \tag{6c}$$

though steady-state has not been explicitly specified. Setting (6c) to zero results in

$$\frac{1}{\psi_{ss}} = \frac{t_c}{d_c} V_{ss} + 1. \tag{6d}$$

Taking the ratio of displacement and time constants to be the characteristic velocity of the weakening process, $d_c/t_c = V_c$, the steady-state relation is

$$\psi_{ss} = \frac{V_c}{V_c + V_{ss}}. \tag{6e}$$

Since the expectation is that the number of bonds decreases as the shear-rate increases, the inverse relationship (6e) is adequate for the conceptual model and matches the data reasonably well (Fig. 2a). The evolution equation (6d) without the time constant is

$$\frac{d\psi}{dt} = \frac{V_c}{d_c}(1 - \psi) - \frac{V\psi}{d_c}. \tag{6f}$$

For equations (6a) and (6f) at velocity $V << V_c$ the fault is strong $\mu = \mu_O$, and at $V >> V_c$ the fault is weak at the residual strength $\mu = \mu_w$. V_c is the velocity at which the fault has lost half its strength, $\mu = (\mu_O - \mu_w)/2$.

3.3. *Discussion*

Two numerical simulations of rate steps using (6a) and (6f) are shown in Figure 3. As the relation was constrained by the data, obviously, there is qualitative agreement with the observations, exponential shear-weakening at high slip speed (Fig. 2b) and rapid strength recovery when the loading rate reverts to values less than V_c, consistent with the duration of strength recovery shown in Figure 2b. In addition, some predicted properties are consistent with the laboratory observations. For example, knowledge of the slip-weakening distance ($d_c \approx 0.5$ m) and characteristic velocity ($V_c = 0.004$ m/s) results in a prediction for the characteristic time controlling the strength recovery. The prediction (121 s) is similar to the characteristic time of the observed recovery (Fig. 2b). Because (6) has at least a vague theoretical basis following a limited understanding of the underlying physics of the gel, additional implications can be tested in subsequent experiments, for example, at higher slip rate where the relationship predicts constant

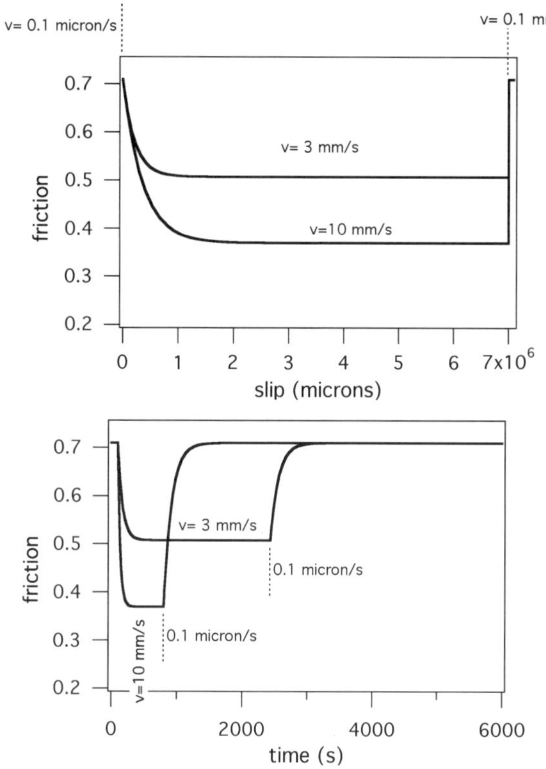

Figure 3

Simulations of unexpected weakening using step increases and decreases in sliding velocity for two different-sized rate steps (0.1 microns/s to 3 mm/s, 0.1 microns/s to 10 mm/s) in friction vs displacement and friction vs time with equations (6a) and (6f). The simulation steps down after slip of 10 microns, then back up after slip of 7 meters, $d_c = 0.5$ m, $t_c = 121$ s, $\mu_0 = 0.71$, $\mu_w = 0.23$, and $V_c = 4.1$ mm/s.

strength or at higher temperature where the time constant of the reaction is expected to decrease. One such prediction is evident in the simulations in Figure 3; the apparent slip-weakening distance increases with increasing slip speed. In an iterative way, trial constitutive relations can be used to identify new experiments that will help to further constrain the underlying physics of weakening which in turn should lead to improved constitutive descriptions.

A number of different mechanisms have been invoked to explain recent laboratory observations of high speed weakening. In some cases, such as melting, weakening is clearly related to shear heating (e.g., Hirose and Shimamoto, 2005). However, in other cases, even when a phase change occurs, the weakening is not obviously related to the reaction or to temperature change (e.g., Han *et al.*, 2007; Mizoguchi *et al.*, 2006). In situations where the weakening is not due directly to shear-heating, equations (6) may provide an adequate empirical representation of lab observations. An important caveat is that (6) has no direct effect and will lead to ill-posedness in spatially dimensioned numerical simulations (RICE *et al.,* 2001), thus it requires additional modification for adequate simulation of dynamic rupture.

4. Faulting with Mixed Brittle and Ductile Behavior

As there is recent renewed interest in faulting at transition zone conditions (e.g., DRAGERT *et al*., 2004; ITO *et al*., 2006), let us consider an example constitutive relation based on lab observations of mixed brittle and ductile behavior. For typical continental crust composition, plastic deformation is favored over brittle fracture and frictional slip at low strain rate when source temperatures and pressures exceed 300C and 100 MPa, (e.g., GOETZE and EVANS, 1979). Therefore, as the brittle ductile transition is approached from above following geothermal and lithostatic gradients, shear zones will become increasingly rate-strengthening at the lowest slip speeds. This is seen in the high temperature quartzofeldspathic friction studies (BLANPIED *et al.,* 1991, 1995; Chester and Higgs, 1992; Chester, 1994, 1995) and for mafic materials at somewhat higher temperatures (BOETTCHER *et al*., 2007). Even when fault strength is pressure dependent, ductile processes such as pressure solution and dislocation motion affect the strength of load-bearing contacts and for quartz and granite in the highest temperature regime and particularly at the lowest sliding rates there is strong rate-strengthening. Thus, constitutive relations for the transition zone should include a transition to positive rate dependence at low slip rates.

While separate faulting studies of transition zone behavior at hydrothermal conditions in quartz and granite (BLANPIED *et al.,* 1991, 1995; CHESTER and HIGGS, 1992; CHESTER, 1994, 1995) require rate-strengthening at low slip rates, the behavior at slip rates above 10 microns/s is unknown and the effective pressures tested are limited. In the absence of a comprehensive study of rate dependence over a wide range of sliding rates at varying effective pressure, we may examine insights from faulting in an analogue material. SHIMAMOTO (1986) studied the brittle-ductile transition in halite at room temperature. Halite deforms by ductile processes (dislocation creep, pressure solution), solution mass transfer processes (crack healing, asperity contact overgrowth), and brittle processes (fracture, frictional slip) at room temperature, as sliding rate and confining pressure are varied. Consequently, the brittle-ductile transition can be studied in halite, avoiding the many experimental difficulties that arise in high temperature faulting experiments. SHIMAMOTO's (1986) experiments on 0.3-mm thick halite shear zones were conducted dry to minimize the influence of pressure solution and solution mass transfer. True ductile deformation was found at all sliding rates at 250 MPa confining pressure, and predominantly brittle behavior was found at 10 MPa confining stress. Interestingly, an additional transition to rate-strengthening behavior occurs at slip rates above 1 micron/s.

The transition to rate strengthening at higher slip rates in SHIMAMOTO's (1986) experiments is one of many experimental examples of what DIETERICH (1978) called a 'high speed cutoff' on rate-weakening. Such a limit is observed even at room temperature for fault slip on granite surfaces (OKUBO and DIETERICH, 1986) and has long-standing physical (DIETERICH, 1978; ESTRIN and BRECHET, 1996; NAKATANI, and SCHOLZ, 2006) and observational (OKUBO and DIETERICH, 1986; KILGORE *et al.*, 1993; NAKATANI and SCHOLZ, 2004, 2006) basis. The transition to rate-strengthening behavior is thought to result from a theoretical limit on the maximum rate of fault normal yielding at asperity contacts (DIETERICH, 1978; ESTRIN and BRECHET, 1996; NAKATANI and SCHOLZ, 2006), as follows. During steady-state slip at constant slip rate V, asperity contact area is constant, however there is dynamic balance between fault normal yielding of contacts which tends to increase contact area, and shear-induced dilatancy that decreases contact area. Starting from the steady-state condition, if sliding velocity were then set immediately to zero, the rate of fault normal yielding at contacts could be measured directly (NAKATANI and SCHOLZ, 2004, 2006). At specified normal stress and surface roughness there is an upper limit on the rate of fault normal yielding. This limit is most easily envisioned by considering the condition of zero applied normal stress. Here we assume that asperity contacts have finite area at zero load as expected for asperity contacts on naturally rough fault surfaces. Under zero load there is no tendency to yield and the minimum observed sliding velocity at which there is no fault normal yielding, the shear cutoff velocity, is 0. At higher normal stress the cutoff velocity is non-zero depending on the specific mechanism of yielding, the normal stress, the tendency to dilate with shear, and the instantaneous area of asperity contact (NAKATANI and SCHOLZ, 2004, 2006). Following this reasoning, the cutoff is expected to increase with increasing normal stress, though the dependence of the cutoff on normal stress will depend on the surface roughness, the specific (nonlinear) elastic properties of asperity contacts (see BROWN and SCHOLZ, 1985), magnitude of shear dilatancy, and the detailed stress sensitivity of the yielding mechanism. Such a reduction of the cutoff velocity with decreasing normal stress is seen in experiments by KILGORE *et al.* (1993).

For Shimamoto's halite experiments, at the lower intermediate pressures (35 to 100 MPa), there is a regime of negative rate-dependence bounded at low and high speeds by rate strengthening (Fig. 4). That there should be a regime of negative rate-dependent friction limited at both higher speed (the high speed cutoff) and lower slip speeds (the brittle ductile transition) has been suggested on theoretical grounds by ESTRIN and BRECHET (1996) both for materials such as metals which are nominally ductile at room temperature and low strain rate, and for brittle silicates. They termed the characteristic shape of friction vs velocity seen in Figure 4 as 'N-shaped'.

Some other minerals exhibit related behavior at modest temperature and pressure, particularly weak phases. For example, the rate dependence of serpentine changes sign as slip speed is varied (REINEN *et al.*, 1994; MOORE *et al.*, 1997). Qualitatively, the rate dependence of serpentine at elevated temperature and pressure (Fig. 5) (MOORE *et al.*, 1997) resembles the transitional behavior seen in Shimamoto's experiments on halite. At the very least serpentine can be considered another analogue material for low temperature lab simulation of the brittle ductile transition (see ESCARTIN *et al.*, 1997).

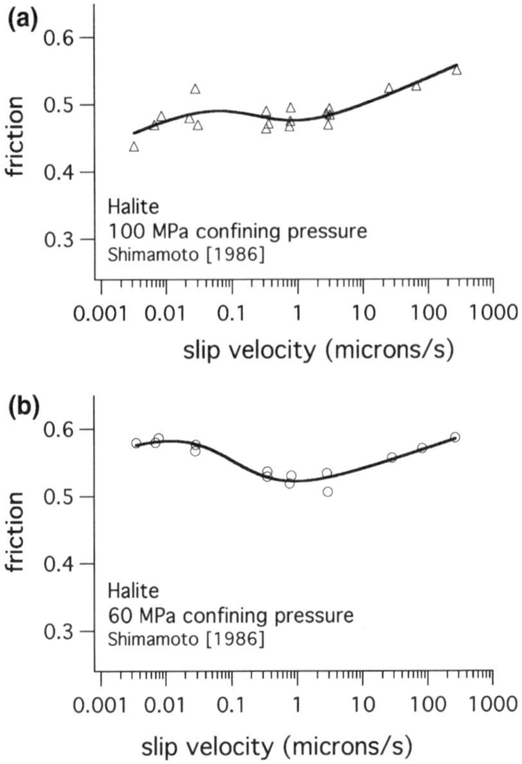

Figure 4

Fault rate dependence over a wide range of sliding velocity at the brittle-ductile transition in an analogue material. These are data from shear of 0.3 mm layers of dry halite from SHIMAMOTO (1986). The published shear stress τ and confining pressure σ_3 data were converted fault normal stress and friction using the relationships $\tau = 0.5\sigma_\Delta \sin 2\phi$, and $\sigma_n = \sigma_3 + 0.5\sigma_\Delta(1 - \cos 2\phi)$, where σ_Δ is the differential stress $\sigma_\Delta = \sigma_1 - \sigma_3$ and $\phi = 35°$ is the angle between the σ_1 direction and the fault plane. Shown also are fits to these data using equation (8b). For a) $\mu_0 = 0.43$, $a = 0.018$, $b = 0.1064$, $V_0 = 0.232$ μm/s. For b) $\mu_0 = 0.477$, $a = 0.014$, $b = 0.15$, $V_0 = 0.106$ μm/s.

In the following section I construct a state variable equation to represent transition zone behavior where the sign of the rate dependence changes with slip rate. This approach follows from DIETERICH (1978) and the resulting equation is similar and related to expressions derived using theoretical considerations by NAKATANI and SCHOLZ (2006) in the context of their detailed experimental program to find the high speed cutoff.

4.1. Some equations

The constraints used to construct the state variable relation are:

1) Contact area increases with time (time strengthening),
2) Transition to rate strengthening at low slip speed,
3) Transition to rate strengthening at high slip speed.

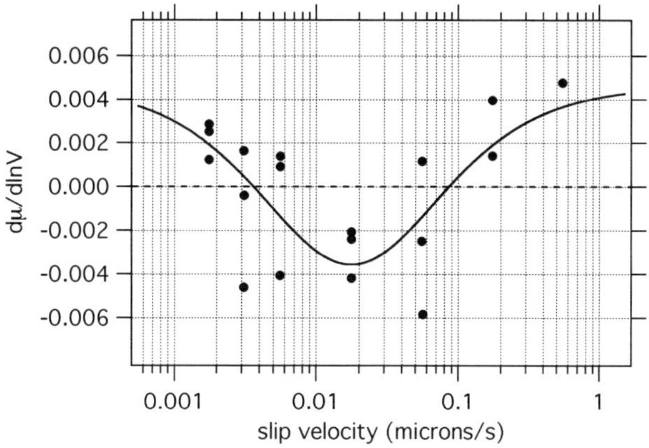

Figure 5
Variation of the steady-state rate dependence of chyrsotile serpentine with sliding velocity at temperatures between 107 and 281 °C and effective normal stresses between 46.5 and 139.5 MPa, from MOORE *et al.* (1997). The conditions are consistent with those at depths between 3 and 9 km in the continental crust assuming hydrostatic pore fluid pressure and normal stress equal to overburden. Shown also is a fit to these data using equation (8c) resulting in $a = 0.0046$, $b = 0.032$ and $V_O = 0.018$ μm/s.

The limited brittle faulting model consists of a first-order fault strength μ_0, a second-order direct dependence on slip rate and a second-order dependence on state, equation (0), following (RUINA, 1983)

$$\tau = \mu\sigma_e = \sigma_e\left(\mu_o + a\ln\frac{V}{V_o} + \mathrm{b}\psi\right).$$

The state variable ψ represents the inelastic portion of the shear load-bearing surface area in the fault zone — the portion of contacting area that can change during sliding at constant normal stress (4a) if the sliding rate is varied. Following the construction of RUINA's (1983) slowness equation (4) and choosing g_1 to allow state (contact area) to increase with time $g_1 = (V_o \exp[-\psi])/d_c$ is

$$\frac{d\psi}{dt} = \frac{V_o}{d_c}\exp(-\psi) + g_2. \tag{7a}$$

To produce rate-strengthening at both high and low velocity introduce the cutoff velocities V_L and V_H. If at steady-state, the state variable is $\psi_{ss} = \ln([V_H + V]/[V_L + V])$ (Fig. 6a) then at slip speeds well below V_L and well above V_H, $d\psi_{ss}/d\ln V \approx 0$, producing pure rate-strengthening in (0) at the high and low speeds. Substituting the desired steady-state into (7a)

$$0 = \frac{V_o}{d_c}\exp(-\psi) + g_2, \tag{7b}$$

leads to

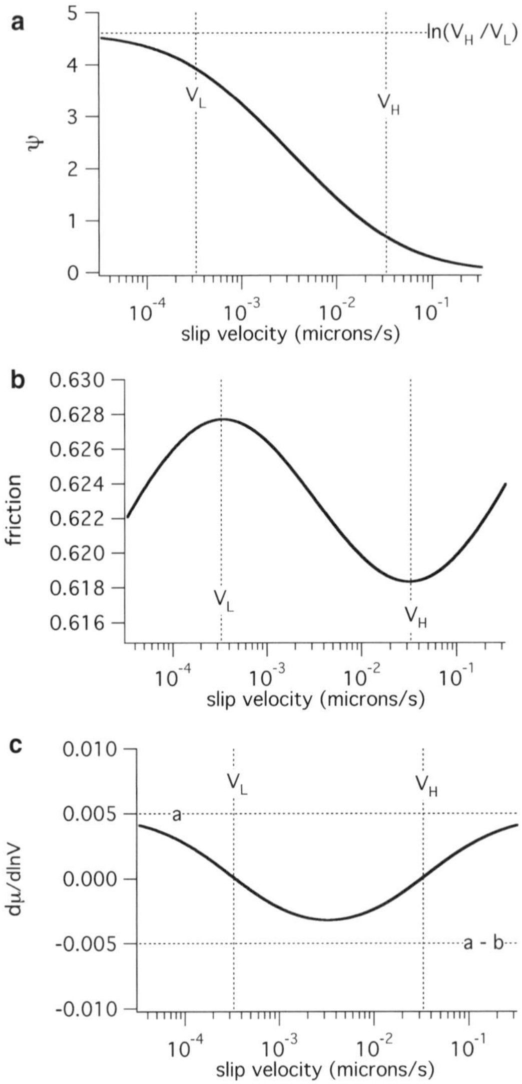

Figure 6
Steady-state properties of trial constitutive relation for periodic slow slip, equations (0) and (7) with $\mu_0 = 0.6$, $a = 0.005$, $b = 0.01$, $V_L = 0.00033$ μm/s, and $V_H = 0.033$ μm/s, a) state, b) strength, c) rate dependence.

$$\frac{d\psi}{dt} = \frac{V_o}{d_c}\left(\exp(-\psi) + \frac{V + V_L}{V + V_H}\right). \tag{7c}$$

This is Ruina's slowness equation with high and low speed cutoffs on state (see NAKATANI and SCHOLZ (2006) for a related expression). The steady-state strength of (0) and (7c) is

$$\mu_{ss} = \mu_o + a\ln\frac{V}{V_0} + b\ln\frac{V_H + V}{V + V_L}, \tag{7d}$$

(Fig. 6b). The steady-state rate dependence is

$$\frac{d\mu_{ss}}{d\ln V} = a + \mathrm{b}V\left[\frac{1}{(V_H + V)} - \frac{1}{(V_L + V)}\right], \tag{7e}$$

(Fig. 6c).

4.2. Relation between Transition Zone Behavior and Dilatancy

In room temperature rock fracture tests anitigorite and lizardite serpentines are markedly weaker than nonhydrous silicate rocks such as granite and dunite. The weakness may be a key to understanding the mixed brittle response of serpentinite and other phyllosilicates at room temperature (ESCARTIN *et al.*, 1997). Serpentines have a low dependence of failure strength on fault normal stress (pressure dependence) which indicates low bulk dilatancy (see BRACE *et al.*, 1966; BRACE, 1978; ESCARTIN *et al.*, 1997). As serpentine has [001] cleavage, shear is accommodated preferentially along these cleavage surfaces resulting in crystallographically-controlled (non-opening) mode II shear fractures. Deformed serpentinite has scarce open mode I cracks that produce the strong pressure dependence in quartzofeldspathic rock. Dilatancy that does occur in serpentinite is restricted to rigid motion due to roughness along the shear planes (ESCARTIN *et al.*, 1997) and in dilatant jogs within regions of shear stepover. For pre-existing faults, the analogous pressure dependence is the coefficient of friction μ. Assuming the same relation between pressure dependence and dilatancy, comparison of μ of the typically strong quartzofeldspathic rocks with that of weak serpentines and other phyllosilicates suggests that dilatancy during frictional slip is similarly and relatively reduced for fault slip in phyllosilicate-rich rocks.

While the standard interpretation of the high speed cutoff on rate-weakening is that at velocities above V_H asperity contact area does not change with sliding speed (NAKATANI and SCHOLZ, 2006), it can be equivalently considered a cutoff on shear-induced dilatancy. Strength loss during rock failure or frictional slip requires a net reduction of contacting area across the fault, which cannot occur unless dilatant volume is generated during the strength loss. If rocks with low-pressure dependence such as serpentinite are indeed less dilatant, the amount and extent of rate-weakening will be limited and quite different than in rocks that are highly pressure dependent at room temperature.

Similar arguments apply to halite at room temperature and can be extended also to quartzofeldspathic materials at elevated temperature and confining pressure. As the data from SHIMAMOTO (1986) show clearly, as confining pressure is increased there is a systematic reduction in the pressure dependence. For example, this can be seen by comparing the friction coefficient at different confining pressures in Figures 4a and 4b, and is more clearly shown in Figure 1 in SHIMAMOTO (1986). The standard interpretation of the reduction of pressure dependence is that dilatancy is suppressed as confining pressure increases. Microfracturing that occurs at the higher confining pressures is manifest increasingly as shear fracture and slip on cleavage surfaces with little accompanying dilatancy. Thus, we may speculate, perhaps expect, that the transition zone behavior of many other brittle materials reflects natural limits on shear-induced weakening (dilatancy) exhibited in SHIMAMOTO's (1986) halite and MOORE *et al.*'s (1997) serpentinite experiments.

4.3. An Alternative Transition Zone State Variable

In SHIMAMOTO's (1986) and MOORE *et al.*'s (1997) experiments the transitions in rate dependence are in fairly close proximity in velocity. Under these special circumstances an approach that is different from (7) can be used to construct the state variable. Again, as with (7), the state variable ψ represents the inelastic portion of the shear-load bearing contact area in the fault zone. The alternative approach is to choose functions g_1 and g_2 that have limits. To illustrate, assuming that contacting area increases exponentially with time of static contact and decreases exponentially with slip leads to equation (6g),

$$\frac{d\psi}{dt} = \frac{V_0}{d_c}(1 - \psi) - \frac{V\psi}{d_c}, \tag{8a}$$

though the physical interpretation is different. Here, reduction of state with slip represents reduction of contact area with accompanying dilatancy and the increase in state with time represents yielding or chemical reactions that restore strength. The steady-state strength of (0) and (8a) is

$$\mu = \mu_o + a \ln \frac{V}{V_0} + \mathrm{b} \frac{V_0}{V_0 + V}, \tag{8b}$$

(Fig. 7b). The steady-state rate dependence is

$$\frac{d\mu_{ss}}{d \ln V} = a - \mathrm{b} \frac{V/V_0}{(1 + V/V_0)^2} \tag{8c}$$

(Fig. 7c). When strength is plotted against log velocity, the steady-rate dependence is symmetric about V_0 (Fig. 7c), and at V_0 rate dependence is at its minimum, $a - b/4$. When $b/4 > a$ the rate dependence at V_0 is negative, indeed for there to be net strength loss $b/4 > a$.

4.4. Discussion

Equations (0) and (8) well fit the analogue observations of transition zone faulting in halite (Shimamoto, 1986) (Fig. 4) and the mixed brittle ductile behavior or serpentinite (Fig. 5). However, the positive rate dependencies at high and low velocities are not well constrained by the observations; neither for halite nor serpentinite has the detailed response to changes in sliding velocity been studied. Thus, the rate-strengthening at low

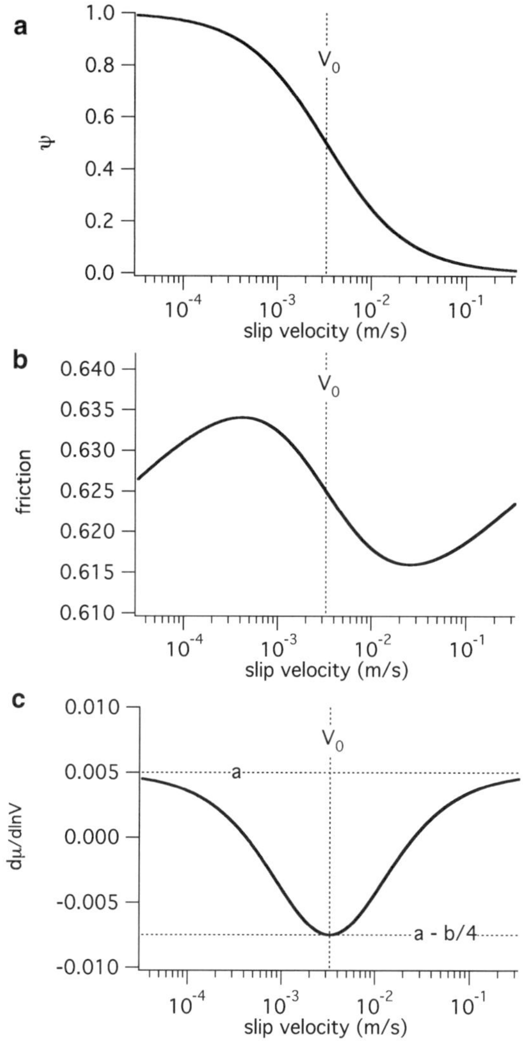

Figure 7
Steady-state properties of trial constitutive relation for periodic slow slip, equations (0) and (8) with $a = 0.005$, $b = 0.05$, $V_0 = 0.0033$ μm/s, a) state, b) strength, c) rate dependence.

slip speed may have a different positive rate dependence than the high speed regime and (0) and (8) may be a oversimplification of the laboratory data.

Slider block simulations conducted with equations (0) and (7) or (8) show periodic slip if the spring constant is sufficiently low. Considering simulations with (0) and (8), periodic slip arises when elastic stiffness is significantly lower than $(b/4 - a)/d_c$. When V_0 is submicron/s, conditions are truly quasi-static at the highest slip speeds, and the 'slip' portion of the limit cycle can be simulated without consideration of inertia or radiation damping (e.g., Rice, 1993). Such behavior may be related to natural episodic slow slip (e.g., SHIBAZAKI and IIO, 2003). An example for loading rate $V_L = 0.001$ μm/s, stiffness $k = 2.76 \times 10^{-6}$ MPa/μm, and $V_0 = 0.0033$ μm/s is shown in Figure 8. The particular choice of the constitutive parameters a, b and σ_e results in a limited rate-weakening region, a maximum slip sped of about 0.1 μm/s, per event slip of 15 mm, stress drop of 0.04 MPa, recurrence of 8 months, and event duration of 4.5 days. Because the fault is rate-strengthening at the lowest slip rates, significant intervent slip can occur; in this example it exceeds 0.25 of the total slip. Such slip-partitioning may be consistent with natural fault slip in the transition zone. Natural slow slip events may account for only 2/3 of the plate motion at the source in Cascadia (e.g., DRAGERT *et al.*, 2004).

5. *Conclusions*

An approach to constructing constitutive equations for modeling seismic fault slip is described. It is a variation on RUINA's (1983) state variable derivation, however it attributes physical meaning and attempts to identify the particular processes associated with state. The intention is to use laboratory, and theory as constraints on the state variable and on the steady-state values of state. The familiar state variable evolution equations developed by RUINA (1980, 1983) and widely used in earthquake modeling can be constructed using the alternative approach developed here. The method of construction is demonstrated by applying it to laboratory observations that are not well modeled by the existing rate and state constitutive equations, unexpected weakening of silica-rich rocks observed by GOLDSBY and TULLIS (2002) at high slip speed, and the mixed brittle and ductile behavior seen in transition zone experiments of SHIMAMOTO (1986). The constitutive equations derived in this paper are included as illustrations but are adequate representations of the particular example data sets and may be useful in quantitative fault models.

For unexpected weakening, the constructed state variable is the percentage of strong bonds between contacting asperities and particles within the fault zone. Evolution of the state variable balances a reduction of state with slip, which represents the shear-induced production of a weak material, with a dependence on time representing chemical reactions that restore strength. For transition zone behavior the state variable is inelastic contact area. Two approaches for constructing the state variable to match the laboratory observations are described. The first explicitly limits the range of rate-weakening slip speed using a high and a low 'cutoff' velocity (DIETERICH, 1978). The second approach instead balances the reduction of state using a slip-weakening term representing dilatancy against a time-strengthening term representing yielding or chemical reactions that restore strength. Both the apparent slip-weakening and time-strengthening terms have limits that are manifested as limits on state at high and low slip speeds.

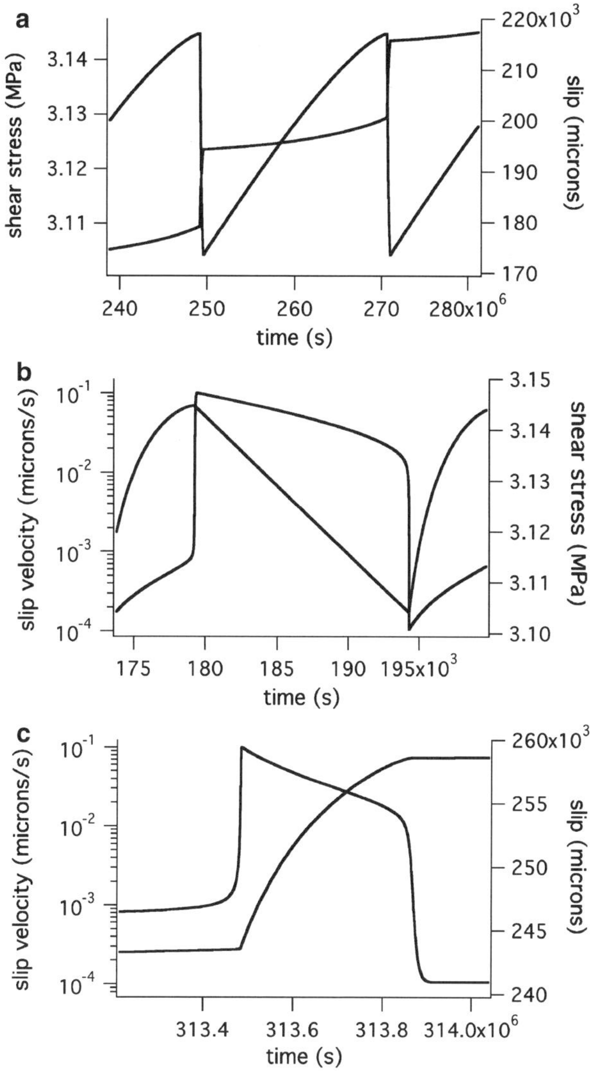

Figure 8
Numerical simulation of equations (0) and (8) with a single degree of freedom slider block. In this simulation $V_L = 0.001$ μm/s, $k = 2.76 \times 10^{-6}$ MPa/μm, $V_0 = 0.0033$ μm/s, $\mu_0 = 0.6$, $a = 0.008$, $b = 0.05$, $d_c = 20$ microns, and $\sigma_n = 5$ MPa. a) Shear stress and slip vs time. b) Slip velocity and shear stress vs time. c) Detailed view of slip and slip velocity during a 'rapid' slip event.

Acknowledgements

Discussion or correspondence with D. Goldsby, T. Tullis, D. Lockner, T. Shimamoto, G. DiToro, J. D. Weeks, M. Nakatani, J. Dieterich, G. Hirth and J. Rice are gratefully acknowledged. Unpublished notes by Jim Rice of his work with G. Perrin on the quadratic state relation were particularly helpful and influential. The paper was enhanced in response to comments by J. Gomberg, M. Boettcher, two anonymous reviewers and the associated editor Y. Ben Zion. The first anonymous reviewer found conceptual and mathematical errors in the original draft. The associate editor and reviewer guided the correction of these errors through an extra round of review. Many thanks to them. This research was supported by the US Geological Survey and the Southern California Earthquake Center. Research at SCEC is funded by NSF Cooperative Agreement EAR-0106924 and USGS Cooperative Agreement 02HQAG0008. This is SCEC contribution 1263.

References

Ampuero, J.-P. and Rubin, A. M. (2008), *Earthquake nucleation on rate and state faults—Aging and slip laws*, J. Geophys. Res. *113*, doi:10.1029/2007JB005082.

Blanpied, M. L., Lockner, D. A., and Byerlee, J. D. (1991), *Fault stability inferred from granite sliding experiments at hydrothermal conditions*, Geophys. Res. Lett. *18*, 609–612.

Blanpied, M. L., Lockner, D. A., and Byerlee, J. D. (1995), *Frictional slip of granite at hydrothermal conditions*, J. Geophys. Res. *100*, 13,045–13,064.

Blanpied, M. L., Marone, C. J., Lockner, D. A., Byerlee, J. D., and King, D. P. (1998), *Quantitative measure of the variation in fault rheology due to fluid-rock interactions*, J. Geophys. Res. *103*, 9691–9712.

Boettcher, M. S., Hirth, J.G., and Evans, B. (2007), *Olivine friction at the base of oceanic seismogenic zones*, J. Geophys. Res. *112*, B0125, doi:10.1029/2006JB004301.

Brace, W. F. (1978), *Volume changes during fracture and frictional sliding*, Pure Appl. Geophys. *116*, 603–614.

Brace, W. F., Paulding, B. W., and Scholz, C. H. (1966), *Dilatancy in the fracture of crystalline rock*, J. Geophys. Res. *71*, 3939–3953.

Brown S. R. and Scholz, C. H. (1985), *The closure of random elastic surfaces in contact*, J. Geophys. Res. *90*, 5531–5545.

Brune, J. N. (1970), *Tectonic stress and the spectra of seismic shear waves from earthquakes*, J. Geophys. Res. *75*, 4997–50090.

Chester, F. M. (1994), *Effects of temperature on friction: constitutive equations and experiments with quartz gouge*, J. Geophys. Res. *99*, 7247–7262.

Chester, F. M. (1995), *A rheologic model for wet crust applied to strike-slip faults*, J. Geophys. Res. *100*, 13,033–13,044.

Chester, F. M. and Higgs, N. G. (1992), *Multimechanism friction constitutive model for ultrafine quartz gouge at hypocentral conditions*, J. Geophys. Res. *97*, 1859–1870.

Dieterich, J. H. (1972), *Time-dependent friction in rocks*, J. Geophys. Res. *77*, 3690–3697.

Dieterich, J. H. (1978), *Time-dependent friction and the mechanics of stick slip*, Pure Appl. Geophys. *116*, 790–806.

Dieterich, J. H. (1979), *Modeling of rock friction 1. Experimental results and constitutive equations*, J. Geophys. Res. *84*, 2161–2168.

Dieterich, J. H. and Kilgore, B.D. (1994), *Direct observation of frictional contacts: New insights for sliding memory effects*, Pure Appl. Geophys. *143*, 283–302.

Dragert, H., Wang, K., and Rogers, G. (2004), *Geodetic and seismci signatures of episodic tremor and slip in the northern Cascadia subduction zone*, Earth Planets Space, *56*, 1143–1150.

DiToro, G., Goldsby, D. L., and Tullis, T. E. (2004), *Friction falls towards zero in quartz rock as slip velocity approaches seismic rates*, Nature, *47*, 436–439.

Escartin, J., Hirth, G., and Evans, B. (1997), *Nondilatant brittle deformation of serpentinites; implications for Mohr-Coulomb theory and the strength of faults*, J. Geophys. Res. *102*, 2897–2913.

Estrin, Y. and Brechet, Y. (1996), *On a model of frictional sliding*, Pure Appl. Geophys. *4*, 745–762.

GOETZE, C. and EVANS, B. (1979), *Stress and temperature in the bending lithosphere as constrained by experimental rock mechanics*, Geophys. J. Roy. Astron. Soc. *59*, 463–478.

GOLDSBY, D. L. and TULLIS, T. E. (2002), *Low frictional strength of quartz rocks at subseismic slip rates*, Geophys. Res. Lett. *29*, 4 pp.

HAN, R., SHIMAMOTO, T., HIROSE, T., REE, J.-H., and ANDO, J-I. (2007), *Ultralow friction of carbonate faults caused by thermal decomposition*, Science *316*, 878–881.

HIROSE, T. and SHIMAMOTO, T. (2005), *Growth of molten zone as a mechanism of slip weakening of simulated faults in gabbro during frictional melting*, J. Geophys. Res. 110.

IDE, S., BEROZA, G. C., SHELLY, D. R., and UCHIDE, T. (2007), *A scaling law for slow earthquakes*, Nature *447*, 76–79.

ITO, Y., OBARA, K., SHIOMI, K., SEKINE, S., and HIROSE, H. (2006), *Slow earthquakes coincident with episodic tremors and slow slip events*, Science *26*, 503–506.

JOHNSON, T. L., WU, F. T., and SCHOLZ, C. H. (1973), *Source parameters for stick-slip and for earthquakes*, Science *129*, 278–280.

KILGORE, B. D., BLANPIED, M. L., and DIETERICH, J. H. (1993), *Velocity dependent friction of granite over a wide randge of conditions*, Geophys. Res. Lett. *20*, 903–906.

LIU, Y. and RICE, J. R. (2005), *Aseismic slip transients emerge spontaneously in 3D rate and state modeling of subduction earthquake sequences*, J. Geophys. Res. *110*, B08307, doi:10.1029/2004JB003424.

MARONE, C., RALEIGH, C. B., and SCHOLZ, C. H. (1990), *Frictional behavior and constitutive modeling of simulated fault gouge*, J. Geophys. Res. *95*, 7007–7025.

MCGARR, A. (1994), *Some comparisons between mining-induced and laboratory earthquakes*, Pure Appl. Geophys, *142*, 467-489.

MCGARR, A. (1999), *On relating apparent stress to the stress causing earthquake fault slip*, J. Geophys. Res. *104*, 3003–3011.

MCGARR, A. and FLETCHER, J. (2003), *Maximum slip in earthquake fault zones, apparent stress, and stick-slip friction*, Bull. Seismol. Soc. Am. *93*, 2355-2362.

MCGARR, A., FLETCHER, J., and BEELER, N. M. (2004), *Attempting to bridge the gap between laboratory and seismic estimates of fracture energy*, Geophys. Res. Lett. *31*, 4 pp.

MIZOGUCHI, K., HIROSE, T., SHIMAMOTO, T., and FUKUYAMA, E. (2007), *Moisture-related weakening and strengthening of a fault activated at seismic slip rates*, Geophys. Res. Lett. *33*, doi:10.1029/2006GL026980.

MOORE, D.E., LOCKNER, D.A., MA, S., SUMMERS, R., and BYERLEE, J.D. (1997), *Strengths of serpentinite gouges at elevated temperatures*, J. Geophys. Res. *102* (B7), 14,787–14,801.

NADEAU, R. M. and JOHNSON, L. R. (1998), *Seismological studies at Parkfield VI: Moment release rates and estimates of source parameters for small repeating earthquakes*, Bull. Seismol. Soc. Am. *88*, 790–814.

NAKATANI, M. and SCHOLZ, C. H. (2004), *Frictional healing of quartz gouge under hydrothermal conditions: 1. Experimental evidence for solution transfer healing mechanism*, J. Geophys. Res. 109, doi:10.1029/2001JB001522.

NAKATANI, M. and SCHOLZ, C. H. (2006), *Intrinsic and apparent short-time limits for fault healing: Theory observations and implications for velocity dependent friction*, J. Geophys. Res. *111*, doi:10.1029/2005JB004096.

OBARA, K., (2002), *Nonvolcanic deep tremor associated with subduction in southwest Japan*, Science *296*, 1679–1681.

OKUBO, P. G. and DIETERICH, J. H. (1981), *Fracture energy of stick-slip events in a large scale biaxial experiment*, Geophys. Res. Lett. *8*, 887–890.

OKUBO, P. G. and DIETERICH, J. H. (1986) *State variable fault constitutive relations for dynamic slip*. In *Earthquake Source Mechanics* (ed. S. Das, J. Boatwright, and C. H. Scholz) Geophys. Mono. 37, AGU, pp. 25-36.

PERRIN, G., RICE, J. R., and ZHENG, G. (1995), *Self-healing Slip Pulse on a Frictional Surface*, J. Mech. Phys. Sol. *43*, 1461–1495.

RICE, J. R. (1983), *Constitutive relations for fault slip and earthquake instabilities*, Pure Appl. Geophys. *121*, 443-475.

RICE, J. R. (1993), *Spatio-temporal complexity of slip on a fault*, J. Geophys. Res. *98*, 9885–9907.

RICE, J. R., LAPUSTA, N., and RANJITH, K. (2001), *Rate- and state-dependent friction and the stability of sliding between elastically deformable solids*, J. Mech. Phys. Sol. *49*, 1865–1898.

REINEN, L. A., WEEKS, J. D., and TULLIS, T. E. (1994), *The frictional behavior of lizardite and antigorite serpentinites: Experiments, constitutive models, and implications for natural faults*, Pure Appl. Geophys. *143*, 317–358.

RIMSTIDT, J. D. and BARNES, H. L. (1980), *The kinetics of silica-water reactions*, Geochimica et Cosmochimica Acta *44*, 1683–1699.

RUINA, A. L. (1980), *Friction laws and instabilities, a quasistatic analysis of some dry frictional behavior*, Ph.D. Thesis, Brown University, 99 pp.

RUINA, A. L. (1983), *Slip instability and state variable friction laws*, J. Geophys. Res. *88*, 10359–10370.

SHIBAZAKI, B. and IIO, Y. (2003) *On the physical mechanism of silent slip events along the deeper part of the seismogenic zone*, Geophys. Res. Lett. *30*, 1489–1493, doi:10.1029/2003GL017047.

SHIMAMOTO, T. (1986), *Transition between frictional slip and ductile flow for halite shear zones at room temperature*, Science *231*, 711–714.

STUMM, W. and MORGAN, J.J. *Aquatic Chemistry* (John Wiley and Sons, New York, (1981)) pp. 780.

TOLSTOI, D.M. (1967), *Significance of the normal degree of freedom and natural normal vibrations in contact friction*, Wear *10*, 199-213.

TSUTSUMI, A. and SHIMAMOTO, T. (1997), *High-velocity frictional properties of gabbro*, Geophys. Res. Lett. *24*, 699–702.

WONG, T.-F. (1986), *On the normal stress dependence of the shear fracture energy*. In *Earthquake Source Mechanics, Geophys. Monogr. Ser.*, Vol. 37 (eds. S. Das et al.), pp. 1–11, (AGU, Washington, D.C. 1986).

(Received June 5, 2008, accepted March 20, 2009)
Published Online First: July 29, 2009

To access this journal online:
www.birkhauser.ch/pageoph

Pure appl. geophys. 166 (2009) 1799–1815
0033–4553/09/101799–17
DOI 10.1007/s00024-009-0517-y

© Birkhäuser Verlag, Basel, 2009

Pure and Applied Geophysics

Nonplanar Faults: Mechanics of Slip and Off-fault Damage

JAMES H. DIETERICH and DEBORAH ELAINE SMITH

Abstract—Stress interactions and sliding characteristics of faults with random fractal waviness in a purely elastic medium differ both qualitatively and quantitatively from those of faults with planar surfaces. With nonplanar fault models, solutions for slip diverge as resolution of the fractal features increases, and the scaling of fault slip with fault rupture dimension becomes nonlinear. We show that the nonlinear scaling of slip and divergence of solutions arise because stresses from geometric interactions at irregularities along nonplanar faults grow with increasing slip and produce backstresses that progressively impede slip. However, in real materials with finite strength, yielding will halt the growth of the interaction stresses, which will profoundly affect slip of nonplanar faults. We infer that in the brittle seismogenic portion of the Earth's crust, off-fault yielding occurs on pervasive secondary faults. Predicted rates of stress relaxation with distance from major faults with random fractal roughness follow a power-law relationship that is consistent with reported clustering of background seismicity up to 15 kilometers from faults.

Key words: Faults, fault mechanics, seismicity, fault roughness, damage, fractal, fault slip.

1. Introduction

Major faults in the brittle crust are complex structures consisting of a variety of components and geometric features. Outside the narrow zone of highly sheared fault gouge and ultra-cataclasite that defines the fault proper are extended regions of damage in the host rock mass (CHESTER *et al.*, 1993; 1996). The damage characteristically decreases in density with increasing distance from faults and consists of secondary faults, mesoscale fractures and microfractures (CHESTER and CHESTER, 1998; CHESTER, 1999). Also, faults are not geometrically flat surfaces and they do not exist in isolation, but form branching structures and networks. These features are evident over a wide range of length scales and have fractal-like characteristics. The waviness and roughness of individual faults can be approximately represented as mated surfaces with random fractal topographies (SCHOLZ and AVILES, 1986; POWER and TULLIS, 1991; SAGY *et al.*, 2007; CANDELA *et al.*, 2009). Fault step-overs have fractal characteristics (OKUBO and AKI, 1987) as well as fault system geometry (KING, 1983; BONNET *et al.*, 2001; BEN-ZION and SAMMIS, 2003). Mature fault networks appear to consist of a continuum of fractures at all scales of

Department of Earth Sciences, University of California Riverside, Riverside, CA 92521, U.S.A.
E-mail: dieterichj@ucr.edu

observation from major faults through secondary faults to local fractures and microcracking. Indeed, there is a hierarchy of structures whereby the secondary faults that comprise the damage zone for a major fault in turn are associated with their own damage zones with increased density of lesser faults and fractures, and so on to smaller length scales.

Previous studies that have examined various aspects of the interplay between fault slip and fault geometry include investigation of slip of wavy faults (SAUCIER *et al.*, 1992; CHESTER and CHESTER, 2000) and slip through idealized fault bends (NIELSEN and KNOPOFF, 1998). The importance of yielding to accommodate slip past geometric complexities was recognized by KING and ABELE (1985) and KING (1986), and to avoid pathological stress conditions NIELSEN and KNOPOFF (1998) and DUAN and OGLESBY (2005; 2006) implemented viscoelastic stress relaxation processes into their simulations of rupture through fault bends and across fault step-overs.

Here we examine slip of fault surfaces with a random fractal roughness. Our focus is on the interactions among slip, off-fault stressing resulting from geometric incompatibilities during slip, and their possible relation to off-fault damage. Particular attention is directed to investigating models with large numbers of fault elements to permit representation of geometric complexity over a wide range of spatial scales. Although the model geometry is much simpler than the complicated network geometries encountered in nature, it incorporates the types of interactions and the range of scales that characterize more complicated fault systems.

2. *Model*

2.1. *Fault Geometry*

The simulations employ faults with surface profiles that have a random fractal roughness, which are generated using the random mid-point displacement method (FOURNIER *et al.*, 1982; PEITGEN *et al.*, 1992). The mean amplitude of deviations from a planar surface at some length scale is given by

$$h = \beta x^H, \tag{1}$$

where x is a length interval along the fault and β is the rms slope, at the reference length $x = 1$ relative to the local fault orientation. For self-affine fractal surfaces, the exponent H can take values from 0–1. Profiles are self-similar when $H = 1$. For natural fault surfaces the exponent H takes values in the range 0.5–1.0 (BROWN and SCHOLZ, 1985; OKUBO and AKI, 1987; POWER and TULLIS, 1991; SAGY *et al.*, 2007). From fault maps we estimate that β generally falls in the range 0.02–0.10 at length scales $L \geq 1$ km. Figure 1 illustrates fault profiles with different values of β. Tests were also conducted using fault profiles created with a power-law filter method that is somewhat more accurate than the midpoint displacement method in generating power-law distributions. However, the test calculations

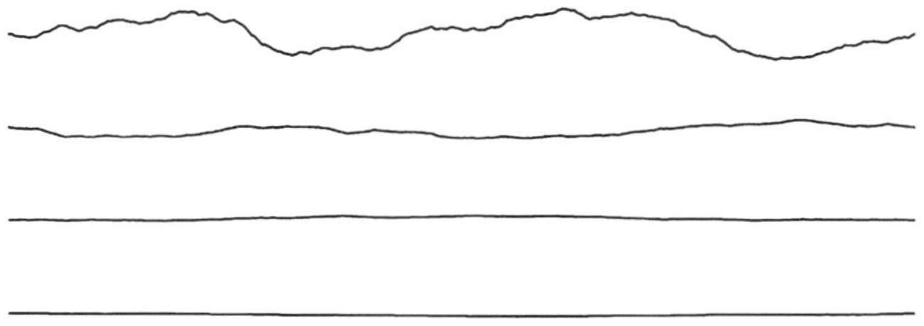

Figure 1
Examples of random fractal fault profiles generated with $H = 1.0$.

indicate that the results presented here are quite insensitive to the method used to generate fault profiles.

The dimensions of the fault elements used in the computations set the minimum length scale that can be represented as a random fractal topography. This study examines effects arising from fault geometry, but does not address issues relating to changes of configuration caused by slip. Consequently, we assume slip is small compared to the fault element size. If slip exceeds element length, irregularities move over one another, and other sliding processes come into play that fall in the realm of friction mechanics.

2.2. *Formulation of Elastic Model*

We employ the boundary element method in plane strain to compute stress changes and slip along the modeled faults. The calculations assume elastic quasi-static deformation of a homogeneous elastic medium. During slip shear and normal stress, τ and σ respectively, at the center of each fault element are related by

$$\tau = \mu\sigma, \tag{2}$$

where μ is the coefficient of friction. Hence, the Coulomb stress

$$S = \tau - \mu\sigma \tag{3}$$

is set to zero during slip. All results reported here were obtained with $\mu = 0.6$.

Shear and normal stress at the center of each fault element due to fault slip anywhere in the model are found by summing the contributions from each element,

$$\tau_i = T_{ij}d_j,\ \sigma_i = N_{ij}d_j, \quad \text{and}, \quad i,j = 1,2,\ldots,n, \tag{4}$$

where (4) employs the summation convention, n is the number of fault elements, d_j is the array of segment slips, and the coefficients T_{ij} and N_{ij} are obtained from the solutions for the elastic dislocations using the relevant distance and segment orientation information. The method used to generate the interaction matrices is based on DIETERICH (1992).

The change of Coulomb stress, resolved on the fault, arising from fault slip is obtained by combining (3) and (4)

$$S_{F_i} = C_{ij} d_j, \quad \text{where} \quad C_{ij} = T_{ij} - \mu N_{ij}. \tag{5}$$

In this model, slip occurs in response to a uniform externally applied load which changes the Coulomb stress on each fault element S_{E_i}. Hence, to model the slip that occurs in response to the stressing increment, the total Coulomb stress on each element from external loading and elastic interactions of fault elements is zero ($S_i = S_{E_i} + S_{F_i} = 0$). This condition with equation (5) gives

$$-S_{E_i} = C_{ij} d_j. \tag{6}$$

The system of equations (6) is solved for the unknown fault slip. In all calculations described below slip is driven by a spatially uniform increment of shear stress on a plane parallel to the overall orientation of the fault. Hence, for a fault oriented perpendicular to the y-coordinate with slip parallel to the x-coordinate $\Delta\sigma_{\mathrm{xy}} \neq 0$, $\Delta\sigma_{\mathrm{yy}} = 0$ and the Coulomb stress that drives fault slip is $S_E = \Delta\sigma_{\mathrm{xy}}$ for planes parallel to the overall orientation of the fault. However, the resolved Coulomb stress on a fault element, S_{E_i} depends on orientation of the element relative to the overall fault orientation.

In the following we present results for slip and fault length in terms of the normalized quantities

$$L' = L/l \quad \text{and} \quad d' = \frac{Gd}{S_E l}, \tag{7}$$

where L' and d' are normalized fault rupture length and slip (at midpoint of the rupture), respectively; L and d are unnormalized rupture length and slip, respectively; and l is the minimum length for resolution of the fractal geometry, which is generally the boundary element length. Hence, L' is the ratio of outer to inner scale lengths that define the fractal geometry. Calculations were also conducted with renderings of the fractal roughness at truncation lengths 2 to 16 times larger than the element dimensions. In all cases essentially identical results were obtained for normalized slip, which confirms the appropriateness of normalizing slip by inner scale length l. Because the model employs linear elasticity, displacements scale as the ratio of change of elastic stresses to elastic modulus.

3. Results

A rather extensive series of calculations (>30,000) were carried out to quantify the effects of fractal geometry on fault slip and stress interactions. We find that sliding characteristics and stress interactions of fractally rough faults in an elastic medium differ both qualitatively and quantitatively from those of faults with mathematically planar surfaces. Examples of slip and stress change on planar and rough faults in plane strain are compared in Figure 2. The results for the planar fault closely agree with the analytic solution of STARR (1928)—slip results in a uniform decrease of shear stress that equals the driving stress increment, and slip follows an elliptical distribution with maximum slip in the center

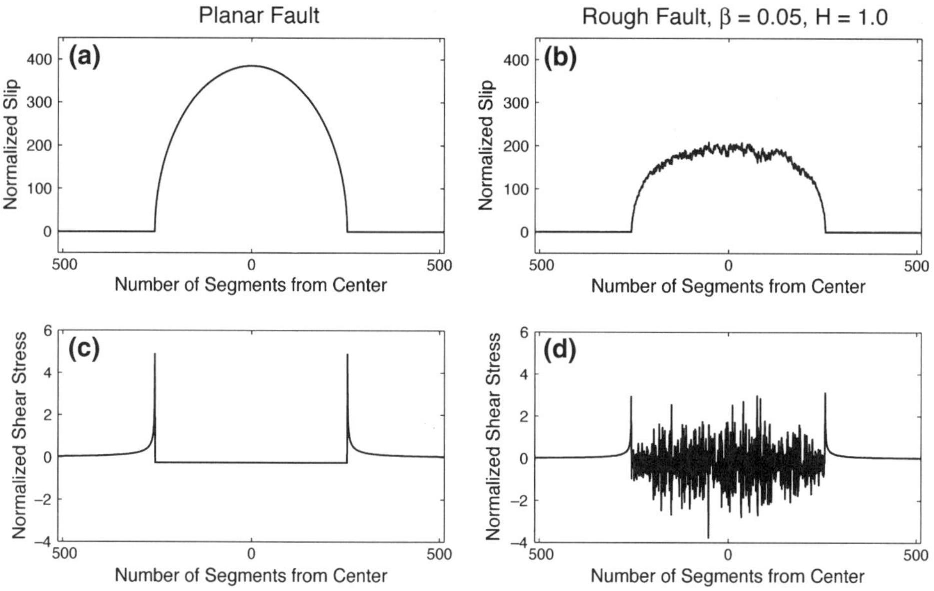

Figure 2
Examples of slip and shear stress change with planar and fractal faults. The generating parameters for the fractal fault are $H = 1$ and $\beta = 0.05$. The driving stress for slip in each model is $S_E/G = 1$. (a) and (b) normalized fault slip d' for planar and fractal faults, respectively. (c) and (d) shear stress change divided by G for planar and fractal faults, respectively.

$$d = \frac{S_E L}{\eta G}, \tag{8}$$

where L is the length of the fault segment that slips, G is the shear modulus (Poisson's ratio set at 0.25) and $\eta = 4/3$. In contrast, the fractal fault has a highly heterogeneous shear-stress drop and slip distribution. Additionally, the average total slip of the fractal fault is less than that of the planar fault.

Figure 3 illustrates certain characteristics of the stress changes that occur in response to slip of both planar and rough faults. Off-fault stress changes are obtained using the elastic dislocation solutions in a manner analogous to the calculation of the interaction coefficients in equation (5). In these examples, the amplitude and distribution of slip along the fault were specified to enable direct comparison of stress patterns in the planar and fractal fault models. In the case of slip on a fault patch (Figs. 3a and 3b), identical elliptical distributions of slip as given by the solution of STARR (1928) were specified for both the planar and fractal fault models. In both of those models slip generated similar overall patterns of stress change, except in the region adjacent to the slipped portion of the fault where highly heterogeneous stresses. The models for uniform global slip with planar and fractal geometries generate fundamentally different stress patterns. Uniform slip on a planar fault in an infinite medium (Fig. 3c) results in no change of stress, while

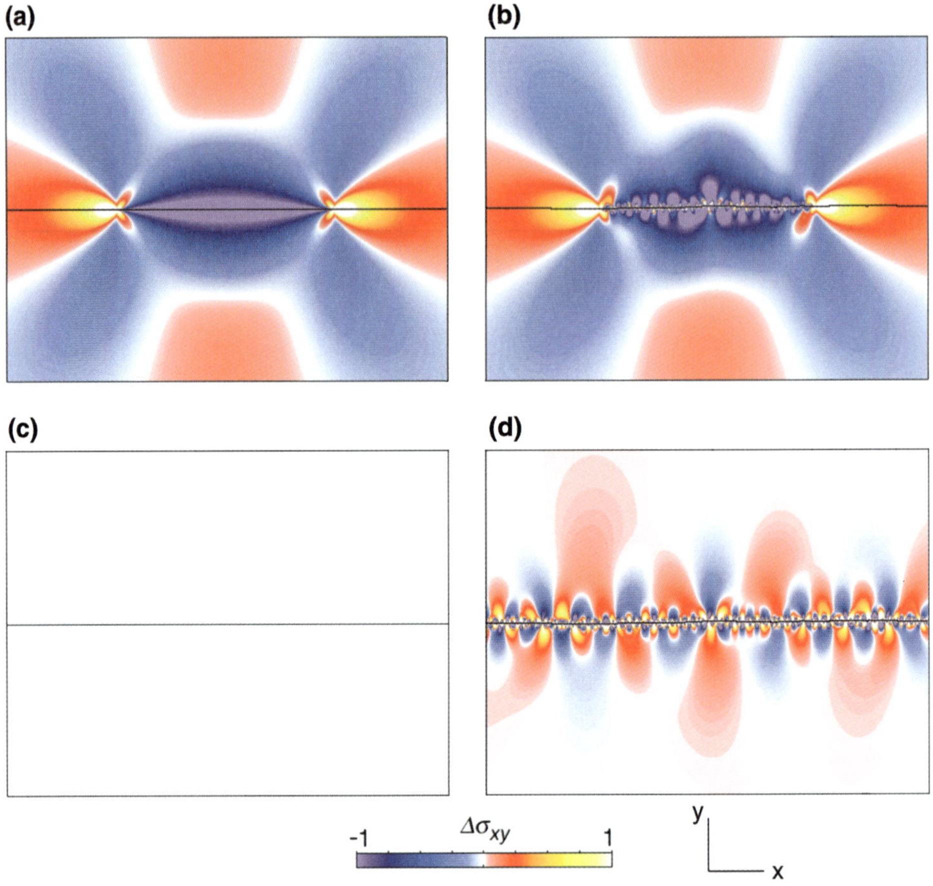

Figure 3
Shear stress changes (arbitrary scale) resulting from slip of planar fault (a) and (c) and fault with random fractal roughness (b) and (d). Fractal fault parameters are $H = 1.0$ and $\beta = 0.02$. (a) and (b) stress changes resulting from elliptical distribution of slip on confined fault segment. (c) and (d) stress changes resulting from uniform slip.

uniform slip of the fractal fault (Fig. 3d) results in highly heterogeneous stresses that show alternating regions of stress increase and decrease at different scales that reflect the random fractal character of the fault geometry.

The differences in stress patterns illustrated in Figure 3 reflect important underlying differences in the interactions that control slip of planar and nonplanar faults. Slip of planar faults can entirely restore the system to the stress state that existed prior to the increment of external stressing. However, slip of nonplanar faults cannot restore the system to the prior state. Although slip of the nonplanar fault returns fault stress to an equilibrium state with respect to the coefficient of friction, slip through geometric irregularities creates a highly heterogeneous state of stress. Those stresses grow with increasing slip to maximum levels that are set by β. The effects of the growth of

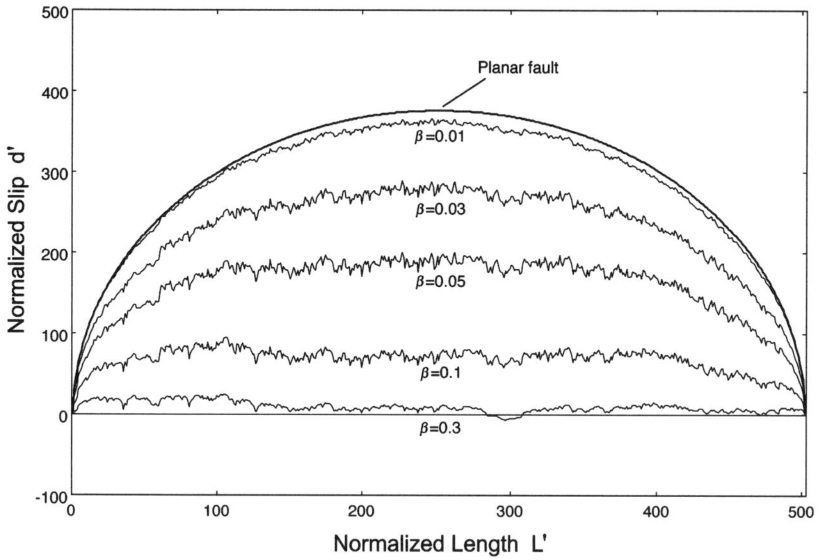

Figure 4
Effect of fault roughness amplitude factor β on slip of random fractal faults. In each calculation the same random fault profile was used with different roughness amplitudes. The fault profiles for the examples illustrated here were generated with $H = 1$. Equivalent results were obtained with $H = 0.5$ and 0.75. To satisfy the Coulomb stress condition, at $\beta = 0.3$ the model solution requires negative slip in one region. A positivity criterion in the slip solution would result in a locked fault in this region.

heterogeneous stresses on system behavior are discussed in greater detail below, following presentation of results for fault slip.

The effect of the amplitude of fault roughness on slip was explored in a series of calculations, wherein the roughness amplitude factor β was systematically varied with fault length held fixed. As one might expect, slip decreases with increasing β. Figure 4 presents results from one such set of calculations with $H = 1.0$. Similar results were obtained with $H = 0.5$, 0.75, and 1.0 and with total number of fault elements from $N = 500$ to $N = 8192$. The number of fault elements in this example is fixed at $N = 500$, and the random profiles have identical shapes but with different amplitudes as set by β. Figure 4 illustrates a number of features that are not observed in the slip of a planar fault surface. First, total slip decreases with increasing roughness. Fault slip with very low roughness ($\beta = 0.01$) deviates only slightly from that of the planar fault. However, with increasing β, total slip decreases and at $\beta = 0.3$, slip is greatly impeded. Second, fluctuations of slip amplitude occur at all wavelengths. The amplitudes of the slip fluctuations when normalized by total slip are roughly proportional to β. Third, with increasing β, the overall shape of the slip distribution along the fault progressively departs from the elliptical shape function of the planar fault. At high values of β, the average slip along the central portion of the fault flattens to a constant value.

Several series of calculations were also carried out wherein the total length of the fault that was allowed to slip L' was varied while element dimensions and amplitude of

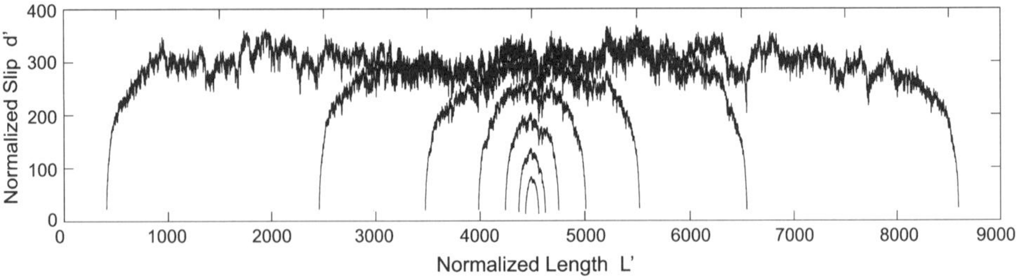

Figure 5
Effect of the ratio of outer to inner scales (L') on slip of random fractal faults. In these simulations the fault geometry was fixed ($H = 1.0$, $\beta = 0.05$) and fault sections of different length were allowed to slip. Equivalent results were obtained with $H = 0.5$ and 0.75.

the roughness β were fixed. Hence, in these simulations the ratio of outer to inner scales increases with the length of the section of the fault that is allowed to slip. The calculations use values of L' from 11 to 8192, β from 0.01 to 0.30 and H of 0.5, 0.75 and 1.0. An example of one such series is illustrated in Fig. 5. At sufficiently small L' the slip of fractal faults agrees rather well with that of a planar fault—the amplitude of slip increases approximately linearly with L' in rather good quantitative agreement with equation (8), and it has an elliptical distribution. However, with increasing L', slip diverges from the linear scaling of the planar fault, and the distribution of slip along the fault loses the elliptical shape characteristic of planar faults. Eventually, at a sufficiently large system size, slip attains a stable maximum value and no longer changes with L'. In summary, the effects of varying L' (Fig. 5) are comparable to the effects of varying β (Fig. 4). With increasing length (L') or roughness (β), the amplitude of slip increasingly deviates from linear scaling of planar faults, and the distribution of slip along the fault progressively flattens from an elliptical distribution to a constant value.

Figure 6 summarizes the results of calculations for slip over a range of values for L', β, with H fixed at 1.0. Similar results with slightly offset curves are obtained with $H = 0.50$ and $H = 0.75$. In Fig. 6, each data point for slip at the mid-point of the fault represents the average of $\geq$100 simulations that used different random number seeds to generate the random fractal profiles. For comparison, the slip as a function of L for a planar fault is plotted as well. Note that curves for slip as a function of L' obtained with different β have identical shapes and asymptotically approach the linear scaling curve of the planar faults at small L', and at large L' slip asymptotically approaches a fixed maximum value of slip. The normalized and unnormalized maximum slip scale by

$$d'_{\max} = c\beta^{-2},\ d_{\max} = c\beta^{-2} l \frac{S_E}{G}. \tag{9}$$

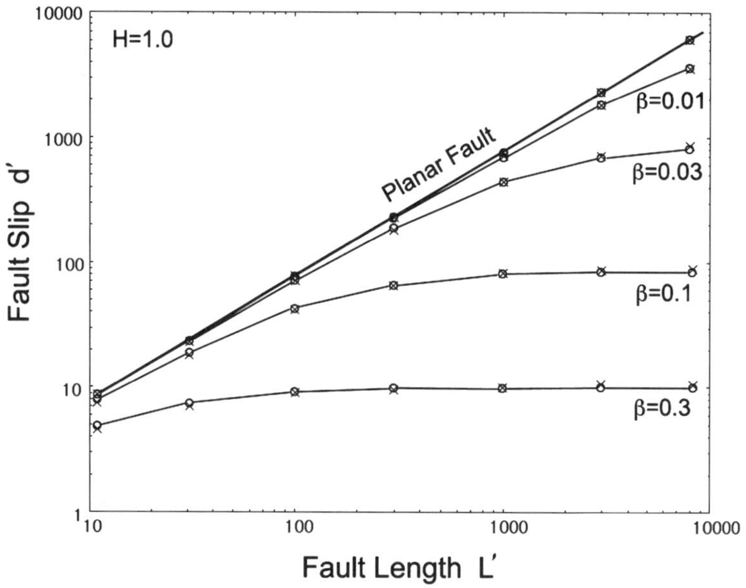

Figure 6
Summary plot of the effects of fault roughness and rupture length on fault slip (measured at the center of the rupture segment). The numerical data in this plot are derived from simulations of the type illustrated in Figure 5. Because the solutions for slip have considerable spatial variability arising from the randomness of the fault profiles, the numerical data (open circles) are the mean values of slip from 200 simulations with different realizations of the fault geometry. Results with H of 0.5, 0.75, and 1.0 differ only in slight shifts of the positions of the curves. The x symbols give values predicted by a simple analytic model described below.

The factor c, which is evaluated from the numerical results, has approximate values of $0.76 \pm .05$, $0.87 \pm .05$ and $0.91 \pm .05$, for $H = 0.50$, 0.75 and 1.0, respectively. The fault rupture length where slip reaches 90% of $d'_{\max}$ is

$$L'_{\max} \approx 1.2c\beta^{-2}. \tag{10}$$

Similarly, the rupture length at which linear scaling breaks down (slip is 90% of that of a planar fault under identical loading) is

$$L'_{\text{planar}} \approx 0.15c\beta^{-2}. \tag{11}$$

4. Discussion

4.1. Non-Linear Scaling of Slip

The reduction of slip in the fractal models relative to that of planar faults apparently occurs because geometric irregularities in fault orientation support stresses that oppose the applied Coulomb stress S_E that drives slip. Because stresses in the model arise from linear elastic interactions, those backstresses that oppose slip must increase linearly with slip. This indicates that the saturation of slip at $d_{\max}$ occurs when the average backstress reaches the applied stress S_E. That is

$$S_{back} = S_E \frac{d}{d_{\max}}, \tag{12}$$

where S_{back} is the spatial average of the backstress along the fault and d is slip at the center of the fault section. Combining equations (8) and (12) gives

$$S_{back} = \frac{G\beta^2}{cl} d. \tag{13}$$

If the proposal that elastic backstresses generated at geometric irregularities control the nonlinear scaling of slip is correct, then planar fault models that incorporate the average slip-dependent backstresses to simulate the average effect of the geometric interactions should have the same nonlinear scaling of slip as the rough fault models. An approximate solution for this problem can be obtained from the relationships for slip of a planar crack (equation 8) and dependence of backstress and slip (equation 13). Each of these effects involves linear changes of stress with slip and together are equivalent to a combined system stiffness linking driving stress and slip

$$d = S_E/(K_1 + K_2), \tag{14}$$

where K_1 is the effective stiffness of a freely slipping crack, which is given approximately by equation (8)

$$K_1 = \frac{S_E}{d} = \frac{\eta G}{L}, \tag{15}$$

and K_2 is the backstress effect of fault irregularities given by equation (13)

$$K_2 = \frac{G\beta^2}{cl}. \tag{16}$$

These yield the following approximate solution for slip at the center of a fault section as a function of length and roughness β

$$d = \frac{S_E/G}{\left(\frac{\eta}{L} + \frac{\beta^2}{cl}\right)}, \tag{17}$$

or in terms of normalized slip and length

$$d' = \frac{1}{\left(\frac{\eta}{L'} + \frac{\beta^2}{c}\right)}. \tag{18}$$

Hence, when L' is sufficiently small, the backstress term (K_2) is negligible compared to the K_1 term, and there is a near linear scaling of slip with L'. However, with continuing

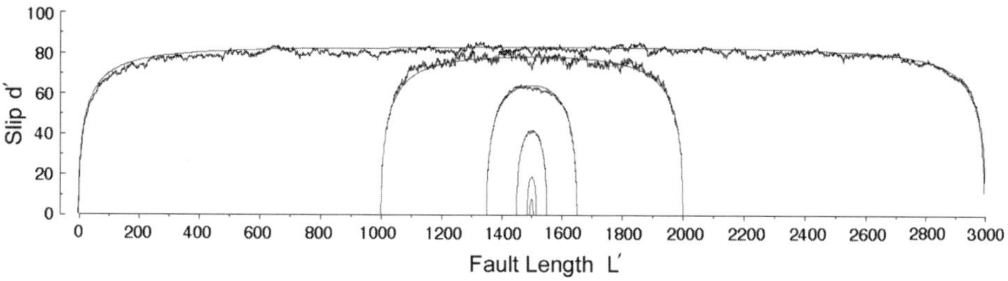

Figure 7
Comparison of slip on fractal fault (irregular curves) and numerical model of a planar fault with backstress (smooth curves). As in Figure 6 the fractal fault curves are averages of 200 simulations. The backstress parameter c is set at 0.91, which is the empirically evaluated value for the surface parameters in this set of simulations ($H = 1.0$, $\beta = 0.10$).

increase of L', the backstress eventually becomes non-negligible, and slip asymptotically approaches $d'_{\max}$. The accuracy of equations (17) and (18) is limited by the factor η, which is set by the shape of distribution of slip along the fault. As shown by Figure 5, the shape of the slip distribution changes from elliptical at small L to a more complex and flattened form at large L. However, a rather good fit of this model to the numerical results is obtained with $\eta = 4/3$, which is appropriate for the elliptical slip distribution that holds when the term with η dominates. Predictions of slip with this model are indicated by the x symbols in Figure 6.

Using a similar approach, boundary element models of planar faults with up to 8192 fault elements were implemented with the backstress equation (13) to simulate the average effect of fault roughness on slip. To solve for slip of a planar fault with backstress, the system of equations for slip (equation 6) is modified to

$$-S_{E_i} = C^*_{ij} d_j, \quad \text{where} \begin{array}{ll} C^*_{ij} = C_{ij} - \frac{G\beta^2}{cl}, & \text{if } i = j \\ C^*_{ij} = C_{ij}, & \text{if } i \neq j \end{array}. \tag{19}$$

As shown by Fig. 7, the simulations closely match both the amplitude of slip as a function of L' and the spatial distribution of slip obtained with the rough fault models. To facilitate comparisons between the planar fault with backstress and the fractal fault, the curves for slip as a function of position from the fractal fault simulations in Figure 7 represent averages of 200 realizations of the random fractal roughness.

4.2. *Stress Changes and Off-fault Damage*

The nonlinear scaling and saturation of slip as fault length increases (Figs. 5, 6, 7) are qualitatively different from anything one would predict from analysis of a single planar fault, and they are not readily apparent in coarsely resolved systems with low amplitude roughness. These characteristics have important practical implications for modeling slip of geometrically complex faults in elastic media. In particular, solutions for slip diverge as model resolution of fractal geometry increases. This effect arises because the backstress that opposes slip, increases as element length l decreases (equation 13), which results in ever-decreasing slip as resolution of geometric features increases. Also, the model is not consistent with the scaling of fault slip by length from earthquake observations, which appear to be well represented by linear elastic models of planar faults. Together, these characteristics indicate that the elastic model of slip as rendered in these computations is an incomplete representation of the stress interactions that control slip of faults in nature. We have demonstrated that nonlinear scaling and retardation of slip arise from backstresses that grow as slip increases. However, growth of stresses cannot occur without limit in real materials with finite strength. Hence, we conclude that yielding and stress relaxation are an essential process during slip of nonplanar faults and may significantly alter fault slip processes.

To avoid pathological stress conditions, NIELSEN and KNOPOFF (1998) and DUAN and OGLESBY (2005; 2006) implemented stress relaxation processes into their simulations of rupture through fault bends and across fault step-overs. Those studies assume a viscoelastic relaxation of stresses at the fault surface, which consists of a monotonic time-dependent exponential decay of fault stresses toward a prescribed background stress state.

However, in the upper portion of the Earth's crust where faulting dominates, we surmise that off-fault yielding will occur primarily through brittle failure processes. Consider for example a nearly planar surface with local deviations from planarity at some wavelength of one part per hundred ($\beta = 0.01$). Slip of a mated fault over those irregularities will generate shear strains on the order of 0.01. Assuming a shear modulus of 30 GPa, the resulting shear stresses will be on the order of 300 MPa, which equals or exceeds the brittle failure strength of rocks at 10 km depth. Because faults universally appear to have waviness with slope deviations well in excess of 0.01 at the kilometer scale, we conclude that widespread off-fault damage by brittle failure must accompany fault slip throughout the depth range of brittle faulting in the crust. Hence, we conclude that nonplanar faults cannot exist in isolation in brittle rocks—faults will always be associated with a system of secondary fractures over a wide range of length scales. The damage will take the form of a system of secondary faults and fractures, and it will extend to distances that are proportional to the long-wavelength irregularities in the geometry of the primary fault.

Nonlinear scaling of slip arises as a consequence of the linear growth of backstresses with slip. Conversely, if backstresses reach the yield limit and do not change with additional slip, then linear scaling of slip with fault length (or width) should be restored. In a related study (DIETERICH *et al.*, 2008; SMITH and DIETERICH, 2008), we have implemented models of stress relaxation due to slip of secondary faults as a bulkstress relaxation process. The model of stress relaxation is based on laboratory observations of rate- and state-dependent fault friction, and it provides a means to incorporate time- and stress-dependencies that are consistent with earthquake laboratory observations of fault slip. We find that stress relaxation eliminates the various pathologic characteristics seen in purely elastic models with fractal faults, and because it halts the growth of backstress with slip, the scaling of slip with fault length is linear.

Some characteristics of the off-fault stress relaxation and associated damage can be assessed with the current model without explicitly modeling the secondary fractures. At any point in the model the state of stress is taken as the sum of tectonic stressing, stress interactions from slip of the explicitly modeled fault, and stress relaxation from continuing motion on the secondary system of fractures. We assume that once a system of brittle damage has formed, the sum of these stressing sources must balance everywhere to bring the long-term growth of stresses to a halt. That is

$$\dot{\sigma}_T + \dot{\sigma}_F + \dot{\sigma}_R = 0, \qquad (20)$$

where $\dot{\sigma}_T$ is the externally imposed tectonic stressing rate on the through-going fault, $\dot{\sigma}_F$ is the long-term stressing rate from all fault interactions in the model, and $\dot{\sigma}_R$ is the long-term stress relaxation rate. Stresses will fluctuate with time due to intermittent fault slip, but the fluctuations will be around a stable long-term mean.

The long-term average of stress relaxation can be directly determined at any point from equation (20) if $\dot{\sigma}_E$ and $\dot{\sigma}_F$ are known. The externally imposed stressing rate $\dot{\sigma}_E$ is a model input, and $\dot{\sigma}_F$ is readily computed from the boundary element model given the long-term fault slip rate. Figure 8 illustrates results of such calculations for rate of off-fault stress relaxation as a function of distance from a rough fault. The rate of stress relaxation falls off rapidly with distance and obeys a power law of the form

$$\dot{S}_R = kG\beta y^{-(2-H)}, \qquad (21)$$

where y is distance from the fault normalized by l, and k is a constant that depends on fault slip rate and the constant c. Minimum distance y is set by inner scale length l. The scalar stress quantity plotted in Figure 8 is the spatial average of the Coulomb stress relaxation rate taken on optimally oriented plans at the normalized distance $y' = y/l$. Similarly, we observe that the absolute values of the components of the stress rate tensor for relaxation likewise exhibit the same power law fall-off as equation (21).

Rates of occurrence of aftershocks and background seismicity characteristically fall off with distance from major through-going faults and support this model. Studies of background seismicity in the vicinity of active faults in southern California (Powers and Jordan, 2008; Hauksson, in press 2009) find that numbers of earthquakes as a function distance from major faults follow a power-law decay to distances of 15 km with values of the decay exponent of around -1.5. Hauksson (2009) reports region-wide average values of the decay exponent of -1.73 for 13 fault segments with Quaternary slip rates $\geq$6 mm/yr and -1.38 for 95 fault segments with Quaternary slip rates $\leq$6 mm/yr. Both the power-law form of the distribution and the distances over which the effect operates are in broad agreement with our model. The average values of the distance decay exponents for seismicity reported in those studies are consistent with the predicted rates of stress relaxation from equation (21) with values from $H = 0.5$ to 1.0. However, some caution should be exercised in drawing detailed quantitative comparisons between the model and

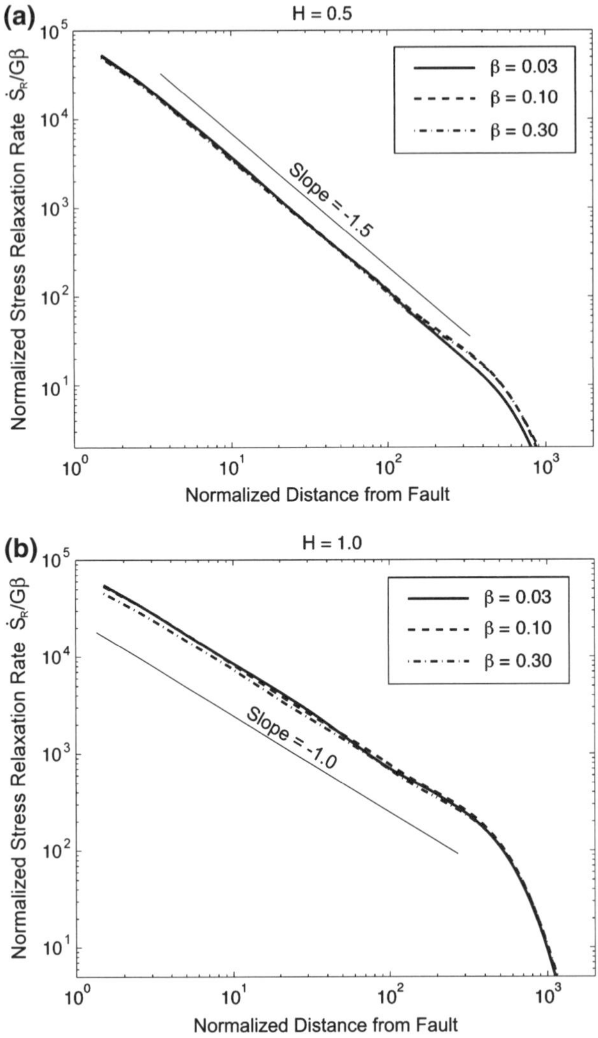

Figure 8

Long-term rate of stress relaxation as a function of normalized distance ($y' = y/l$) from faults with random fractal roughness with $H = 0.5$ (Fig. 8a) and $H = 1.0$ (Fig. 8b). Stress scale is arbitrary and depends on fault slip rate. The simulations employ a model of a periodic fault of length $L' = 1000$, with spatially uniform slip rate. Periodicities to $\pm 10\ L'$ were used to compute the interaction matrices of equation (4). The fall-off from power-law scaling beginning at normalized distances of $y' \approx 500$ is set by the outer scale length of the fractal fault geometry L'.

seismicity data. The model results presented here are based on 2D simulations, and depending on kilometer scale fault structures in the third dimension, the decay of stress relaxation rates with distance in the models may be described by a different exponent. Also, rates of stress relaxation may not scale linearly with earthquake rates if magnitude frequency distributions of earthquakes are functions of distance from the major faults.

5. *Conclusions*

In summary, computations with purely elastic models of fault slip without stress relaxation and subjected to uniform tectonic stressing have numerous characteristics that are not seen in planar fault models. These include divergence of slip solutions as resolution of fractal-like geometric features increases and breakdown of linear scaling of fault slip with fault dimensions. These effects arise from geometric interactions that can be represented as backstresses that increase with slip and progressively inhibit fault slip.

However, such stress cannot grow without limit. Yielding and stress relaxation are inevitable consequences of the deformation that arises in response to slip of faults with fractal roughness and finite material strength. We infer that in the brittle seismogenic portion of the Earth's crust yielding occurs on pervasive secondary faults.

A variety of different earthquake and faulting processes may cause off-fault damage in nature. First, off-fault damage may be a relic of the localization process that forms faults. Laboratory experiments (LOCKNER *et al.*, 1991; RECHES and LOCKNER, 1994) field observations (MARTEL *et al.*, 1988) and models (LYAKHOVSKY *et al.*, 1997; LYAKHOVSKY, 2001) indicate that fault formation operates through a process in which diffuse small-scale fractures progressively coalesce to form larger through-going faults. Second, damage may arise as a consequence of high dynamic stresses during earthquake slip. Such stresses are highly localized around the rupture front and appear to be sufficiently intense to create a damage zone in the vicinity of the tip of propagating earthquake rupture (ANDREWS, 2004; 2005; RICE *et al.*, 2005). Third, as demonstrated in this study, yielding will occur as a result of geometric incompatibilities that develop during slip of nonplanar faults. The effects discussed here are unavoidable consequences of fault geometry that operate to distances comparable to the long-wavelength features of fault geometry. Earthquake data compiled by POWERS and JORDAN (2008) and HAUKSSON (2009) demonstrate that stress relaxation through background seismicity extends to distances of at least 15 kilometers from major faults. The predicted power-law decay of the rate of stress relaxation with distance is broadly consistent with the seismicity data.

Acknowledgements

We thank Charlie Sammis for helpful suggestions. This material is based on work supported by the National Science Foundation under Grant No. EAR-0636064. The research was also supported by the southern California Earthquake Center (SCEC). SCEC is funded by NSF Cooperative Agreement EAR-0106924 and USGS Cooperative Agreement 02HQAG0008. The SCEC contribution number for this paper is 1280.

References

Andrews, D. (2004), *Rupture with dynamically determined breakdown displacement*, Bull. Seismol. Soc. Am. *94*, 769–775.

Andrews, D. (2005), *Rupture dynamics with energy loss outside the slip zone*, J. Geophys. Res.-Sol. Earth *110*, B01307.

Ben-Zion, Y. and Sammis, C. G. (2003), *Characterization of fault zones*, Pure Appl. Geophys. *160*, 677–715.

Bonnet, E., Bour,O., Odling, N. E., Davy, P., Main, I., Cowie, P., and Berkowitz, B. (2001), *Scaling of fracture systems in geological media*, Rev. Geophys. *39*, 347–383.

Brown, S. R. and Scholz, C. H. (1985), *Broad bandwidth study of the topography of natural rock surfaces*, J. Geophys. Res.-Sol. Earth Planets *90*, 2575–2582.

Caine, J., Evans, J., and Forster, C. (1996), *Fault zone architecture and permeability structure*, Geology *24*, 1025–1028.

Candela, T., Renard, F., Bouchon, M., Marsan, D., Schmittbuhl, J., and Voisin, C. (2009), *Characterization of fault roughness at various scales: Implications of three-dimensional high resolution topography measurement*, Tectonophysics.

Chester, F., *Field guide to the Punchbowl Fault Zone at Devil's Punchbowl Los Angeles County Park, California* (Texas A&M University, Texas 1999).

Chester, F. and Chester, J. (1998), *Ultracataclasite structure and friction processes of the Punchbowl fault, San Andreas system, California*, Tectonophysics *295*, 199–221.

Chester, *et al.* (1996).

Chester, F., Evans, J., and Biegel, R. (1993), *Internal structure and weakening mechanisms of the San-Andreas Fault*, J. Geophys. Res.-Sol. Earth *98*, 771–786.

Chester, F. M. and Chester, J. S. (2000), *Stress and deformation along wavy frictional faults*, J. Geophys. Res.-Sol. Earth *105*, 23421–23430.

Dieterich, J. H. (1992), *Earthquake nucleation on faults with rate-dependent and state-dependent strength*, Tectonophysics *211*, 115–134.

Dieterich, J. H., Richards-Dinger, K., and Smith, D. E. (2008), *Large-scale simulations of fault system seismicity*. Paper presented at 6th ACES International Workshop on *Earthquake Simulation*, Cairns, Australia, May 11–16, 2008.

Duan, B. and Oglesby, D. D. (2005), *Multicycle dynamics of nonplanar strike-slip faults*, J. Geophys. Res. *110*, B03304.

Duan, B. and Oglesby, D. D. (2006), *Heterogeneous fault stresses from previous earthquakes and the effect on dynamics of parallel strike slip earthquakes*, J. Geophys. Res. *111*, B05309.

Fournier, A., Fussell, D., and Carpenter, L. (1982), *Computer rendering of stochastic-models*, Communications of the ACM *25*, 371–384.

Hauksson, E. (2009), *Spatial Separation of Large Earthquakes, Aftershocks, and Background Seismicity: Analysis of Interseismic and Coseismic Seismicity Patterns in Southern California*, Pure Appl. Geophys., *Special Frank Evison Issue*, in press.

King, G. C. P. (1983), *The accommodation of large strains in the upper lithosphere of the Earth and other solids by self-similar fault systems – The geometrical origin of the b-value*, Pure Appl. Geophys. *121*, 761–815.

King, G. C. P. (1986), *Speculations on the geometry of the initiation and termination processes of earthquake rupture and its relation to morphology and geological structure*, Pure Appl. Geophys. *124*, 567–585.

King, G. C. P. and Nabélek, J. (1985), *Role of fault bends in the initiation and termination of earthquake rupture*, Science *228*, 984–987.

Lockner, D., Byerlee, J., Kuksenko, V., Ponomarev, A., and Sidorin, A. (1991), *Quasi-static fault growth and shear fracture energy in granite*, Nature *350*, 39–42.

Lyakhosvsky, V., Ben-Zion, Y., and Agnon, A. (1997), *Distributed damage, faulting, and friction*, J. Geophys. Res. *102*, 27635–27649.

Lyakhovsky, V. (2001), *Scaling of fracture length and distributed damage*, Geophys. J. Internatil. *144*, 114–122.

Martel, S., Pollard, D., and Segall, P. (1988), *Development of simple strike-slip fault zones, Mount Abbot quadrangle, Sierra-Nevada, California*, Geol. Soc. Am. Bull. *100*, 1451–1465.

Nielsen, S. B. and Knopoff, L. (1998), *The equivalent strength of geometrical barriers to earthquakes*, J. Geophys. Res.-Sol. Earth *103*, 9953–9965.

Okubo, P. G. and Aki, K. (1987), *Fractal geometry in the San-Andreas fault system*, J. Geophys. Res.-Sol. Earth Planets *92*, 345–355.

Peitgen, H., Jürgens, H., and Saupe, D., *Chaos and Fractals: New Frontiers of Science* (Springer-Verlag, New York 1992).

Power, W. L. and Tullis, T. E. (1991), *Euclidean and fractal models for the description of rock surface roughness*, J. Geophys. Res. *96*, 415–424.

Powers, P. M. and Jordan, T. H. (2008), *Distribution of seismicity across strike-slip faults in California, EOS Trans. AGU*, *89*, Fall Meet. Suppl., Abstract S21B-1831.

Reches, Z. and Lockner, D. (1994), *Nucleation and growth of faults in brittle rocks*, J. Geophys. Res.-Sol. Earth *99*, 18159–18173.

Rice, J., Sammis, C., and Parsons, R. (2005), *Off-fault secondary failure induced by a dynamic slip pulse*, Bull. Seismol. Soc. Am. *95*, 109–134.

Sagy, A., Brodsky, E., and Axen, G. (2007), *Evolution of fault-surface roughness with slip*, Geology, *35*, 283–286.

Saucier, F. E., Humphreys, E., and Weldon, R. (1992), *Stress near geometrically complex strike-slip faults — Application to the San-Andreas Fault at Cajon Pass, Southern California*, J. Geophys. Res.-Sol. Earth *97*, 5081–5094.

Scholz, C. H. and Aviles, C. A. (Eds.), *The Fractal Geometry of Faults and Faulting*, 147–156 pp. (Am. Geophys. Union, Washington D.C. 1986).

Smith, D. E. and Dieterich, J. H. (2008), *Rate-state modeling of stress relaxation in geometrically complex systems*. Paper presented at Seismol. Soc. Am. Annual Meeting, Santa Fe, NM, April 16–18, 2008.

Starr, A. (1928), *Slip in a crystal and rupture in a solid due to shear*, Proc. Cambridge Philosoph. Soc. *24*, 489–500.

(Received February 9, 2009, revised March 17, 2009)
Published Online First: August 1, 2009

To access this journal online:
www.birkhauser.ch/pageoph

Pure appl. geophys. 166 (2009) 1817–1851
0033–4553/09/101817–35
DOI 10.1007/s00024-009-0521-2

© Birkhäuser Verlag, Basel, 2009

Pure and Applied Geophysics

Characterization of Fault Roughness at Various Scales: Implications of Three-Dimensional High Resolution Topography Measurements

THIBAULT CANDELA,[1] FRANÇOIS RENARD,[1,2] MICHEL BOUCHON,[3] ALEXANDRE BROUSTE,[4] DAVID MARSAN,[5] JEAN SCHMITTBUHL,[6] and CHRISTOPHE VOISIN[3]

Abstract—Accurate description of the topography of active fault surfaces represents an important geophysical issue because this topography is strongly related to the stress distribution along fault planes, and therefore to processes implicated in earthquake nucleation, propagation, and arrest. To date, due to technical limitations, studies of natural fault roughness either performed using laboratory or field profilometers, were obtained mainly from 1-D profiles. With the recent development of Light Detection And Ranging (LIDAR) apparatus, it is now possible to measure accurately the 3-D topography of rough surfaces with a comparable resolution in all directions, both at field and laboratory scales. In the present study, we have investigated the scaling properties including possible anisotropy properties of several outcrops of two natural fault surfaces (Vuache strike-slip fault, France, and Magnola normal fault, Italy) in limestones. At the field scale, digital elevation models of the fault roughness were obtained over surfaces of 0.25 m^2 to 600 m^2 with a height resolution ranging from 0.5 mm to 20 mm. At the laboratory scale, the 3-D geometry was measured on two slip planes, using a laser profilometer with a spatial resolution of 20 μm and a height resolution less than 1 μm.

Several signal processing tools exist for analyzing the statistical properties of rough surfaces with self-affine properties. Among them we used six signal processing techniques: (i) the root-mean-squares correlation (RMS), (ii) the maximum-minimum height difference (MM), (iii) the correlation function (COR), (iv) the RMS correlation function (RMS-COR), (v) the Fourier power spectrum (FPS), and (vi) the wavelet power spectrum (WPS). To investigate quantitatively the reliability and accuracy of the different statistical methods, synthetic self-affine surfaces were generated with azimuthal variation of the scaling exponent, similar to that which is observed for natural fault surfaces. The accuracy of the signal processing techniques is assessed in terms of the difference between the "input" self-affine exponent used for the synthetic construction and the "output" exponent recovered by those different methods. Two kinds of biases have been identified: Artifacts inherent to data acquisition and intrinsic errors of the methods themselves. In the latter case, the statistical results of our parametric study provide a quantitative estimate of the dependence of the accuracy with system size and directional morphological anisotropy. Finally, based on this parametric study, we used the most reliable

[1] Laboratoire de Géodynamique des Chaînes Alpines, CNRS-OSUG, University Joseph Fourier - Grenoble I, BP 53, 38041 Grenoble, France. E-mail: Thibault.Candela@bvra.e.ujf-grenoble.fr; francois.renard@ujf-grenoble.fr

[2] Physics of Geological Processes, University of Oslo, Oslo, Norway.

[3] Laboratoire de Géophysique Interne et Tectonophysique, CNRS-OSUG, University Joseph Fourier - Grenoble I, Grenoble, France.

[4] Laboratoire Manceau de Mathématiques, CNRS, Université of Le Mans, Université du Maine, Le Mans, France. E-mail: Alexandre.Brouste@univ-lemans.fr

[5] Laboratoire de Géophysique Interne et Tectonophysique, CNRS, University of Savoie, Le Bourget du Lac, France. E-mail: david.marsan@univ-savoie.fr

[6] UMR 7516, Institut de Physique du Globe de Strasbourg, Strasbourg, France.

techniques (RMS-COR, FPS, WPS) to analyze field data. These three methods provide complementary results. The FPS and WPS methods determine a robust characterization of the fault surface roughness in the direction of striations and perpendicular to them. The RMS-COR method allows investigation of the azimuth dependence of the scaling exponent. For both field and laboratory data, the topography perpendicular to the slip direction displays a similar scaling exponent $H_{\perp} = 0.8$. However, our analysis indicates that for the Magnola fault surface the scaling roughness exponent parallel to the mechanical striation is identical at large and small scales $H_{//} = 0.6$–0.7, whereas for the Vuache fault surface it is characterized by two different self-affine regimes at small and large scales. We interpret this cross-over length scale as a witness of different mechanical processes responsible for the creation of fault topography at different spatial scales.

Key words: Fault, 3-D laser scanner, fault-surface roughness, self-affine surface, roughness exponent.

1. Introduction

Knowledge of the detailed fault geometry is essential to understand certain major processes involved in faulting such as grain comminution or asperities abrasion during slip, geometrical heterogeneity of the stress field that controls earthquake nucleation (Campillo *et al.*, 2001; Voisin *et al.*, 2002a), rupture propagation, and arrest (Voisin *et al.*, 2002b). Establishing correlations between geometrical properties of fault roughness (Power *et al.*, 1987, 1988; Power and Tullis, 1991; Schmittbuhl *et al.*, 1993; Lee and Bruhn, 1996; Power and Durham, 1997; Renard *et al.*, 2006; Sagy *et al.*, 2007), seismic behavior of faults (Okubo and Aki, 1987; Parson, 2008), frictional strength and critical slip distance (Scholz, 2002), wear processes during fault zone evolution (Power *et al.*, 1988) represent a fundamental issue to understand seismic faulting.

At the laboratory scale, Amitrano and Schmittbuhl (2002) highlight a complex coupling between fault gouge generation and fault roughness development. At larger scale, asperities control the slip distribution of earthquake (Peyrat *et al.*, 2004). Indeed asperities on active fault planes concentrate the stress (Marsan, 2006; Schmittbuhl *et al.*, 2006) and therefore may control earthquake nucleation (Lay *et al.*, 1982; Scholz, 2002) and the propagation of the rupture to its ultimate arrest (Aki, 1984). High resolution relocations of earthquakes using the multiplet technique have shown streaks of earthquake along several faults in California (Rubin *et al.*, 1999). This pattern has been interpreted as resulting from the presence of an organized large scale roughness (asperities) resisting slip (Schaff *et al.*, 2002).

Despite recent progress in seismology, the imaging of fault planes over a large range of scales at depth is not yet available. A quasi-unique access to high resolution description of the fault plane comes from exhumed fault scarp observations. This requires, of course, that the main morphological patterns of faults mapped at the surface of the Earth persist at least across the seismogenic zone. Due to technical limitations, prior comparative studies of natural fault roughness were mainly based on 1-D profilometry (Power *et al.*, 1987, 1988; Power and Tullis, 1991; Schmittbuhl *et al.*, 1993; Lee and Bruhn, 1996; Power and Durham, 1997). These 1-D measurements have shown that fault roughness can be characterized by a scale invariance property described by a self-affine geometry (see Section 2 for the definition of self-affinity) with a roughness scaling exponent close to 0.8 for profiles oriented in a direction perpendicular to the striations observed on the fault plane. Such striations are aligned in the direction of slip. The value of 0.8 is similar to what was measured for tensile cracks (POWER *et al.*, 1987; SCHMITTBUHL *et al.*, 1995b; BOUCHAUD, 1997). Moreover, the influence of slip was also quantified: The fault surfaces have smaller roughness amplitude along the slip direction than perpendicular to it (POWER *et al.*, 1988; POWER and TULLIS, 1991; LEE and BRUHN, 1996; POWER and DURHAM, 1997). The compiled fault roughness statistics of several studies (POWER and TULLIS, 1991; LEE and BRUHN, 1996; BEN-ZION and SAMMIS, 2003) suggest a change in scaling properties between large and short length scales. However, considering the noise in their data, these authors underlined that it was not possible to decipher whether this variation was related to small-scale surface weathering of the fault scarp or to the faulting process itself.

With the recent development of high resolution distance meters, it is now possible to use accurate statistical approaches to quantify fault roughness. Indeed, portable 3-D laser scanners (also called LiDAR, Light Detection And Ranging) allow mapping fault surface outcrops over scales of 0.5 mm to several tens of meters. The accuracy of the measurements enables a reliable quantification of the data. RENARD *et al.* (2006) and SAGY *et al.* (2007) demonstrated precisely the scaling invariance and anisotropy properties of fault topography using ground based LIDAR and laboratory profilometers. In these previous studies, statistical analysis of fault roughness was carried out with a single signal processing tool. However, SCHMITTBUHL *et al.* (1995a,b) recommend the simultaneous use of different methods in order to appreciate the confidence in the measured scaling invariance properties.

In the present study, we use new roughness data to extend the type of measurements made by RENARD *et al.* (2006) and SAGY *et al.* (2007) and propose a parametric study of the statistical results of fault topography, using multiple signal processing tools. In order to investigate the reliability and accuracy of the different signal processing methods, synthetic self-affine surfaces were generated with azimuthal variation of the scaling exponent. These synthetic rough surfaces are completely characterized by two different self-affine exponents prescribed in perpendicular directions. When comparing these synthetic surfaces to natural faults, one should keep in mind that any self-affine model can only describe a real surface to a finite degree of accuracy, and only within a finite range of scales. After this systematic study, we used the most reliable and accurate techniques to investigate the scaling properties and anisotropy of several outcrops of two natural fault surfaces that have been measured using 3-D laser scanners in the field and a laser profilometer in the laboratory.

This paper is organized as follows. In Section 2, following a brief explanatory discussion of the self-affine notion, the generation process of synthetic self-affine surfaces with a directional morphological anisotropy is described. In Section 3, statistical methods to define the self-affine properties are reviewed. Section 4 is devoted to the systematic study of the accuracy of the methods. Section 5 is devoted to the acquisition of data on natural fault. In Section 6, analysis of the roughness, covering six orders of magnitude of length scales, is performed on several outcrops of two natural faults. Finally, in Section 7, we conclude by linking our findings on the statistical properties of natural fault topography to the results of earlier studies, with the ultimate goal of developing a more mutually consistent description of fault asperities geometry.

2. *Generation of Self-affine Surfaces*

2.1. *Self-affinity in 1-D and 2-D*

Surface roughness analysis provides a statistical characterization of a surface which is simpler and easier to use than a complete deterministic description. In geophysics, BROWN and SCHOLZ (1985) and POWER *et al.* (1987) demonstrated the self-similar property of natural fault surfaces at field scale. Coming from statistical physics, a more general scaling transformation has been proposed: self-affinity (MANDELBROT, 1985, 1986; VOSS, 1985) that was successfully used for the quantitative description of fault roughness (SCHMITTBUHL *et al.*, 1993, RENARD *et al.*, 2006).

A self-affine 1-D profile remains unchanged under the scaling transformation $\delta x \rightarrow \lambda \delta x$, $\delta z \rightarrow \lambda^H \delta z$ for 1-D profiles (Fig. 1) extracted from a surface (MEAKIN, 1998). Here, δx is the coordinate along the profile and δz is the roughness amplitude. For a self-affine profile, the scaling exponent H, also called Hurst exponent, lies in the range $0 \leq H < 1$. Accordingly, self-affinity implies that a profile appears less rough as the scale increases. In other words, if a profile is self-affine, a magnified portion of it will appear statistically identical to the entire profile if different magnifications are used in the x and z directions (Fig. 1).

For 2-D surfaces, this self-affinity property can be described for sets of 1-D parallel profiles extracted from the surface. Moreover, if the surface is striated along some given orientation, anisotropic scaling behavior can emerge if H varies for different directions in the plane of the surface. An anisotropic self-affine surface $Z(x, y)$ with coordinates (x, y) obeys the property: $Z\left(\lambda^{1/H_{//}}x, \lambda^{1/H_{\perp}}y\right) = \lambda Z(x,y)$, where λ is a positive dilation factor, $H_{//}$ and $H_{\perp}$ are the Hurst exponents, comprised between 0 and 1, in two perpendicular directions of the surface. $H_{//}$ is defined along a direction parallel to the main striations, and $H_{\perp}$ is defined along a direction perpendicular to the striation (Fig. 1b).

2.2. *Synthetic Anisotropic Self-affine Surfaces*

To calculate synthetic fault surfaces (Fig. 1b), we used a Fourier based method to simulate a matrix scaling random Gaussian field on a 2-D grid, where an anisotropy matrix $E = \begin{pmatrix} 1/H_{//} & 0 \\ 0 & 1/H_{\perp} \end{pmatrix}$ was introduced when calculating the 2-D Gaussian random field. The eigenvalues of this matrix correspond to the inverse of the two roughness exponents $H_{//}$, and $H_{\perp}$ that characterize the self-affine properties of the

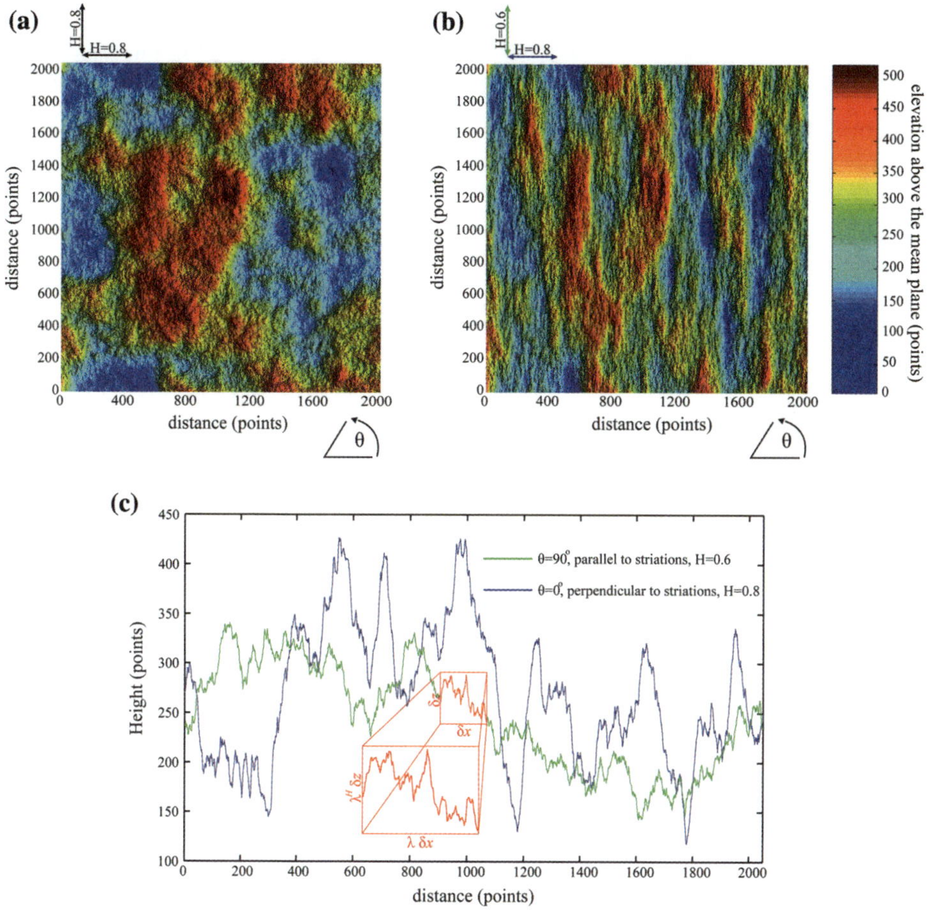

Figure 1

Digital Elevation Models (DEM) of 2D synthetic self-affine surfaces (up) and 1-D profiles (down) generated using the algorithm of the appendix. (a) Surface with an isotropic self-affine property characterized by a Hurst exponent of 0.8. (b) Anisotropic self-affine surface with two Hurst exponents ($H_{//} = 0.6$ and $H_{\perp} = 0.8$) in perpendicular directions. (c) Representative 1-D profiles extracted in two perpendicular directions of surface (b). Inset: magnified portion of a profile along the $H_{//}$ direction (parallel to the striations), which has a statistically similar appearance to the entire profile when using a rescaling $\delta x \to \lambda\, \delta x$, $\delta z \to \lambda^H \delta z$.

generated surface in two perpendicular directions (BIERME *et al.* 2007; CLAUSEL and VEDEL, 2008). The code to generate an anisotropic 2-D self-affine surface, written in Matlab©, is given in Appendix A and can be run easily on a desktop computer.

In the following sections, we decompose the signal processing analysis of rough surfaces in two stages. Firstly, we present the six signal processing tools used to estimate the self-affine property of an isotropic surface with a single Hurst exponent (Fig. 1a), as observed for example for fresh mode I brittle fractures in rocks (POWER *et al.*, 1987;

Schmittbuhl *et al.*, 1995b; Bouchaud, 1997). For this, we have synthesized several isotropic surfaces with an exponent in the range [0.1–0.9] and grid sizes in geometrical series: 129 × 129 points, 513 × 513 points, 2049 × 2049 points. Secondly, we analyze synthetic anisotropic surfaces (Fig. 1b) with $H_{//}$ in the range [0.7–0.9] and $H_{\perp}$ in the range [0.4–0.9].

3. Statistical Signal Processing Methods

We have used six different methods that characterize the amplitude of the roughness at various spatial wavelengths. All these methods, presented in the following subsections, are based on the analysis of 1-D profiles (Fig. 1c) that are extracted from the 2-D Digital Elevation Model (DEM) of 2-D surfaces (Fig. 1a, b). For each surface, a set of 1-D parallel profiles in a specific direction are extracted, detrended and then analyzed. Then, the properties are averaged over all the 1-D profiles to characterize the 2-D surface in the chosen direction. To study the azimuthal dependence of the statistical properties of the surfaces, we have extracted profiles in several directions, following a 360° rotation. Estimating the statistical properties of the surface in various directions (Renard *et al.*, 2006) allows characterizing a morphological anisotropy.

For the application to natural fault surfaces, we also tested how the noise in the data and the presence of missing points could affect the estimation of fault surface. Indeed, the raw scanner data consist of clouds of points, with x, y, and z coordinates, sampled more or less regularly. Sometimes, data are missing (vegetation on the fault plane, low reflectivity of the scanner light beam), and the surface is incomplete. An interpolation is then necessary, which induces a bias in the estimation of scaling exponents that also need to be estimated.

3.1. Root-mean-squares Correlation (RMS) and Maximum-minimum Height Difference (MM) Methods

Consider a 1-D profile $L(x)$ that is divided into windows of width δx and indexed by the position of the first point x_0 of the band. The standard deviation $\sigma(\delta x)$ of the height $L(x)$ and the height difference $h(\delta x)$ between the maximum and minimum height are computed for each band, and then averaged over all the possible bands of fixed width δx spanning the profile, by varying the origin x_0. We then obtain $\langle\sigma(\delta x)\rangle$ and $\langle h(\delta x)\rangle$, where both quantities follow a power law for a self-affine profile: $\langle\sigma(\delta x)\rangle \propto \delta x^H$ and $\langle h(\delta x)\rangle \propto \delta x^H$ (Schmittbuhl *et al.* 1995a).

Note that this technique is useful when H is not too close to 0 or 1, where a significant error can be measured (see Figs. 3a, b – 4a, b). Usually, levelling off of $\sigma(\delta x)$ at small δx values is due to the noise in the data (see Figs. 7c, d), and leveling-off at large δx is due to the finite size of the profile.

3.2. Height-height Correlation Function (COR) Method

For a signal $L(x)$, we consider the height-height correlation function defined by $C(\Delta x) = \left[\left\langle (L(x) - L(x + \Delta x))^2 \right\rangle\right]^{1/2}$, which estimates the average height-height difference between two points of the profile separated by a distance Δx. For a self-affine profile, the correlation function follows a powerlaw such that $C(\Delta x) \propto \Delta x^H$ where H is the Hurst exponent.

3.3. Standard Deviation of the Correlation Function (RMS-COR) Method

For a profile $L(x)$ containing N points, the height difference ΔL between each pair of points separated by a distance Δx is calculated. The window size Δx is varied between the sampling distance and the size of the system and, for a given Δx, the standard deviation of the height difference $\sigma(\Delta L_{\Delta x})$ is calculated. For a self-affine surface this measurement follows a powerlaw such that $\sigma(\Delta x) \propto \Delta x^H$. This method was successfully applied to characterize the self-affine properties of the Vuache fault plane (RENARD *et al.* 2006).

3.4. Fourier Power Spectrum (FPS) Method

The Hurst exponent H can be estimated from the Fourier power spectrum which has a power law form for a 1-D self-affine profile (BARABASI and STANLEY, 1995; MEAKIN, 1998). For each parallel profile, the Fourier power spectrum $P(k)$, *i.e.*, the square of the modulus of the Fourier transform, is calculated as a function of the wavenumber k. Then the spectrum of the entire surface is calculated by stacking all 1-D Fourier transforms to reduce the noise associated with individual profiles. For each profile of length L containing N increments, the spatial frequencies range between $1/L$ and the Nyquist frequency $N/2L$ (*i.e.*, the reciprocal of the interval between data points). When plotting the power spectrum as a function of k in log-log space, a self-affine function reveals a linear slope, which is itself a function of the Hurst exponent H through $P(k) \propto k^{-1-2H}$.

3.5. Average Wavelet Coefficient Power Spectrum (WPS) Method

The average wavelet coefficient method consists of decomposing the input signal into amplitudes that depend on position and scale. The wavelet transform of each 1-D profile $L(x)$ is defined as $W_{a,b} = \frac{1}{\sqrt{a}} \int_{-\infty}^{+\infty} \psi\left(\frac{x-b}{a}\right) |L(x)|\, dx$ where ψ is the wavelet function. Then the wavelet coefficients are averaged over the translation factor b for each length scale a: $W_a = \langle W_{a,b} \rangle_b$. If the profile is self-affine, the wavelet transform verifies statistically that, for any positive dilatation factor λ, $W_{a,b}[L(\lambda x)] = \lambda^H W_{a,b}$. Accordingly, the averaged wavelet coefficients scale as $W_a \propto a^{H+1/2}$. A wide range of wavelet functions can be used. For a simple and efficient implementation, we chose the Daubechies wavelet of order 12 as suggested in SIMONSEN *et al.* (1998).

4. Quantitative Estimation of the Accuracy of Roughness Analysis Methods

4.1. Synthetic Isotropic and Anisotropic Rough Surfaces

Figures 1a and 1b display the topography of synthetic rough surfaces where the data set includes 2049 × 2049 points regularly spaced on a grid. Figure 1a shows an isotropic rough surface, whereas Figure 1b shows an anisotropic surface, with corrugations elongated parallel to the direction of smaller Hurst exponent (analogue to the direction of slip on a natural fault surface) and covering a wide range of scales.

The roughness amplitude of a profile parallel to the striation direction (green curve in Fig. 1c) is smaller than that of a perpendicular profile. The profile extracted along the direction with the smallest exponent (green curve) appears more jagged at small scales compared to a perpendicular profile, showing the different effects of the anisotropy of the surface on the waviness and amplitude of the profiles.

The outputs of the statistical methods described in Section 2.3 are represented on Figure 2. Each curve is calculated by averaging the outputs of all possible parallel 1-D profiles extracted from the anisotropic surface of Figure 1b. The results are represented in a log-log plot, allowing visualizion of the linear portion of the curve that characterizes a power-law distribution (Fig. 2). This linear portion is binned in a small number of increments, and a power-law fit is performed to extract the Hurst exponent that characterizes the self-affinity of the profile. The best fits are performed for each curve and a value of the "output" self-affine exponent is then calculated for all the six signal processing methods.

Using the RMS correlation function, we have also extracted sets of parallel profiles in several directions, at an angle θ to the direction of the striations. For each set of profile, we have calculated the correlation function and estimated the value of H. The angular dependence of H could be represented on a polar plot (inset in Fig. 2d) (Renard *et al.*, 2006). The anisotropy of such plot characterizes the anisotropy of the surface: An isotropic surface is represented as a circle of radius H, whereas an anisotropic one has a more complex elliptical shape.

4.2. Isotropic Surfaces: Effect of Size and Input Exponent on the Output Estimation of the Hurst Exponent

The comparison between the input Hurst exponent used to calculate an isotropic synthetic surface and the output Hurst exponent estimated using the six different methods is represented on Figure 3, for different system sizes. The RMS, MM, COR and RMS-COR methods are all mainly sensitive to the value of the input self-affine exponent (the typical trend of the curve is not parallel to the diagonal). Small self-affine exponents are systematically overestimated whereas large exponents are underestimated. In contrast, the error for the WPS method is mainly a function of the system size (the response is more or less parallel to the diagonal). The FPS method appears the most accurate

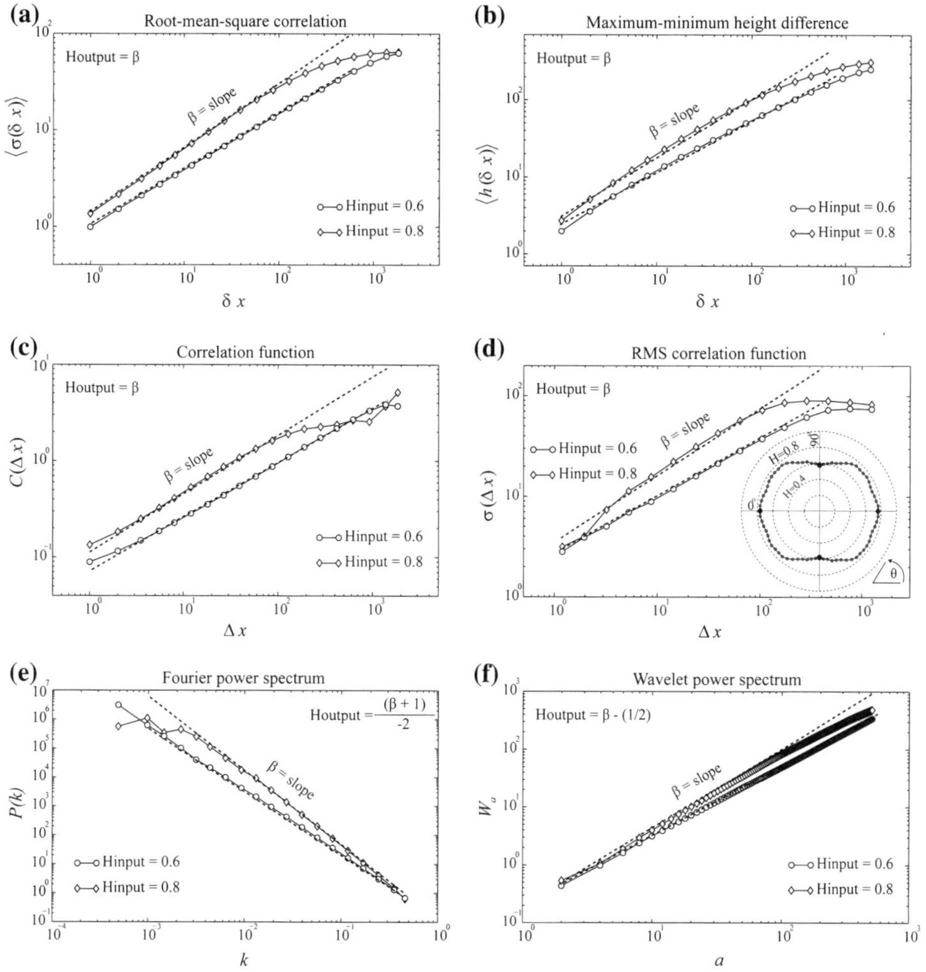

Figure 2

Outputs of the six signal processing techniques applied on the data of the anisotropic self-affine surface shown in Figure 1b. (a) Root-mean-squares correlation (RMS), (b) maximum-minimum height difference (MM), (c) correlation function (COR), (d) RMS correlation function (RMS-COR), (e) Fourier power spectrum (FPS), (f) Wavelet power spectrum (WPS). The dashed lines represent fits of the curves to obtain the output β. The inset in (d) displays a polar plot of H obtained by the RMS-COR method and allowing then to determine the azimuth dependence of H. The black points correspond to the Hurst exponents for the two profiles shown on this plot.

technique, with only slight sensitivity to the input self-affine exponent and size effects. This conclusion should however be interpreted cautiously as the algorithm used to generate the synthetic surface is based on a Fourier transform approach. The conclusion of this comparison test is that the FPS, WPS, and RMS-COR methods are the most reliable because they have a small dependence on the value of the input Hurst exponent and a slight dependence on system size.

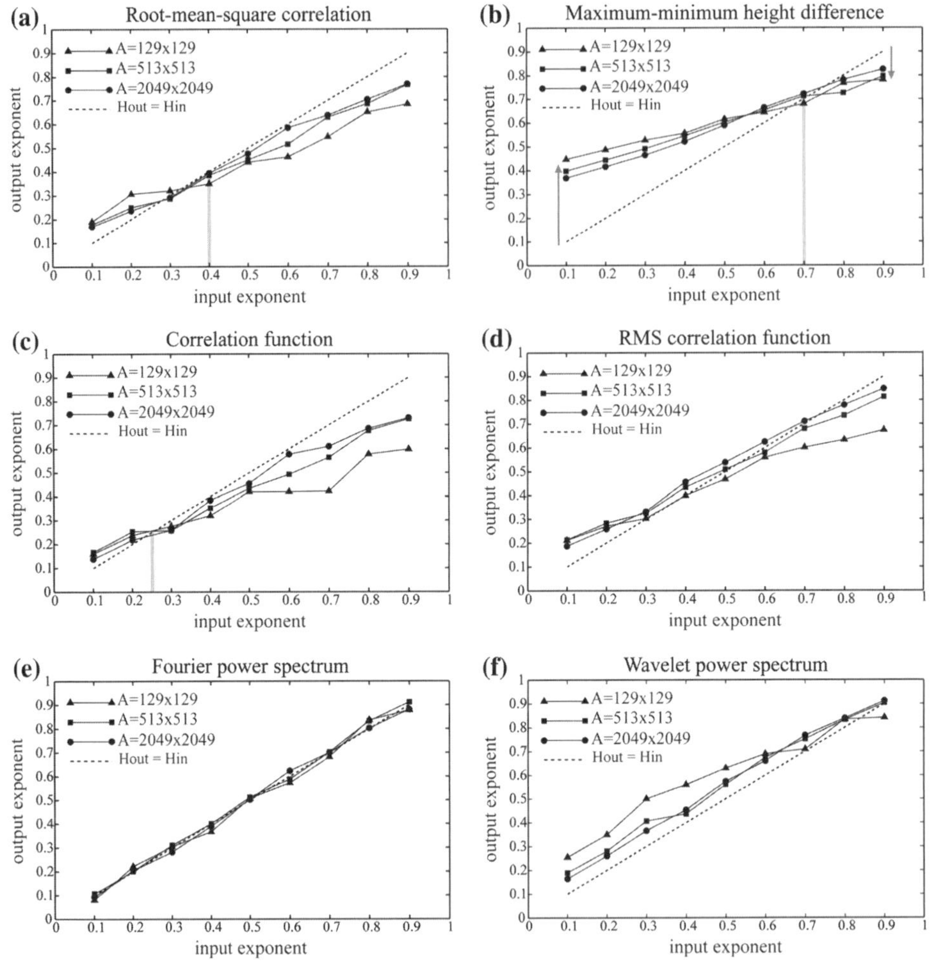

Figure 3

Comparisons between the "input" Hurst exponents introduced in the construction of isotropic fractional Brownian surfaces and the "output" exponent measured on these surfaces using the six independent signal processing techniques. The three outputs illustrate effects of system size. (a) Root-mean-squares correlation (RMS), (b) maximum-minimum height difference (MM), (c) correlation function (COR), (d) RMS correlation function (RMS-COR), (e) Fourier power spectrum (FPS), (f) Wavelet power spectrum (WPS). The gray line in (a), (b), and (c) indicates for which input exponent the error in the estimation is closest to zero.

4.3. Anisotropic Surfaces: Interaction between the Two Input Roughness Exponents

For synthetic self-affine anisotropic surfaces, we have calculated the error on the estimation of the two Hurst exponents. For this, we have built surfaces (2049 $\times$ 2049 points, similar to Fig. 1b) for which the Hurst exponents $H_{\mathrm{input}//}$ and $H_{\mathrm{input}\perp}$ in two perpendicular directions were varied in the range [0.4–0.9]. We have then used the six signal processing techniques to estimate the values of these same exponents. The absolute error in the estimation of each Hurst exponent (Fig. 4) depends on the input value of these parameters and also on the amplitude of their difference ($H_{input//} - H_{input\perp}$).

This error is particularly large for the RMS (up to 20%), MM (up to 25%), and COR (up to 35%) methods. When the input anisotropy ($H_{input//} - H_{input\perp}$) increases, the absolute error on the two output exponents increases accordingly. The absolute error is smaller in the direction of the smallest exponent (analogue to the direction of striation on a natural fault surface) than perpendicular to it. Moreover, it is also noteworthy to mention that these three techniques show significant errors in the estimation for input exponents close to 0.8–0.9 even if the anisotropy is minimal, demonstrating the limited reliability of these methods to detect an exponent close to one.

The RMS-COR analysis is also sensitive to the input anisotropy (Fig. 4d), however such an effect is not strongly pronounced (the absolute errors are smaller, up to 15%). For this method, the error does not depend on the values of the two input Hurst exponents. For example, with a synthetic surface defined by $H_{input//} = 0.8$ and $H_{input\perp} = 0.6$, the absolute error in the estimation of each Hurst exponent is almost identical. As shown in Figure 4d, when anisotropy is small, the errors do not increase significantly for input values close to 1 unlike the three previous methods.

The FPS and WPS analyses are only slightly sensitive to the "input" anisotropy and the estimated Hurst exponents do not depend on the input exponent values. Our analysis clearly shows that the FPS, the WPS and, to a lesser extent, the RMS-COR methods are the most reliable. More precisely, the RMS-COR and the WPS techniques slightly underestimate and overestimate, respectively, the roughness exponent compared to the FPS method.

We have also analyzed the azimuthal dependence of the Hurst exponent for synthetic anisotropic self-affine surfaces. Comparisons of the "output" anisotropy estimated using the RMS-COR method and the "input" anisotropy are represented on Figure 5. We have used this technique because it does not require interpolation of the profiles, whereas the FPS and WPS methods would need regularly spaced data points. A significant directional morphological anisotropy of surfaces is visible on these polar plots of H even if a low "input" anisotropy is imposed, thus demonstrating the reliability of the RMS-COR method to detect a slight morphological anisotropy. Remarkably, following a 360° rotation, the azimuth variation of H is not progressive. When departing a few degrees from the direction of the smallest "input" exponent, the "output" exponent is already very close to the largest "input" exponent. This property of anisotropic self-affine surface is not well understood yet.

A tentative way to calculate the error on the anisotropy that is made when estimating the anisotropy of the surface $|H_{//} - H_{\perp}|$ is represented in Figure 6. This plot indicates the error on the estimation of the anisotropy of the surface, and therefore provides some bounds on the accuracy of the determination of this property. Most methods underestimate the anisotropy, except the Fourier power spectrum which slightly overestimates it. For the RMS, MM, and COR methods, when the "input" anisotropy $|H_{//} - H_{\perp}|$ increases, the absolute error on the "output" anisotropy increases accordingly.

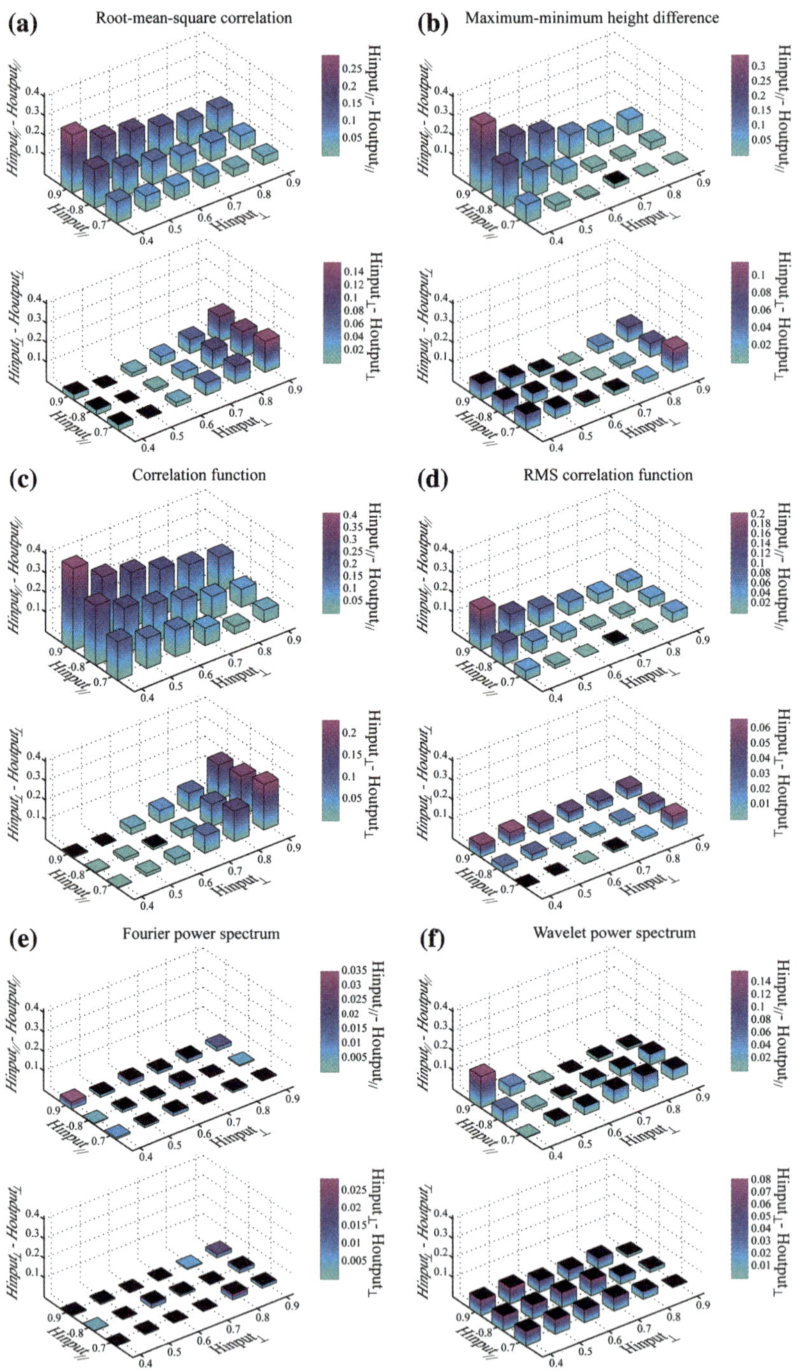
(a)
Root-mean-square correlation
(b)
Maximum-minimum height difference
(c)
Correlation function
(d)
RMS correlation function
(e)
Fourier power spectrum
(f)
Wavelet power spectrum
Hinput// - Houtput//
Hinput⊥ - Houtput⊥
Hinput//
Hinput⊥

◄

Figure 4

Intrinsic errors in the estimation of the Hurst exponents for anisotropic synthetic surfaces characterized by two Hurst exponents $H_{\text{input}//}$ and $H_{\text{input}\perp}$ in perpendicular directions (2049 × 2049 points, similar to Figure 1b). For each signal processing method, histogram plots are represented where the horizontal axes contain the Hurst exponents $H_{\text{input}//}$ and $H_{\text{input}\perp}$ used as inputs to generate the synthetic surface. The vertical axis represents the difference between the input exponent and the estimated output Hurst exponent, using the six different methods. For each method, two histogram plots are represented: the upper one shows the difference ($H_{\text{input}//} - H_{\text{output}//}$) and the lower one the difference ($H_{\text{input}\perp} - H_{\text{output}\perp}$). The black top surfaces on the histogram bars indicate a negative difference (overestimation of the output exponent) and the colored ones a positive difference (underestimation of the output exponent). (a) Root-mean-squares correlation (RMS), (b) maximum-minimum height difference (MM), (c) correlation function (COR), (d) RMS correlation function (RMS-COR), (e) Fourier power spectrum (FPS), (f) Wavelet power spectrum (WPS).

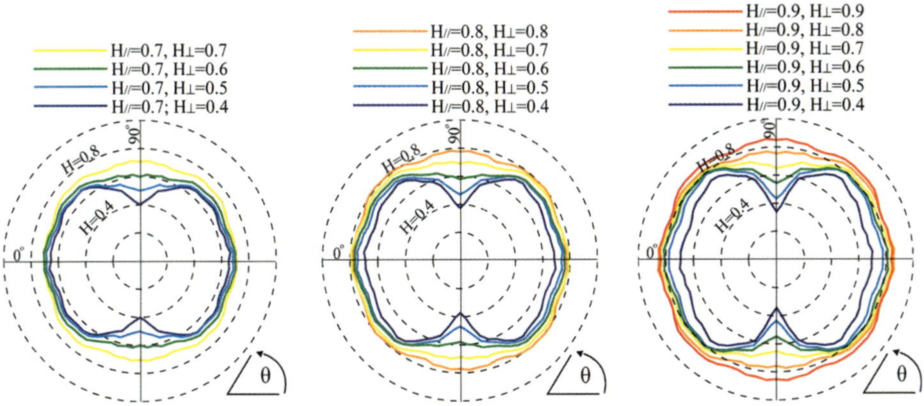

Figure 5

Polar plots of the angular dependence of the two Hurst exponents $H_{//}$ and $H_{\perp}$ for synthetic anisotropic surfaces with principal directions oriented at angles $\theta_{//} = 0°$ and $\theta_{\perp} = 90°$. The Hurst exponents $H_{//,\perp}(\theta)$, as defined by the slope $\beta = H$ in Figure 2d, were calculated on series of 1-D profiles extracted at an angle θ on 2049 × 2049 points surfaces. Three series of simulations are represented here for three values of $H_{//}$ in the range [0.7–0.9]. For each polar plot, the different curves correspond to successive values of $H_{\perp}$. The dashed lines correspond to the values of the output Hurst exponent measured with the RMS correlation method (center dashed circle: $H = 0.2$, outer dashed circle: $H = 1$).

Moreover, this absolute error is similar for all surfaces with the same "input" anisotropy, regardless of the values of the two "input" self-affine exponents.

The determination of the "output" anisotropy with the RMS-COR, FPS, and WPS methods is less sensitive to the "input" anisotropy, except for the highest anisotropy, thus demonstrating the robustness of these three methods to determine the azimuth dependence of the statistical properties of an anisotropic self-affine surface. More precisely, estimates reported for the WPS method are somehow systematically lower than those of the two other techniques.

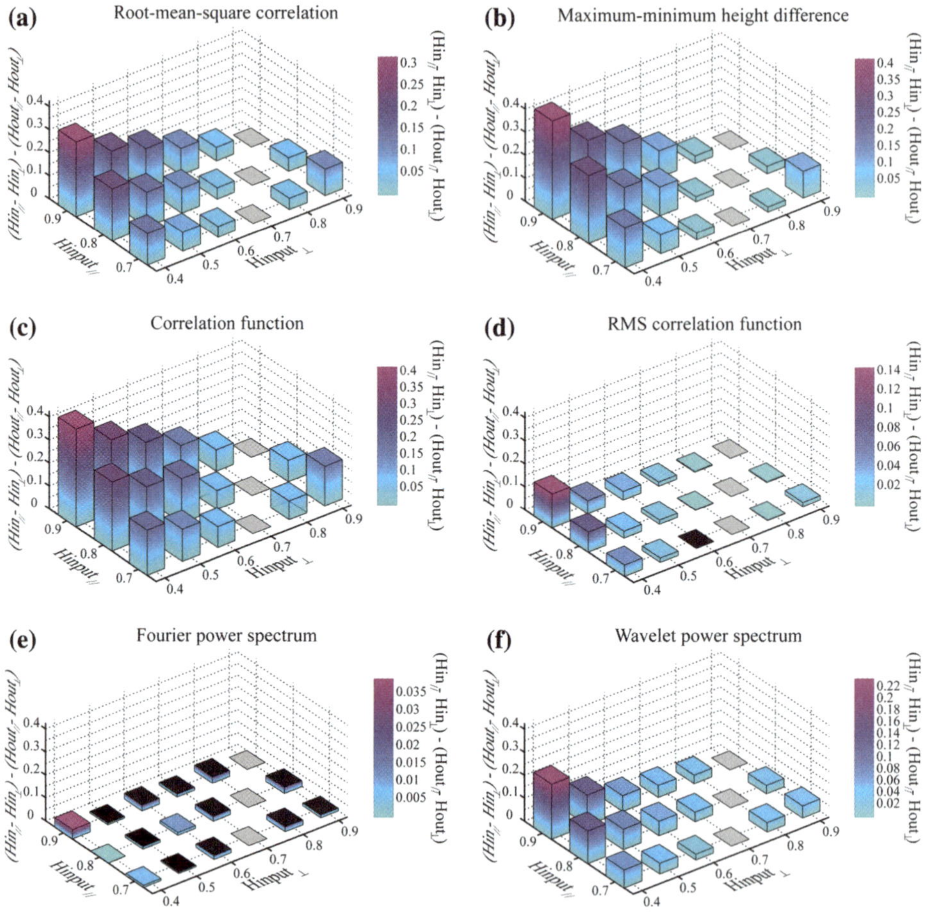

Figure 6

Quantification of the intrinsic errors on the estimation of the anisotropy of the surface ($H_{//} - H_{\perp}$). The difference between the "input" anisotropy (difference between the two "input" Hurst exponents) introduced in synthetic surfaces (2049 × 2049 points, similar to Fig. 1b), minus the "output" anisotropy is represented for the six signal processing methods: (a) Root-mean-squares correlation (RMS), (b) maximum-minimum height difference (MM), (c) correlation function (COR), (d) RMS correlation function (RMS-COR), (e) Fourier power spectrum (FPS), (f) wavelet power spectrum (WPS). Bars with black top surface indicate an overestimation ("input" anisotropy - "ouput" anisotropy < 0) and colored top surfaces indicate an underestimation ("input" anisotropy - "ouput" anisotropy > 0). Gray bars indicate isotropic surfaces ($H_{//} = H_{\perp}$), thus without errors.

4.4. *Effect of Noise*

In all physical measurements, noise is present in the data because of the limited resolution of the measuring device. Such noise is usually described using Gaussian statistics with a zero mean and a constant variance. We have analyzed how the presence of noise in synthetic data could alter the estimation of the Hurst exponent. For this, we have calculated synthetic anisotropic surfaces and added a Gaussian noise with a standard deviation equal to 1/200 of the standard deviation of the rough surface (Figs. 7a, b). This artificial alteration of the synthetic surface mimics measurement biases obtained on natural data, for example due to electronic noise in the measurement device or due to weathering of the fault surface. We have then estimated the Hurst exponents using the six signal processing methods and compared the results with the noise-free analysis. The results confirm that adding noise to the synthetic data induces a leveling-off of the curves at small length scales (Renard *et al.*, 2006; Sagy *et al.*, 2007) and therefore a possible underestimation of the Hurst exponent, for all six signal processing methods (Figs. 7c-h).

Despite the fact that the Gaussian white noise added is isotropic, each plot (Figs. 7c-h) indicates that the effect of noise is slightly dependent on the azimuth of the profile: The underestimation of the Hurst exponent is more pronounced along striations than perpendicularly to them. Indeed, the addition of noise in the rough signal preferentially alters the roughness at small scales, and therefore has a stronger effect on the profiles parallel to the striations because they are characterized by a smaller amplitude at large length scales compared to the profiles perpendicular to the striations.

For the RMS, MM and RMS-COR methods, the noise does affect not only the small length scales but also the large length scales. Indeed, such an effect is strongly pronounced for these three methods and, slopes of the curves in Figs. 7c-f lead to a significant underestimation of the actual value of the self-affine exponents. Notably, the polar plot of H from a surface with added noise obtained with the RMS-COR technique (see Fig. 7f) shows errors of 10% and 20% on the Hurst exponent measured in directions perpendicular and parallel to striations, respectively.

Conversely, the COR, FPS, and WPS techniques are less sensitive to the addition of noise. At large scale, the noise appears as a negligible correction, and even if the curves are affected at small scales, the estimation of the self-affine exponent is less affected.

4.5. Effect of Missing Data

When considering natural fault measurements, local weathering and/or the presence of vegetation may form patches of missing data. To study their influence on estimation of fault surface properties, we generated incomplete data sets removing an increasing percentage of clusters of points from a synthetic surface that initially contained 513×513 points (Fig. 8).

For the FPS and WPS methods, the incomplete cloud of points was interpolated across the gaps (Fig. 8b), using a linear fit. However, for the RMS-COR method, the biased data can be used without interpolation of the holes (Fig. 8c).

Typically, in our natural data sets 5% of interpolated holes is the maximum percentage of spurious points removed from the raw scanner data. The results (Figs. 8d-f) indicate that the RMS-COR, FPS and WPS analysis show an error of only 4% on the Hurst exponent estimated on a surface with 40% holes compared to a complete surface. Therefore, 5% of missing points does not significantly affect the measurement of surface properties, regardless of the technique employed.

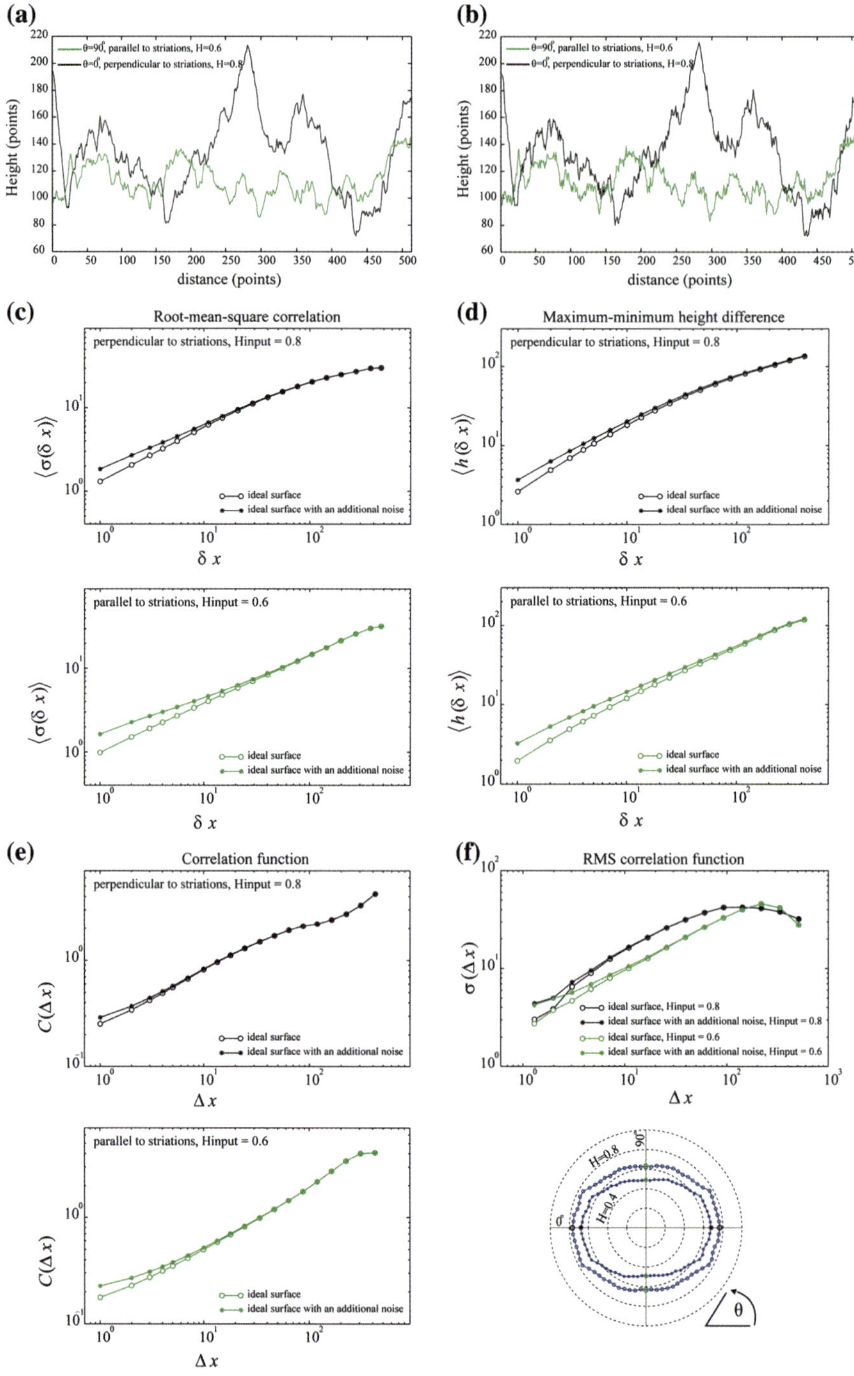
(a)
(b)
θ=90°, parallel to striations, H=0.6
θ=0°, perpendicular to striations, H=0.8
Height (points)
distance (points)
(c)
Root-mean-square correlation
perpendicular to striations, Hinput = 0.8
parallel to striations, Hinput = 0.6
⟨σ(δ x)⟩
δ x
ideal surface
ideal surface with an additional noise
(d)
Maximum-minimum height difference
⟨h(δ x)⟩
(e)
Correlation function
C(Δ x)
Δ x
(f)
RMS correlation function
σ(Δ x)
ideal surface, Hinput = 0.8
ideal surface with an additional noise, Hinput = 0.8
ideal surface, Hinput = 0.6
ideal surface with an additional noise, Hinput = 0.6
90°
0°
H=0.8
H=0.4
θ

◀

Figure 7

Influence of an additional noise on the self-affinity property of an anisotropic synthetic surface (513 × 513 points) characterized by two Hurst exponents ($H_{//} = 0.6$ and $H_{\perp} = 0.8$) in perpendicular directions. The Gaussian white noise is characterized by a standard deviation two hundred times smaller than the roughness amplitude of the synthetic surface. (a) Example of 1-D profiles extracted in two perpendicular directions of an "ideal" surface. (b) The same profiles but with an additional noise. Note that altered profiles appear more jagged or rougher at small scale compared to the noise-free profiles. Analyses of those altered surfaces are performed with the six independent self-affine methods: (c) Root-mean-squares correlation (RMS), (d) maximum-minimum height difference (MM), (e) correlation function (COR), (f) RMS correlation function (RMS-COR), (g) Fourier power spectrum (FPS), (h) wavelet power spectrum (WPS). For each method, except the RMS-COR function, two plots are represented: The upper plot shows the difference between the noise-free surface and the altered surfaces in the direction of the largest exponent and the lower one in the direction of the smallest exponent. For the RMS-COR technique, the upper plot represents the difference between the noise-free and the altered surface in both directions: perpendicular and parallel to striations. The lower subplot displays two polar plots of H obtained with the noise-free and the biased surfaces. Note that the flattening of the scaling behavior at large scales is related to a finite size effect.

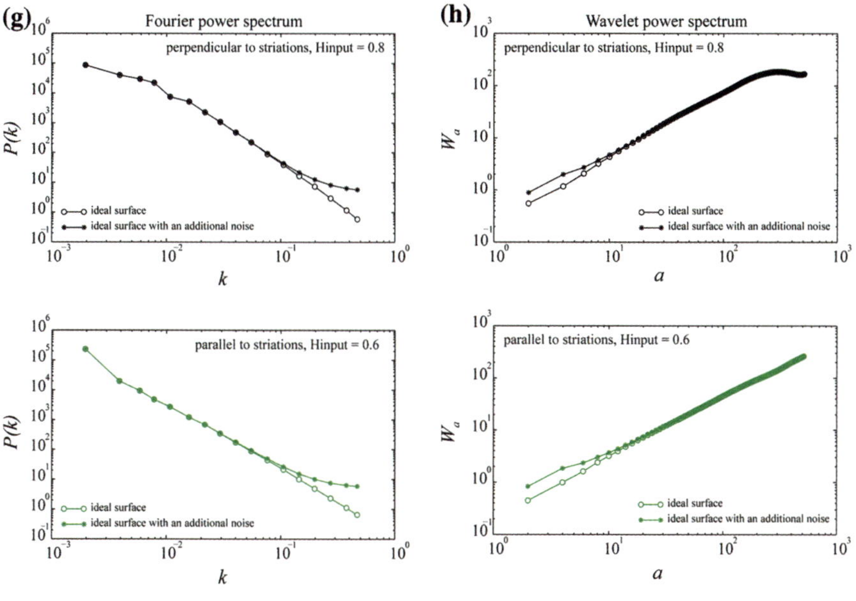

Figure 7
contd.

5. *Acquisition of Roughness Data on Natural Faults at Various Scales*

5.1. Acquisition of the Data on the Field and in the Laboratory

The roughness data of several fault samples were acquired at various scales using five different scanning devices (Table 1). At the laboratory scale, we used a home-made laser profilometer (Meheust, 2002), where a sample, set on a 2-axis moving table, is scanned

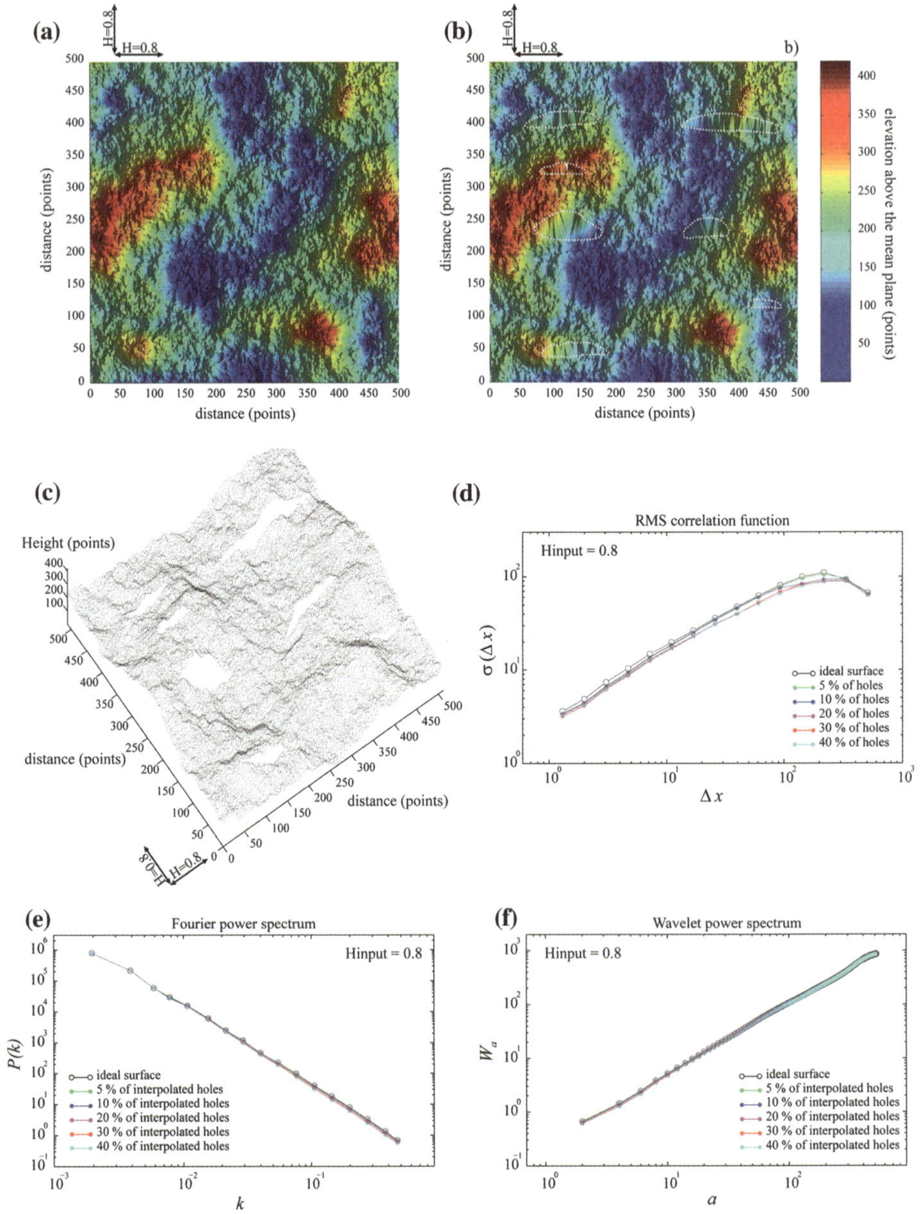

Figure 8

Influence of the interpolation of missing data on the estimation of the Hurst exponent. (a) Synthetic surface (513 × 513points) with H = 0.8. (b) Same surface with 5% of holes (white dashed lines) that have been interpolated. (c) 3-D view of the surface in a) with the holes. (d) RMS correlation function (RMS-COR), (e) Fourier power spectrum (FPS), (f) wavelet power spectrum (WPS). The different curves on each plot present the result of the analysis for five different percentages of missing points (5%, 10%, 20%, 40%).

Table 1

Characteristics of the field and laboratory scanner devices

3-D scanner device	S10	GS100	LMS Z420i	Ilris-3D	Lab. Profilometer
Company	Trimble	Trimble	Riegl	Optech	Univ. Strasbourg
Resolution (dx)	0.5 mm	10 or 20 mm	20 mm	20 mm	20 or 24 μm
Noise on the data	0.9 mm	4.5 mm	10.2 mm	20 mm	<1 μm
Acquisition speed	70 pts/s	5000 pts/s	5000 pts/s	2500 pts/s	60 pts/s

by measuring the distance between the sample and a laser head. The horizontal scanning steps are either 20 or 24 micrometers and the vertical resolution is better than 1 micrometer.

On field outcrops we measured several surfaces with four different LIDARs, where two main technologies were used. The S10 system (Table 1) contains a laser source and two cameras; the distance between the laser head and a surface point is measured by triangulation. The maximum shooting distance is around 15 m and the resolution in the distance measurement is close to 0.5 mm. Surfaces of several square meters can be measured with this system. The main drawback of this system is that it must be operated during night time otherwise the day light may blind the cameras.

The other three LIDAR systems (Table 1) are based on the same technology and were built by three different companies: A light pulse is sent from a laser head and the time of flight to the target point is measured, allowing calculation of the distance, knowing light velocity. The whole target surface is scanned by rotating the laser head at constant angular velocity. The main advantages of this technology is that fast scanning rates can be achieved (up to 5000 points/s), the shooting distance can be as large as 1500 m and the system can be operated even under daylight. However, compared to the S10 system the measurement accuracy is lower, between 1 and 2 cm. Note also that if the laser wavelength is in the infra-red range, absorption by water present on the target surface might also alter the quality of the data.

We have used these scanning measurement devices on two faults in limestones, where outcrop fault planes were scanned at various scales. Hand samples of slip surfaces were also collected and measured with the laboratory profilometer.

5.2. *The Magnola Normal Fault*

The Magnola fault outcrop, in the Fuccino area, is part of the extensive fault system in the Central Apennines (Italy). This 15-km long normal fault shows microseismic activity and offsets limestone beds with a vertical displacement larger than 500 meters and a slight shear component witnessed by mechanical striations dipping at a 85° angle on the fault plane. This site is characterized by recent exhumation of the fault (Palumbo *et al.*, 2004; Carcaillet *et al.*, 2008) with less alteration by weathering than older exhumed portions of the fault. Recent earthquakes have exposed a ~10-m thick band of fresh limestone (Fig. 9a), where mechanical striations and grooves at all scales are still visible.

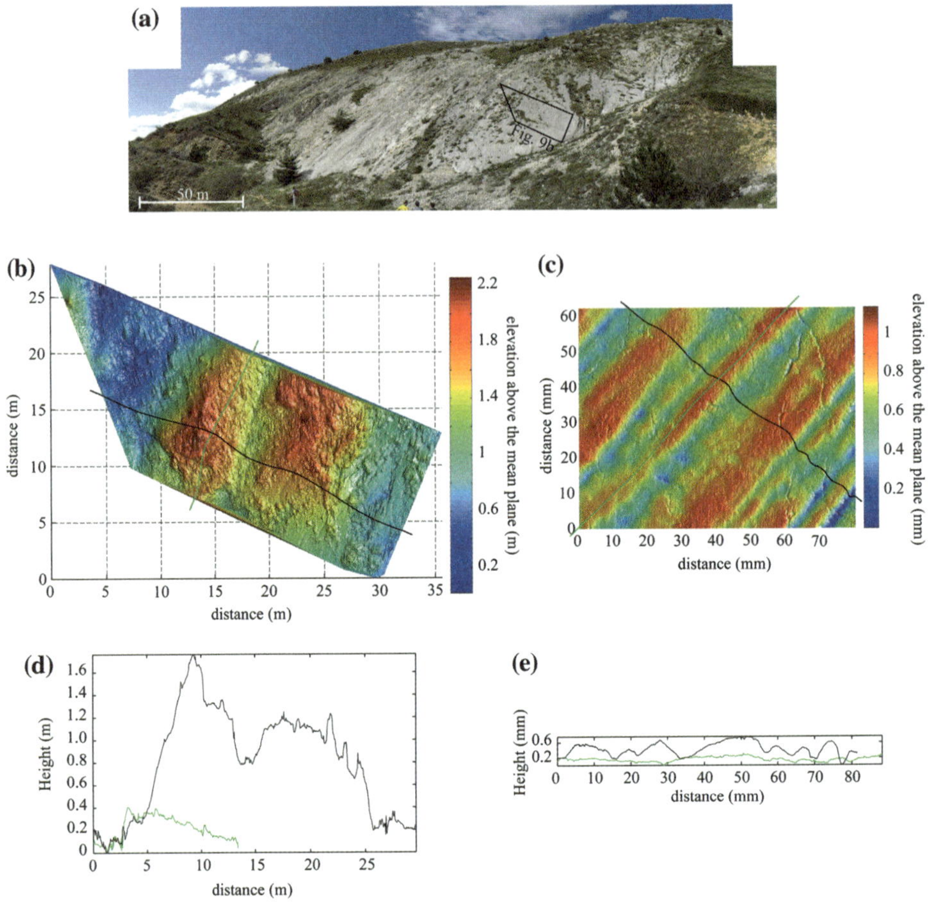

Figure 9

3-D scanner data of the Magnola fault slip surface at different scales. (a) Photograph of the fault surface showing significant weathering and covering by vegetation. The black polygon corresponds to the surface shown below. (b) Digital Elevation Model (DEM) of the surface A32 (Table 2). The LIDAR data contain 799,830 points, sampled on a roughly regular grid of $\sim 40 \times \sim 40$ mm. The measurements were performed on a $\sim 20 \times \sim 20$ mm grid and then averaged on a coarser grid. The resolution of the elevation is 20 mm. The fault surface shows elongated bumps (red) and depressions (blue), which have a pitch of approximately 85°, and indicated a main normal slip motion. The corrugation, with maximum amplitude of around 2.2 m, can be observed at all scales down to the measurement resolution. (c) DEM of hand sample M2 (Table 2) that contains 3999×3120 points on a regular mesh of 20×20 μm^2. The surface contains grooves (blue) and ridges (red) aligned parallel to slip and with maximum amplitude of around 1.2 mm. (d) and (e) Example of 1-D profiles extracted from the DEM of surface A32 (d) and hand sample M2 (e), in directions perpendicular (black profile) and parallel (green profile) to striations. The position of extracted profiles is shown by a black line (perpendicular to striations) and a green line (parallel to striations) on each scan surface. Note the vertical exaggeration (x10) in order to highlight the roughness of parallel and normal profiles at all scales for field and laboratory data. The reduction of the amplitude of parallel profiles compared to normal profiles is an expected consequence of striations.

Table 2

Fault surfaces analyzed in this study

Fault	Surface area, dx	Scanner
Vuache, SURF1	17 × 10 m, 20 mm	GS100
Vuache, SURF7	24 × 11 m, 30 mm	GS100
Vuache, SURF6	45 × 9 m, 20 mm	LMS Z420i
Vuache, SMALL	10 × 9 cm, 24 μm	Lab. Profilometer
Vuache, SURF-JPG	0.5 × 0.5 m, 1 mm	S10
Magnola, A32	35 × 15 m, 20 mm	Optech
Magnola, M2	7.2 × 4.5 cm, 20 μm	Lab. Profilometer

We have scanned several subsurfaces of the fault wall (Table 2, Fig. 9b) and collected one hand sample from the roof of the fault for laboratory measurements (Table 2, Fig. 9c). This sample, that shows mechanical striations, was excavated from below the ground surface to get a slip surface preserved from climatic weathering. The larger outcrop surfaces show evidence of erosion and some karstic water outlets provided holes for vegetation. Small bushes and grass outcrops had to be removed either directly from the fault plane, or from the LIDAR data. The result was incomplete data sets of the fault surface, with missing points in the records. Nevertheless, elongated bumps and depressions at large scales (Fig. 9b), and grooves and ridges at small scales (Fig. 9c) aligned parallel to slip can be observed. Note that 1-D profiles (Figs. 9d, e) extracted from the Digital Elevation Model (DEM) of both the field surface (Fig. 9d) and hand sample (Fig. 9e) appear to be rough at all scales.

5.3. The Vuache Strike-slip Fault

The Vuache fault is an active strike-slip fault system in the western part of the French Alps (Thouvenot, 1998). The fault outcrop we analyzed lies on a short segment that connects to the main Vuache fault and is no longer active. This fault was analyzed in Renard *et al.* (2006) and we present here new data of large- and small-scale slip surfaces (Fig. 10).

This fault has a mainly strike-slip component, witnessed by large elongated bumps and depressions associated with linear striations of smaller size observed at all scales up to the resolution of the scanners LIDAR (Fig. 10c-e). Conversely, the laboratory data show that the surface below the centimeter scale appears more polished and only smooth decimeter ridges persist (Fig. 10f). Moreover, Figure 10h shows that 1-D profiles extracted from the hand sample fault plane also appear clearly smooth at small scale. This observation is confirmed when we perform an isotropic dilation of profiles (Fig. 10h). In a different way, we observe a roughness at all scales on profiles extracted from field surfaces (Fig. 10g).

The fault offsets meter-scale beds of limestones and the fault plane was exhumed ten years ago by the activity of a quarry. As a consequence, the LIDAR measurements could be performed on fresh surfaces, where weathering was minimum and no vegetation had

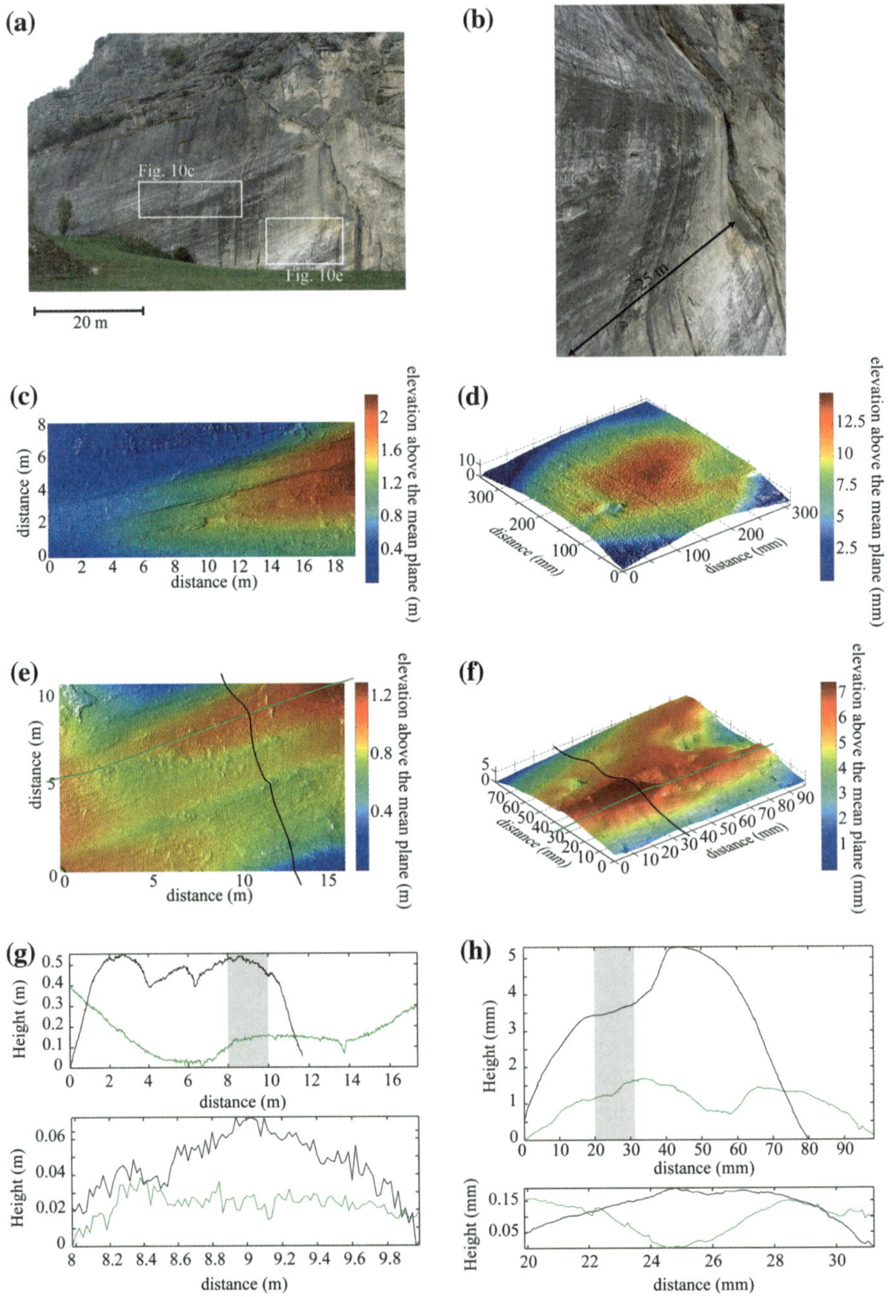

developed on the fault plane. For these surfaces, the data recovery was excellent, greater than 99.5%. We therefore obtained the topography of the surfaces without holes in the data, making the signal processing results reliable.

◀ Figure 10

3-D data of the Vuache fault slip surface at all scales. (a) Photograph of the outcrop, where the white boxes correspond to the surfaces shown in (c) and (e). (b) Lateral photograph of the slip plane that highlights its remarkable waviness. (c-e) Fault surface topography of SURF1 and SURF6 (Table 2). Each surface contains approximately 450,000 points sampled on a constant grid of 20 × 20 mm. The resolution of the elevation is 10.2 mm for (c) and 4.5 mm for (e), respectively. The surfaces show large elongated bumps (red) and depressions (blue) with maximum amplitude of around 2 m associated with linear striations of smaller size (grooves and ridges). Both geometrical patterns with a pitch of 15–25° indicate a strike-slip motion with a small normal component. (d) DEM of the bumpy zone SURF-JPG (Table 2) that contains 107,606 points sampled on a regular grid of 1 × 1 mm. Note the vertical exaggeration. The resolution of the elevation is 0.9 mm. (f) DEM of the hand sample SMALL (Table 2) that contains 4099 × 3333 points on a constant grid of 24 × 24 μm^2. The resolution of the elevation is less than one micrometer. The scans clearly show a smoothing of the roughness from large to small scales. Large fault surfaces measured on the field have asperities over the entire range of observed scales down to the measurement resolution. Conversely, the laboratory data indicate that the surface below the centimeter scale appears more polished and only smooth decimeter ridges persist. (g) and (h) Example of 1-D profiles extracted from the DEM of surface SURF6 (e) and hand sample SMALL (f), in directions perpendicular (black profile) and parallel (green profile) to striations. The position of extracted profiles is shown by a black line (perpendicular to striations) and a green line (parallel to striations) on each scan surface. For each surface, lower subfigures represent zooms on the portions of profiles (located by grey zones on the upper profiles) which have been detrended and dilated with identical ratios over the horizontal and the vertical directions.

6. *Roughness Results and Interpretation*

6.1. *Magnola Fault*

The roughness analysis results of the Magnola fault outcrop and hand sample are shown on Figure 11. To focus our analysis on the scaling properties of amplitude of the roughness versus wavelength, *i.e.*, the Hurst exponent, we chose to normalize the Fourier power spectrum and the averaged wavelet coefficients spectrum by their respective prefactors (Figs. 11a-d). Indeed, for a self-affine profile, the power spectrum and averaged wavelet coefficient behave respectively as: $P(k) \propto A_k k^{-1-2H}$ and $W_a \propto A_a a^{H+1/2}$ where prefactors A_k and A_a define the overall amplitude of the roughness. We will see later in the paper how to characterize and compare these prefactors.

On each plot (Figs. 11a-d) both LIDAR data (upper curve), acquired with the Optech scanner (Table 2), and laboratory profilometer data (lower curve) are represented, showing a scaling behavior over 5.5 orders of magnitude of length scales (50 μm to approximately 20 m). Each curve represents an average over a large set of parallel 1-D profiles extracted from the DEM of the fault surface shown on Figure 9. The level of noise for the field LIDAR scanner (Table 1) is estimated as the height of the flat part of the spectrum at small length scales and is indicated by the black arrows (Fig. 11). The flattening of the scaling behavior at large scales is related to a finite size effect.

The FPS and WPS techniques performed along and perpendicular to the slip direction (Figs. 11a-d) indicate that the power laws can easily be connected for the field and laboratory data, demonstrating the robustness of the scaling behavior. Moreover, our

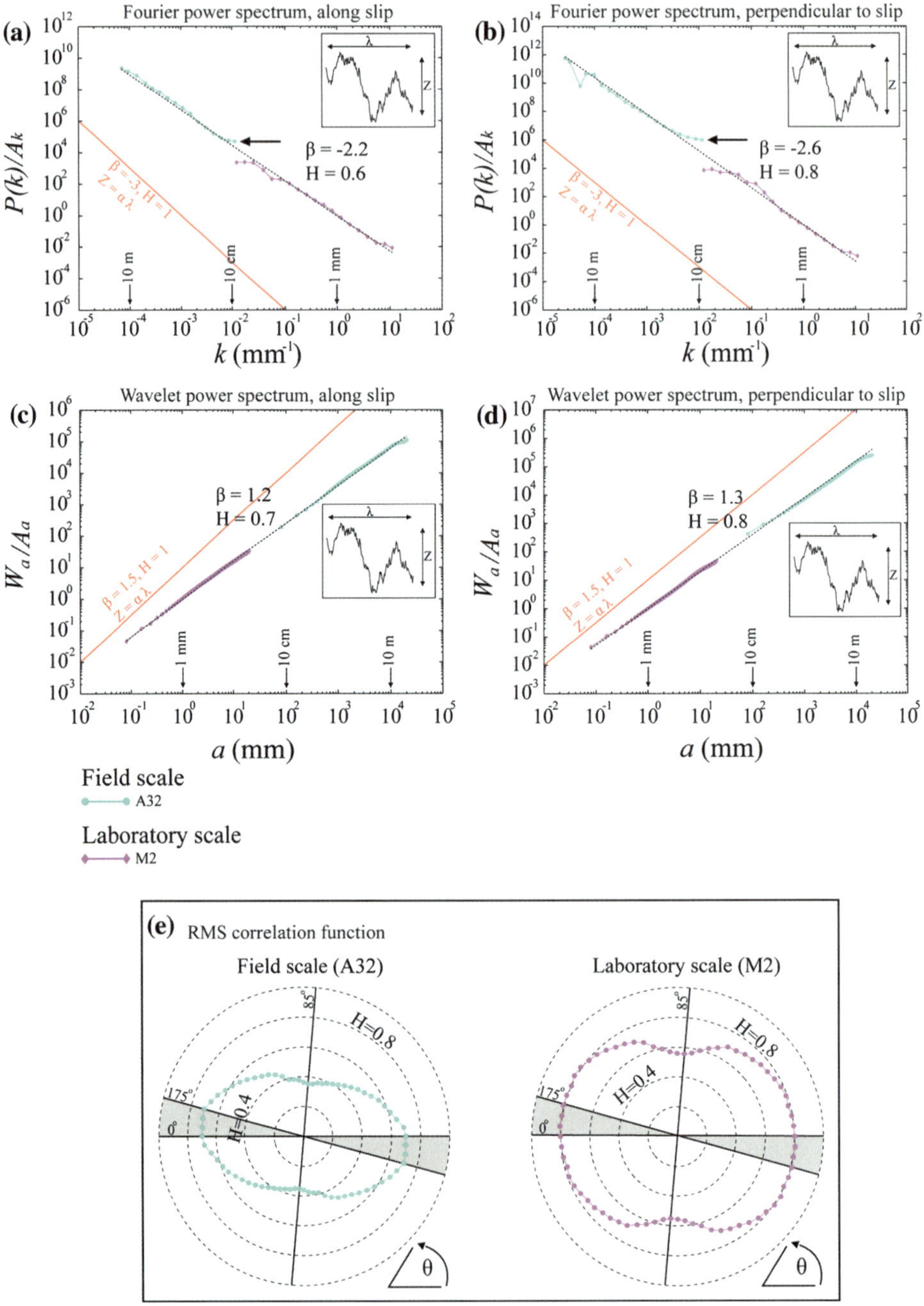

results highlight a significant directional morphological anisotropy over a wide range of scales: The profiles parallel to the slip direction are rougher than perpendicular ones (Power *et al.*, 1988; Power and Tullis, 1991; Lee and Bruhn, 1996; Power and Durham,

◀

Figure 11

Roughness scaling analysis from the best preserved outcrops of the Magnola fault plane, scanned using ground-based LIDAR (surface A32, Fig. 9b), or using the laboratory profilometer (hand sample M2, Fig. 9c). (a-b) Fourier Power Spectrum (FPS) normalized, and (c-d) wavelet power spectrum (WPS) normalized along two directions, parallel and perpendicular to the direction of slip, are represented in log-log plots. Power-law fits (black dotted line) are performed for each curve and the corresponding slopes (β) and roughness exponents (H) are indicated above the spectra. The inset displays an example of the amplitude (Z) and the wavelength (λ) of a rough profile. Contours (red lines) of constant amplitude (Z) to wavelength (λ) ratio, reflecting a self-similar behavior, are provided to allow easier interpretation of the spectra. Black arrows indicate the level of noise of the LIDAR. (e) Surface anisotropy revealed by the angular variation of the Hurst exponent determined by the RMS correlation function method. The polars plot of H on the left and the right sides correspond to data of the field surface and hand sample shown on Figure 9, respectively.

1997; RENARD *et al.*, 2006; SAGY *et al.*, 2007). These two methods estimate a similar Hurst exponent perpendicular to the slip orientation ($H_{\perp} = 0.8$) across the entire range of scales investigated in this study, a property similar to fresh mode I fracture surfaces (POWER *et al.*, 1987; SCHMITTBUHL *et al.*, 1995b; BOUCHAUD, 1997). However, the FPS technique indicates a greater anisotropy, $|H_{//} - H_{\perp}| = 0.2$, where $H_{//}$ represents the Hurst exponent in the direction of slip, than quantified by the WPS method ($|H_{//} - H_{\perp}| = 0.1$). Moreover, the WPS method overestimates the self-affine exponent along the slip direction ($H_{//} = 0.7$) compared to the FPS method ($H_{//} = 0.6$). An attempted explanation of these last two differences is given by our parametric study of synthetic rough surfaces: The WPS method slightly overestimates the roughness exponents when measuring couples of Hurst exponents in perpendicular directions with range of values similar to those of natural fault surface (0.6 to 0.9). Notably, the exponents accuracy with anisotropic surface of 2049×2049 points and for two Hurst exponents in perpendicular direction of 0.8 and 0.6 is numerically estimated as –0.01 and –0.06 for the wavelet method, respectively (Fig. 4f). For example, an amount of –0.06 should be added to the measured minimal exponent with the WPS analysis to obtain the actual one. Accordingly, on natural fault surface with two perpendicular exponents of 0.8 and 0.6 calculated by the FPS, the estimated Hurst exponent in direction parallel to slip is systematically lower than with the WPS method (Figs. 11, 13). Moreover, since the error on the output Hurst exponent is greater in direction of slip than perpendicularly to it, consequently the output anisotropy decreases, as observed on natural fault surface (Figs. 11, 13).

Our estimations obtained on the hand sample with the RMS-COR function (see Figs. 11e) show that the minimum Hurst exponent ($H = 0.6$) is at 85°, in the direction of slip, and the maximum Hurst exponent ($H = 0.8$) is almost in the perpendicular direction. These two extreme values of the self-affine exponent also correspond to those determined at all scales by the FPS and WPS methods. At the field scale (see Fig. 11e), the minimum ($H = 0.4$) and maximum ($H = 0.7$) Hurst exponents are observed in directions similar to that for the hand sample. However, the results obtained with the RMS-COR method suggest that the roughness exponent of the Magnola fault surface is smaller at the field scale compared to the laboratory scale, regardless of the azimuth. We ascribe this variation to natural weathering (pitting) of the exposed fault surface at short wavelengths, as POWER and TULLIS (1991).

Indeed, the section of the Magnola fault surface (Fig. 9b) shows an increase of the roughness amplitude at short wavelengths created by weathering. Other sections of the Magnola fault surface, that are not presented in this study, display evidence of strong alteration at short length scales leading to a significant reduction of the Hurst exponent, regardless of azimuth. Conversely, the clean hand sample (Fig. 9c), that shows mechanical striations, should represent the actual topography of the fault surface at short length scales related to the faulting process before the action of climatic weathering. This hypothesis is supported by the fact that the fresh slip plane, scanned in laboratory, displays the same self-affine RMS-COR regimes in the directions parallel and perpendicular to slip as those estimated using the FPS and WPS methods. The increase of roughness at short wavelengths on the field surface due to the erosion appears to be similar to the effect of an additional noise tested on "ideal" synthetic surfaces (Fig. 7). In both cases, the Hurst exponent decreases in all directions. Our statistical study led on synthetic surfaces shows that the noise effect is more pronounced when using the RMS-COR technique. To summarize, the roughness scaling estimated using the RMS-COR calculated on the weathered field surface exhibits a decrease of the Hurst exponent in all directions, which is not observed with the FPS and WPS techniques.

On the polar plot of H obtained at the laboratory scale (see Fig. 11e), when departing a few degrees from the direction of slip, the Hurst exponent is close to the value of the maximum Hurst exponent measured in the direction perpendicular to mechanical striations. Such behavior is also visible on "ideal" synthetic surfaces (Fig. 5). In other words, the azimuth of the maximum Hurst exponent is not well-defined (gray shadows on Fig. 11e), while the minimum exponent corresponds to a specific orientation. Note that this property is less visible on the polar plot of the altered field fault section (see Fig. 11e) where the angular variation of H is more progressive.

To obtain a full description of the fault asperity geometry, the prefactor of the scaling function also has to be characterized. For instance, using the standard deviation $\sigma(\delta x)$ of the height $L(x)$, the prefactor can be designed as: $\sigma(\delta x) = l_r^{1\text{-H}} (\delta x)^{\text{H}}$ where the prefactor l_r denotes a length scale, also known as the topothesy of the fault roughness (SIMONSEN *et al.*, 2000; SCHMITTBUHL *et al.*, 2006, 2008). The topothesy corresponds physically to the length scale for which the slope of the rough profile is equal to 1: $\sigma(l_r) = l_r$. In other words, l_r is the theoretical length scale over which the rough profile has a mean slope of 45°. The smaller l_r, the flatter the profile appears on a macroscopic scale.

Figures 12a and b show respectively the scaling of the RMS along the slip and perpendicular to it, of the field surface topography and of the hand sample surface topography. From the best fit of these curves, we estimate the roughness exponent and the topothesy (Fig. 12c) of these two surfaces, along two perpendicular directions. For both scales, the local slope of the surface is always significantly smaller than 45 degrees for the range of scales that we explored, i.e., from 40 mm to 30 m for the field surface and from 0.04 to 90 mm for the hand sample surface, as shown on Figures 12a, b. In other

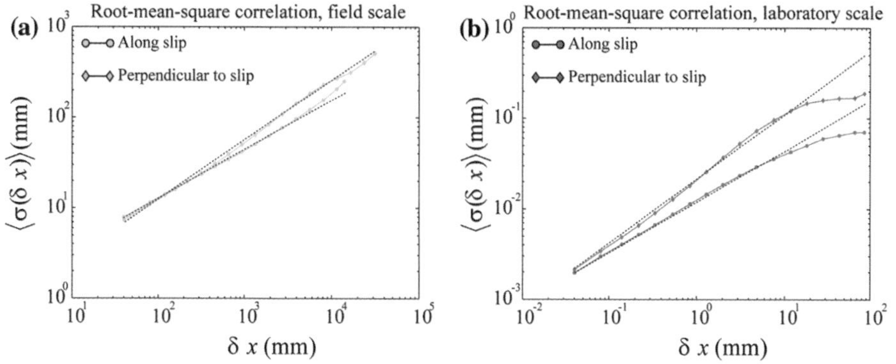

Topothesy of the Magnola fault surface

(c)

		Field scale	Laboratory scale
Magnola fault surface	Perpendicular to slip	0.12 mm	3.1x10^{-7} mm
	Parallel to slip	0.82 mm	4.6x10^{-5} mm

Figure 12
Scaling of the root-mean-squares correlation function (RMS) for the Magnola fault surface topography at field scale (a) and laboratory scale (b) along slip direction and perpendicular to it. The best fits (black dotted lines) of the form $\sigma(\delta x) = l_r^{1\text{-H}}\,(\delta x)^{\text{H}}$ define the roughness exponent H and the topothesy l_r (c).

words, the estimated prefactors are very small, indicating that these surfaces are rather smooth.

6.2. *Vuache Fault*

As explained for the Magnola fault results, the Fourier power spectrum and the averaged wavelet coefficients are normalized by their respective prefactors A_k and A_a in order to directly compare the different slopes of the scaling laws.

The FPS and WPS techniques highlight a significant directional morphological anisotropy over six orders of spatial scales (Figs. 13a-d). Profiles parallel to slip have a smaller Hurst exponent than perpendicular ones (Power *et al.*, 1988; Power and Tullis, 1991; Lee and Bruhn, 1996; Power and Durham, 1997; Renard *et al.*, 2006; Sagy *et al.*, 2007). These two methods, applied to series of profiles perpendicular to the direction of slip, indicate that the power laws of individual surfaces can easily be connected across the wide range of scales investigated (Figs. 13b, d), and the value of $H_\perp = 0.8$ is similar to that which was measured on the Magnola normal fault surface. However, in the slip parallel direction there is a slight change in the magnitude of $H_{//}$ located in the length scale range between 5 mm and 2 cm (Figs. 13a, c). $H_{//} = 0.75$ is larger below this length scale range than at larger length scales, where $H_{//} = 0.65$ and $H_{//} = 0.7$ for the FPS and WPS methods, respectively. The smoother aspect of the Vuache fault surface in the direction of slip compared to the perpendicular direction is an obvious and expected consequence of

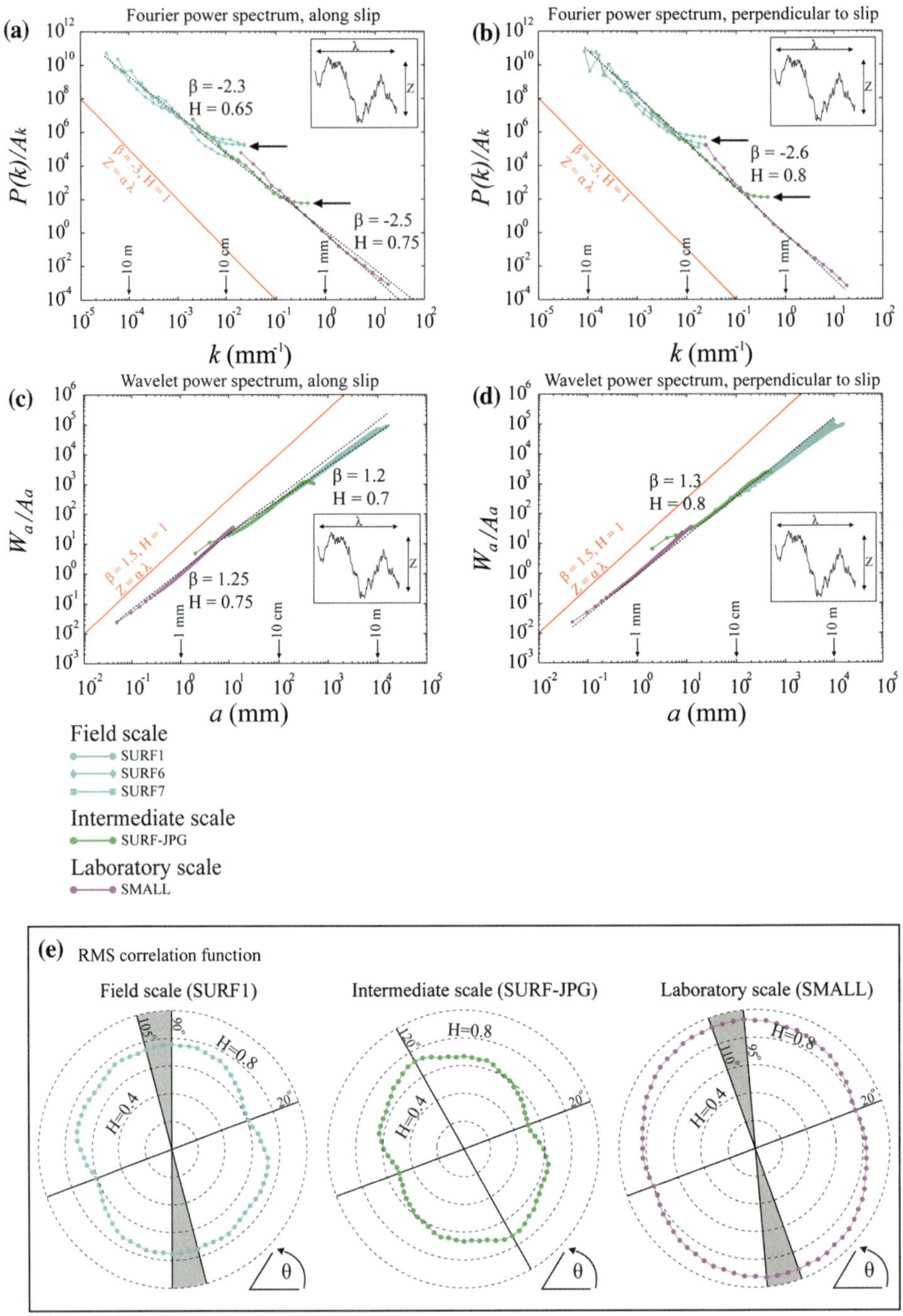
(a) Fourier power spectrum, along slip
P(k)/Ak
k (mm-1)
β = -2.3
H = 0.65
β = -2.5
H = 0.75
β = -3, H = 1
Z = α λ
10 m
10 cm
1 mm
(b) Fourier power spectrum, perpendicular to slip
β = -2.6
H = 0.8
(c) Wavelet power spectrum, along slip
Wa/Aa
a (mm)
β = 1.2
H = 0.7
β = 1.25
H = 0.75
β = 1.5, H = 1
(d) Wavelet power spectrum, perpendicular to slip
β = 1.3
H = 0.8
Field scale
SURF1
SURF6
SURF7
Intermediate scale
SURF-JPG
Laboratory scale
SMALL
(e) RMS correlation function
Field scale (SURF1)
Intermediate scale (SURF-JPG)
Laboratory scale (SMALL)
H=0.8
H=0.4
105°
90°
20°
120°
110°
95°
θ

◀ Figure 13
Roughness scaling analysis from five surfaces (Fig. 10, Table 2) of the Vuache fault, covering 6 orders of magnitude of frequencies or wavelengths (40 μm to approximately 40 m). The data collected contain four surfaces (blue and green curves) that have been scanned using LIDAR and one hand sample (magenta curve) measured by laboratory profilometer. (a-b) Fourier power spectrum (FPS) normalized and (c-d) wavelet power spectrum (WPS) normalized along two directions, parallel and perpendicular to the direction of slip, are represented in log-log plots. Power-law fits (black dotted line) are shown for each curve and the corresponding slopes (β) and roughness exponents (H) are indicated next to the spectra. Each curve is an average over a series of parallel profiles extracted from the DEM shown on Figure 10. The level of noise for the Lidar scanners is estimated as the height of the flat part of the spectrum at small length scales and is indicated by the black arrows. The flattening of the scaling behavior at large scales is related to a finite size effect. Contours (red lines) of constant amplitude (Z) to wavelength (λ) ratio, reflecting a self similar surface, are also indicated as a guide for the eye. The inset displays an example of the amplitude (Z) and the wavelength (λ) of a rough profile. (e) Roughness anisotropy revealed by the angular variation of the Hurst exponent determined by the RMS-COR method. The polar plots of H on the left, in the center and on the right correspond to data of the field surface, the intermediate scale section and hand sample, respectively (see Fig. 10).

striations, however the smallest directional morphological anisotropy at the laboratory scale compared to the field scale is novel in this study. At the field scale, the morphology of the elongated bumps and depressions along the slip direction is different from the grooves and striations observed at the laboratory scale (Fig. 10). As the fault Vuache outcrop is quite fresh, and was preserved from the climatic erosion, we propose that this cross-over in the slope at length scale of several millimeters is significant. We interpret this cross-over length scale as a witness of different mechanical processes responsible for the creation of fault topography at different spatial scales.

Our results obtained at the field and intermediate scales with the RMS-COR function (see left and center plots on Fig. 13e) show that the minimum Hurst exponent ($H_{//} \in [0.55–0.6]$) is oriented at 20° with respect to the horizontal, indicating the fault had a normal component besides a strike-slip component. The maximum value $H = 0.75$ is located for an almost perpendicular direction. These two extreme values of the self-affine exponent are slightly lower when estimated using the RMS-COR function than when using the FPS or the WPS methods. This slight underestimation with the RMS-COR technique is consistent with our results on synthetic surfaces for the accuracy in this range of parameters (Fig. 4). Indeed, our parametric study with synthetic anisotropic surfaces shows that the estimation of H calculated with the RMS-COR technique slightly underestimates its actual value (Fig. 4d). At the laboratory scale (see right plot on Fig. 13e), $H_{//} = 0.8$ and $H_{\perp} = 0.9$ are located in orientations similar to those for the three larger surfaces measured on the field. However, the polar plot of H calculated using the RMS-COR function suggests an increase of the scaling exponents in all directions, while the estimation using the FPS or WPS techniques underlined that the exponent increased only along the slip direction. In addition, the two exponents $H_{//}$ and $H_{\perp}$ for the hand sample are overestimated when calculated with the RMS-COR method compared to the FPS and WPS techniques. The latter observation cannot be explained by the results of our parametric study on synthetic surfaces. As a consequence of the extremely smooth

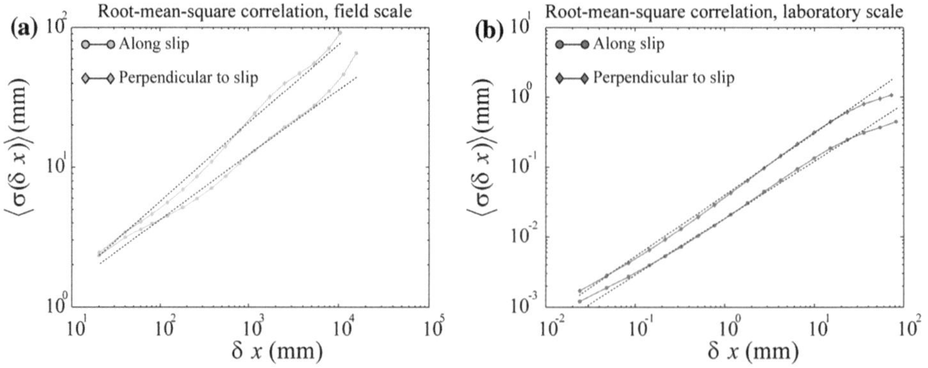

(c) Topothesy of the Vuache fault surface

		Field scale	Intermediate scale	Laboratory scale
Vuache fault surface	Perpendicular to slip	0.27 mm	1×10^{-3} mm	7.5×10^{-13} mm
	Parallel to slip	0.63 mm	1.4×10^{-3} mm	1.1×10^{-9} mm

Figure 14
Scaling of the root-mean-square correlation function (RMS) for the Vuache fault surface topography at field scale (a) and laboratory scale (b) along slip direction and perpendicular to it. The best fit of the form $\sigma(\delta x) = l_r^{1-H}\,(\delta x)^H$ defines the roughness exponent H and the topothesy l_r (c).

aspect of the hand sample at small scales (Figs. 10f, h), the RMS-COR method could lose its robustness. However, a new reliable result is that the directional morphological anisotropy calculated by the RMS-COR function significantly decreases at the laboratory scale, as observed with the FPS and WPS techniques.

As observed on the polar plot of H calculated on the Magnola fault surface (Fig. 11e), the azimuth of the maximum Hurst exponent is less constrained (gray shadows on Fig. 13e) than the single specific orientation of the striations. Remarkably, despite the weak anisotropy of the hand sample topography, the slip direction is always significant, demonstrating the accuracy and reliability of the RMS-COR method.

We also estimated the topothesy of the Vuache fault surface (Fig. 14) when computing the standard deviation $\sigma(\delta x)$ of the height $L(x)$. As for the Magnola fault data, the prefactor l_r always lies outside the range of scales that we explored, *i.e.*, from 20 mm to 15 m for the field surface and from 0.024 mm to 80 mm for the laboratory surface, as shown on Figures 14a-b. In other words, the topothesy of the measured fault planes represents a theoretical value that is very small (see Fig. 14c) and that confirms that fault surfaces are rather smooth. Moreover, it is of interest to note that at the field scale, the topothesy value of the Vuache fault surface (Fig. 14c) is similar to that of the Magnola fault (Fig. 12c). However, at laboratory scale this quantity is significantly lower for the Vuache hand sample compared to that of the Magnola. This characteristic highlights the smoothing of the Vuache fault plane at small scale as already observed on the DEM (Fig. 10f), the 1-D profiles of the Vuache hand sample (Fig. 10 h), and the roughness scaling analysis (Fig. 13).

7. *Discussion and Conclusion*

The six statistical tools used in this study have different responses to the effect of two kinds of biases, the intrinsic errors of the methods (Figs. 3, 4, 5, 6) and the artifacts inherent in data acquisition (Figs. 7, 8). Using a parametric approach, where we varied the size of the surface and its anisotropy, we selected the three most reliable and accurate methods (RMS-COR, FPS, WPS) for roughness analysis of natural fault topography (Figs. 11, 13). The FPS and WPS techniques allow obtaining a robust description of the fault asperity geometry in the direction of slip and perpendicular to it (Figs. 11a-d, 13a-d). Additionally, the RMS-COR technique is robust to characterize the angular variation of H (Figs. 11e, 13e). The Hurst exponent estimation at various scales for the natural slip surfaces displays the same trends and provides a consistent and robust characterization of their scaling regimes. We emphasize that the slight variations in the results given by each of these methods fall within the range of error estimated by our parametric study (see Section 4).

One of the most robust results of our scaling analysis is that the FPS and the WPS methods estimate a same Hurst exponent equal to 0.8 in the direction perpendicular to slip, over approximately 6 orders of magnitude of length scales for two different fault surfaces (Figs. 11b,d and 13b,d). However, in the slip direction two different scaling behaviors are highlighted: The Magnola fault surface manifests an identical scaling regime at large and small scales (Figs. 11a, c). Conversely, the scaling property of the Vuache fault roughness exhibits a cross-over in the slope at length scale of several millimeters (Figs. 13a, c). In other words, the scaling property of this fault surfaces is similar at large scales but changes at small scales. The statistical analyses (Fig. 13) and the scan of the Vuache fault surface (Fig. 10) clearly show a smoothing of the roughness below a length scale of several millimeters.

The scaling regime of 0.8 measured in the direction perpendicular to slip is a classical result already observed for tensile cracks (POWER *et al.*, 1987; SCHMITTBUHL *et al.*, 1995b; BOUCHAUD, 1997), indicating that the topography of the fault surface in the direction perpendicular to slip has not registered the effect of shear. Along slip, the general interpretation is that mechanical wear processes, such as frictional ploughing, cause striations that reduce the amplitude of the large scale roughness (POWER *et al.*, 1987, 1988; POWER and TULLIS, 1991; POWER and DURHAM, 1997; SAGY *et al.*, 2007) and accordingly the Hurst exponent. Nevertheless, our scaling analysis seems to indicate different mechanical processes responsible for the creation of fault topography at different spatial scales.

Prior comparative studies of natural fault roughness based on 1-D profilometry (POWER and TULLIS, 1991; LEE and BRUHN, 1996) suggest a change in scaling properties between large and short length scales. However, due to technical limitations, their measures were not sufficiently accurate to decipher if this variation was related either to small-scale surface weathering of the fault scarp or to the faulting process itself.

From laboratory experiments CHEN and SPETZLER (1993) suggest that the break in slope at length scales of several millimeters is caused by a change in the dominant mode of deformation from small-scale intergranular cracking to intragranular cracking at a larger scale. We think this interpretation does not apply to the Magnola and Vuache faults because the grain scale of these limestones is very small (< 0.1 mm), well below the observed cross-over length scale.

Recently, SAGY *et al.* (2007) observed that faults with large cumulated slip display surfaces with elongate, quasi elliptical bumps at field scale and are polished at small scales. Conversely, faults with a small cumulative slip are rough on all scales. SAGY and BRODSKY (2009) proposed that the waviness of the large slip fault surface reflects variations of thickness of the cohesive layer under the slip surface formed as boudinage structures (JOHNSON and FLETCHER, 1994; SMITH, 1977; TWISS and MOORES, 1992; GOSCOMBE *et al.*, 2004). Therefore, they evoke two different deformation mechanisms between large and small scales: abrasion caused by frictional sliding at the origin of the smoothing at small scales, and "boudinage" creating elongated bumps and depression at large scales.

From our study of natural fault roughness, we observe large elongated bumps and depressions in slip direction on two different fault planes (Figs. 9b, 10a-e). There is no evidence that the small segment, polished at small scale (Figs. 10f, h), that connects to the main Vuache fault has accumulated more slip than the Magnola fault surface that is rough at all scales (Figs. 9b, c). Therefore, we propose that large elongated bumps and depressions of several meters in length with maximum amplitude of around 2 m may reflect the processes of lateral growth and branching that link together several fault surfaces, during all the stages of the evolution of the fault zone, as suggested by LEE and BRUHN (1996) and LIBICKI and BEN-ZION (2005).

At small scales, two different kinds of scaling regime are observed on the two fault surfaces, (Figs. 11, 13), both being linked to mechanical wear process. Frictional sliding is expressed through ploughing of small asperities and is responsible for the small-scale abrasional striations on the Magnola fault surface (Fig. 9c). This mechanism is also responsible for the polishing of the Vuache fault surface below the centimeter scale (Figs. 10f, h). One should keep in mind that only one hand sample was measured for each surface and therefore it is possible that differences of the scaling behavior between the two fault planes reflect spatial heterogeneity of core fault at small scales. Notably, on the Vuache fault, although the surface appears polished at the laboratory scale on the whole surface, zones with striations due to ploughing elements could be present. A more extensive study of fault roughness in several different faults should therefore produce more information on the mechanisms at work during faulting.

Acknowledgments

We would like to thank Jim Dieterich and an anonymous reviewer for helpful comments. We acknowledge Benoit Dubacq for technical help on *Matlab* and Sebastien Boutareaud for encouragement and constructive comments that enhanced this work. Anne-Marie Boullier, Jean-Pierre Gratier, Mai-Linh Doan, Emily Brodsky are decidedly thanked for their fruitful discussions concerning the Vuache fault surface. We also acknowledge a research grant from the program TUNES at the University Joseph Fourier – Grenoble I.

Appendix A

```
function RoughSurf = Synthetic2DFault(N,H1,H2)

% This Matlab(c) function creates a self-affine 2D surface,
% with a directional anisotropy (courtesy of Hermine Bierme,
% Univ. Paris V, France), when H1 is different from H2.
% Input parameters:
% N = size of the surface: (2^(N+1)+1) x (2^(N+1=+1)
% Typically N must be between 8 and 11 when running on a
desktop computer.
% H1, H2: Hurst exponents in two perpendicular directions
% must be positive, smaller than 1.
% Output result:
% RoughSurf: rough surface of size (2^(N+1)+1) x (2^(N+1=+1),
% with two perpendicular directions of self-affinity
% characterized by Hurst exponents H1 and H2, l1 = 1/H1 and
% l2 = 1/H2 represent the eigen values of the anisotropy
% diagonal matrix
l1 = 1/H1;
l2 = 1/H2;
X = (-2*2^N:2:2*2^N)/(2^(N+1));
X(2^N+1) = 1/2^N;
Y = (-2*2^N:2:2*2^N)/(2^(N+1));
Y(2^N+1) = 1/2^N;
XX = X(ones(1,2*2^N+1),:);
YY = Y(ones(1,2*2^N+1),:)';
clear X Y

% rho is the pseudo-norm associated to the eigen values l1 and
l2
% rho(x,y)=(abs(x)^(2/l1) + abs(y)^(2/l2) )^(1/2)
rho = sqrt(abs(XX).^(2/l1)+abs(YY).^(2/l2));
clear XX YY

%phi is the spectral density of the field built from rho
phi = rho.^(1 + (l1+l2)/2);
clear rho

% W = Fourier transform of the anisotropic Gaussian field
Z = randn(2*2^N+1,2*2^N+1);
W = fftshift(fft2(Z))./phi;
clear Z
T = real(ifft2(ifftshift(W)));
RoughSurf = T-T(2^N+1,2^N+1);

% Plot the 2D rough surface
imagesc(RoughSurf);
axis equal
axis tight
```

References

AMITRANO, D. and SCHMITTBUHL, J. (2002), *Fracture roughness and gouge distribution of a granite shear band*, J. Geophys. Res. *107*(B12), 2375, doi: 10.1029/2002JB001761.

AKI, K. (1984), *Asperities, barriers, characteristic earthquakes and strong motion prediction*, J. Geophys. Res. *89*, 5867–5872.

BARABASI, A.-L. and STANLEY, H. E., *Fractal Concepts in Surface Growth* (Cambridge University Press. 1995).

BIERME, H., MEESCHAERT, M. M., and SCHEFFLER, H.-P. (2007), *Operator scaling stable random fields*, Stoch. Proc. Appl. *117*, 312–332.

BEN-ZION, Y. and SAMMIS, C.G. (2003), *Characterization of fault zones*, Pure Appl. Geophys. *160*, 677–715, doi: 10.1007/PL00012554.

BOUCHAUD, E. (1997), *Scaling properties of cracks*, J. Phys. Condens. Matter *9*, 4319–4344.

BROWN, S. R. and SCHOLZ, C. H. (1985), *Broad bandwidth study of the topography of natural rock surfaces*, J. Geophys. Res. *90*, 2575–2582.

CAMPILLO, M., FAVREAU, P., IONESCU, I. R., and VOISIN, C. (2001), *On the effective friction law of an heterogeneous fault*, J. Geophys. Res. *106*, 307–322.

CARCAILLET, J., MANIGHETTI, I., CHAUVEL, C., SCHLAGENHAUF, A., and NICOLE, J. M. (2008), *Identifying past earthquakes on an active normal fault (Magnola, Italy) from the chemical analysis of its exhumed carbonate fault plane*, Earth Planet. Sci. Lett. *271*, 145–158, doi:10.1016/j.epsl.2008.03.059.

CHEN, G. and SPETZLER, H. A. (1993), *Topographic characteristics of laboratory-induced shear fractures*, Pure Appl. Geophys. *140*, 123–135.

CLAUSEL, M. and VEDEL, B. (2008), *Explicit constructions of operator scaling stable random Gaussian fields*, preprint to Adv. Appl. Prob.

GOSCOMBE, B. D., PASSCHIER, C., and HANDA, M. (2004), *Boudinage classification: end-member Boudin types and modified boudin structures*, J. Struct. Geol. *26*, 739–763.

JOHNSON, A. M. and FLETCHER, R. C., *Folding of Viscous Layers: Mechanical Analysis and Interpretation of Structures in Deformed Rock* (Columbia University Press. 1994).

LAY, T., KANAMORI, H., and RUFF. (1982), *The asperity model and the nature of large subduction zone earthquakes*, Earth. Predict. Res. *1*, 3–71.

LEE, J.-J. and BRUHN, R. (1996), *Structural anisotropy of normal fault surfaces*, J. Struct. Geol. *18*, 1043–1059.

LIBICKI, E. and BEN-ZION, Y. (2005), *Stochastic Branching Models of Fault Surfaces and Estimated Fractal Dimensions*, Pure Appl. Geophys. *162*, 1077–1111, doi: 10.1007/s00024-004-2662-7.

MANDELBROT, B. B. (1985), *Self-affine fractals and fractal dimension*, Phys. Scripta. *32*, 257–260.

MANDELBROT, B. B., *Fractals in Physics* (Elsevier, Amsterdam. 1986).

MARSAN, D. (2006), *Can coseismic stress variability suppress seismicity shadows? Insights from a rate-and-state friction model*, J. Geophys. Res. *111*, B06305, doi:10.1029/2005JB004060.

MEAKIN, P., *Fractals: Scaling and Growth far From Equilibrium* (Cambridge Univ. Press, New York. 1998).

MEHEUST, Y. (2002), *Ecoulements dans les fractures ouvertes*, Ph. D. Thesis, Univ. Paris VI and Ecole Normale Supérieure, Paris.

OKUBO, P. G. and AKI, K. (1987), *Fractal geometry in the San Andreas fault system*, J. Geophys. Res. *92*, 345–355.

PALUMBO, L., BENEDETTI, L., BOURLES, D., CINQUE, A., and FINKEL, R. (2004), *Slip history of the Magnola fault (Apennines, Central Italy) from ^{36}Cl surface exposure dating: Evidence for strong earthquakes over the Holocene*, Earth Planet. Sci. Lett. *225*, 163–176.

PARSONS, T. (2008), *Persistent earthquake clusters and gaps from slip on irregular faults*, Nature Geosci. *1*, 59–63, doi:10.1038/ngeo.2007.36.

PEYRAT, S., OLSEN, K. B., and MADARIAGA, R. (2004), Which dynamic rupture parameters can be estimated from strong ground motion and geodetic data?, Pure Appl. Geophys. *161*, 2155–2169, doi: 10.1007/s00024-004-2555-9.

POWER, W. L., TULLIS, T. E., and WEEKS, J. D. (1988), *Roughness and wear during brittle faulting*, J. Geophys. Res. *93*, 15,268–15,278.

POWER, W. L., TULLIS, T. E., BROWN, S. R., BOITNOTT, G. N., and SCHOLZ, C. H. (1987), *Roughness of natural fault surfaces*, Geophys. Res. Lett. *14*, 29–32.

POWER, W. L. and TULLIS, T. E. (1991), *Euclidean and fractal models for the description of rock surface roughness*, J. Geophys. Res. *96*, 415–424.

POWER, W. L. and Durham, W. B. (1997), *Topography of natural and artificial fractures in granitic rocks: Implications for studies of rock friction and fluid migration*, Int. J. Rock Mech. Min. Sci. *34*, 979–989.

RENARD, F., VOISIN, C., MARSAN, D., and SCHMITTBUHL, J. (2006), *High resolution 3D laser scanner measurements of a strike-slip fault quantify its morphological anisotropy at all scales*, Geophys. Res. Lett. *33*, L04305, doi: 10.1029/2005GL025038.

RUBIN, A. M., GILLARD, D., and GOT, J. -L. (1999), *Streaks of microearthquakes along creeping faults*, Nature *400*, 635–641.

SAGY A., BRODSKY, E. E., and AXEN, G. J. (2007), *Evolution of fault-surface roughness with slip*, Geology *35*, 283–286.

SAGY, A. and BRODSKY, E. E. (2009), *Geometric and rheological asperities in an exposed fault zone*, J. Geophys. Res. *114*(B02301), doi:10.1029/2008JB005701.

SCHAFF, D. P., BOKELMANN, G. H. R., BEROZA, G. C., WALDHAUSER, F., and ELLSWORTH, W. L. (2002), *High-resolution image of Calaveras fault seismicity*, J. Geophys. Res. *107*(B9), 2186, doi:10.1029/2001JB000633.

SCHMITTBUHL, J., CHAMBON, G., HANSEN, A., and BOUCHON, M. (2006), *Are stress distributions along faults the signature of asperity squeeze*?, Geophys. Res. Lett. *33*, L13307, doi:10.1029/2006GL025952.

SCHMITTBUHL, J., GENTIER, S., and ROUX, R. (1993), *Field measurements of the roughness of fault surfaces*, Geophys. Res. Lett. *20*, 639–641.

SCHMITTBUHL, J., VILOTTE, J., and ROUX, S. (1995a), *Reliability of self-affine measurements*, Phys. Rev. E. *51*, 131–147.

SCHMITTBUHL, J., SCHMITT, F., and SCHOLZ, C. (1995b), *Scaling invariance of crack surfaces*, J. Geophys. Res. *100*, 5953–5973.

SCHMITTBUHL, J., STEYER, A., JOUNIAUX, L., and TOUSSAINT, R. (2008), *Fracture morphology and viscous transport*, Int. J. Rock. Mech. Min. Sci. *45*, 422–430.

SCHOLZ, C. H., *The Mechanics of Earthquake and Faulting* (Cambridge Univ. Press, New York. 2002).

SIMONSEN, I., HANSEN, A., and NES, O.M. (1998), *Using wavelet transforms for Hurst exponent determination*, Phys. Rev. E. *58*, 2779–2787.

SIMONSEN, I., VANDEMBROUCQ, D., and ROUX, S. (2000), *Wave scattering from self-affine surfaces*, Phys. Rev. E. *61*, 5914–7.

SMITH, R. B. (1977), *Formation of folds, boudinage, and mullions in non-Newtonian materials*, Geol. Soc. Amer. Bull. *88*, 312–320.

STEIN, M. L. (2002), *Fast and exact simulation of fractional Brownian sufaces*, J. Comput. Graph. Statist. *11*, 587–599.

THOUVENOT, F. (1998), *The ML 5.3 Epagny (French Alps) earthquake of 1996 July 15: A long-awaited event on the Vuache Fault*, Geophys. J. Int. *135*, 876–892.

TWISS, R. J. and MOORES, E. M., *Structural Geology* (W. H. Freeman, New York. 1992).

VOISIN, C., CAMPILLO, M., IONESCU, I., HASSANI, R., and NGUYEN, Q.-L. (2002a), *Process and signature of initiation on a finite fault system: A spectral approach, Geophys.* J. Int. *148*, 120–131.

VOISIN, C., IONESCU, I., and CAMPILLO, M. (2002b), *Crack growth resistance and dynamic rupture arrest under slip dependent friction*, Phys. Earth Planet. Inter. *131*, 279–294.

VOSS, R.F., *Scaling Phenomena in Disorder Systems* (Plenum, New York. 1985).

(Received September 30, 2008, accepted March 17, 2009)
Published Online First: June 29, 2009

To access this journal online:
www.birkhauser.ch/pageoph

Pure appl. geophys. 166 (2009) 1853–1884
0033–4553/09/101853–32
DOI 10.1007/s00024-009-0510-5

© Birkhäuser Verlag, Basel, 2009

Pure and Applied Geophysics

Spatio-temporal Slip, and Stress Level on the Faults within the Western Foothills of Taiwan: Implications for Fault Frictional Properties

Ya-Ju Hsu,[1,2] Jean-Philippe Avouac,[2] Shui-Beih Yu,[1] Chien-Hsin Chang,[3] Yih-Min Wu,[4] and Jochen Woessner[2,5]

Abstract—We use preseismic, coseismic, and postseismic GPS data of the 1999 Chi-Chi earthquake to infer spatio-temporal variation of fault slip and frictional behavior on the Chelungpu fault. The geodetic data shows that coseismic slip during the Chi-Chi earthquake occurred within a patch that was locked in the period preceding the earthquake, and that afterslip occurred dominantly downdip from the ruptured area. To first-order, the observed pattern and the temporal evolution of afterslip is consistent with models of the seismic cycle based on rate-and-state friction. Comparison with the distribution of temperature on the fault derived from thermo-kinematic modeling shows that aseismic slip becomes dominant where temperature is estimated to exceed 200° at depth. This inference is consistent with the temperature induced transition from velocity-weakening to velocity-strengthening friction that is observed in laboratory experiments on quartzo-feldspathic rocks. The time evolution of afterslip is consistent with afterslip being governed by velocity-strengthening frictional sliding. The dependency of friction, μ, on the sliding velocity, V, is estimated to be $\partial\mu/\partial \ln V = 8 \times 10^{-3}$. We report an azimuthal difference of about 10–20° between preseismic and postseismic GPS velocities, which we interpret to reflect the very low shear stress on the creeping portion of the décollement beneath the Central Range, of the order of 1–3 MPa, implying a very low friction of about 0.01. This study highlights the importance of temperature and pore pressure in determining fault frictional sliding.

Key words: The Chi-Chi earthquake, fault friction, fault rheology, stress, fault slip distribution.

1. *Introduction*

Fault slip occurs as steady aseismic creep or as recurring transient slip, the latter of which might be seismic or aseismic. This process leads to an alternation between long periods of slow elastic strain accumulation, the interseismic period, and short periods of strain release by transient slip events and large earthquakes. An unanswered first-order

[1] Institute of Earth Sciences, Academia Sinica, Nankang, Taipei, Taiwan.
E-mail: yaru@earth.sinica.edu.tw
[2] Division of Geological and Planetary Sciences, California Institute of Technology, California, U.S.A.
[3] Central Weather Bureau, Taipei, Taiwan.
[4] Department of Geosciences, National Taiwan University, Taipei, Taiwan.
[5] Swiss Seismological Service, ETH Zürich, Zürich, Switzerland.

question regarding the 'seismic cycle' process is whether asperities, defined loosely here as zones with large coseismic slip (e.g., larger than a given fraction of the peak slip), that broke during large earthquakes, coincide with fault patches that remain locked during the interseismic period. If so, geodetic monitoring of interseismic strain could be used to assess the location and, possibly either the time of a future earthquake or the amount of coseismic slip, hence the magnitude (SHIMAZAKI and NAKATA, 1980).

The fault slip accrues with time depending primarily on the fault friction law and its eventual spatio-temporal variability. Fault friction can be rate-weakening, allowing for stick-slip motion or rate-strengthening promoting aseismic slip (SCHOLZ, 1990). For quartz-feldspathic rocks, rate-strengthening seems to be promoted at high temperature (T > 200–250°C), hence at depths deeper than typically 15 km, or for poorly consolidate alluvium or fault gouge at shallow depths (BLANPIED *et al.*, 1995; MARONE, 1998). However, the relation between fault friction and the mode of slip is not trivial. For example, earthquakes can only nucleate in areas obeying a rate-weakening friction law but can propagate dynamically in rate-strengthening areas (LAPUSTA *et al.*, 2000; TSE and RICE, 1986). Furthermore, a fault portion might obey a rate-strengthening friction law but might seem locked if it lies in the stress shadow of a fully locked zone. In that case, it might slip by aseismic afterslip when the stress shadow effect is removed as the shadowing asperities break. For example, it is possible that the shallow portion of subduction zones would dominantly slip aseismically, but would appear locked in the interseismic period as it lies in the shadows of deeper locked patches (BURGMANN *et al.*, 2005). This mechanism might also explain why shallow creep is rarely observed, except in the case of fault segments which dominantly creep at all depths such as the creeping segment of the San Andreas fault (BURFORD and HARSH, 1980; LISOWSKI and PRESCOTT, 1981; TITUS *et al.*, 2005) or along the southern portion of the Longitudinal Valley fault in Taiwan (ANGELIER *et al.*, 1997; LEE *et al.*, 2001b, 2003; LIU and YU, 1990; YU and LIU, 1989; YU and KUO, 2001).

Another issue, which has been a controversial issue for decades, concerns the stress level under which active faults operate (SCHOLZ, 2000). Deep borehole measurements in a stable tectonic area (BRUDY *et al.*, 1997; ZOBACK and TOWNEND, 2001) suggest that the differential stress within the brittle crust is generally high, near the threshold of Coulomb failure as obtained from laboratory estimates of the coefficient of friction of rocks which generally range 0.6–0.85 (BYERLEE, 1978). On the other hand, there is evidence that the shear stress acting on major crustal faults is rather small, typically about 10 times less than expected from rock friction experiments. This conclusion is reached particularly for the San Andreas fault zone where frictional heating seems very limited (SHAW, 1995) and where the principal horizontal stress is nearly perpendicular to the fault plane (HAUKSSON, 1990; JONES, 1988; MOUNT and SUPPE, 1987; ZOBACK *et al.*, 1987). A low stress level on fault zones has also been deduced from the influence of coseismic stress change due to the Landers earthquake (HARDEBECK and HAUKSSON, 2001) as well as 2002 Denali earthquake (WESSON and BOYD, 2007) on the stress orientations derived from earthquake focal mechanisms. There is more consensus that thrust faults and detachments below

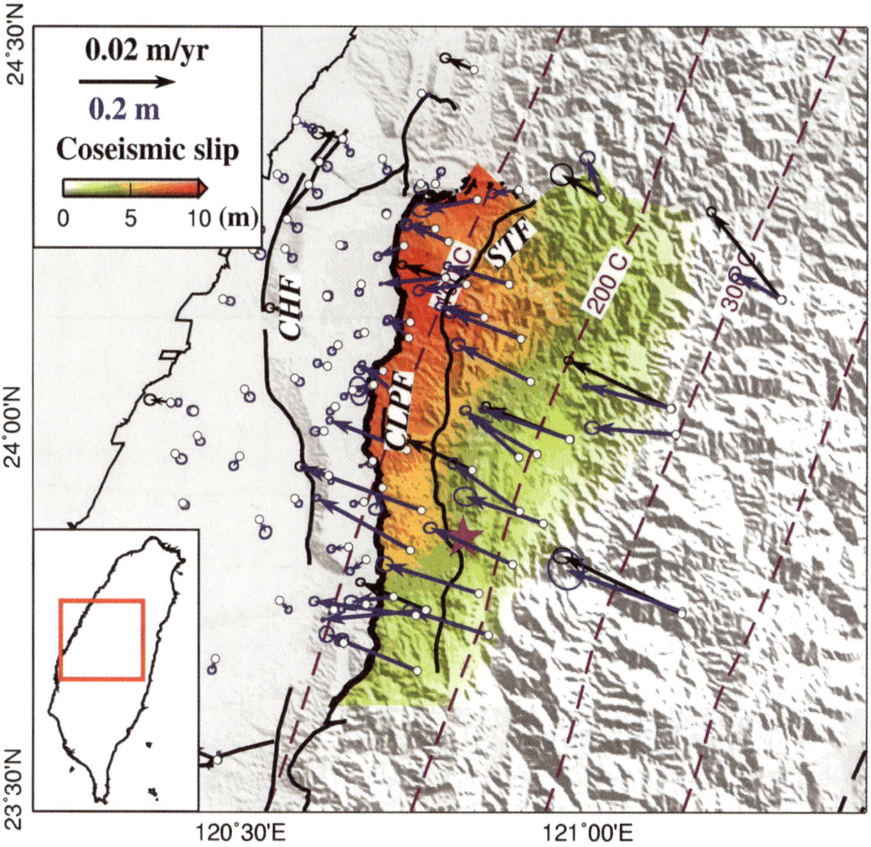

Figure 1
GPS velocities across the western foothills of central Taiwan and coseismic slip distribution of the 1999, M_w 7.6, Chi-Chi earthquake. Black arrows show interseismic GPS velocities (1993–1999), with 95% confidence ellipses, relative to the permanent GPS stations on Penghu Island before the Chi-Chi earthquake (Yu *et al.*, 2001). Blue arrows indicate postseismic displacement over 15 months following the Chi-Chi earthquake (Yu *et al.*, 2003). Distribution of coseismic slip, derived from inversion of GPS displacements is shown in color. The pink star denotes the epicenter of the Chi-Chi earthquake. Isotherms on the fault (purple dash lines) are derived from the thermo-kinematic model (Simoes *et al.*, 2007a). Major active faults are indicated as black lines and the 1999 Chi-Chi rupture is shown as a thick black line. CHF: Changhua fault, CLPF: Chelungpu fault, STF: Shuantung fault.

orogenic wedges, including Taiwan, need to be weak as required from their generally low taper angle (*e.g.*, Davis *et al.*, 1983; Suppe, 2007).

We discuss these questions based on the analysis of geodetic strain across the western foothills of Taiwan where a M_w 7.6 earthquake occurred in 1999 (Figs. 1 and 2). We take advantage of the wealth of geodetic, seismological and geological data available from this area to investigate the contribution of interseismic slip, afterslip and coseismic slip to the long-term fault slip and the resulting stress transfer.

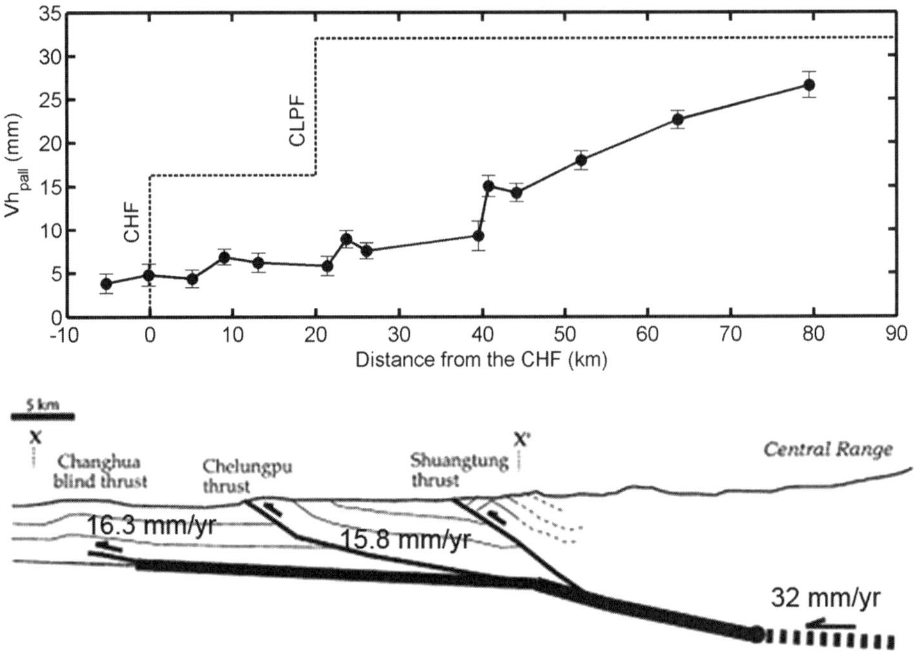

Figure 2

A schematic of interseismic velocity profile parallel to the plate convergence direction and a simplified geological cross section across the western foothills of Taiwan. Studies on the deformed strata terrace inferred slip rates of 16.3 ± 4.1 mm/yr on the Changhua fault (CHF) and 15.8 ± 5.1 mm/yr on the Chelungpu fault (CLPF) (SIMOES *et al.*, 2007b). The geological cross section is constrained from numerous seismic profiles in that the faults merge at depth into a single major décollement (YUE *et al.*, 2005). The geodetic data show that the active faults are locked at the surface but some aseismic deformation must occur at depth below the Central Range, probably along the downdip continuation of the décollement (CARENA *et al.* 2002). The shortening rates across the Changhua and the Chelungpu faults are approximately equal to the geodetic shortening rate across the western foothills, suggesting that little deformation occurs elsewhere, other than on these two major faults.

In the following, we first give an overview of the seismotectonic context of the study area. Next, we analyze the pre-, co-, and postseismic data in a self-consistent manner to assess how fault patches with interseismic, coseismic and postseismic slip pave the fault to contribute to a quasi-uniform slip distribution in the long term and the significance of the difference of azimuth between pre- and postseismic GPS velocities.

2. *Seismotectonic Setting of the Western Foothills of Taiwan*

Three major active faults are recognized along the western foothills of central Taiwan. Seismic profiles and bedding dip angles measured at the surface suggest that they all sole into a single décollement at a depth of about 6 km (YUE *et al.*, 2005) (Fig. 2). The most frontal fault is the Changhua fault which is a blind thrust fault responsible for the formation of Pakuashan anticline at its tip. The shortening rate across Pakuashan is estimated to be 16.3 ± 4.1 mm/yr from the study of dated growth strata deposited over the last 300 ka (SIMOES *et al.*, 2007b). The most internal fault is the Shuangtung fault. The Shuangtung fault also deforms recent terraces, but based on its geomorphic expression, it seems to be a second-order feature compared to the two other faults. The Chelungpu fault lies between the Changhua fault and the Shuangtung fault. The shortening rate across the Chelungpu fault is estimated to be 15.8 ± 5.1 mm/yr from the deformation of dated Holocene terraces (SIMOES *et al.*, 2007b).

GPS data acquired over the years preceding the Chi-Chi earthquake (YU *et al.*, 2001) indicate a shortening rate across the western foothills approximately equal to the sum of the long-term shortening rates across Pakuashan anticline and Chelungpu fault (Fig. 2). It follows that the ~30 mm/yr shortening rate across the western foothills has to be taken up by either aseismic or seismic slip on the Chelungpu and Changhua faults and their downdip continuation as a décollement (or as a shear zone with finite thickness).

The M_w 7.6, 1999 Chi-Chi earthquake ruptured the Chelungpu thrust fault along the western foothills of Taiwan extending approximately 100-km-long with up to 10–15 m of coseismic slip (JI *et al.*, 2001; JOHNSON *et al.*, 2001; SHIN and TENG, 2001). GPS data acquired over the years preceding the Chi-Chi earthquake (YU *et al.*, 2001) indicate that the area with significant coseismic slip is primarily locked before the main shock (DOMINGUEZ *et al.*, 2003; HSU *et al.*, 2003; LOEVENBRUCK *et al.*, 2001). GPS data acquired after the main shock shows postseismic deformation reflecting dominantly afterslip downdip from the coseismic rupture area (HSU *et al.*, 2002, 2007; YU *et al.*, 2003).

GPS measurements also reveal a systematic counter-clockwise rotation of postseismic surface displacements relative to interseismic velocities at the sites within the Central Range (Fig. 1). The azimuthal differences between preseismic and postseismic GPS velocities are only marginally significant at each single site where both pre- and post-Chi-Chi measurements are available. The fact that the postseismic azimuth is systematically rotated counter-clockwise at all GPS sites located within the range is an indication that this rotation is probably not fortuitous, however. The possibility that the difference of azimuth can be a reference frame issue can be excluded since the interseismic and postseismic GPS data were processed using identical procedures and the same reference frames (International Terrestrial Reference Frame 1997, ITRF97) (YU *et al.*, 2001, 2003). We analyze below the significance of this observation.

3. Modeling of Geodetic Strain

3.1. Method and Data

The geodetic data used in this study include coseismic displacements (YU *et al.*, 2001), the postseismic GPS displacements recorded over 15 months after the mainshock

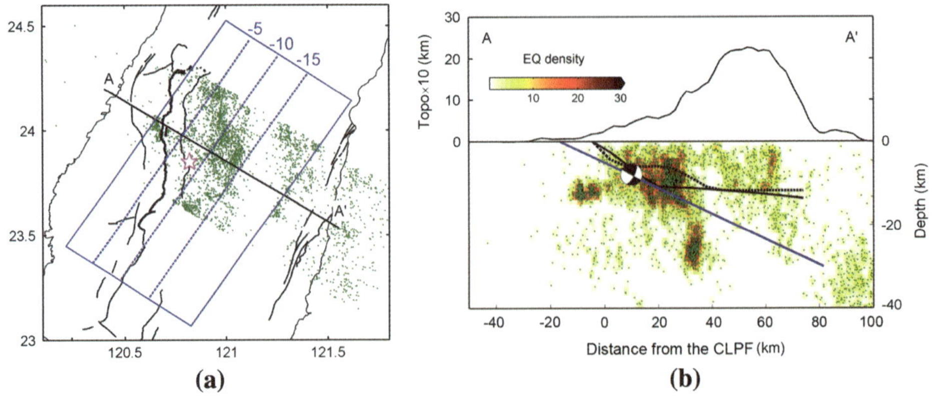

Figure 3

Geometries of the Chelungpu fault, and of the simplified planar fault used to model interseismic strain from a forward slip modeling approach. (a) The green circles denote aftershocks within a 60-km wide swath in the first 15 months after the Chi-Chi main shock. A-A' and blue rectangle indicate a NW-SE trending transect of the modeled Chelungpu fault and the simplified planar fault. Blue dash lines denote iso-depth contours of the simplified planar fault. The depth in kilometer is shown as a blue number. (b) The depth profiles along A-A' indicated in (a). The black solid and dash lines indicate the fault geometries determined by inversions and Yue *et al.* (2005), respectively. The blue line denotes the fault geometry of the simplified planar fault used to model interseismic strain. Color scale indicates the earthquake density (number of event/ km^2). The beach ball shows the focal mechanism of the Chi-Chi main shock.

(Yu *et al.*, 2003), and GPS site velocities collected before the Chi-Chi earthquake that consists of 12 campaign-mode sites and two permanent sites (Yu *et al.*, 2001).

For the purpose of this study, we model the interseismic GPS velocities (1993–1999) and the postseismic GPS displacements (over the 15 months following the Chi-Chi earthquake) in a self-consistent manner, *i.e,* assuming the same geometry for the décollement and using the same modeling technique.

We first utilized the 3-D fault geometry constrained by surface geology, seismic profiles, well data and geological balanced cross sections proposed by Yue *et al.* (2005) (black dashed line in Fig. 3b). The fault geometry is well constrained beneath the western foothills from reliable seismic data and surface structural constraints. It is assumed that the fault extends eastward beneath the Central Range as a décollement at a relatively shallow depth as suggested from relocated seismicity (Carena *et al.*, 2002). Because the geometry of the fault beneath the Central Range is not that well constrained, we also searched for the geometry that best fits the geodetic data (black solid line in Fig. 3b). It should be pointed out that the model assumes that deformation at depth can be represented by fault slip on a localized décollement. In reality, it is probable that deformation is more distributed than we assume and that some amount of pure shear must be taking place to allow shortening and thickening of the crust beneath the range. The décollement is therefore a convenient representation of deformation at depth although not necessarily a very realistic representation.

The fault is subdivided into ~ 150 small rectangular patches with sizes ranging from 5 km $\times$ 5 km to 8 km $\times$ 8 km. This representation of the fault was found to be a reasonable compromise so that the fault model follows reasonably well the sinuous fault trace as observed in the field, and its spatial extent is large enough to allow modeling coseismic, postseismic, and interseismic slip. The Green functions, *i.e.,* the surface displacements at the sites of observations related to unit slip vectors with either pure dip-slip or pure strike-slip components, are computed from the theory of elastic dislocation in an elastic half space (OKADA, 1985). The rigidity modulus is fixed to a standard value of 30 GPa. Because the preseismic and postseismic displacements show relatively stable azimuths (HSU *et al.*, 2007; YU *et al.*, 2001), we ignore temporal variations of rake during interseismic or postseismic deformation. Furthermore, we ascertained that the interseismic data can be fitted reasonably well without introducing spatial variations in rake.

The data are inverted using a weighted least-squares approach by minimizing the following functional:

$$F(\mathbf{s}, \alpha, \beta, \mathbf{m}) = \left\| \sum^{-1/2} (G(\mathbf{m})\mathbf{s} - \mathbf{d}) \right\|^2 + \alpha^{-2} \left\| \nabla^2 \mathbf{s} \right\|^2 + \beta^{-2} \|\mathbf{s}\|^2, \tag{1}$$

where $\Sigma^{-1/2}$ is the inverse square-root of the data covariance matrix; $G(\mathbf{m})$ are Green's functions which depend on the fault geometry parameters $\mathbf{m}$; $\mathbf{s}$ is slip; $\mathbf{d}$ is the observed displacements and ∇^2 is the finite-difference approximation of the Laplacian smoothing operator. To regularize the inversions, we use two damping parameters which characterizes the weights put on the model smoothness, α, and on minimizing the geodetic moments, β, and they are determined by cross validation (MATTHEWS and SEGALL, 1993). The fit to the data is quantified from the mean of the normalized square residuals defined as:

$$\chi_r^2 = \frac{1}{N} \left\| \sum^{-1/2} (G(\mathbf{m})s - \mathbf{d}) \right\|^2 = \frac{1}{N} \sum_{i=1}^{N} \left(\frac{d_i^{obs} - d_i^{pred}}{\sigma_i} \right)^2, \tag{2}$$

where N is the number of data points; σ_i is the 67% uncertainty on the i-th data point; $d_i^{\text{obs}}, d_i^{\text{pred}}$ are the observed and predicted values on the i-th data point. A value of 1 of χ_r^2 means that the model fits the data within uncertainties on average. Note that although this quantity, χ_r^2, resembles the reduced chi-square, it does not have the same statistical significance since the number of model parameters is not considered.

3.2. Coseismic Slip Model

The coseismic slip model obtained using the 3-D fault geometry proposed by YUE *et al.* (2005) yields χ_r^2 of 599 (black dashed line, Fig. 3b). The weighted root-mean-square misfit (wrms) of the horizontal and vertical displacements are 0.15 m and 0.21 m, respectively, about 10–20 times the estimated uncertainties. If the fault geometry beneath

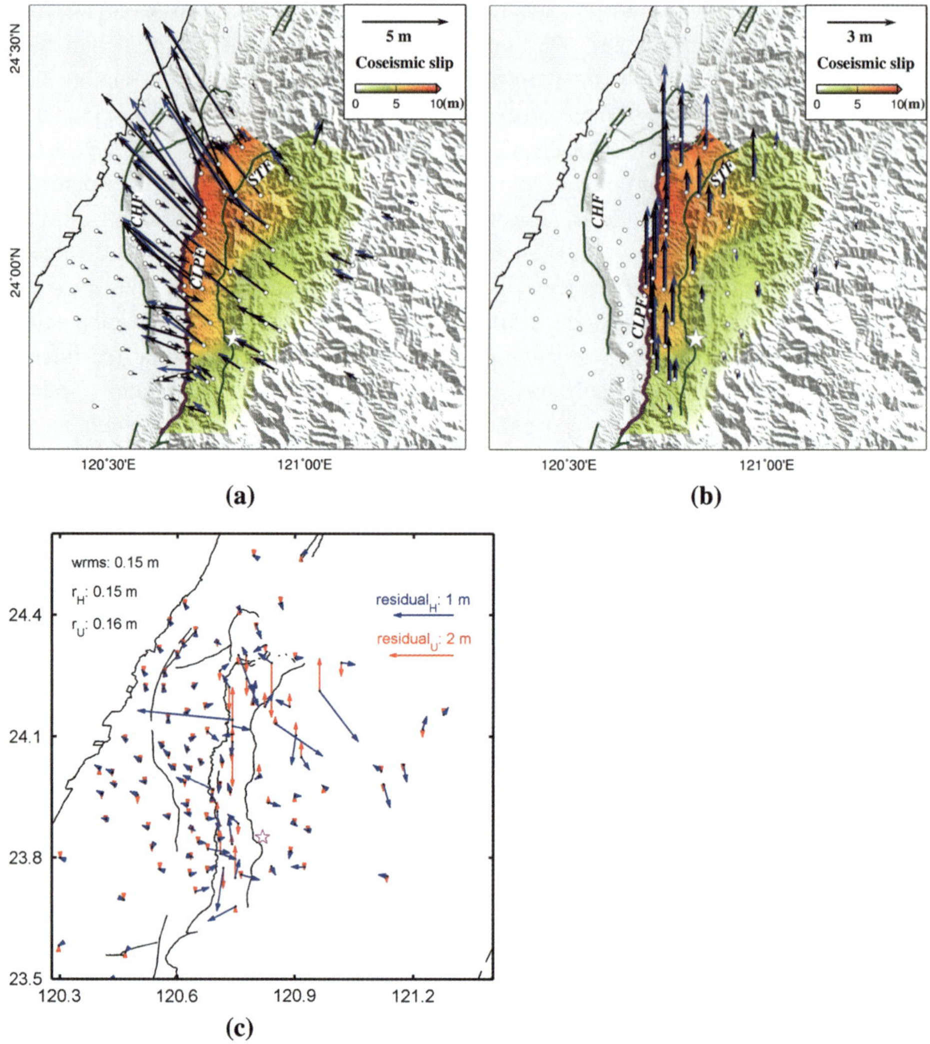

Figure 4

Coseismic slip distribution of the 1999 M_w 7.6 Chi-Chi earthquake. (a) Black and blue arrows indicate observed coseismic horizontal displacements (Yu *et al.*, 2001) and predictions from our best-fitting coseismic slip model. Color indicates coseismic slip. White star indicates the epicenter of the Chi-Chi earthquake. (b) Vertical displacements. Major faults are indicated as green lines and the 1999 Chi-Chi rupture is shown as a purple line. (c) Residuals of horizontal (blue) and vertical (red) displacements.

the Central Range is optimized (black solid line, Fig. 3b), the misfit, χ_r^2, is reduced to 499 (Fig. 4). Figure 4c shows the residuals between the observed displacements and those computed from the optimal coseismic model. Although the residuals are larger than the data uncertainties on average, these residuals manifest no systematic pattern that could

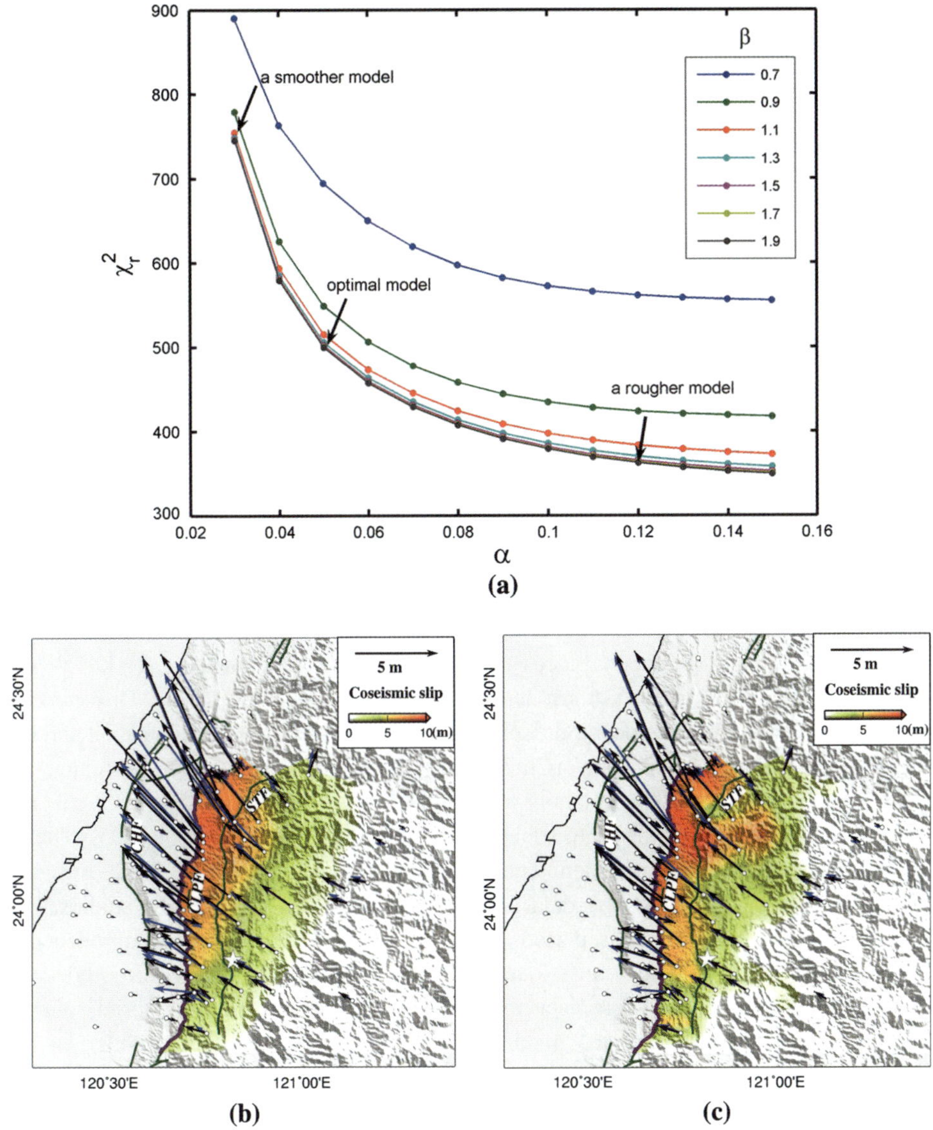

Figure 5

(a) Plot showing how the value of the χ_r^2 (see equation (2)) evolves as a function of the weight put on the model smoothness (proportional to $1/\alpha$) and on minimizing the geodetic moment (proportional to $1/\beta$). (b) Black and blue arrows indicate observed coseismic horizontal displacements (Yu *et al.*, 2001) and predictions from the smoother coseismic slip model indicated in (a). (c) Same as (b) but for the rougher slip model indicated in (a).

inspire improvement of the fit further. Figure 5a shows how the fit to the data evolves as a function of the weight put on the model smoothness (proportional to $1/\alpha$) and on minimizing the geodetic moment (proportional to $1/\beta$). Figures 5b and 5c show the slip

Table 1

The reduced-chi-squares (χ_r^2), *geodetic moment and the rate of moment deficit in coseismic, postseismic, and interseismic slip models*

	Coseismic model	Postseismic model	Interseismic model		
			Model 1	Model 2	Model 3
χ_r^2	499	8.5	24.5	10.9	2.7
Geodetic moment/ moment rate	3.6×10^{20} (N m)	3.9×10^{19} (N m)	1.2×10^{18} (N m/yr)	2.3×10^{18} (N m/yr)	3.7×10^{18} (N m/yr)

distributions obtained for alternative models smoother and rougher, respectively, than the preferred model of Figure 4. It ensures that even if the slip distribution is allowed to be rougher than that of the preferred model, the fit to the data does not improve substantially. There is therefore no obvious way to improve the fit to the data within the framework of our modeling which assumes that the surface deformation is due to slip on a single fault embedded in a homogeneous elastic half space. The three models shown in Figures 4 and 5 are only marginally different. We conclude that the preferred model is a robust first-order representation of the main characteristics of the coseismic slip distribution.

The optimal fault geometry places the décollement at a depth of ~ 12 km depth, considerably deeper than the 5–6 km depth proposed by Yue *et al.* (2005). This value is more consistent with the hypocentral depth of the Chi-Chi earthquake main shock and the depth distribution of aftershocks (Chang *et al.*, 2007). Our fault geometry in fact approaches Yue *et al.*'s alternative fault model which was proposed to better reconcile the structural constraints with the main shock hypocenter (Fig. 21, Yue *et al.*, 2005). The fit to the geodetic data remains, deficient, corresponding to a wrms of 0.15 (Fig. 4c), however this model does fit the data better than the models in previous studies. For example, Johnson *et al.* (2001) reported a wrms of 0.19 m obtained from a homogeneous elastic half-space model and Johnson and Segall (2004) obtained wrms values between 0.155 m and 0.17 m from layered elastic models, with eventual lateral material contrast across the fault. Reconciling the geodetic data and the structural constraints on the Chelungpu fault geometry would probably require allowing for variations of elastics properties in 3-D and eventual anelastic deformation off the main fault zone. Nonelastic deformation of the footwall has indeed been documented in the field and is an expected feature given the nonplanar geometry of the fault (Chen *et al.*, 2007b). These sophistications are beyond the scope of this study. Also it is possible that the uncertainties assigned to the coseismic displacements may be underestimated.

Most of coseismic slip is shallow with a maximum near the surface and a gradual downdip decrease to negligible slip at about 10 km depth (Fig. 4). The maximum slip in our best-fitting model is 10 m, located at the northwestern edge of the fault. The slip potency is 11873 km^2-m, corresponding to a moment of 3.6×10^{20} N-m (Table 1), which is somewhat larger than the value of 2.7×10^{20} N-m estimated from the joint inversion of geodetic and seismological data (JI *et al.*, 2003), but close to the scalar moment of 3.4×10^{20} N-m of the Harvard CMT solution.

Our coseismic slip distribution is quite similar to those obtained in the previous studies that were based on the same GPS data set (JOHNSON *et al.*, 2001; JOHNSON and SEGALL, 2004) or jointly with remote sensing data (DOMINGUEZ *et al.*, 2003; LOEVENBRUCK *et al.*, 2004) or seismological data (JI *et al.*, 2003), which all yielded misfits similar to those obtained with our modeling. In these other models, the fault geometry is similar with the strike approximately N-S and dip 30° to the east. All these models use a homogeneous earth model, except in the cases of JI *et al.* (2003) which assume a layered earth model and JOHNSON and SEGALL (2004) which explores different layered structures with lateral contrast. Despite different assumptions and data sets used in these models, they all show slip distributions similar to the one we have obtained with a maximum coseismic slip generally between 10 and 15 m on the shallow northern portion of the Chelungpu fault.

3.3. *Afterslip Model*

We compute the afterslip model corresponding to cumulative slip over 15 months using the same procedures as that used to derive the coseismic slip model. Figure 6 shows the corresponding afterslip distribution. The fit to the surface postseismic displacements corresponds to a χ_r^2 of 8.5 (Table 1). The less satisfactory fit is likely because our model accounts only for afterslip while other processes such as poroelastic effect and postseismic viscous relaxation probably also contribute. However, it is clear that afterslip is the dominant source of the observed postseismic surface strain (HSU *et al.*, 2007). In our model, afterslip over the 15 months following the Chi-Chi earthquake released a cumulative moment of 3.9×10^{19} N-m, constituting about 11% of the coseismic moment.

3.4. *Interseismic Model*

The interseismic and postseismic geodetic data show a gradual increase of horizontal velocities relative to the foreland across the Central Range, consistent with elastic deformation of the crust driven by aseismic creep along a basal décollement, while shallow faults remain locked. To first order, the geodetic data can thus be modeled reasonably well assuming that the shallow portion of the décollement is fully locked (creep rate is zero) until it reaches some depth that is adjusted to best fit the geodetic observations (DOMINGUEZ *et al.*, 2003; HSU *et al.*, 2003; LOEVENBRUCK *et al.*, 2001).

One difficulty arises because both the Changhua and the Chelungpu faults might be considered. In theory, if creep occurs only downdip of where they merge into their common décollement, the models obtained using one, the other or both faults should be equivalent. This is because the geometry of any fully locked fault should not matter since it does not creep and hence does not contribute to interseismic strain. This is

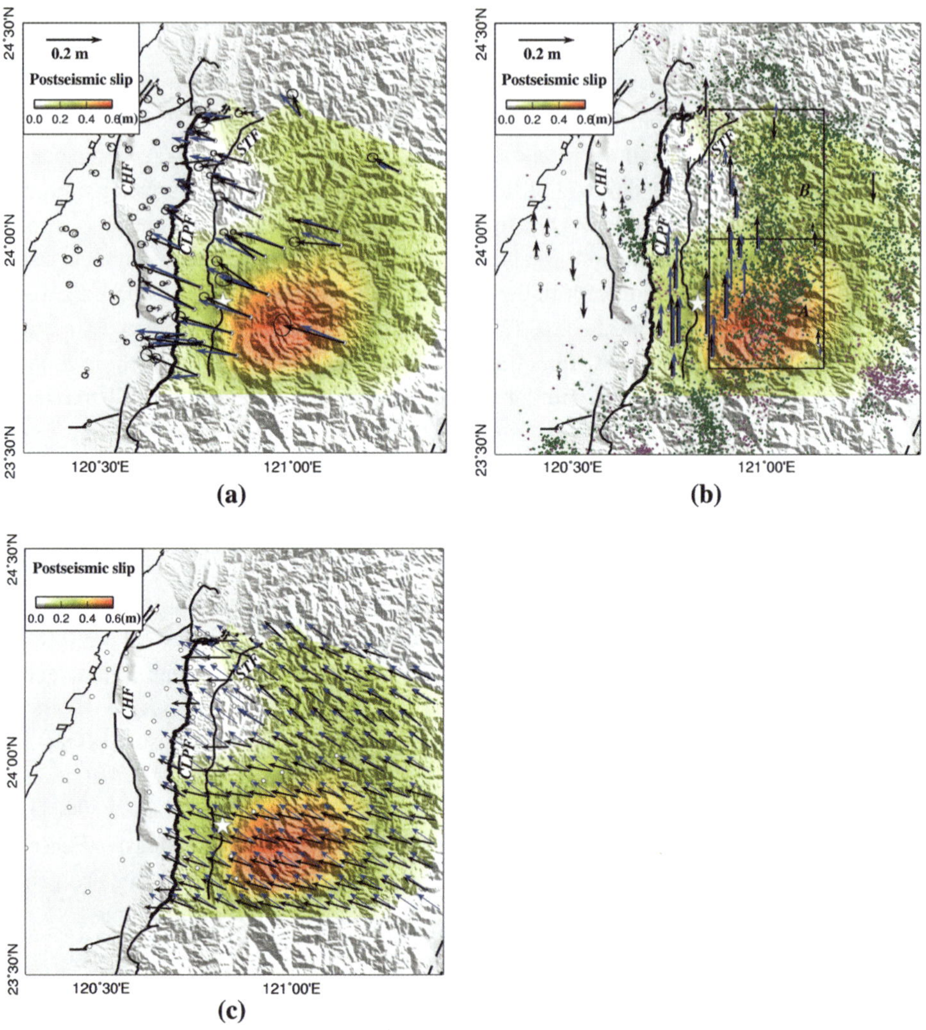

Figure 6

Postseismic slip distribution of the Chi-Chi earthquake (a) Black and blue arrows indicate observed (Yu *et al.*, 2003) and predicted postseismic horizontal displacements. Color indicates cumulative postseismic slip over a 15-month period following the main shock. White star shows the epicenter of the Chi-Chi earthquake. (b) Vertical displacements. Black boxes indicate the two regions (*A*: south, *B*: north) within which the stress tensor is determined from earthquake focal mechanisms. Purple circles denote the seismicity between 1992 and 1999.7, before the Chi-Chi main shock. Green circles denote the seismicity between 1999.7 and 2001, after the Chi-Chi main shock. (Data from the Central Weather Bureau in Taiwan) (c) Azimuths of postseismic (black vectors) and interseismic (blue vectors) fault slip vectors.

approximately the case here and we therefore ignore the Changhua fault in the modeling of the interseismic strain. This is equivalent to assuming that the Changhua fault is fully locked entirely to where it merges with the Chelungpu fault. Note however that when it comes to estimating the rate of accumulation of moment deficit in the interseismic period, it is important to take into account real fault geometries and how slip rate on the décollement is partitioned between slip on the Changhua fault and on the Chelungpu fault.

Either a backslip (Savage, 1983) or a forward slip modeling approach can be adopted to provide for the interseismic slip rate distribution. In the present context, the main advantage of the backslip model is that the area that needs to be modeled is small. However, the backslip model assumes that no elastic strain accumulates over the long term; an assumption that holds only if the creeping fault is strictly planar. Comparisons with finite-element modeling of the seismic cycle on a thrust fault in which distributed inelastic deformation and erosion is allowed, have shown that both the forward slip and the backslip approaches are reasonable first-order approximations, given that the geometry of the creeping fault is close enough to planar (Vergne *et al.*, 2001). The most important difference between the two approaches lies in their opposite edge effects. With the backslip model, areas of the fault that are not explicitly modeled (at depths or laterally) are implicitly assumed to creep at the long-term slip rate; while they are implicitly assumed to be locked in the forward slip model. Here, we use both approaches and compare their results. We present in fact three interseismic models. The first two are obtained from a backslip modeling approach employing a fault geometry consistent with that used for the modeling of coseismic and postseismic deformation. In these two models, the long-term slip rate is assumed to be constant (30 mm/yr) and the rake is uniform, set to an optimum value of 304° obtained from a grid search. Because the Chelungpu fault is non-planar, the backslip approach is not well justified. The long-term elastic strain accumulation leads to very large stresses, violating the assumption that long-term deformation does not lead to strain. We therefore also produce a model from a forward slip modeling approach. In that case, we additionally optimize the fault geometry to fit the geodetic data best.

In Model 1, we test the hypothesis that the slip distribution due to the Chi-Chi earthquake would mirror the interseismic slip rate pattern. To do so, we consider a backslip model in which the fault moves backward at a rate proportional to the sum of the coseismic slip and afterslip models. In that case, the only parameter is the 'long-term' slip rate. A long-term slip rate of 30 mm/yr yields the best fit to the data (Fig. 7a) that corresponds to a χ_r^2 of 24.5 (Table 1). The fit is not very satisfactory, particularly because the model predicts too much creep along the northern and southern segments of the Chelungpu fault (Fig. 7a). We conclude that the coseismic plus postseismic slip model is not exactly proportional to the interseismic slip deficit.

Model 2 (Fig. 7b) was obtained by inverting for the best-fitting backslip distribution. Here we used a fault geometry slightly different from the one for the coseismic and postseismic models. To avoid the change of the fault strike on the northern Chelungpu fault, we connect the Chelungpu fault to the Sanyi fault northward. The optimal slip model corresponds to a χ_r^2 of 10.9 (Table 1). The slip distribution obtained from this approach shows strong locking of most of the area that ruptured during the Chi-Chi

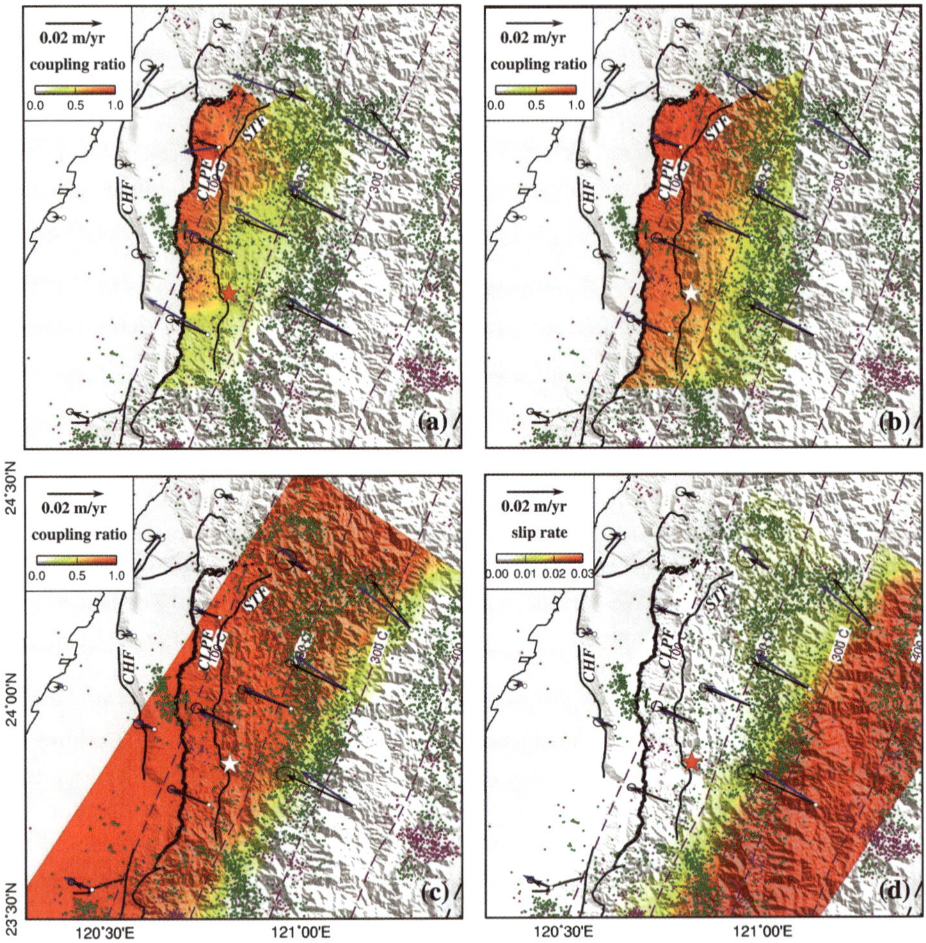

Figure 7

Interseismic coupling, observed and predicted interseismic horizontal surface velocities. The coupling ratio is defined as 1 minus the ratio of the interseismic creep rate to the long-term creep rate (30 mm/yr). Purple and green dots show earthquakes with $M_L > 3$ recorded between 1993 and 1999.7, and between 1999.7 and 2001 (data from the Central Weather Bureau), respectively. The models in (a) and (b) were obtained from a backslip modeling approach (SAVAGE, 1983) based on the coseismic fault geometry. The model in (a) assumes a backslip distribution exactly matches the sum of coseismic slip and afterslip distributions on the Chelungpu fault. The model in (b) was obtained from least-squares adjustment of the backslip distribution. The model in (c) was obtained from the inversion of fault creep rate on the décollement, using a forward slip modeling approach. The corresponding creep rate distribution is shown in (d). The reduced chi-square values corresponding to these models are listed in Table 1. Note that, as noticed in an earlier study (DOMINGUEZ *et al.*, 2003), that seismicity follows quite closely the downdip edge of the locked fault zone. Isotherms on the fault (purple dash lines) are derived from the thermo-kinematic model (SIMOES *et al.*, 2007a).

earthquake. The main differences compared with the previous model are that the slip distribution is less compact and that locking is stronger on the southern part of the Chelungpu fault where only limited moment was released in 1999. The misfits at the northernmost sites are probably due to the fact that the fault area beyond the northern limit of the model is implicitly assumed to be creeping, while it is probably partly locked in reality.

Model 3 (Fig. 7c) was obtained from the inversion of the surface velocities from a forward slip modeling approach. The geometry of the décollement and its continuation beneath the Central Range is approximated using a single planar fault (blue rectangle in Fig. 3a). We search for the optimal fault geometry where interseismic slip occurred as a uniform rate of 30 mm/yr in the direction of 304°. The best-fitting model corresponds to a décollement striking 34°, and dipping 17° to the east (Fig. 3). The fault extends somewhat deeper with a dip more steeply than the décollement inferred from the modeling of the coseismic deformation, but both meet with the ramp of the Chelungpu fault at a depth of 14 km (Fig. 3b). We impose a uniform slip rate of 30 mm/yr in the direction of 304° on the fault deeper than the locking depth (14 km) and predicted surface velocities on GPS sites. We then invert for a slip model on the westward continuation of the décollement using the residual velocity field. The optimal interseismic slip distribution is the sum of the uniform slip model and residual slip model. The fit to the data corresponds to a χ_r^2 of 2.7. The spatial distribution is similar to that obtained in Model 2. The result in Model 3, like Model 2, also requires strong locking at the south, although the width of the locked zone is significantly narrower than that to the north. The coupling pattern resembles the distribution of coseismic slip, although it extends somewhat deeper where most afterslip occurred.

Seismicity in the interseismic and postseismic periods appears to be clustered along the downdip edge of the locked fault zone (Fig. 7). That is, the seismicity clearly follows the boundary between the locked fault zone and the creeping décollement. This was noticed first by Dominguez *et al.* (2003) who suggested that, as observed in the Himalaya (Bollinger *et al.*, 2004; Cattin and Avouac, 2000), seismicity would be triggered by stress build up near the up-dip edge of the creeping zone.

4. Estimation of the Stress Level Associated to the Décollement beneath the Central Range

The interseismic model predicts slip on the décollement striking 304°, approximately parallel to the convergence between the Philippine Sea Plate and South China, while the postseismic model yields slip striking about 275°, if assumed uniform (Fig. 8). Slip on the basal décollement (or equivalently a shear zone with a finite width) should be colinear with the shear stress acting on it. Following a reasoning similar to that of Hardebeck and Hauksson (2001), who analyzed the stress change due to the Landers earthquake using earthquake focal mechanisms, the difference of azimuth between interseismic velocities and postseismic displacements, if significant, suggests a coseismic stress change of the order of magnitude of the preseismic stress on the décollement.

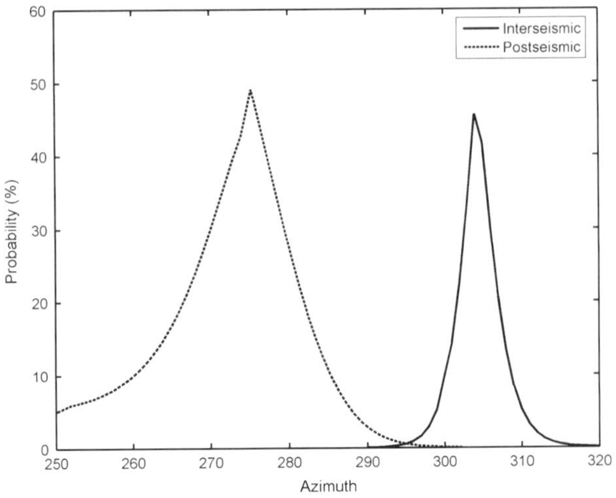

Figure 8
Azimuth of slip on the décollement inferred from the inversion of the interseismic and of the postseismic data. These inversions are conducted assuming a uniform rake. The t-test is used to estimate the probability distribution.

4.1. Significance of the Difference between Preseismic and Postseismic Azimuths

In order to test the significance of the difference of rakes between preseismic and postseismic models, we have run inversions of the interseismic GPS data and of the postseismic data assuming a uniform rake that was varied between 250° and 320° (Fig. 8). Using the t-test, we estimate the probability of rejection of any particular value of the slip azimuth in view of the fit obtained to either the interseismic or postseismic data. The uniform rake that best fit the interseismic data is 304° and the one that best fit the postseismic data is 275°. The t-test shows that the probability that the post- and the interseismic data reflect a slip distribution with the same rake, assumed uniform, is less than 5%. This shows that the rakes required to fit the postseismic and the interseismic data do differ significantly. If we believe that the assumption of uniform rake is valid and that formal uncertainties on the GPS data are reasonably well estimated, the difference of slip rakes in the interseismic and postseismic periods is found to be significant from 0 at the 95% confidence level. However, it should be realized that the rake might not be uniform and that interseismic and postseismic slip is not occurring at the same location. It is difficult to demonstrate rigorously that interseismic and postseismic rakes differ significantly when considered at the same location on the fault, due to the lack of spatial resolution in the interseismic model. Now, if we consider the afterslip model in which the rake is allowed to be nonuniform (Fig. 6c), it shows a rake of about 285° in the area with maximum afterslip (area *A* in Fig. 6b). It is meaningful to compare the interseismic and postseismic rake there because this area beneath the Central Range is with significant slip in the interseismic period as well as in the postseismic period. We therefore conclude that in area *A* the coseismic

Table 2

The stress tensors obtained in area A and area B (Fig. 9) are characterized by strikes and plunges of the three principal stress axes, σ_1, σ_2 and σ_3, in degrees

	Number of events	σ_1		σ_2		σ_3		φ^a	Azi^b_{stress}	Azi^c_{slip}
		strike	plunge	strike	plunge	strike	plunge	$\frac{\sigma_2-\sigma_3}{\sigma_1-\sigma_3}$		
South[1](*A*)	**59**	**287°**	**19°**	**196°**	**2°**	**100°**	**71°**	**0.48**	**286° ± 6°**	**285°**
North[1] (*B*)	**33**	**134°**	**3°**	**296°**	**87°**	**44°**	**1°**	**0.56**	**206° ± 61°**	**300°**
South[2](*A*)	40	288°	24°	28°	22°	155°	57°	0.66	298° ± 10°	285°
North[2] (*B*)	41	129°	4°	5°	82°	219°	7°	0.27	275° ± 66°	300°
South[3] (*A*)	42	287°	27°	21°	10°	129°	61°	0.68	291° ± 4°	285°
North[3](*B*)	70	129°	6°	343°	83°	220°	4°	0.40	252° ± 38°	300°

[a] The parameter, φ, is the ratio of the principal stress difference.
[b] The azimuth of the shear stress computed on a horizontal décollement with 95% confidence interval based on the bootstrap result of 1000 resampling focal mechanism data sets.
[c] The azimuth of postseismic slip on a horizontal décollement.
[1] Data from CWB and BATS with magnitude larger than 4 and focal depth less than 35 km.
[2] Data from Wu *et al.* (2008) with magnitude larger than 4 and focal depth less than 20 km.
[3] Data from Wu *et al.* (2008) with magnitude larger than 4 and focal depth less than 35 km.

stress change seems to have induced a notable change of slip azimuth. By contrast, Figure 6c shows no significant difference of slip azimuth in area *B* (as defined in Fig. 6b); while this is the area where the difference between interseismic and postseismic azimuths of displacements at the GPS sites is most significant. One reason might be that the assumption of a uniform slip azimuth does not allow to fit favourable of the observed azimuths of interseismic displacements in area *B* as seen in Figure 7. It suggests that the interseismic fault slip azimuth might trend more northerly in this area than predicted by the uniform rake model. If so, the azimuth of slip on the décollement would also have rotated counter-clockwise in area *B* due to the Chi-Chi earthquake. However, we are unable to estimate well that rotation based on the modeling presented here.

4.2. Comparison of Fault-Slip and Shear Stress Azimuths

We now test the hypothesis that slip on the décollement is parallel to the shear stress acting on it. To do so, we have analyzed earthquake focal mechanisms determined from first motion data by the Central Weather Bureau (CWB) (http://www.cwb.gov.tw/) and from the modeling of broadband waveforms recorded by the Broadband Array in Taiwan for Seismology (BATS, http://www.earth.sinica.edu.tw/). Because of the scarcity of earthquakes in the Chi-Chi rupture area prior to 1999, we could only carry this analysis for the postseismic period. We used the algorithm based on minimizing the misfit of the rake angle between the assumed and inverted stress tensors for both fault plane solutions to determine the orientation of the principal stress (Michael, 1984; 1987). The result of the stress inversion is given in Table 2. We consider two areas (labels *A* and *B* in Fig. 9). The focal mechanisms within area *A* are consistent with a relatively homogeneous stress

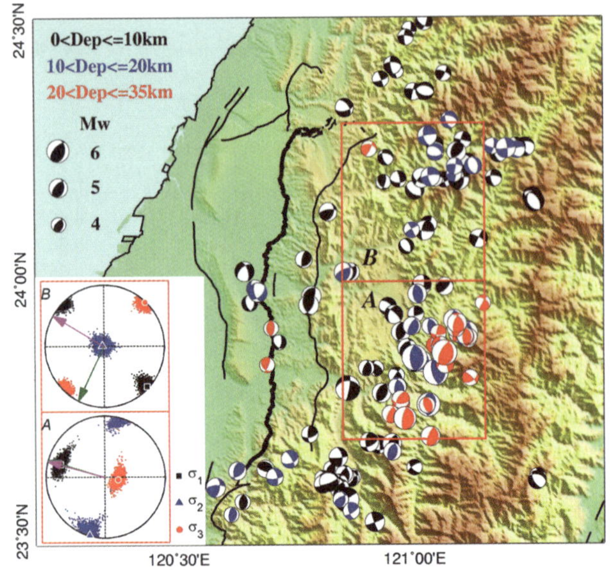

Figure 9

Focal mechanisms of aftershocks with $M_w > 4$, over the first 15 months following the Chi-Chi main shock. The color code indicates focal depths. Major faults are shown in black and the 1999 Chi-Chi rupture is shown in a thicker black line. Red boxes indicate the two regions (*A*: south, *B*: north) within which the stress tensors are determined from earthquake focal mechanisms. The resulting stress tensors are shown in the bottom left inset. Different symbols representing three principal axes plotted in equal-area projection of the lower hemisphere. The dots show the distributions of stress axes within 95% confidence region. Magenta and green vectors are azimuths of postseismic slip and of the shear stress computed on the flat décollement at a depth of about 12 km.

field. To test the sensibility of the result to the assumed depth range and to the focal mechanism data sets, we have varied the depth range and considered another data set obtained from first motions analysis determined from seismic and accelerometric records (Wu *et al.*, 2008). The stress tensor in area *A* only varies slightly (Table 2), with the azimuth of shear stress on the décollement always close to N285°. The focal mechanisms within area *B* are more diverse and the results are more sensitive to the data set used in the inversion. It could result from a depth variation or a stronger heterogeneity of the stress tensor in area *B* than in area *A*. In any case, the focal mechanisms are generally consistent with the vertical axis being close to a principal stress direction.

Next, we compare the azimuth of the shear stress computed from the stress tensors with the azimuth of slip on the décollement (Table 2). Our analysis demonstrates that the shear stress and shear strain azimuths on the décollement are coherent. For area *A*, the result indicates that the azimuth of slip on the décollement inferred from the geodetic data is reasonably consistent with the stress tensor obtained from earthquakes with focal depths less than 35 km, which occurred within 15 months after the Chi-Chi earthquake (the same period as the time span covered by the postseismic GPS displacements analyzed here). In addition, if we constrain the postseismic slip azimuth on the décollement to be equal to the azimuth of shear stress, it yields a value of χ_r^2 about 8.7, which approaches the value of 8.5 obtained for the best-fitting slip azimuth. For area *B*, the hypothesis is contradicted. One possible explanation would be that the stress field, as revealed by the focal mechanisms, is quite heterogeneous and not representative of the shear stress on the décollement. In addition, because the intermediate principal stress component is nearly vertical, the azimuth of the shear stress on the décollement is very sensitive to the assumed attitude of the décollement (taken horizontal here). Alternatively, our estimate of the azimuth of postseismic slip on the décollement might be erroneous due to the scarcity of pre-Chi-Chi GPS data in this area. Therefore, we only focus on area *A* in the following section.

4.3. *Implications for Shear Stress Level*

Here, we assume that the difference between interseismic and postseismic rakes reflects a change of the stress acting on the décollement, due to the fact that the postseismic stress, $\underline{\underline{\sigma}}^{\text{post}}$, is the sum of the preseismic stress, $\underline{\underline{\sigma}}^{\text{pre}}$, and the coseismic stress change, $\underline{\underline{\Delta\sigma}}$. In theory, if $\underline{\underline{\Delta\sigma}}$ is known, it is then possible to place constraints on the pre- and postseismic stress acting on the décollement, based on the assumption that the rotation of the shear stress is equal to the rotation of the slip azimuth. It might be assumed that the normal stress, i.e. the vertical stress, is only slightly affected by the coseismic stress change. In that case, a simple vector diagram can be constructed to estimate the magnitude of the pre- and postseismic shear stresses on the décollement. We estimate the coseismic stress change beneath the Central Range using the coseismic slip model (Fig. 4). The coseismic stress change in the box *A* (Fig. 9) is estimated by taking the average stress change of 1000 points within a 1000 km^3 cube centered at the grid point to avoid biased results from short-wavelength variations of the stress field. The averaged coseismic shear stress at 12 km depth on the décollement is of the order of 1 MPa in the direction of 330° within area *A*. The azimuths of preseismic and postseismic shear stress are 304° and 285°, respectively. Based on the vector diagram, we estimate that the preseismic and postseismic shear stress in area *A* are 2.2 ± 2.2 MPa and 1.3 ± 1.4 MPa, respectively. The uncertainties given here were estimated assuming that the uncertainties on the azimuths of preseismic, coseismic, and postseismic shear stresses are 5°, 5°, and 10°, respectively, and that the uncertainty on the amplitude of coseismic stress is 1 MPa.

Another approach consists in considering that the postseismic stress tensor acting on the décollement is correctly represented by the stress tensor determined from focal mechanisms. In that case, the normal stress change is also taken into account in the computation. If $\underline{\underline{\Delta\sigma}}$ is known and if the orientation of the principal postseismic stress and the stress ratio $\phi\,(= \sigma_2 - \sigma_3/\sigma_1 - \sigma_3)$ are constrained from the focal mechanisms of the aftershocks, the postseismic differential stress, $\sigma_1 - \sigma_3$, can then be determined from the condition that preseismic shear stress be colinear with the preseismic slip on the décollement. We assume a postseismic stress tensor with principal stress orientations and ratio ϕ constrained from the Chi-Chi aftershocks (bold text in Table 2) and a differential

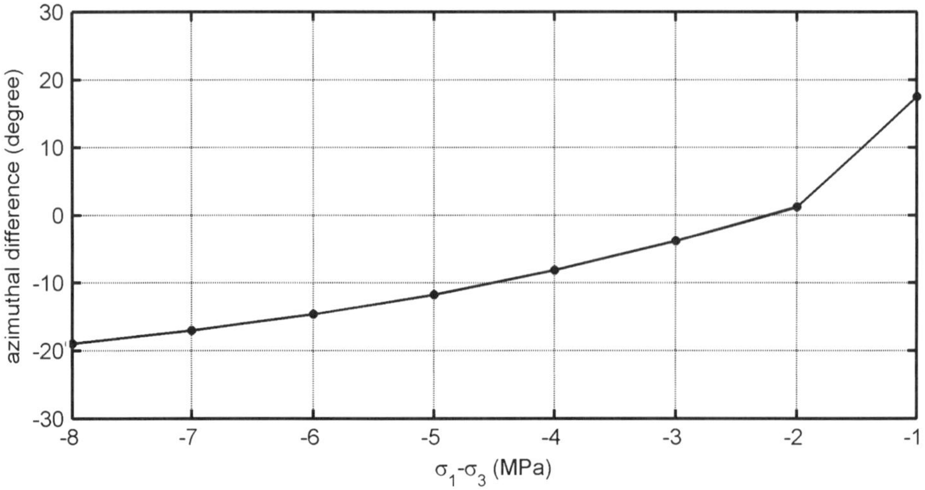

Figure 10
Difference between the azimuths of preseismic shear stress and interseismic slip as a function of differential postseismic stress for area *A*.

stress between 1 and 8 MPa. We then solve for the value of the postseismic differential stress that yields the smallest magnitude azimuthal difference between the preseismic shear stress and the preseismic slip on the décollement. Note that the colinearity is not guaranteed. The result shows that the azimuthal difference between the preseismic shear stress and the preseismic slip on the décollement is the minimum for the postseismic differential stress of 2 MPa in area *A* (Fig. 10). According to this reasoning, the low shear stress acting on the décollement would result from both a geometric effect, one principal stress component being near vertical, and from a low deviatoric stress within the wedge.

5. *Discussion*

5.1. Comparison of Interseismic, Coseismic and Postseismic Slip Models

Figures 4, 6, and 7d allow comparison of the distributions of coseismic slip, afterslip and interseismic slip rates. The three distributions complement each other and the following points can be made:

1. The area that ruptured in 1999 was strongly locked before the main shock. There is no evidence that the rupture propagated into the area that was creeping before the earthquake.
2. Not all of the area that was strongly locked before 1999 ruptured during the Chi-Chi earthquake. The interseismic model shows strong locking at shallow depths on the

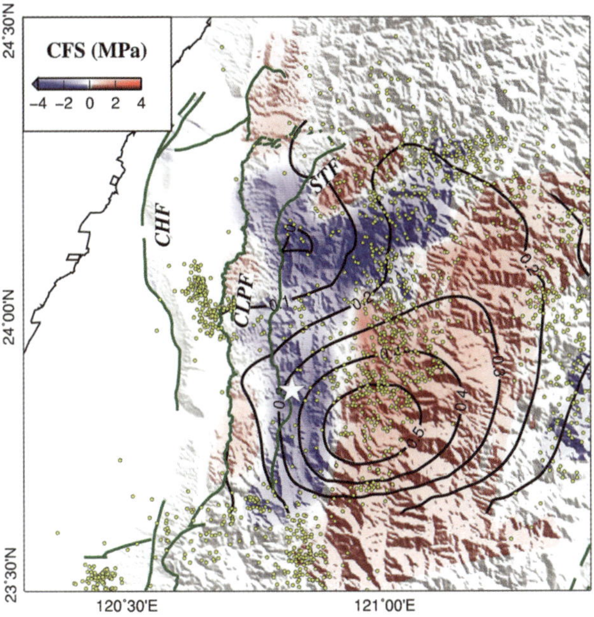

Figure 11
Comparison of afterslip distribution and coseismic Coulomb stress change on the décollement. The Coulomb stress changes (color) were computed assuming an average fault strike of 0°, dip angle of 10°, and a rake of 90° at 15 km depth. Black contours with interval of 0.1 m indicate postseismic slip. White star denotes the epicenter. Yellow dots indicate aftershocks occur at depths ranging from 10 and 20 km over a 15-month period after the main shock.

southern segment of the Chelungpu fault where the earthquake did not produce much slip in 1999, was not creeping before 1999, and did not experience much afterslip.

3. Afterslip occurred dominantly downdip from the rupture area of the Chi-Chi earthquake in a transition zone that was creeping before 1999, so it was probably characterized by a rate-strengthening friction. On this creeping zone, the coseismic deformation has induced a significant increase of Coulomb stress of about 1 MPa at 15 km depth based on a fault geometry with N-S trending, 10° east dipping as illustrated in Figure 11. The area with the increase of Coulomb stress is consistent with the aftershock distribution at 10–20 km depth.

To first-order, this simple picture compares well to theoretical models of the seismic cycle based on the depth variation of frictional properties of faults (Lapusta *et al.*, 2000; Scholz, 1998; Tse and Rice, 1986) (Fig. 12). Steady aseismic slip occurs at deep depths in the interseismic period until the instability results in a coseismic rupture at shallow depths. The relaxation of the coseismic stress causes afterslip downdip from the coseismic rupture in the postseismic period. However, afterslip does not completely fill the gap between coseismic and interseismic slip distributions. Given the relatively rapid decay of afterslip rate, the misfit is not due to the short period of afterslip considered here.

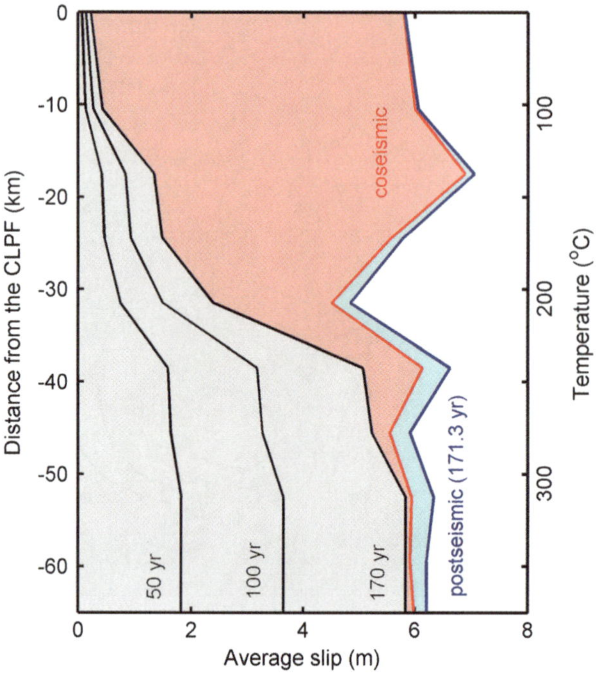

Figure 12
Average slip as a function of distance and temperature over the seismic cycle of the Chelungpu fault. The plot was obtained from determining the average coseismic slip, afterslip and interseismic slip within a 5-km wide swath along a profile parallel to 340° (close to AA' in Fig. 3). Interseismic slip is from Model 3 in Figure 7d.

The Chi-Chi earthquake has the unusual property of having released most of its moment at relatively shallow depths. Indeed, in examples where coseismic slip could be constrained from geodetic and remote-sensing data, the maximum slip is generally found at a depth of about 5–10 km. (e.g., Avouac *et al.*, 2006; Fialko *et al.*, 2005; Konca *et al.*, 2007; Simons *et al.*, 2002). In addition, the Chi-Chi earthquake only released slight moment on the southern segment. Other earthquakes rupturing the western foothills might release a larger proportion of moment at depths or on the southern segment. Consequently it is possible that the area of the Chelungpu fault that remains locked in the interseismic period produces a variety of coseismic ruptures. Each earthquake would have a different coseismic slip distribution however the various types of earthquakes would collectively add to aseismic slip (interseismic or postseismic) and even out the cumulative slip distribution. Another possibility is that aseismic transients, not observed during the period covered by our geodetic data, might occur in the transition zone. In any case, the Chelungpu fault must produce earthquakes with different slip distributions than the Chi-Chi event since there is no indication that the slip rate on the fault decreases southwards as does the coseismic slip.

Figure 3b shows that many aftershocks occurred at depth deeper than the brittle-ductile transition inferred from the geodetic data. These deep aftershocks may reflect heterogeneities of crustal rheology or a transient deepening of the brittle-ductile transition depth due to the large strain rates generated by the main shock rupture below the décollement. Transient deepening of the brittle-ductile transition has been simulated in numerical experiment (BEN-ZION and LYAKHOVSKY, 2006), and advocated for in some observational studies (ROLANDONE *et al.*, 2004; SCHAFF *et al.*, 2002).

Interestingly, the Chi-Chi earthquake nucleated near the downdip edge of the locked fault zone (see epicenter location in Fig. 7d), as is also observed in numerical experiments (LAPUSTA *et al.*, 2000). It should be noticed that, in these models, temperature is thought to be the key parameter controlling the transition from rate-weakening at shallow depths to rate-strengthening at greater depths. We assess this possibility in the following section.

5.2. Effect of Temperature on Fault Friction

Temperature is thought to be a controlling parameter determining the mode of slip on subduction megathrust (HYNDMAN *et al.*, 1997; OLESKEVICH *et al.*, 1999; PEACOCK and HYNDMAN, 1999) as well as on intracontinental faults (BLANPIED *et al.*, 1995). To assess the temperature on the Chelungpu fault and its downdip continuation as a décollement, we use the thermal model (SIMOES *et al.*, 2007a) that has been derived from the modeling of thermochronological and thermometric data along three sections across Taiwan. We find that aseismic slip in the interseismic and postseismic periods is dominant where temperature exceeds approximately 200°C (Figs. 7 and 12), a similar feature as also reported in the Nepal Himalaya based on the same approach (AVOUAC, 2003). This temperature is consistent with laboratory experiments on granite at hydrothermal conditions indicating the transition from velocity-weakening friction to velocity-strengthening friction occurred around 250 $\sim$ 350°C, probably due to the onset of crystal plasticity of quartz (BLANPIED *et al.*, 1995; SCHOLZ, 1998). Temperature is thus likely a controlling parameter of downdip variations of frictional properties along the Chelungpu fault. Other parameters such as the lithology or fluid must also play a role, in particular to explain lateral variations, or shallow creep, but they are not quantified here.

5.3. No Evidence for Shallow Creep

At shallow depths, continental faults can sometimes exhibit a rate-strengthening frictional behavior as suggested from experimental results and the observation of shallow aseismic slip in some cases (MARONE *et al.*, 1990; MARONE, 1998). It is then possible that fault patches with rate-strengthening friction might not creep in the interseismic period if they lie in the stress shadow of deeper locked asperities (BURGMANN *et al.*, 2005). In that case, they would be expected to slip aseismically as a result of afterslip following the rupture of the shadowing asperities, or to eventually produce spontaneous aseismic transients (PERFETTINI and AMPUERO, 2008). The observation of a few centimeters of aseismic slip on Changhua fault triggered by the Chi-Chi earthquake (PATHIER *et al.*, 2003) suggest that this is a plausible mechanism in the context of the western foothills of Taiwan. However, in the case of the Chi-Chi earthquake, there is no shallow deficit of slip that may diagnostic a shallow rate-strengthening zone (FIALKO *et al.*, 2005). We observe no shallow afterslip following the main shock. Therefore we exclude that there would be a significant proportion of the shallow Chelungpu fault that would be rate-strengthening but would appear locked in the interseismic period. On the other hand, some afterslip seems to overlap with the area that is locked in the interseismic period (Fig. 7c). Coseismic slip and afterslip near the hypocentral region in the southern CLPF may be an indication of a mosaic of fault patches with rate-weakening and rate-strengthening friction.

5.4. *Implications for Fault Rheology*

Our analysis reveals that the coseismic stress change related to the Chi-Chi earthquake has induced a difference in azimuth of geodetic displacements within the Central Range measured before and after the Chi-Chi earthquake. We infer that the shear-stress acting on the décollement is of the same order of magnitude as the coseismic stress change beneath the Central Range, estimated to about 1 MPa (Fig. 11). The modeling of that effect, assuming that deformation across the Central Range is governed by aseismic creep along a horizontal décollement, requires a very low shear stress on the décollement, of the order of only 2 MPa. The low shear stress is consistent with the fact that the focal mechanisms require one principal stress component to be near vertical. Given the $\sim$ 10 km depth of that décollement and assuming a crustal density of $\sim$ 2.9, this low shear stress implies a very low effective friction of $\sim$ 0.01. This estimate is consistent with the inference of the low taper angle between the topographic slope and the décollement, which constitutes about 8°, requiring an effective basal friction lower than 0.08 (CARENA *et al.*, 2002; SUPPE, 2007).

Deformation at 10–20 km depth beneath the Central Range, where temperature is estimated to be between 200 and 300°C (SIMOES *et al.*, 2007a), might occur either in the brittle or ductile regime. In any case, the exceptionally low shear stress suggests that fluid is probably playing a key role. The presence of fluids is suggested by the high conductivity measured from magnetotelluric sounding (CHEN and CHEN, 2002) and a localized zone with high V_p/V_s ratio at $\sim$ 10 km depth beneath the Central Range (WU *et al.*, 2007). In addition, microstructures and abundant quartz veins within the Central Range slates are evidence of fluid-assisted deformation (CHAN *et al.*, 2005). Finally, fluids must be released as the Chinese continental margin underthrusts the western foothills and experiences dehydration and metamorphism. In the brittle regime, fluids can contribute to a low effective friction if the pore pressure is high and enhances stable frictional sliding. Through chemical effects, fluids would also enhance pressure solution deformation (CHESTER, 1995), as well as diffusion and dislocation creep (KOHLSTEDT, 1995)

As reported, frictional afterslip can explain the temporal evolution of geodetic deformation following the Chi-Chi earthquake (HSU *et al.*, 2007; PERFETTINI and AVOUAC, 2004) and we shall therefore start by considering this mechanism. Rock friction experiments have shown that that the relationship between stress and sliding velocity during rate-strengthening friction can generally be written

$$\tau_{ss} = \sigma_n \mu^* + (a - b)\sigma_n \ln(V/V^*), \tag{3}$$

where τ_{ss} is the driving shear stress, σ_n is the normal stress, $(a-b) > 0$ is a rheological parameter, V is the sliding velocity, and $\mu^*$and V^* are the reference values. This phenomenological law seems to hold for temperatures reaching about 300°C for quartzo-feldspathic rocks, even when deformation is assisted by fluid or distributed within some gouge zone with finite thickness (BLANPIED *et al.*, 1995; CHESTER, 1995; MARONE, 1998).

In the steady-state approximation and assuming that the creeping velocity just before the coseismic stress change applied is equal to the asymptotic creeping velocity, V_o, afterslip rate should decay according to PERFETTINI and AVOUAC (2004)

$$V(t) = V_0 \frac{d \exp(t/t_r)}{1 + d(\exp(t/t_r) - 1)}, \tag{4}$$

where d is the velocity jump due to coseismic stress change; t_r is the relaxation time associated with frictional afterslip and t is time after the main shock.

We have extracted the slip history (HSU *et al.*, 2007) in two areas (*A* and *B* in Fig. 9) to test the model predictions. In both cases, the temporal decay (Fig. 13) is consistent

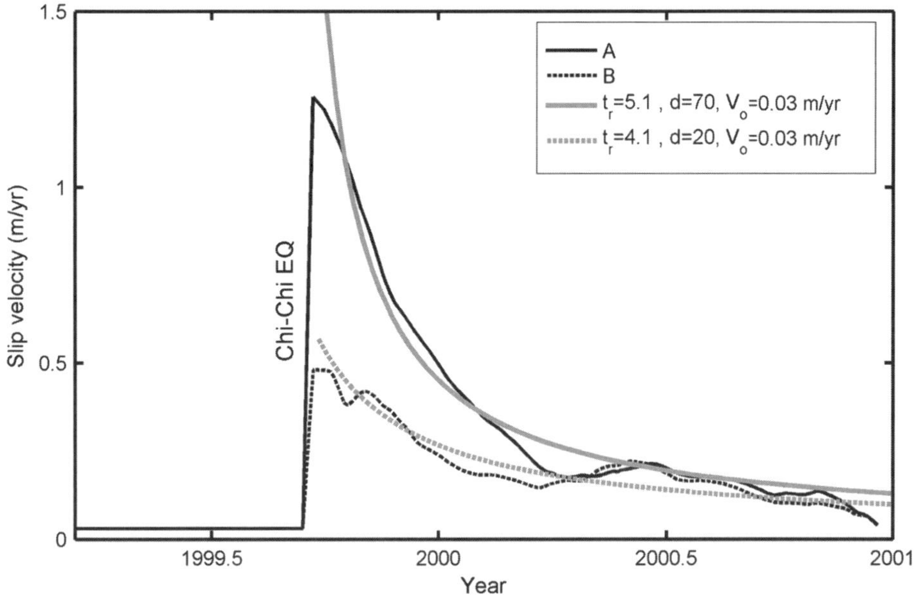

Figure 13
Temporal evolution of slip (black curves) due to afterslip computed at two locations *A* and *B* as indicated in Figure 6b, based on the afterslip model of HSU *et al.*, 2007. They are compared to the predictions of afterslip (grey curves) from a 1-D rate-strengthening frictional model (PERFETTINI and AVOUAC, 2004) (equation (4)).

with that predicted from equation (4). The two parameters, the relaxation time (t_r) and the velocity jump (d), are solved by grid search. The value of $(a - b) \cdot \sigma_n$ can then be estimated from $(a - b) \cdot \sigma_n = \Delta CFS/\log(d) = t_r/T_{\text{inter}}\Delta CFS$ (PERFETTINI and AVOUAC, 2004). The coseismic Coulomb stress change (ΔCFS) is estimated to $\sim$10 MPa, and the average time interval between two earthquakes (T_{inter}) similar to the Chi-Chi event is estimated to $\sim$200 years. We infer $(a - b) \cdot \sigma_n$ to 0.3–2.4 MPa and 0.2–3.3 MPa in area A and B, respectively. Altogether, we estimate $(a - b) \cdot \sigma_n$ to about 0.2–3.3 MPa. Assuming a hydrostatic pore pressure, a crustal density of 2.9, and a décollement depth of 10 km, it would imply a value of $(a - b)$ of about 8×10^{-3}. This value is in agreement with estimates derived from laboratory experiments which range from 5×10^{-3} to 3.0×10^{-2} (MARONE, 1998), or from postseismic observations, which fall in the range 10^{-2}–10^{-4} (HEARN *et al.*, 2002; HSU *et al.*, 2006; JOHNSON *et al.*, 2006; MIYAZAKI *et al.*, 2004; PERFETTINI and AVOUAC, 2004; PERFETTINI *et al.*, 2005)

5.5. *Return Period of Seismic Ruptures across the Western Foothills of Central Taiwan*

Interseismic strain implies a rate of accumulation of moment deficit which can be computed from integrating the slip-rate deficit over the locked fault zones. This rate is estimated to between 1.2×10^{18} and 3.7×10^{18} N-m/yr (Table 1). If a large earthquake like the Chi-Chi earthquake were the only type of event releasing the slip deficit on the fault portion that is locked in the interseismic period, it would take about 100 $\sim$ 300 years to accumulate enough deficit to produce an earthquake as large as the Chi-Chi earthquake. A similar estimate, although less well-constrained by the geodetic data, was obtained by SIMOES *et al.* (2007b). However, as mentioned above, it is clear that the real behavior is not the simple regular repetition of the Chi-Chi type earthquakes. First, the slip distributions during successive ruptures of the Chelungpu fault cannot be similar to that observed in 1999 since they must add to an approximately uniform cumulative slip distribution. If coseismic slip at one point on the fault is relatively constant from one event to another as is often observed (SIEH, 2000), seismic ruptures should be more frequent on the southern segment than on the northern segment of the Chelungpu fault. Second, the earthquakes should be about equally distributed on the Chelungpu and Changhua faults given their similar long-term slip rates.

No earthquake as large as the Chi-Chi earthquake is known to have occurred in the area over the last few centuries. Paleoseismic investigations have revealed prehistorical ruptures with slip similar to those observed in 1999 at a few sites, however, it seems clear that these paleoseismic events occurred at different times and at different places (CHEN *et al.*, 2004, 2007a; LEE *et al.*, 2001a; STREIG *et al.*, 2007). Clearly, the Chelungpu fault does not act with the repetition of a characteristic earthquake similar to the Chi-Chi event. The longest record of paleoseismic investigations, which was carried at Chushan on the southern segment of the Chelungpu fault, has revealed five different events with slip similar to that observed in 1999 over the last 2000 years. Similarly, at Shi-Jia site, three events occurred over the last

1200 years (STREIG *et al.*, 2007). Thus the return period of large events on the Chelungpu fault seems to conflict with estimates derived from the long term slip rate and the pattern of interseismic locking. There is possibility that the paleoseismic data do not cover a long enough period to be representative of the long-term behavior of the Chelungpu fault. Another possibility is that we would underestimate grossly the fraction of aseismic slip taken up by transient aseismic slip events, and that no such aseismic transients have occurred during the period covered by the geodetic data prior to the Chi-Chi earthquake.

6. Conclusion

The study of coseismic, postseismic and interseismic slip distributions and the comparison with the thermal structure across the western foothills of Taiwan shed light on the parameters determining the mode of slip on crustal faults and fault rheology. We find that slip on the Chelungpu fault is dominantly aseismic at depths exceeding about 10 km, where the temperature increases beyond about 200°C, and is dominantly seismic at shallower depths. The downdip transition from rate-strengthening to rate-weakening friction is likely governed by temperature as it corresponds to the same range of temperature as that observed in laboratory experiments on quartzo-feldspathic rocks. Afterslip following the Chi-Chi earthquake occurred along the edge of the zone that is creeping in the interseismic period and evolved as predicted from rate-strengthening friction with the frictional parameter $d\tau_{ss}/d \ln V$ estimated to be about 1.5 MPa. Furthermore, the absence of any significant shallow afterslip on the Chelungpu fault indicates that the shallow portion of the fault, which was locked before the Chi-Chi earthquakes, primarily obeys a rate-weakening friction law.

We explore the possibility of assessing fault-stress level by comparing fault-slip rakes in the pre- and postseismic period. Indeed, the azimuth of frictional sliding in the creeping area downdip of the seismogenic zone seems to have changed due to the coseismic stress change induced by the Chi-Chi earthquake. It is difficult to demonstrate that the data really require a significant change of fault-slip rake, however, if accepted, the observation would suggest an extremely low shear stress level on this creeping basal décollement of the order of 2 MPa. This low shear stress level is consistent with the low taper angle of the orogenic wedge and the inference that the orogenic wedge does not seem to suffer significant internal deformation as it is thrust up the basal décollement. It implies that the wedge would be overcritical, the topographic slope being steeper than that which is required for the wedge to be at the verge of brittle 'passive' failure, *i.e.* by thrust faulting (DAHLEN, 1990). The low basal shear stress might reflect a high pore pressure and/or a shear zone deforming by pressure-solution creep.

The rupture area of the Chi-Chi earthquake lies mainly within the area of the Chelungpu fault that was locked in the interseismic period but coseismic, interseismic and postseismic slip do not add to a uniform fault slip, if only the Chi-Chi-like rupture is assumed. The pattern of locking is probably relatively stationary, as would be expected if it were to relate primarily to temperature or lithology along the fault. By contrast, the seismic release process is probably quite variable with the irregular occurrences of seismic ruptures with different slip distributions. This is not unexpected, given that such a complexity can arise even on a simple fault system with geometric fault irregularities or strength heterogeneities (BEN-ZION, 2001).

Acknowledgments

We thank the editor, Dr. Y. Ben-Zion, Dr. J. C. Savage, and an anonymous reviewer for their constructive comments. We are grateful to many colleagues at the Institute of Earth Sciences, Academia Sinica, as well as the Central Weather Bureau for kindly provide geodetic and earthquake focal mechanism data. We have benefited from stimulating discussions with Mark Simons. This work was supported by the National Science Foundation grant EAR-0537625 to Caltech, the Gordon and Betty Moore Foundation, and the National Science Council of the Republic of China grant NSC 95-2119-M-001-064 –MY3. This is a contribution of Caltech Tectonics Observatory number 113 and the Institute of Earth Sciences, Academia Sinica, IESAS1335.

REFERENCES

ANGELIER, J., CHU, H. T., and LEE, J. C. (1997), *Shear concentration in a collision zone: Kinematics of the Chihshang Fault as revealed by outcrop-scale quantification of active faulting, Longitudinal Valley, eastern Taiwan*, Tectonophysics *274*, 117–143.

AVOUAC, J. P. (2003) *Mountain building, erosion and the seismic cycle in the Nepal Himalaya.* In Adv. Geophys. (ed. R. Dmowska, pp. 1–79 Elsevier, Amsterdam, 2003).

AVOUAC, J. P., AYOUB, F., LEPRINCE, S., KONCA, O. and HELMBERGER, D. V. (2006), *The 2005, M_w 7.6 Kashmir earthquake: Sub-pixel correlation of ASTER images and seismic waveforms analysis*, Earth Planet. Sci. Lett. *249*, 514–528.

BEN-ZION, Y. (2001), *Dynamic ruptures in recent models of earthquake faults*, J. Mech. Phys. Solids *49*, 2209–2244.

BEN-ZION, Y. and LYAKHOVSKY, V. (2006), *Analysis of aftershocks in a lithospheric model with seismogenic zone governed by damage rheology*, Geophys. J. Int. *165*, 197–210.

BLANPIED, M. L., LOCKNER, D. A., and BYERLEE, J. D. (1995), *Frictional slip of granite at hydrothermal conditions*, J. Geophys. Res. *100*, 13045–13064.

BOLLINGER, L., AVOUAC, J. P., CATTIN, R., and PANDEY, M. R. (2004), *Stress buildup in the Himalaya*, J. Geophys. Res. *109*, doi:10.1029/2003JB002911.

BRUDY, M., ZOBACK, M. D., FUCHS, K., RUMMEL, F. and BAUMGARTNER, J. (1997), *Estimation of the complete stress tensor to 8 km depth in the KTB scientific drill holes: Implications for crustal strength*, J. Geophys. Res. *102*, 18453–18475.

BURFORD, R. O. and HARSH, P. W. (1980), *Slip on the San Andreas fault in central California from alignment array surveys*, Bull. Seismol. Soc. Am. *70*, 1233–1261.

BURGMANN, R., KOGAN, M. G., STEBLOV, G. M., HILLEY, G., LEVIN, V. E., and APEL, E. (2005), *Interseismic coupling and asperity distribution along the Kamchatka subduction zone*, J. Geophys. Res. *110*, 10.1029/2005JB003648.

BYERLEE, J. (1978), *Friction of Rocks*, Pure Appl. Geophys. *116*, 615–626.

CARENA, S., SUPPE, J. and KAO, H. (2002), *Active detachment of Taiwan illuminated by small earthquakes and its control of first-order topography*, Geology *30*, 935–938.

CATTIN, R. and AVOUAC, J. P. (2000), *Modeling mountain building and the seismic cycle in the Himalaya of Nepal*, J. Geophys. Res. *105*, 13389–13407.

CHAN, Y. C., OKAMOTO, K., YUI, T. F., IIZUKA, Y., and CHU, H. T. (2005), *Fossil fluid reservoir beneath a duplex fault structure within the Central Range of Taiwan: Implications for fluid leakage and lubrication during earthquake rupturing process*, Terra Nova *17*, 493–499.

CHANG, C. H., WU, Y. M., ZHAO, L., and WU, F. T. (2007), *Aftershocks of the 1999 Chi-Chi, Taiwan, earthquake: The first hour*, Bull. Seismol. Soc. Am. *97*, 1245–1258.

CHEN, C. C. and CHEN, C.-S. (2002), *Sanyi-Puli conductivity anomaly in NW Taiwan and its implication for the tectonics of the 1999 Chi-Chi earthquake*, Geophys. Res. Lett. *29*, 1166.

CHEN, W. S., LEE, K. J., LEE, L. S., PONTI, D. J., PRENTICE, C., CHEN, Y. G., CHANG, H. C., and LEE, Y. H. (2004), *Paleoseismology of the Chelungpu Fault during the past 1900 years*, Quaternary International *115*, 167–176.

CHEN, W. S., YANG, C. C., YEN, I. C., LEE, L. S., LEE, K. J., YANG, H. C., CHANG, H. C., OTA, Y., LIN, C. W., LIN, W. H., SHIH, T. S., and LU, S. T. (2007a), *Late Holocene paleoseismicity of the southern part of the Chelungpu fault in central Taiwan: Evidence from the Chushan excavation site*, Bull. Seismol. Soc. Am. *97*, 1–13.

CHEN, Y. G., LAI, K. Y., LEE, Y. H., SUPPE, J., CHEN, W. S., LIN, Y. N. N., WANG, Y., HUNG, J. H., and KUO, Y. T. (2007b), *Coseismic fold scarps and their kinematic behavior in the 1999 Chi-Chi earthquake Taiwan*, J. Geophys. Res. *112*, doi:10.1029/2006JB004388.

CHESTER, F. M. (1995), *A rheologic model for wet crust applied to strike-slip faults*, J. Geophys. Res. *100*, 13033–13044.

DAHLEN, F. A. (1990), *Critical taper model of fold-and-thrust belts and accretionary wedges*, Annu. Rev. Earth Planet. Sci. *18*, 55-99.

DAVIS, D., SUPPE, J. and DAHLEN, F. A. (1983), *Mechanics of fold-and-thrust belts and accretionary wedges*, J. Geophys. Res. *88*, 1153–1172.

DOMINGUEZ, S., AVOUAC, J. P., and MICHEL, R. (2003), *Horizontal coseismic deformation of the 1999 Chi-Chi earthquake measured from SPOT satellite images: Implications for the seismic cycle along the western foothills of central Taiwan*, J. Geophys. Res. *108*, doi:10.1029/2001JB000951.

FIALKO, Y., SANDWELL, D., SIMONS, M., and ROSEN, P. (2005), *Three-dimensional deformation caused by the Bam, Iran, earthquake and the origin of shallow slip deficit*, Nature, *435*, 295–299.

HARDEBECK, J. L. and HAUKSSON, E. (2001), *Crustal stress field in southern California and its implications for fault mechanics*, J. Geophys. Res. *106*, 21859–21882.

HAUKSSON, E. (1990), *Earthquakes, faulting and stress in the Los Angeles Basin*, J. Geophys. Res. *95*, 15,365–315,394.

HEARN, E. H., BURGMANN, R., and REILINGER, R. E. (2002), *Dynamics of Izmit earthquake postseismic deformation and loading of the Duzce earthquake hypocenter*, Bull. Seismol. Soc. Am. *92*, 172–193.

HSU, Y.-J., SIMONS, M., YU, S.-B., KUO, L.-C., and CHEN, H.-Y. (2003), *A two-dimensional dislocation model for interseismic deformation of the Taiwan mountain belt.*, Earth Planet. Sci. Lett. *211*, 287–294.

HSU, Y. J., BECHOR, N., SEGALL, P., YU, S.-B., KUO, L. C., and MA, K. F. (2002), *Rapid afterslip following the 1999 Chi-Chi, Taiwan earthquake*, Geophys. Res. Lett. *29*, 10.1029/2002GL014967.

HSU, Y. J., SIMONS, M., AVOUAC, J. P., GALETZKA, J., SIEH, K., CHLIEH, M., NATAWIDJAJA, D., PRAWIRODIRDJO, L., and BOCK, Y. (2006), *Frictional afterslip following the 2005 Nias-Simeulue earthquake, Sumatra,* Science *312*, 1921–1926.

HSU, Y. J., SEGALL, P., YU, S. B., KUO, L. C., and WILLIAMS, C. A. (2007), *Temporal and spatial variations of postseismic deformation following the 1999 Chi-Chi, Taiwan earthquake*, Geophys. J. Int. *169*, 367–379.

HYNDMAN, R. D., YAMANO, M., and OLESKEVICH, D. A. (1997), *The seismogenic zone of subduction thrust faults*, The Island Arc, *6*, 244–260.

JI, C., HELMBERGER, D. V., SONG, T.-R. A., MA, K.-F., and WALD, D. J. (2001), *Slip distribution and tectonic implications of the 1999 Chi-Chi, Taiwan earthquake*, Geophys. Res. Lett. *28*, 4379–4382.

JI, C., HELMBERGER, D. V., WALD, D. J., and MA, K. F. (2003), *Slip history and dynamic implications of the 1999 Chi-Chi, Taiwan, earthquake*, J. Geophys. Res. *108*, doi:10.1029/2002JB001764.

JOHNSON, K. J., HSU, Y.-J., SEGALL, P., and YU, S.-B. (2001), *Fault geometry and slip distribution of the 1999 Chi-Chi, Taiwan earthquake imaged from inversion of GPS data*, Geophys. Res. Lett. *28*, 2285–2288.

JOHNSON, K. M. and SEGALL, P. (2004), *Imaging the ramp-decollement geometry of the Chelungpu fault using coseismic GPS displacements from the 1999 Chi-Chi, Taiwan earthquake*, Tectonophysics *378*, 123–139.

JOHNSON, K. M., BURGMANN, R., and LARSON, K. (2006), *Frictional properties on the San Andreas fault near Parkfield, California, inferred from models of afterslip following the 2004 earthquake*, Bull. Seismol. Soc. Am. *96*, S321–S338.

JONES, L. M. (1988), *Focal mechanisms and the state of stress on the San Andreas fault in southern California*, J. Geophys. Res. *93*, 8869–8891.

KOHLSTEDT, D. L., EVANS, B., and MACKWELL, S. J. (1995), *Strength of the lithosphere: Constraints imposed by laboratory experiments*, J. Geophys. Res. *100*, 587–517,602.

KONCA, A. O., HJORLEIFSDOTTIR, V., SONG, T. R. A., AVOUAC, J. P., HELMBERGER, D. V., JI, C., SIEH, K., BRIGGS, R., and MELTZNER, A. (2007), *Rupture kinematics of the 2005 M-w 8.6 Nias-Simeulue earthquake from the joint inversion of seismic and geodetic data*, Bull. Seismol. Soc. Am., *97*, S307–S322.

LAPUSTA, N., RICE, J. R., BEN-ZION, Y., and ZHENG, G. T. (2000), *Elastodynamic analysis for slow tectonic loading with spontaneous rupture episodes on faults with rate- and state-dependent friction*, J. Geophys. Res., *105*, 23765–23789.

LEE, J.-C., CHEN, Y.-G., SIEH, K., MUELLER, K., CHEN, W.-S., CHU, H.-T., CHAN, Y.-C., RUBIN, C., and YEATS, R. (2001a), *A vertical exposure of the 1999 surface rupture of the Chelungpu fault at Wufeng, western Taiwan: Structural and paleoseismic implications for an active thrust fault*, Bull. Seismol. Soc. Am. *91*, 914–929.

LEE, J. C., ANGELIER, J., CHU, H. T., HU, J. C., and JENG, F. S. (2001b), *Continuous monitoring of an active fault in a plate suture zone: a creepmeter study of the Chihshang Fault, eastern Taiwan*, Tectonophysics *333*, 219–240.

LEE, J. C., ANGELIER, J., CHU, H. T., HU, J. C., JENG, F. S., and RAU, R. J. (2003), *Active fault creep variations at Chihshang, Taiwan, revealed by creep meter monitoring, 1998–2001*, J. Geophys. Res. *108*, doi:10.1029/2003JB002394.

LISOWSKI, M. and PRESCOTT, W. H. (1981), *Short range distance measurements along the San Andreas fault system in central California*, Bull. Seismol. Soc. Am., *71*, 1607–1624.

LIU, C. C. and YU, S.-B. (1990), *Vertical crustal movements in eastern Taiwan and their tectonic implications*, Tectonophysics *183*, 111–119.

LOEVENBRUCK, A., CATTIN, R., LE PICHON, X., COURTY, M. L., and YU, S. B. (2001), *Seismic cycle in Taiwan derived from GPS measurements*, Comptes Rendus De L Academie Des Sciences Serie Ii Fascicule a-Sciences De La Terre Et Des Planetes, *333*, 57–64.

LOEVENBRUCK, A., CATTIN, R., LE PICHON X., DOMINGUEZ, S., and MICHEL, R. (2004), *Coseismic slip resolution and postseismic relaxation time of the 1999 Chi-Chi, Taiwan, earthquake as constrained by geological observations, geodetic measurements and seismicity*, Geophys. J. Int. *158*, 310–326.

MARONE, C., RALEIGH, C. B., and SCHOLZ, C. (1990), *Frictional behavior and constitutive modelling of simulated fault gouge*, J. Geophys. Res. *95*, 7007–7025.

MARONE, C. (1998), *Laboratory-derived friction laws and their application to seismic faulting*, Annu. Rev. Earth Planet. Sci. *26*, 643–696.

MATTHEWS, M. V. and SEGALL, P. (1993), *Estimation of depth-dependent fault slip from measured surface deformation with application to the 1906 San Francisco earthquake*, J. Geophys. Res. *98*, 12,153–112,163.

MICHAEL, A. J. (1984), *Determination of stress from slip data - faults and folds*, J. Geophys. Res. *89*, 1517–1526.

MICHAEL, A. J. (1987), *Use of focal mechanisms to determine Stress - A control study*, J. Geophys. Res. *92*, 357–368.

MIYAZAKI, S., SEGALL, P., FUKUDA, J., and KATO, T. (2004), *Space time distribution of afterslip following the 2003 Tokachi-oki earthquake: Implications for variations in fault zone frictional properties*, Geophys. Res. Lett. *31*, doi:10.1029/2003GL019410.

MOUNT, V. S. and SUPPE, J. (1987), *State of stress near the San-Andreas fault - Implications for wrench tectonics*, Geology *15*, 1143–1146.

OKADA, Y. (1985), *Surface deformation to shear and tensile faults in a half space*, Bull. Seism. Soc. Am. *75*, 1135–1154.

OLESKEVICH, D. A., HYNDMAN, R. D., and WANG, K. (1999), *The updip and downdip limits to great subduction earthquakes: Thermal and structural models of Cascadia, south Alaska, SW Japan, and Chile*, J. Geophys. Res. *104*, 14965–14991.

PATHIER, E., FRUNEAU, B., DEFFONTAINES, B., ANGELIER, J., CHANG, C. P., YU, S.-B., and LEE, C.-T. (2003), *Coseismic displacements of the footwall of the Chelungpu fault caused by the 1999, Taiwan, Chi-Chi earthquake from InSAR and GPS data*, Earth Planet. Sci. Lett. *212*, 73–88.

PEACOCK, S. M. and HYNDMAN, R. D. (1999), *Hydrous minerals in the mantle wedge and the maximum depth of subduction thrust earthquakes*, Geophys. Res. Lett. *26*, 2517–2520.

PERFETTINI, H. and AVOUAC, J. P. (2004), *Postseismic relaxation driven by brittle creep: A possible mechanism to reconcile geodetic measurements and the decay rate of aftershocks, application to the Chi-Chi earthquake, Taiwan*, J. Geophys. Res. *109*, doi:10.1029/2003JB002488.

PERFETTINI, H., AVOUAC, J. P., and RUEGG, J. C. (2005), *Geodetic displacements and aftershocks following the 2001 Mw = 8.4 Peru earthquake: Implications for the mechanics of the earthquake cycle along subduction zones*, J. Geophys. Res. *110*, doi:10.1029/2004JB003522.

PERFETTINI, H., and AMPUERO, J. P. (2008), *Dynamics of a velocity strengthening fault region: Implications for slow earthquakes and postseismic slip*, J. Geophys. Res. *113*, doi:10.1029/2007JB005398.

ROLANDONE, F., BURGMANN, R., and NADEAU, R. M. (2004), *The evolution of the seismic-aseismic transition during the earthquake cycle: Constraints from the time-dependent depth distribution of aftershocks*, Geophys. Res. Lett. *31*, 10.1029/2004GL021379.

SAVAGE, J. C. (1983), *A dislocation model of strain accumulation and release at a subduction zone*, J. Geophys. Res. *88*, 4984–4996.

SCHAFF, D. P., BOKELMANN, G. H. R., BEROZA, G. C., WALDHAUSER, F., and ELLSWORTH, W. L. (2002), *High-resolution image of Calaveras Fault seismicity*, J. Geophys. Res. *107*, 10.1029/2001JB000633.

SCHOLZ, C. *The Mechanics of Earthquakes and Faulting*, 439 pp, (Cambridge University Press, New York 1990).

SCHOLZ, C. H. (1998), *Earthquakes and friction laws*, Nature *391*, 37–42.

SCHOLZ, C. H. (2000), *Evidence for a strong San Andreas fault*, Geology, *28*, 163–166.

SHAW, B. E. (1995), *Frictional weakening and slip complexity in earthquake faults*, J. Geophys. Res. *100*, 18239–18251.

SHIMAZAKI, K. and NAKATA, T. (1980), *Time-predictable recurrence model for large earthquakes*, Geophys. Res. Lett. *7*, 279–282.

SHIN, T. C. and TENG, T. L. (2001), *An overview of the 1999 Chi-Chi, Taiwan, earthquake*, Bull. Seismol. Soc. Am. *91*, 895–913.

SIEH, K. (2000), *The repetition of large earthquake ruptures*, paper presented at Hokudan International Symp. and School on Active (Letter Press Co. Ltd., FaultingHokudan, Japan, 2000).

SIMOES, M., AVOUAC, J. P., BEYSSAC, O., GOFFE, B., FARLEY, K. A., and CHEN, L. (2007a), *Mountain building in Taiwan: A thermokinematic model*, J. Geophys. Res. *112*, doi: 10.1029/2006JB004824.

SIMOES, M., AVOUAC, J. P., and CHEN, Y. G. (2007b), *Slip rates on the Chelungpu and Chushiang thrust faults inferred from a deformed strath terrace along the Dungpuna river, west central Taiwan*, J. Geophys. Res. *112*, doi:10.1029/2005JB004200.

SIMONS, M., FIALKO, Y., and RIVERA, L. (2002), *Coseismic deformation from the 1999 M-w 7.1 Hector Mine, California, earthquake as inferred from InSAR and GPS observations*, Bull. Seismol. Soc. Am. *92*, 1390-1402.

STREIG, A. R., RUBIN, C. M., CHEN, W. S., CHEN, Y. G., LEE, L. S., THOMPSON, S. C. MADDEN, C., and LU, S. T. (2007), *Evidence for prehistoric coseismic folding along the Tsaotun segment of the Chelungpu fault near Nan-Tou, Taiwan*, J. Geophys. Res. *112*, doi:10.1029/2006JB004493.

SUPPE, J. (2007), *Absolute fault and crustal strength from wedge tapers*, Geology *35*, 1127–1130.

TITUS, S. J., DEMETS, C., and TIKOFF, B. (2005), *New Slip rate estimates for the creeping segment of the San Andreas fault, California*, Geology *33*, 161–240.

TSE, S. T., and RICE, J. R. (1986), *Crustal earthquake instability in relation to depth variation of frictional slip properties*, J. Geophys. Res. *91*, 9452–9472.

VERGNE, J., CATTIN, R., and AVOUAC, J.-P. (2001), *On the use of dislocations to model interseismic strain and stress build-up at intracontinental thrust faults*, Geophys. J. Int. *147*, 155–162.

WESSON, R. L., and BOYD, O. S. (2007), *Stress before and after the 2002 Denali fault earthquake*, Geophys. Res. Lett. *34*, doi:10.1029/2007GL029189.

WU, Y. M., CHANG, C. H., ZHAO, L., SHYU, J. B. H., CHEN, Y. G., SIEH, K., and AVOUAC, J. P. (2007), *Seismic tomography of Taiwan: Improved constraints from a dense network of strong motion stations*, J. Geophys. Res. *112*, doi:10.1029/2007JB004983.

WU, Y. M., ZHAO, L., CHANG, C. H., and HSU, Y. J. (2008), *Focal-mechanism determination in Taiwan by genetic algorithm*, Bull. Seismol. Soc. Am. *98*, 651–661.

YU, S.-B., KUO, L.-C., HSU, Y.-J., SU, H.-H., LIU, C. C., HOU, C.-S., LEE, J.-F., LAI, T.-C., LIU, C. C., LIU, C.-L., TSENG, T.-F., TSAI, C.-S., and SHIN, T.-C. (2001), *Preseismic deformation and coseismic displacements associated with the 1999 Chi-Chi, Taiwan, earthquake*, Bull. Seismol. Soc. Am. *91*, 995–1012.

YU, S.-B., HSU, Y.-J., KUO, L.-C., CHEN, H.-Y., and LIU, C.-C. (2003), *GPS measurements of postseismic deformation following the 1999 Chi-Chi earthquake*, J. Geophys. Res. *108*, 2520.

YU, S. B. and LIU, C. C. (1989), *Fault creep on the central segment of the Longitudinal Valley Fault, eastern Taiwan*, Proc. Geol. Soc. China *32*, 209–231.

YU, S. B., and KUO, L. C. (2001), *Present-day crustal motion along the Longitudinal Valley Fault, eastern Taiwan*, Tectonophysics *333*, 199–217.

YUE, L. F., SUPPE, J., and HUNG, J. H. (2005), *Structural geology of a classic thrust belt earthquake: the 1999 Chi-Chi earthquake Taiwan (M-w = 7.6)*, J. Struct. Geol. *27*, 2058–2083.

ZOBACK, M. D., ZOBACK, M. L., MOUNT, V. S., SUPPE, J., EATON, J. P., HEALY, J. H., OPPENHEIMER, D., REASENBERG, P., JONES, L., RALEIGH, C. B., WONG, I. G., SCOTTI, O., and WENTWORTH, C. (1987), *New Evidence on the state of stress of the San-Andreas fault system*, Science *238*, 1105–1111.

ZOBACK, M. D., and TOWNEND, J. (2001), *Implications of hydrostatic pore pressures and high crustal strength for the deformation of intraplate lithosphere*, Tectonophysics *336*, 19–30.

(Received September 15, 2008, accepted March 3, 2009)
Published Online First: July 29, 2009

To access this journal online:
www.birkhauser.ch/pageoph

Pure appl. geophys. 166 (2009) 1885–1908
0033–4553/09/101885–24
DOI 10.1007/s00024-009-0520-3
© Birkhäuser Verlag, Basel, 2009
Pure and Applied Geophysics

Landslides, Ice Quakes, Earthquakes: A Thermodynamic Approach to Surface Instabilities

Klaus Regenauer-Lieb,[1,2] David A. Yuen,[3] and Florian Fusseis[1]

Abstract—The total rate of rock deformation results from competing deformation processes, including ductile and brittle mechanisms. Particular deformation styles arise from the dominance of certain mechanisms over others at different ambient conditions. Surprisingly, rates of deformation in naturally deformed rocks are found to cluster around two extremes, representing coseismic slip rates or viscous creep rates. Classical rock mechanics is traditionally used to interpret these instabilities. These approaches consider the principle of conservation of energy. We propose to go one step further and introduce a nonlinear far-from-equilibrium thermodynamic approach in which the central and explicit role of entropy controls instabilities. We also show how this quantity might be calculated for complex crustal systems. This approach provides strain-rate partitioning for natural deformation processes occurring at rates in the order of 10^{-3} to 10^{-9} s^{-1}. We discuss these processes using examples of landslides and ice quakes or glacial surges. We will then illustrate how the mechanical mechanisms derived from these near-surface processes can be applied to deformation near the base of the seismogenic crust, especially to the phenomenon of slow earthquakes.

Key words: Seismology, geodynamics, instabilities, thermodynamics, entropy production, numerical modelling.

Notation

Variable

S	*Entropy*
s	*Specific entropy*
Q	*Heat*
T	*Absolute temperature*
ρ	*Density*
$e_{tot,\ kin}$	*Specific total, kinetic energy*
u	*Specific internal energy*
V	*Volume of the reference volume element (RVE)*
A	*Surface of the RVE*
σ_{ij}	*Total Cauchy stress tensor*
ε_{ij}	*Total strain tensor*
q_i	*Surface heat flow, radiative and conductive*

[1] Multi-scale Earth System Dynamics, School of Earth and Environment, University of Western Australia, Crawley, WA 6009, Australia. E-mail: klaus.regenauer-lieb@csiro.au; fusseis@cyllene.uwa.edu.au
[2] CSIRO Exploration and Mining, PO Box 1130, Bentley, WA 6102, Australia.
[3] Department of Geology and Geophysics and Minnesota Supercomputing Institute, University of Minnesota, Minneapolis, MN 55455, U.S.A.

ψ	*Specific Helmholtz free energy*
ε_{ij}^{el}	*First state variable, the total elastic strain tensor*
α_k	*Other state variables (e.g. total creep strain tensor)*
C_α	*Specific heat for constant* α
λ	*Thermal expansion coefficient*
p	Pressure
ξ	*Fractional volume of one phase in a two phase mixture*
χ	*Taylor-Quinney heat conversion efficiency*
L	*Latent heat release*
κ	*Thermal diffusivity*

1. Introduction: Classical Rock Mechanics and Thermodynamics Approaches

Here we present the hypothesis that far-from-equilibrium, nonlinear thermodynamics provides a unified framework for earthquake modeling. The rapid movements of large masses of the crust have a direct impact on human communities and a deep understanding of these phenomena requires researchers to take a fresher view into the dynamical properties of rock deformation near the surface. The strain energy of a deforming rock is generally released through a number of different creep mechanisms. These deformation mechanisms span a wide range and involve elastic straining, dislocation- and diffusion creep, metamorphic reactions, phase transitions, chemical reactions, fluid-rock interaction, cleavage fracturing, cavitation and frictional sliding.

Observational evidence for the different mechanisms come from a wide range of disciplines, from extrapolation of experimental data, microstructure, microgeochemistry (isotope measurements, 'geospeedo-metry'), remote sensing (GPS/InSAR) and geophysical measurements (borehole strainmeters). Total rates of natural deformation processes have been bracketed (Fig. 1) either at seismic rates (e.g., fracture propagation, frictional sliding/melting, which happens within seconds) or at strain rates of 10^{-10} to 10^{-15} s^{-1} where point/line-defect diffusion controls the viscous (i.e., temperature-activated) response of a rock. A geological system where deformation takes place at all strain rates is the seismogenic crust. Deformation occurs on the time scale of the earthquake cycle on the order from tens to hundreds of years. During interseismic periods various 'slow' diffusion processes ultimately control most deformation, whereas during earthquakes 'rapid', unstable brittle processes control the deformational response of the fault to coseismic stress release.

Consequently, during the earthquake cycle, deformation rates vary over about 14 orders of magnitude. Within the earthquake cycle "slow earthquakes" release strain energy over a time scale of several hours to days, often as a precursor to larger earthquakes (e.g., LINDE *et al.*, 1996). This time scale suggests that slow earthquakes happen at intermediate rates between those achieved during interseismic and coseismic periods.

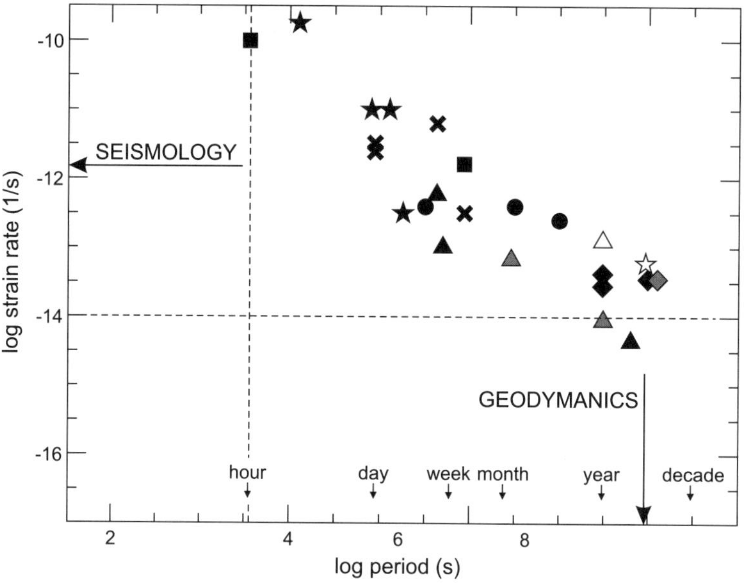

Figure 1
Measured strain rates by: GPS (gray), InSAR(white), Strainmeters (black). Triangles show postseismic deformation from the Landers earthquake, squares show slow earthquakes, open squares single station observations. GPS and InSAR data have higher sensitivity at long periods (months to decades) while borehole strain meters are better at short periods. While some data are derived close to their detection limit, the plots illustrate the point that there is no separation between the classical geodynamic strain rates (dashed line plate motion) and seismological strain rate. Geological deformation takes place at all rates. Source: PBO Steering Committee, The Plate Boundary Observatory: Creating a four-dimensional image of the deformation of western North America, White paper providing the scientific rationale and deployment strategy for a Plate Boundary Observatory based on a workshop held October 3-5, 1999, modified after: http://www.nap.edu/books/0309065623/xhtml/images/p20007dc2g215001.jpg.

In classical rock mechanics widely different micromechanical deformation mechanisms are used to explain the wide range of strain-rates observed in nature. However, the processes are normally not treated in a fully coupled way. There also does not exist a unified theory linking the scales. Rock mechanical properties are assigned to a given micromechanical group (see Appendix A). Interactions are often parameterized in experimentally derived state-dependent laws considering averaging variables (RICE 1971). In earthquake modeling, the use of experimentally constrained (DIETERICH, 1979a, b) rate and state-dependent friction law has proven to be very useful (TSE and RICE, 1986). These rock-mechanical descriptions are in themselves very similar to an isothermal thermomechanic approach but they are normally not derived from thermomechanics *a priori*, see ZIEGLER (1983) for a full thermomechanic approach. This work has now progressed further and an excellent approach to plasticity theory based on thermodynamic principles has been published recently (HOULSBY and PUZRIN, 2007). We argue, however, to go one step further for earthquake modeling. Such full nonisothermal thermodynamic formulations have only been available since the early 90's, being originally motivated by the advent of thermal infrared imaging of deformation processes (CHRYSOCHOOS and DUPRE, 1991). The approach has, however, a huge potential beyond this application. It offers an extension of thermomechanics into the nonisothermal domain. The approach also offers the possibility to formulate a unified multi-scale thermodynamical framework for coupling mechanical and chemical simulations (POULET and REGENAUER-LIEB, 2009; RAMBERT *et al.*, 2007).

For earthquake modeling the clear advantage is that dynamic properties of rocks are not tied to specific experiments but property variations can be predicted in a truly time- or state- dependent manner based on the rate of entropy production. In this contribution we will elaborate further what this theory might hold for the application to earthquake modeling. We relate this concept to earlier case studies of thermal-mechanical instabilities, flowing of large ice sheets and unstable landslides.

2. *Instabilities and their Mathematical Expression*

CLAUSIUS (1865) introduced the concept of entropy as a simplifying concept for the understanding of thermal processes and not as a complication. Quite the opposite appears to be the case today. In geosciences we seem to have forgotten Clausius's message and try to avoid entropy production as a simplifying method in our calculations.

There is considerable confusion about entropy in the present literature. Entropy appears to be a concept that is only understood by information theorists or theoretical physicists/chemists. In the following we try to describe it as a simplifying concept that allows us to understand geomechanical problems with a thermodynamic foundation. In this attempt we avoid to go into subjects such as stochastic geometry and information theory (ATTARD, 2006) and focus on the potential application to earthquake simulation. In its most basic form the definition of entropy was put forward in order to formalize the simple concept that heat flows spontaneously from a hot to a cold object. No process can take place whose net effect is only to transfer heat from a cold object to a hot object. We hence revisit Clausius original definition by which entropy is defined as a quantity to describe what was later known as the second law of thermodynamics.

$$dS \geq \frac{\delta Q}{T}, \tag{1}$$

where S is the entropy of the system, d represents an infinitesimally small change of a state function, δ represents that of a path function. Q is the heat and T the absolute temperature. Maximum entropy defines a thermodynamic system at equilibrium. This entropy state is used later for the definition of a representative volume element (RVE). For reversibility the inequality in turns into equality. We also introduce the specific entropy which is

$$dS = \int_V \rho s \, dV. \tag{2}$$

Now consider a representative volume element in the current configuration and thermodynamic equilibrium. Its internal energy in motion is the sum of its specific internal energy u plus its specific kinetic energy e_{kin}.

$$\int_V \rho \, e_{tot} \, dV = \int_V \rho \, u \, dV + \int_V \rho \, e_{kin} \, dV \tag{3}$$

In the following we use the classical approach of creeping flows conveniently used for geodynamic modeling. We neglect the effect of the kinetic energy. This clearly does not apply to earthquake mechanics, however, as a first step towards modeling the conditions leading to an earthquake we wish to investigate competing strain rate processes prior to the development of kinetic energy. As a caveat we emphasize that this approach is meaningful only for investigation of the early mechanisms underlying the physics of slow earthquakes, ice quakes and landslides.

2.1. Local Equilibrium: The Small Scale

The time scale of interest defines the choice of the suitable size of the representative local equilibrium volume element v in a material reference frame (current volume) for a given observation time. This is the crucial step in the thermodynamics equilibrium assumption. Consider two thermodynamics processes happening at two very different time and length scales, i.e., the time t_1 to reach local equilibrium in the reference volume considerably smaller than the size of the system under study and the time t_2 required to reach the equilibrium in the entire system, $t_1 << t_2$. Following Onsager's regression hypothesis (ONSAGER, 1931) the time evolution of the fluctuation of a given physical value in an equilibrium system obeys the same laws on the average as the change of the corresponding macroscopic variable in a nonequilibrium system.

For the particular time scale of interest we define the smallest volume that allows a calculation of an average continuum property (RICE 1971) as the arbitrary reference volume in local thermodynamic equilibrium. Even at this smallest scale of the RVE we place ourselves in the framework of continuum mechanics where every point of the material can itself be imaged as a continuum. In other words, we assume that the RVE contains a sufficient number of discrete entities such that the laws of thermodynamics apply. This approach can be applied to all scales down to microscale. However, below nanoscale, statistical mechanics would replace the approach presented here. Emergent equilibrium properties of the continuum systems are embedded in a larger system that is not in equilibrium. We obtain for negligible kinetic energy the following relation for the RVE:

$$\int \dot{e}_{tot}\, dV = \int \dot{u}\, dV, \tag{4}$$

where, for convenience, the overdot denotes differentiation with respect to the Lagrangian, material time derivative. For this reference volume the first law of thermodynamics spells out the energy conservation under the assumption of small perturbations, conservation of mass and creeping flow

$$\int_V \rho\dot{u}\, dV = \int_V \sigma_{ij}\dot{\varepsilon}_{ij}\, dV + \int_V r_i\, dV - \int_A q_i\, dA. \tag{5}$$

We assume Einstein's summation convention and identify σ_{ij} as the total Cauchy stress tensor, which is work conjugate to the symmetric strain tensor ε_{ij}. In continuum mechanics work conjugacy implies that the product of the stress with the strain increments gives the total rate of work input to the material, per unit volume. We are using symmetric tensors because these follow from moment equilibrium of the continuum. In a first attempt to use thermodynamics for linking earthquake modeling with geodynamics we do not use the theory of micropolar continua because nonsymmetric tensor formulations would be required. Conservation of energy is achieved if the rate of external work expended on the reference volume is equal to the mechanical work, including the heat produced in the reference volume (first term on the right-hand side) plus a term related through internal heat r_i generation such as radioactive decay, Joule heating or heat generation through chemical reactions (second term on the right-hand side) minus the radiative and conductive heat transfer q_i on the surface of the volume respectively (third term).

Note that the additional heat term r_i appears as a source term on the right-hand side if e.g., radioactive decay is considered without considering its effect on the specific internal energy u. Such a loose formalism may appear convenient (POULET and REGENAUER-LIEB, 2009) if the scale of the radioactive isotopes is not resolved in the mathematical treatment. However, we emphasize that this loose formalism is strictly seen as not satisfactory in a more general approach (HOULSBY and PUZRIN, 2007).

The fundamental thermodynamic energy balance for the specific energy is given in terms of the specific entropy s by

$$u = \psi(T, \varepsilon_{ij}^{el}\{\alpha_k\}) + sT, \tag{6}$$

where ψ is the specific Helmholtz free energy and the state variables are the elastic strain ε^{el} and the absolute Temperature T and α_k other state variables including for instance the plastic strains caused by the individual micromechanical deformation mechanisms.

This equation describes the system in thermodynamic equilibrium given the following condition. *Equilibrium is achieved if the entropy goes to a maximum.* As a universal principle we hence use a strong form of the second law of thermodynamics. i.e. rather than just implying the second law of thermodynamics as a criterion for the direction of heat flow (from hot to cold, therefore the entropy must be positive) we require it to reach a maximum. Maximum entropy implies that the conjugate quantity to entropy, i.e., the stored energy (Helmholtz free energy, Gibbs free energy) goes to a minimum at equilibrium. Therefore for the small volume element we can derive equilibrium material properties such as elastic properties simply from a chemical Gibbs free energy minimizer (SIRET *et al.*, 2008). Note that time-dependent processes do not explicitly enter the discussion yet as we are still discussing equilibrium thermodynamics and time is irrelevant.

In order to illustrate this concept consider the following example: The RVE is a purely isothermal conductive solid with initial thermal insulation at all boundaries. At time $t_0 << t_1 << t_2$ the top and bottom thermal boundaries are removed and the RVE is subject to higher constant temperature from below and a lower constant temperature on the top boundary. Heat flows through the RVE and it is not in equilibrium. According to our choice this RVE can be considered in equilibrium only when it has reached its maximum entropy (linear thermal gradient) at time t_1. There is no more time dependency in the system. We can derive material parameters for this RVE for time scale t_2 by solving for the minimum Helmholtz free energy. Having defined the thermodynamic equilibrium time/length scale we may now wish to proceed to the nonequilibrium processes at the large scale. In the following we will implicitly assume integration over the RVE and drop the volume integral.

2.2. *Nonequilibrium Assumption: The Large Scale*

For the large nonequilibrium system we need to know how the macroscopic system responds over some reference time to small fluctuations of a given physical value at the local equilibrium. The small reference volume is at equilibrium. Equation (6) is appropriate for the small volume element but insufficient for describing the large nonequilibrium system. For this we have to consider the time evolution for which we differentiate Equation (6) with respect to time

$$\dot{u} = \dot{\psi}(T, \varepsilon_{ij}^{el}, \alpha_k) + \dot{s}T + \dot{T}s. \tag{7}$$

The second term now describes the rate of entropy production and the third term the coupled variation of temperature with time. Feedback between the three terms on the right-hand side of Equation (7) underpins the localization phenomena discussed in this paper. This feedback expresses itself as a competition of rates of processes, which happen at vastly different time and length scales. Equation (7) describes a rigorous and consistent framework, within which models can be developed, to describe a wide range of material processes, which we believe to be important for explaining the strain rate cascade in Figure 1.

Equation (7) relates the rate of internal energy to the sum of the time-rate of the Helmholtz free energy plus the rate of the entropy production composed of mechanical and thermal dissipation processes. Feedback between any of the terms can lead to instability. As an example, feedback between thermal and mechanical dissipation processes leads to the well-known thermal runaway instability (e.g., OGAWA, 1987) for instance if production of heat is faster than loss of heat due to diffusion. However, the thermal runaway process is only one of many possible feedbacks. In more general terms Equation (7) defines for the general case, conditions for instability if some critical internal parameter is exceeded. Shear heating may be the first and most obvious feedback mechanism in the strain-rate cascade of Figure 1 but we show later on in the Landslide example that it is overtaken by another critical mechanism once the strain rates have increased sufficiently above geodynamic rates. In the following we highlight the important difference of this nonlinear far-from-equilibrium thermodynamic approach formalized in Equation (7) with theories for localization in the classical rock mechanics formulation (Appendix A) and Prigogines localization theories for chemistry (next section).

Classical rock mechanics deals with isothermal deformation and Equation (7) simplifies to the theory of thermomechanics (see Appendix A). There is no thermal dissipation and this important feedback is suppressed. The time dependence of thermal diffusion is also missing and, if there is no kinetic energy and no chemical diffusion, then thermomechanics reverts to the classical theory of plasticity with a check on thermodynamic consistency. In this case the internal energy rate (rate of mechanical work applied to volume element) is equal to the rate of change of Helmholtz free energy (stored energy) plus the dissipation (entropic work rate or dissipated part of the plastic work rate). Because the time dependence is lost, the entropy must be formulated along a new quasi-steady state upon reaching the plastic yield stress. ZIEGLER (1983) shows that this leads to the principle of *maximimum entropy production* as a thermomechanics consistency check for continuum mechanics. Maximum entropy production is a strong assumption and it is only achieved if the nonlinear thermal mechanical process has reached some form of steady state, i.e., temperature does not change. This assumption is hence not very useful for earthquake modeling and we recommend use of the entire formulation in Equation (7).

2.3. Linear Stability and Localization

A localization phenomenon can be understood mathematically as a bifurcation phenomenon where the velocity field of a smoothly deforming/reacting solid aborts the continuous branch and takes a new discontinuous path. Equation (7) can be analyzed using classical concepts of linear stability and we can predict from this critical quantities of rate of Helmholtz free energy, rate of entropy production or dissipation. We highlight possible caveats owing to the nonlinear nature of the equations. In the thermal runaway example discussed above we have already illustrated the basic concept underlying linear stability for a critical value. The localization problem is understood mathematically as an unstable solution of the system of partial differential equations. In a linear stability analysis this problem is tackled by simplifying to linear ordinary differential equations.

The problem solved hence is a linearized solution to what is in principle a nonlinear matrix problem. An example in mechanics would be, where a tangential stiffness matrix is a function of the nodal displacements. We look for the unknown increment of nodal displacement leading to bifurcation. This tangential stiffness is equal to a residual nodal force. If the determinant of the stiffness matrix becomes zero a bifurcation is detected. It is often pointed out that the neglect of the nonlinear terms allows establishment of a sufficient criterion for instability but not of a necessary criterion. In addition this analysis only allows us to detect a bifurcation but does not allow the prediction of how the system behaves far from this equilibrium point. The analysis is conventionally extended to derive Lyapunov functions.

Any local equilibrium state can be analyzed by means of local linear approximation to the underlying differential equation such as Equation (7). This can be done to assess stability of any equilibrium state in an extended linear thermodynamic framework. This is indeed the great achievement of Prigogine and the group from Brussels (GLANSDORFF *et al.*, 1973). The theory was, however, only designed for investigation of thermodynamic stability of arbitrary mass-action kinetic networks in which the reaction velocity is assumed to be proportional to the concentrations of the involved reactants. Temperature variations were not considered and the framework was reduced to a thermomechanic approximation (see Appendix A). In addition in view of the difficulty of the problem Prigogine restricted his analysis to a linearized system where Lyapunov's method can be used to derive stability far-from-equilibrium. It is important to note that on the basis of this assumption it is a linear non-equilibrium thermodynamic framework. The chosen problem of a mass diffusion is indeed (like the simple thermal diffusion problem discussed above) in its basic nature a linear thermodynamic problem. Prigogine showed the principle of minimum entropy production for the stability of non-equilibrium linear thermodynamics. We seek stability for the competing rate process of non-equilibrium, nonlinear thermodynamics. For this no analytical technique is known and we have to make use of a numerical approach.

We are left with an extrapolation from linear thermodynamics (LAVENDA, 1978). If we expand terms of the nonlinear partial differential equation in a Taylor series, we may postulate that the first term of the expansion is applicable to the more general case, with the caveat that it is a sufficient but not a universal theory. We want to know how the system behaves close to the equilibrium point, e.g., whether it moves towards or away from the equilibrium point, it should therefore be good enough to keep just the linear terms. However, in a marginally stable system the higher order terms may be crucial for the phenomenon of localization and the linear theory breaks down. This means that in most cases the linearized solution will work as a good approximation to the nonlinear case, however, in some it will not work.

In order to evaluate what happens after the instability criteria are fulfilled, we cannot hence use the theory proposed by Prigogine of least rate of dissipation as a generalized basis. It only applies for the case of linear far-from-equilibrium thermodynamics. For the more general case, nonlinear partial differential equations are necessary to describe additional feedback mechanisms. Under these conditions we may not be able to describe the evolution of the system from extremum principles of entropy production. It becomes apparent from this discussion that there exists to date no universal analytically tractable theory for localization for such a case. We are left with a numerical assessment of the case and have to solve explicitly the feedback between entropy production, thermal variation and changes in Helmholtz free energy. For this system we use the Finite-Element method to derive the far-from-equilibrium behavior of the system for small perturbation away from equilibrium of a local equilibrium volume element. In the following we show how the entropy production may be calculated more explicitly.

2.4. *Calculating the Rate of Entropy Production*

The rate of entropy production can be calculated more explicitly from the above. We first expand the rate of Helmholtz free energy production using the chain rule

$$\dot{\psi}_{T,\varepsilon_{ij}^{el},\{\alpha_k\}} = \left(\frac{\partial\psi}{\partial T}\right)_{\varepsilon_{ij}^{el},\{\alpha_k\}} \dot{T} + \left(\frac{\partial\psi}{\partial\varepsilon_{ij}^{el}}\right)_{T,\{\alpha_k\}} \dot{\varepsilon}_{ij}^{el} + \left(\frac{\partial\psi}{\partial\alpha_k}\right)_{T,\varepsilon_{ij}} \dot{\alpha}_k, \tag{8}$$

where the subscripts of the partial differential equation are constant and we use Maxwell's relations (POULET and REGENAUER-LIEB, 2009) which follow from the second law of thermodynamics

$$s = -\left(\frac{\partial\psi}{\partial T}\right)_{\varepsilon_{ij}^{el},\{\alpha_k\}}, \tag{9}$$

$$\sigma_{ij} = \rho\left(\frac{\partial\psi}{\partial\varepsilon_{ij}}\right)_{T,\{\alpha_k\}}, \tag{10}$$

and define the specific heat from

$$c_\alpha \equiv -T\left(\frac{\partial^2\psi}{\partial T^2}\right)_{\varepsilon,\{\alpha_k\}}, \tag{11}$$

also from Equation (9)

$$\dot{s} = -\frac{\partial^2\psi}{\partial T^2}\dot{T} - \frac{\partial^2\psi}{\partial T\partial\alpha_k}\dot{\alpha}_k - \frac{\partial^2\psi}{\partial T\partial\varepsilon_{ij}^{el}}\dot{\varepsilon}_{ij}^{el}, \tag{12}$$

$$T\dot{s} = c_\alpha\dot{T} - T\frac{\partial^2\psi}{\partial T\partial\alpha_k}\dot{\alpha}_k - T\frac{\partial^2\psi}{\partial T\partial\varepsilon_{ij}^{el}}\dot{\varepsilon}_{ij}^{el}, \tag{13}$$

substituting Equations (5) and (7) into Equation (13) we obtain the classical energy equation with additional feedback terms that arise due to the thermal–mechanical couplings.

$$\rho c_\alpha \dot{T} = \sigma_{ij}\dot{\varepsilon}_{ij} - \rho\frac{\partial\psi}{\partial\varepsilon_{ij}^{el}}\dot{\varepsilon}_{ij}^{el} - \rho\frac{\partial\psi}{\partial\alpha_j}\dot{\alpha}_k + \rho T\frac{\partial^2\psi}{\partial T\partial\varepsilon_{ij}^{el}}\dot{\varepsilon}_{ij}^{el} + \rho T\frac{\partial^2\psi}{\partial T\partial\alpha_j}\dot{\alpha}_k + r_i - div\, q_i. \quad (14)$$

We are here describing the additional feedback terms in the order of their appearance. The first term on the right-hand side $\sigma_{ij}\dot{\varepsilon}_{ij}$ minus the second and third term on the right is the deformational power that is converted into heat

$$\sigma_{ij}\dot{\varepsilon}_{ij} - \rho\frac{\partial\psi}{\partial\varepsilon_{ij}^{el}}\dot{\varepsilon}_{ij}^{el} - \rho\frac{\partial\psi}{\partial\alpha_j}\dot{\alpha}_k = \chi_{(t)}\sigma_{ij}\dot{\varepsilon}_{ij}^{diss}. \quad (15)$$

The second term describes the elastic power not converted into heat, and the third term other microstrain processes that are not converted into heat such as those caused by chemical strain or other microstructural modification processes or recoverable processes. All nonthermal physical deformation processes are hidden in the second and third terms. We expect that an explicit treatment will become very important for future studies. The shear heating term is well known in the literature but normally expressed in a simpler fashion shown in Equation (15) through the introduction of the nondimensional factor χ, which is the Taylor-Quinney heat conversion efficiency with permissible values between 0 and 1. In addition only the dissipative strain rates $\dot{\varepsilon}_{ij}^{diss}$ such as those from creep or plastic deformation are considered in classical nonthermodynamic formulations. Since most experimental creep laws are reported for steady state and follow from the postulate of maximum entropy production for quasi-steady state (Appendix A), the heat conversion at steady state is expected to be very efficient. Values for most materials are indeed between 80–95% conversion efficiency (Chrysochoos and Belmahjoub, 1992). The Taylor-Quinney coefficient is often ascribed to unity and neglected.

The fourth term on the right-hand side of Equation (14) describes thermal elastic coupling. In the nonthermodynamic literature this is often neglected or replaced by the scalar or tensor-valued volumetric or scalar linear thermal-elastic expansion coefficient λ and the third term simplifies to

$$\rho T\frac{\partial^2\psi}{\partial T\partial\varepsilon_{ij}}\dot{\varepsilon}_{ij} \equiv \lambda T_{equ}\dot{p}, \quad (16)$$

where p is the pressure and T_{equ} is a thermodynamic equilibrium temperature change. The fifth term on the right-hand side of Equation (14) is the thermal-mechanical coupling term through any additional state variable. Assuming for an example a two-phase thermodynamic system with the fractional volume ξ of one phase the coupling term is equivalent to the latent heat release upon phase change is

$$L = \rho T \frac{\partial^2 \psi}{\partial \xi_i \partial T} \dot{\xi}_i. \tag{17}$$

Equation (14) collapses for this simple case

$$\rho c_p \dot{T} = \chi_{(t)} \sigma_{ij} \dot{\varepsilon}_{ij}^{diss} + \lambda_{th} T_{equ} \dot{p} + L + r_i + \rho c_p \kappa \nabla^2 T \tag{18}$$

where κ is the thermal diffusivity. This equation is also known as the energy equation and it is typically solved for each representative volume element by finite-element analysis. In the numerical solution technique the question whether the volume element for discretization is in thermodynamic equilibrium for the chosen time step needs to be evaluated carefully (REGENAUER-LIEB and YUEN, 2004).

The above-described form of the energy equation is indeed what is used in some modern modeling approaches which we will discuss in the following. If solutions are obtained through full coupling of the energy equation with the momentum and continuity equation (Appendix B), these calculations are consistent with the thermodynamic theory presented here. However, the thermodynamic theory provides additional useful tools for the assessment of instabilities such as extremum principles in entropy production to be described in a forthcoming contribution. Its biggest advantage, however, is the provision of a basic framework for cross-scale modeling, thus allowing a link of geodynamics with seismological simulations. Another advantage, relevant for seismology, is that material properties can be derived as energetically self-consistent time-dependent parameters.

3. *Instabilities and their Natural Expression*

Earthquakes are instabilities that are always manifested in an unexpected manner because of the unpredictable nature of the nonlinear material properties in the shallow portion of the crust, where a critical temperature separates a stable creeping regime from a domain where fractures can readily develop. This temperature is known in the literature (e.g., MORRIS, 2008), as the brittle-ductile transition temperature and holds the key to our understanding of the critical interplay between fracture mechanics and earthquakes. The depth extent of rupture in large continental earthquakes is found to be limited in the temperature regime approximately by the 570 K isotherm (STREHLAU, 1986). This is well within the estimated (REGENAUER-LIEB and YUEN, 2006; 2008) range of onset of creep between 450 and 500 degrees K for quartz. The far-reaching concept that the earthquake mechanism is possibly rooted in the ductile realm was first introduced in the classical work of HOBBS *et al.* (1986) and the recent work of JOHN *et al.* (2009).

Natural expressions of these instabilities can be found in geological field evidence, which includes paleo-earthquakes, ice-quakes, landslides, pseudotachylytes found in fault zones and the phenomenon of grain-size reduction, and laboratory evidence for different types of faulting. In this work we will also discuss the potential danger of ice-quakes caused by these brittle-ductile instabilities, which might exert dramatic influence on raising the sea levels over short time scales. We describe some ongoing laboratory experiments which will call our attention to the potentially important role played by volatiles, strain-localization and also by crustal phase transitions.

4. *Landslides*

Landslides represent an ideal example by which a suite of different dissipative processes is triggered and instability ensues. We present here an example from Vaiont, located in the Dolomite region of the Italian Alps. On October 9, 1963 a catastrophic landslide occurred. It slipped for 45 seconds with a speed of up to 30 m/s smashing a water reservoir which at the time contained 115 million m^3 of water. A wave of water was pushed up the opposite bank and destroyed the village of Casso, located 260 m above the lake level. Other villages further downstream were also erased and 2500 lives were lost.

The slip rates of the landslides were carefully monitored before and during the event and the report MÜLLER (1964) concludes that "...the interior kinematic nature of the mobile mass, after having reached a certain limit velocity at the start of the rock slide, must have been a kind of thixotropy. This would explain why the mass appears to have slid down with an unprecedented velocity which exceeded all expectations. Only a spontaneous decrease in the interior resistance to movement would allow one to explain the fact that practically the entire potential energy of the slide mass was transformed without internal absorption of energy into kinetic energy. Such a behavior of the sliding mass was beyond any possible expectation; nobody predicted it and the author believes that such a behavior was in no way predictable..."

VEVEAKIS *et al.* (2007) were able to unravel the nonlinear physics underlying the unprecedented speed of the event. The authors test the working hypothesis that slip was localized in a clay-rich water-saturated layer. They propose shear heating as the primary mechanism for triggering the long-term phase of accelerating creep, and model the creeping phase using a rigid block moving over a thin zone of high shear strain rates including shear heating. Introducing a thermal softening and velocity strengthening law for the basal material, the authors reformulate the governing equations of a water-saturated porous material, obtaining an estimate for the collapse time of the slide. They were able to calibrate the model with real velocity measurements from the slide. In order to keep the mathematical formulation tractable and to explore the limitations of the shear heating mechanism, the model was kept as simple as possible. In their impressive fit of the predicted velocity to the observed velocity (Fig. 2) they can reproduce the observations from month 5 before the catastrophic slide using the simple shear-heating hypothesis alone (VEVEAKIS *et al.*, 2007). However, the very final phase cannot explain the explosive phase of accelerated creep observed in the Vaiont Landslide. This phase

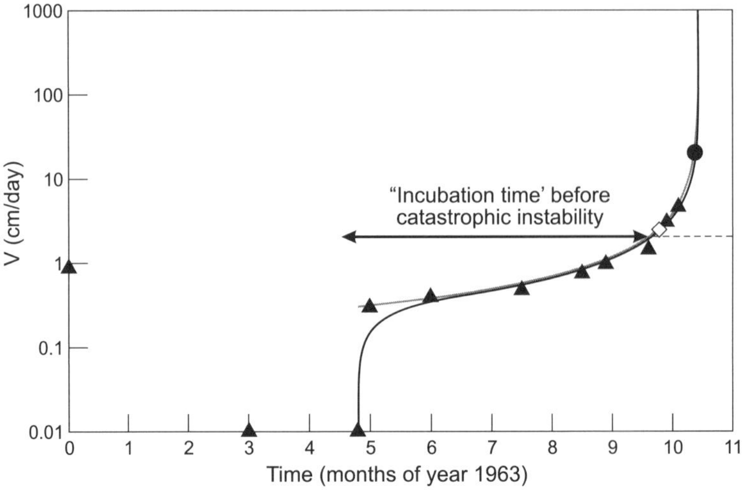

Figure 2
Slip-velocity of the Vaiont 1963 landslide (MULLER, 1964) with a hyperbolic fit (HELMSTETTER *et al.*, 2004) shown by a grey line and the model prediction of VEVEAKIS *et al.* (2007) depicted by a black solid line; modified after VEVEAKIS *et al.* (2007).

occurred during the total loss of strength in the slipping zone in the last minutes prior to the slide.

VEVEAKIS *et al.* (2007) tested the hypothesis that this second phenomenon is explained by the onset of thermal pressurization, triggered by dewatering reactions of the clay due to the temperature rise within the clay-rich layer. Their calibrated data predicted a critical temperature of dewatering of 36°C, which fits experimental data. Given that the ambient average temperature at the base of the slide is expected to be 8°C, this estimate provides an independent verification of the shear-heating hypothesis. As the mechanical dissipation through slow creep in the clay rises slowly it finally approaches a critical temperature, where dehydration reactions start. The associated water release, which increases the pore fluid pressure, subsequently leads to a catastrophic instability. For this to happen the temperature must rise up to the critical pressurization temperature, leading to an explosive pore pressure rise, up to the point of 'full pressurization': The point where pore pressure reaches the value of the total normal stress, leading to fluidization of the shear band.

We conclude from this analysis that the shear-heating hypothesis is a very robust theory to explain landslides. The physics of thermally self-accelerating localized shear in a clayey gouge is indeed a fundamental mechanism. The authors also offer solutions to the fact that not all landslides reach the catastrophic proportions encountered in Vaiont. The sudden, last minute acceleration of the slide is explained with an almost instantaneous rise of the pore pressure because of shear-heating. This critical temperature for dehydration is sufficiently high (36°C) that it may not be reached by all landslides.

A unique feature of the instability is the prolonged time before the actual onset of the catastrophic instability. When consulting Figure 2 one sees that at least 5 months before the fast instability the system appears to have been "simmering", slowly increasing its slip rate, consequently its basal temperature in the basal clay layer. The basic process underlying the instability is inherently an intrinsic thermal process. Once started, it is likely that no amount of engineering intervention (short of freezing the clay) would have been able to stop the landslide.

4.1. Ice Quakes

The concept of a 'incubation time' before onset of catastrophic instability is common in thermal feedback systems, such as combustion, where there are multiple time scales present in the phenomenon. It has been described for thermal runaway instabilities for rocks on lithospheric scale processes (REGENAUER-LIEB and YUEN, 2000). The hyperbolic curves of instability are essentially the same as that of Figure 2, except that the incubation time can be hundreds thousand years before the onset of a catastrophic instability instead of five months. Regenauer-Lieb and Yuen show the same curve as Figure 2 in their Figure 9 (REGENAUER-LIEB and YUEN, 2003), except that the "simmering" time-scale is hundreds of thousands of years. For ice sheets it can be between 100 years and 10 k years before the onset of ice-quakes (YUEN *et al.*, 1986). While the phenomenon of ice-quakes has been postulated by these fundamental analyses as a strict consequence of the underlying fundamental thermodynamics, seismological evidence for 184 of such events has only been provided recently (TSAI and EKSTRÖM, 2007). It is interesting to note that the size of the events appears to peak above the detection threshold, suggesting that the size of glacial earthquakes has a characteristic nature. This could indicate that ice-quakes only occur whenever a critical energy level is reached. This observation supports an energy-based instability, however, follow up numerical studies are certainly required to fully investigate this exciting phenomenon.

SCHUBERT and YUEN (1982) also suggest massive basal melting of the Antarctic ice sheet in the form of an explosive shear-heating instability (similar to the Vaiont event). They note that the present thickness of the Antarctic ice sheet is close to the critical value for instability and perform 1-D models to investigate the effect of increased ice accumulation as finite amplitude perturbation. The difference to the case presented for Vaiont is that Schubert and Yuen do not model direct observational data which was not available at the time of writing (TSAI and EKSTRÖM, 2007). However, in principle they also consider a second style of cascading instability, which is triggered after reaching a critical energy level. This second instability would be a giant melting instability associated with a phase transition (melting). These arguments are however, for the time at best indicative. They may well be robust if the entropy were constrained in a hypothetical 1-D scenario, however, the complexity of higher dimension may throw a spanner in the works and need more processes.

4.2. Instabilities in Crustal Deformation

The Wenchuan earthquake (May 12, 2008) in Sichuan, China (Parsons *et al.*, 2008) has reminded us once again that earthquakes happen unexpectedly and are a very nonlinear process. The earthquake exhibits its inherent stochastic nature with the aftershocks forming a migratory pattern toward the northeast. The unusual rupture mechanism of this earthquake has a complex strike-slip and a strong thrust component in a continental setting. Many of the nonlinear physics associated with earthquake genesis are not yet known. To date, earthquakes have been studied mainly as an elastic-deformation process, driven by kinematic boundary and initial conditions, sometimes with interacting faults (Robinson and Benites, 1995). However, realistic earthquake models require strong phenomenological and thermodynamical foundations. Such a modeling approach should consider thermodynamics within the governing equations, since the non-equilibrium nature of the entire process would influence the entire development of earthquake instabilities.

In the examples listed above we have proposed that multiple instabilities are necessary at multiple scales to fill the gap in strain rates between the fast earthquake event and the slow geodynamic creep event (Fig. 1). In the ice example we have raised concerns that a plausible mechanism in 1-D may become obsolete in 2-D or 3-D because of additional geometric complexities. The Vaiont landslide example provides sound evidence for multiple thresholds (at least two) with multiple critical values with cascading scale. The physics of natural landslides may very well be explained by exceeding the lowest critical value, which is the stability limit within which shear-heating becomes more intense than the diffusion of heat away from the shear zone. However, there is a second threshold more elevated in the energy scale, namely that of reaching a critical temperature for dehydration of clays inside the shear zone. When this critical value is reached the shear-heating instability cascades up in scale to the catastrophic event where kinetic energy must be considered. One could argue that in the deeper crust there are several such critical temperatures associated with individual phase transitions, ultimately leading to the melting instability (Kanamori *et al.*, 1998). We examine in the following section previous modeling and observational evidence for cross-scale instabilities.

In the search for the place where we have the richest instability mechanism available we are immediately attracted to the 'semi-brittle', brittle-ductile transition zone which from laboratory extrapolations is postulated to cover a significant fraction of the crust (Kohlstedt *et al.*, 1995). The brittle-ductile transition is an area where we can have a dual material behavior. The rock can fracture in a violent brittle manner but at the same time it can creep in a ductile manner. By its very nature, deformation within the brittle-ductile transition must hence be able to cross the scales from geodynamics to seismology and observations reported in Figure 1 might be rooted here (Strehlau, 1986). In a thermodynamic sense the brittle-ductile transition also is prominent because it is the area in the crust where the maximum dissipation occurs. Based on the sum of the arguments laid out above we conclude that the brittle-ductile transition holds the key to the mechanisms promoting earthquakes.

Unfortunately, we know very little about the mechanical behavior of the brittle-ductile transition. To date there are only very few modeling approaches oriented in this direction. Pioneering work for investigating the relation between damage by brittle cracking and aseismic creep has been proposed (LYAKHOVSKY *et al.*, 2005; LYAKHOVSKY and BEN-ZION, 2008), the role of crustal phase transition and grain size reduction have been investigated by (GUEYDAN *et al.*, 2001; 2004) and the feedback between brittle instabilities by thermal–expansion and shear-heating and ductile instabilities by shear-heating and accelerated creep have been presented (REGENAUER-LIEB and YUEN, 2006, 2008). The latter point, thermal–mechanical coupling via shear-heating in fault zones has been stressed many times since DAVE GRIGGS (1969) as a key to the understanding of geodynamical processes. However, there is, to date, no clarity on the role of the different feedback mechanisms in the crust. We argue that an understanding of the zone where transitions between brittle and ductile failures occur is critical to understanding both earthquakes on the short time scale and plate tectonics over longer time scales of millions of years. We also performed microscale observations aimed at reconciling the brittle and ductile deformation regimes (FUSSEIS *et al.*, in prep).

We investigated an exposed midcrustal section of the Redbank shear zone in Australia. Microstructural and geochemical analyses have shown that the rocks have been subject to shearing at greenschist-facies metamorphic conditions in the presence of an aqueous fluid phase (FUSSEIS *et al.*, 2009). Deformation was mostly accommodated by viscous-grain-boundary-sliding combined with creep fracturing. We found evidence for the formation of creep failure planes (Fig. 3), which form when a critical density and size of cavities promotes localized failure; cf. (DIMANOV *et al.*, 2003). While cavity growth is controlled by a creep mechanism (viscous grain boundary sliding), cavity interconnection and failure is a spontaneous nonstable process. The fact that the potential failure planes, which are preserved in our samples in a ‘healed condition’, are rather limited in extent suggests that failure in the presented case never reached catastrophic dimensions. However, our observation might be a possible key to bridge the gap in strain rates (Fig. 1) controlling the earthquake cycle at the base of the seismogenic zone, which underlie the Wenchuan earthquake.

5. *Discussion*

Future work in modelling of earthquakes, ice-quakes and landslides shares many of the same characteristics and we should appreciate this commonality. It requires us to put in more feedback processes to make them realistic. If we want to advance a step in joining geological, geophysical and numerical observations for reproducing shorter time scale instabilities, it cannot be as simple as before, as indeed pointed out in the work by LIU *et al.* (2007).

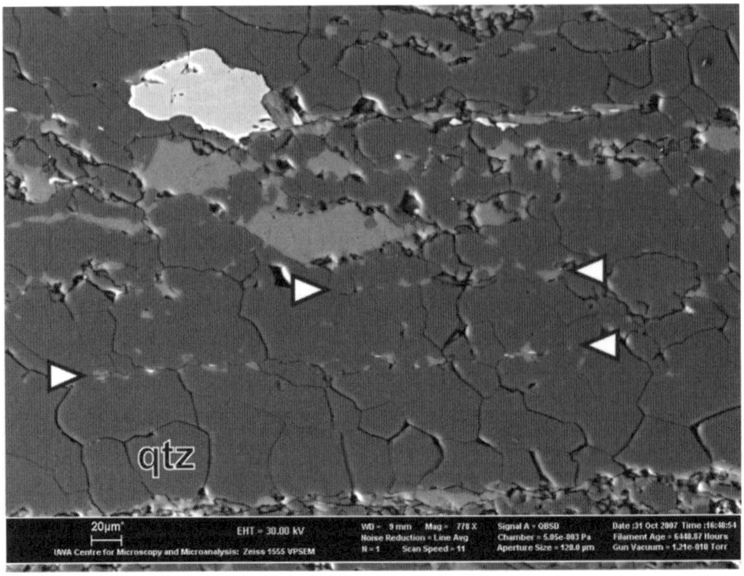

Figure 3
Back Scattered Electron image from the central portion of a natural shear zone shown in the Redbank, Australia. Microcavities (in black) decorate grain boundaries. Aligned grain boundaries mark a possible midcrustal failure plane in a quartz band (between white arrows). The failure plane is also traced by secondary K-feldspar (white) that precipitate during syn-kinematic fluid circulations. See Fusseis *et al.* (2009) for more details.

We propose that future numerical models always should be complemented and benchmarked by direct seismological/geological/geodynamics observations in a method of multi-scale data assimilation. This multi-scale data assimilation is crucial since, as we have pointed out, there is currently no analytical theory available that allows a prediction of the importance of the individual feedback mechanism for localization. It is hence important to build up a catalogue of important threshold values for the onset of the particular feedback and map these feedbacks to a given scale of observation. We have discussed in this paper only numerical models for thermal processes. We have, however, also put forward as a modeling challenge geological evidence for feedback processes that are not entirely thermally based. The microstructural analyses in Figure 3 point to the prominent role of the formation of microcavities in conjunction with fluid flow and dynamic permeability in the ductile crust. This observation may play a fundamental role for the strain-rate gap between geodynamics and earthquakes (Fig. 1). There may be many more processes that are unexamined yet.

Thermodynamics also allows a fair degree of simplification for these processes and provides an alternate route to assess the validity of 'brute force' modeling results. Thermodynamic approaches can be used to combine laboratory work (Rybacki *et al.*, 2008) on natural specimens with forward numerical modelling in fault simulations and micro- to -mesoscale structural observations (Fusseis *et al.* in prep). In this paper we have reviewed what we believe to be solid evidence for thermodynamic instabilities that involve the consideration of the parameter temperature as a free variable. These are landslides and ice-quakes. We have not touched upon the many questions open regarding the generation of pseudotachylytes, melt formation and large-strain localization structure in dynamically recrystallizing faults such as Glarus (SCHMID, 1982) versus pseudotachylites in the Woodroffe thrust in Central Australia (LIN, 2008). We believe that at present they are providing only circumstantial evidence for their underlying process. While pseudotachylites are a very good indication of crustal melting, it is unclear whether the temperature rise is a result or a cause of localization. Melting is possible either because brittle faults are propagating into the ductile realm or they may be evidence for shear-heating instabilities propagating upwards from the ductile region. The very fact that they are happening in the brittle-ductile transition does not give preference to either mechanism.

A new vantage point that is opening a fresh view on ductile instabilities is given by inclusion of crustal phase transformations, mineral reactions and the role of chemical processes that are promoted by the role of volatiles from water and CO_2 and other crustal fluids. The thermodynamic approach proposed here provides a natural inclusion of these processes, which may help to promote shear localization in the ductile realm. To this end we are currently generalizing the approach to include chemical feedbacks (POULET and REGENAUER-LIEB, 2009). In more general terms the entropy production is the product of a thermodynamic force times a thermodynamic rate of flow. In the case that the thermodynamic flow is the flow of chemical species, the length scale for the representative volume element is the length scale of a diffusing species, which can be used to define the volume element.

Acknowledgments

We thank Yehuda Ben-Zion for encouraging us to write this article. We are grateful for stimulating discussions on shear-heating with Boris Kaus, Gabriele Morra, Sergei Medvedev, Ali Karrech and Thomas Poulet. D. A. Yuen acknowledged support from the VLAB grant given by the National Science Foundation. K. Regenauer-Lieb and Florian Fusseis acknowledge support from the Western Australian Government through the Premier's Fellowship Program. This research has been supported by the UWA research grant 12104344 and the Australian Synchrotron Research Program.

Appendix A: Thermomechanics

A much simpler framework has been proposed for Geomechanics (COLLINS and HOULSBY, 1997). The so-called "thermomechanics" approach has initially been proposed as a thermodynamic consistency check for plasticity (ZIEGLER, 1983). However, the additional advantage of the new approach is that geomechanical constitutive laws now can be derived directly from the entropy production of the underlying physical processes. This is a significant step forward, as it provides a unifying framework on the basis of physics to derive yield conditions and flow rules from the entropy production alone. The following paragraphs present a short introduction into the theory of thermomechanics. However, as will be shown in the following, thermomechanics removes key feedback potentials from Equation (7), with a strong simplifying assumption of isothermal deformation, i.e., $\dot{T}s = 0$, hence

$$\dot{u} = \dot{\psi}(T, \varepsilon^{el}, \alpha_j) + \dot{s}T, \tag{A1}$$

because there is no temperature change, there is also no heat produced in the volume and no heat flowing in and out of the reference volume and the first law becomes

$$\rho \dot{u} = \sigma_{ij} \dot{\varepsilon}_{ij}, \tag{A2}$$

since there is no adiabatic temperature change the mechanical dissipation simplifies to

$$\dot{s}T = \tilde{\Phi}, \tag{A3}$$

where the over tilde signals path dependency of the mechanical dissipation potential function. With this and Equation (A1) we have

$$\frac{1}{\rho}\sigma_{ij}\dot{\varepsilon}_{ij} = \frac{\partial \psi}{\partial \varepsilon_{ij}}\dot{\varepsilon}_{ij} + \frac{\partial \psi}{\partial \alpha^k}\dot{\alpha}^k + \frac{\partial \tilde{\Phi}}{\partial \dot{\alpha}^k}\dot{\alpha}^k, \tag{A4}$$

where α^k is the microstrain of the individual process in the representative volume element. The microstrain can be dislocation glide, dislocation climb, diffusion creep, formation of cracks, phase transitions or others. These microstrain processes can be described at the large scale of volume integration by two fundamentally different processes and we obtain

$$\sigma_{ij}\dot{\varepsilon}_{ij} = \sigma_{ij}\dot{\varepsilon}_{ij}^{el} + (1 - \chi_{(t)})\sigma_{ij}\dot{\varepsilon}_{ij}^{noheat} + \chi_{(t)}\sigma_{ij}\dot{\varepsilon}_{ij}^{diss}. \tag{A5}$$

Since the macroscopic strain is the sum of the microstrain processes α^k this formulation leads to the classical additive strain rate decomposition.

$$\dot{\varepsilon}_{ij} = \dot{\varepsilon}_{ij}^{el} + \sum \dot{\varepsilon}_{ij}^{noheat} + \sum \dot{\varepsilon}_{ij}^{diss}. \tag{A6}$$

The deformational work can be reversible and elastic, it can be plastic/viscous both through the storage of energy (e.g., in the surface energy of microcracks) or it can be released as a heat-producing or other dissipative process that does not release heat. The summation over the microstrains can be in serial or parallel. In the summation a serial description implies that the microstrain processes do not depend on another mechanism to be activated. A parallel description implies that the microstrain processes are mutually dependent. In the latter case one over the individual reciprocal strains performs the summation while in the former it is just a simple addition. These are the competing rate processes leading to localization. However, we note that the feedback terms in the energy equations have been removed through the isothermal assumption. Therefore, while this approach is instructive, it does not lend itself to earthquake modeling. We need to consider the full thermodynamic framework.

Appendix B: Coupling Temperature, Momentum and Continuity Equations

A numerical solution technique for Equation (7) requires only modest additional sophistication for modern numerical solution techniques when it is cast in its explicit form in Equation (14). In fact, the use of variational principles to find a minimum in the stored energy potential function is all that is needed. It is important to point out though that it is not enough to just solve Equation (14) separately to the underlying momentum conservation and continuity requirements. If this is done all the possible feedbacks are lost. It is necessary to derive a minimum stored energy solution of the displacements and the force balance for a given boundary condition that satisfies Equation (14). This requires full coupling of the energy equation with the momentum and continuity equations. Modern displacement finite-element methods are built around this method. In the following only a short description of an industry standard code is given, for more in-depth reading please consult the theory manual (ABAQUS/Standard, 2000).

The momentum equation describes the force and moment equilibrium. We write the equation here in integral form for the use of finite elements. Let **f** be the body force at any point within the volume and **n** the unit outward normal to a surface A. Equilibrium is achieved if the surface traction plus the body forces balance each other

$$\int_A \mathbf{n} \cdot \sigma_{ij} dA + \int_V \mathbf{f} dV = 0. \tag{B1}$$

Applying the Gauss theorem to the surface integral we obtain the equation for translational force equilibrium

$$\int_V \nabla \cdot \sigma_{ij} dV + \int_V \mathbf{f} dV = 0. \tag{B2}$$

The equation applies pointwise to an arbitrary volume and the volume integration may be dropped. The equations for rotational equilibrium can be written likewise by considering moments, however, the assumption must be made that the stress tensor is symmetric. At this stage we do not wish to abandon symmetry of stress in our numerical approach. The symmetry assumption of the stress tensor may be revisited at a later stage.

In displacement-interpolation finite-element analysis the force or moment equilibrium is routinely extended to include the continuity equation by the principle of "virtual work" here written as a principle of "virtual power". For this equation Equation (B2) is multiplied by a virtual velocity field, i.e., arbitrary vectorial test functions based on virtual displacements that satisfy the condition of continuity. These virtual displacements are used to verify whether the momentum equation is in local equilibrium.

$$\int_V \left(\nabla \cdot \sigma_{ij} + \mathbf{f}\right) \cdot \delta v \, dV = 0, \tag{B3}$$

$$\int_V \sigma_{ij} \nabla \cdot \delta v \, dV = \int_A \mathbf{n}\sigma_{ij} \cdot \delta v \, dA + \int_V \mathbf{f} \cdot \delta v \, dV = 0, \tag{B4}$$

and with the divergence of the virtual velocity field being the virtual strain rates this simplifies to

$$\int_V \sigma_{ij} \nabla \cdot \delta v \, dV = \int_V \sigma_{ij} \delta \dot{\varepsilon}_{ij} \, dV. \tag{B5}$$

The finite-element approach satisfies the momentum and continuity equation by variational principles through the minimization of the potential energy rate density functionals. According to Equation (6) minimization of the Helmholtz free energy for the smallest element size (RVE in equilibrium) is all that is needed and for any given dissipation potential the equations are closed.

References

ABAQUS/Standard (2000), 384 pp., Hibbit, Karlsson and Sorenson Inc.

Attard, P. (2006), *Theory for non-equilibrium statistical mechanics*, Phys. Chem. Chem. Phys. *8*, 3585–3611.

Chrysochoos, A. and Dupre, J. P. (1991), *Experimental analysis of thermomechanical coupling by infra-red thermography, in Anisotropy and localization of plastic deformation*, Proc. Plasticity '91, The third Internatl. Symp. on plasticity and its Current Applications (eds. Boehler;J-P and A. S. Khan, pp. 540–543, (Elsevier, London 1991)

Chrysochoos, A. and F. Belmahjoub (1992), *Thermographic analysis of thermomechanical couplings*, Archives Mechanics *44*(1), 55–68.

Clausius, R. (1865), *Über verschiedene für die Anwendung bequeme Formen der Hauptgleichungen der mechanischen Wärmetheorie*, Annalen der Physik und Chemie, 125, 353–400.

Collins, I. F., and G. T. Houlsby (1997), *Application of thermomechanical principles to the modelling of geotechnical materials*, Proc. Roy. Soc. London, A *453* 1964, 1975–2001.

Dieterich, J. H. (1979a), *Modeling of Rock Friction .1. Experimental Results and Constitutive Equations*, J. Geophys. Res. *84*(NB5), 2161–2168.

Dieterich, J. H. (1979b), *Modeling of Rock Friction .2. Simulation of Pre-Seismic Slip*, J. Geophys. Res. *84*(NB5), 2169–2175.

Dimanov, A. *et al.* (2003), *Creep of polycrystalline anorthite and diopside*, J. Geophys. Res. Sol. Earth, 108(B1).

Fusseis, F. *et al.* (2009), *Creep cavitation can establish a dynamic granular fluid pump in ductile shear zones*, Nature, doi:10.1038/nature08051.

Fusseis, F. *et al.* (in prep), *Earthquakes triggered by creep failure at the brittle ductile transition*, Geology, in prep.

Glansdorff, P. *et al.* (1973), *Thermodynamics Theory of Structure, Stability and Fluctuations*, Am. J. Phys. *41*(1), 147–148.

Griggs, D. and Baker, D. (1969), *The origin of deep-mantle earthquakes.* In Properties of Matter under Unusual Conditions (ed. H. M. a. S. Fernbach), pp. 23–42, Interscience New York.

GUEYDAN, F. *et al.* (2001), *Grain-size-sensitive flow and shear-stress enhancement at the brittle-ductile transition of the continental crust*, Internatl. J. Earth Sci. *90*(1), 181.

GUEYDAN, F. *et al.* (2004), *Mechanics of low-angle extensional shear zones at the brittle-ductile transition*, J. Geophy. Res. B: Solid Earth, *109*(12), 1.

HELMSTETTER, A., *et al.* (2004), *Slider block friction model for landslides: Application to Vaiont and La Clapiere landslides*, J. Geophy. Res. *109*, B002160.

HOBBS, B. E. *et al.* (1986), *Earthquakes in the ductile regime*, Pure Appl. Geophy. *124*(1/2), 310–336.

HOULSBY, G, and A. PUZRIN, *Principles of Hyperplasticity, An Approach to Plasticity Theory Based on Thermodynamic Principles*, 351 pp. (Springer, Berlin 2007).

JOHN, T. *et al.* (2009), *Generation of intermediate-depth earthquakes by self-localizing thermal runaway*, Nature Geoscience *2*, 137–140.

KANAMORI, H. *et al.* (1998), *Frictional melting during the rupture of the 1994 Bolivian earthquake*, Science *279*(5352), 839–842.

KOHLSTEDT, D. L. *et al.* (1995), *Strength of the lithosphere: Constraints imposed by laboratory measurements*, J. Geophys. Res. *100*(B9), 17587–17602.

LAVENDA, B. H. *Thermodynamics of Irreversible Processes*, 182 pp. (MacMilland Press Ltd, London 1978).

LIN, A. (2009), *Seismic slipping in the lower crust, inferred from granulite-related pseudotachylyte in the Woodroffe thrust central Australia*, Pure Appl. Geophys., in press.

LINDE, A. T. *et al.* (1996), *A Slow Earthquake Sequence on the San Andreas Fault*, Nature *383*(6595), 65–68.

LIU, M. *et al.* (2007), *Parallel computing of multi-scale continental deformation in the western United States: Preliminary results*, Phys. Earth Planet. Inter. *163*, 35–51.

LYAKHOVSKY, V. *et al.* (2005), *A viscoelastic damage rheology and rate- and state-dependent friction*, Geophys. J. Internatl. *161*(1), 179.

LYAKHOVSKY, V. and Y. BEN-ZION (2008), *Scaling relations of earthquakes and aseismic deformation in a damage rheology model*, Geophys. J. Internatl. *172*, 651–662.

MORRIS, J. (2008), *Stronger, tougher steels*, Science *320*, 1023–1024.

MULLER, L. (1964), *The rock slide in the Vaiont valley*, Felsmechanik Ingenieurgeologie, 2, 148–212.

OGAWA, M. (1987), *Shear instability in a viscoelastic material as the cause of deep focus earthquakes*, J. Geophys. Res. *92*(B1), 13801–13810.

ONSAGER, L. (1931), *Reciprocal relations in irreversible processes*, Phys. Rev. *38*(12), 2265.

PARSONS, T. *et al.* (2008), *Stress changes from the 2008 Wenchuan earthquake and increased hazard in the Sichuan basin*, Nature *254*, 509–510.

POULET, T. and REGENAUER-LIEB, K. (2009), *A unified mult-scale thermodynamical framework for coupling geomechanical and chemical simulations*, Tectonophysics, submitted.

RAMBERT, G. *et al.* (2007), *On the direct interaction between heat transfer, mass transport and chemical processes within gradient elasticity*, European J. Mechanics-A/Solids *26*(1), 68–87.

REGENAUER-LIEB, K. and YUEN, D. (2000), *Quasi-adiabatic instabilities associated with necking processes of an elasto-viscoplastic lithosphere*, Phys. Earth Planet. Inter. *118*, 89–102.

REGENAUER-LIEB, K. and YUEN, D. A. (2003), *Modeling shear zones in geological and planetary sciences: Solid- and fluid-thermal-mechanical approaches*, Earth Sci. Revi., *63*, 295–349.

REGENAUER-LIEB, K. and YUEN, D. A. (2004), *Positive feedback of interacting ductile faults from coupling of equation of state, rheology and thermal-mechanics*, Phys. Earth Planet. Inter. *142*(1-2), 113–135.

REGENAUER-LIEB, K., and YUEN, D. (2006), *Quartz Rheology and short time-scale crustal instabilities*, Pure Appl. Geophys. *163*(9), 1–18.

REGENAUER-LIEB, K., and YUEN, D. (2008), *Multiscale brittle-ductile coupling and genesis of slow earthquakes*, Pure Appl. Geophys. *165*(3-4), 523–543.

RICE, J. (1971), *Inelastic constitutive relations for solids: An internal-variable theory and its application to metal plasticity*, J. Mechan. Phys. Sol. *19*(6), 433–455.

ROBINSON, R. and BENITES, R. (1995), *Synthetic seismicity models of multiple interacting faults*, J. Geophys. Res. *100*(B9), 18229–18238.

RYBACKI, E. *et al.* (2008), *High-strain creep of feldspar rocks: Implications for cavitation and ductile failure in the lower crust*, Geophys. Res. Lett. *35*, L04304.

SCHMID, S. M. *Microfabric Studies as Indicators of deformation mechanisms and flow laws operative in Mountain Building*. In Mountain Belts (ed. K. J. Hsu), (Academic Press, London 1982).

SCHUBERT, G. and YUEN, D. (1982), *Initiation of ice ages by creep instability and surging of the East Antarctic ice sheet*, Nature *296*, 127–130.

SIRET, D. *et al.* (2008), *PreMDB, a thermodynamically consistent material database as a key to geodynamic modelling*, Geotechnica Acta, doi:10.1007/s11440-008-0065-0.

STREHLAU, J. *A discussion of the depth extent of rupture in large continental earthquakes.* In *Earthquake Source Mechanics*, Proc. 5th Maurice Ewing Symp. on Earthquake Source Mechanics, May 1985 (ed., AGU, New York 1986).

TSAI, V. and EKSTROM, G. (2007), *Analysis of Glacial earthquakes*, J. Geophys. Res. *112*, F03S22.

TSE, S. T. and RICE, J. R. (1986), *Crustal Earthquake instability in relation to the depth variation of frictional slip properties*, J. Geophys. Res.-Sol. Earth and Planets *91*(B9), 9452–9472.

VEVEAKIS, E. *et al.* (2007), *Thermo-poro-mechanics of creeping landslides: the 1963 Vaiont slide, Northern Italy*, J. Geophys. Res. *112*, F03026.

YUEN, D. A. *et al.* (1986), *Explosive growth of shear-heating instabilities in the down- slope creep of ice sheets*, J. Glaciol. *32*(112), 314–320.

ZIEGLER, H. *An Introduction to Thermomechanics* (North Holland, Amsterdam 1983).

(Received October 1, 2008, accepted March 26, 2009)
Published Online First: June 27, 2009

To access this journal online:
www.birkhauser.ch/pageoph